# Graduate Texts in Contemporary Physics

## Graduate Texts in Contemporary Physics

R. N. Mohapatra: **Unification and Supersymmetry: The Frontiers of Quark-Lepton Physics**

R. E. Prange and S. M. Girvin (eds.): **The Quantum Hall Effect**

M. Kaku: **Introduction to Superstrings**

J. W. Lynn (ed.): **High-Temperature Superconductivity** (in preparation)

H. V. Klapdor (ed.): **Neutrinos**

# Neutrinos

Edited by H.V. Klapdor

With Contributions by
F. T. Avignone, III, R. L. Brodzinski, P. Depommier,
F. von Feilitzsch, G. Gelmini, W. Hillebrandt, H. V. Klapdor,
P. Langacker, S. P. Mikheyev, R. N. Mohapatra, K. Muto,
A. Y. Smirnov, K. Winter

With 164 Figures

Springer-Verlag Berlin Heidelberg New York
London Paris Tokyo

Professor Dr. Hans Volker Klapdor PHYSICS

MPI für Kernphysik, Postfach 103980,
D-6900 Heidelberg 1, Fed. Rep. of Germany

*Series Editors*

Joseph L. Birman
Department of Physics
The City College of the
City University of New York
New York, NY 10031, USA

H. Faissner
Physikalisches Institut
RWTH Aachen
D-5100 Aachen, Fed. Rep. of Germany

Jeffrey W. Lynn
Department of Physics and Astronomy
University of Maryland
College Park, MD 20742, USA

ISBN 3-540-50166-5 Springer-Verlag Berlin Heidelberg New York
ISBN 0-387-50166-5 Springer-Verlag New York Berlin Heidelberg

Library of Congress Cataloging-in-Publication Data. Neutrinos / edited by Hans V. Klapdor ; with contributions by F. T. Avignone III ... [et al.]. p. cm. – (Graduate texts in contemporary physics) Includes index. 1. Neutrinos. I. Klapdor, H. V. (Hans Volker), 1942- . II. Avignone, F. T., 1933- . III. Series. QC793.5.N42N49 1988 539.7'215–dc19 88-28160



Printing: Weihert-Druck GmbH, D-6100 Darmstadt
Binding: J. Schäffer GmbH & Co. KG., D-6718 Grünstadt
2156/3150-543210 - Printed on acid-free paper

# Preface

Since the time of Fermi and Pauli, neutrinos have been central to our understanding of the weak interaction. Nowadays they play a corresponding role within the framework of theories of grand unification (GUTs, SUSYs,...), for which the nature (Dirac, Majorana,...) and mass of the neutrino yield important boundary conditions. Neutrinos with a nonvanishing Majorana mass would correspond to the breaking of baryon-lepton (B–L) number conservation and would rule out, for example, the minimal SU(5) model.

Since, on the other hand, neutrinos are the best candidates for dark matter in the universe, their mass could determine its large-scale structure and evolution, thus yielding, besides the direct relation between the structure of GUTs and the evolution of the early universe, another close connection between the laws of microphysics and cosmology. Neutrinos are also the only direct probes of processes in the cores of collapsing stars, and their properties might be one of the possible solutions to the solar neutrino puzzle.

The purpose of this book is to present, in ten chapters written by experts in their specific areas, 1) our current understanding of the nature and properties of the neutrino in the framework of present theories of particle physics; 2) the impact of neutrino physics on neighboring disciplines such as astrophysics and cosmology; and 3) present experimental efforts to uncover the nature of the neutrino, for example by neutrino oscillations, nuclear double beta decay, high-energy neutrino reactions, and searches for lepton flavor violation.

I would like to express my gratitude to my colleagues who contributed their expertise to this book. The concern and excellent cooperation of Prof. W. Beiglböck and Dr. H.-U. Daniel of Springer-Verlag are gratefully acknowledged.

Heidelberg
November 1988

*H.V. Klapdor*

# Contents

# Neutrino Properties

*F. von Feilitzsch*

Technical University Munich, D-8000 München, Fed. Rep. of Germany

## 1. Introduction

Since the postulation of the neutrino by W. Pauli in 1930 as a "not detectable" neutral particle accompanying the electron in nuclear $\beta$-decay, it continued to be one of the most mysterious particles in physics. It took 26 years before it could be directly observed, via the inverse reaction of the $\beta$-decay,

$$\bar{\nu}_e + p \rightarrow n + e^+$$

by F. Reines and C.L. Cowan in 1956 [Rei59]. In addition to the experimental verification of the existence of the neutrino, the cross section for the inverse $\beta$-decay reaction was measured to be $\sigma = (6.7 \pm 1.5) \cdot 10^{-43} cm^2/fission$ for the neutrino spectrum as it was emitted from the nuclear fission reactor at which the experiment was performed. In the meantime a "world average" cross section from different experiments is

$$\sigma = (6.32 \pm 0.21) \cdot 10^{-43} cm^2/fission$$

for $^{235}U$ fission neutrinos [Mik88]. This extremely low cross section is one reason why the interaction via which the neutrino communicates with matter is called the "weak interaction". Shortly later in 1957, C.S. Wu et al. [Wu57] discovered the parity violation in nuclear $\beta$-decay, proving that this interaction was not symmetric against a mirror transformation in space. The electrons emitted in nuclear $\beta$-decay of $^{60}Co$ are found to be left-handed, which shows that the copiously emitted antineutrino should be right-handed if the angular momentum is conserved in the weak interaction.

From charge conservation we conclude that the neutrino has to be electrically neutral. There remain, however, two major questions:

- Is the neutrino emitted in the $\beta$-decay purely left-handed (i.e. the antineutrino purely right-handed) ?
- does the neutrino have a non-vanishing rest mass?

The first of these questions is related to whether the weak interaction, as indicated by experimental observations, is purely composed by a left-handed vector and axial vector force (V-A interaction). The second question arises from the fact that we do not have any physical need for a neutrino mass to be identical to zero. This is in contrast to the case of the photon, where the renormalizability of quantum electrodynamics requires, and relativity implies, a zero rest mass of the photon.

Experimentally these questions can be tested in various forms. The right-handed contributions to the weak interaction may be investigated by measuring the spin alignment of the charged lepton emitted in weak decays as well as by the search for neutrinoless double $\beta$-decay. In addition the neutrinoless double $\beta$-decay would imply that neutrinos are massive Majorana particles. This process is discussed in more detail in other parts of this book.

The manifestation of neutrino masses depends on whether or not the weak eigenstates are pure or mixed states of the mass eigenstates, i.e. the mixing matrix is purely diagonal or contains off-diagonal elements.

If the weak eigenstate of the neutrino contains several different mass eigenstates, we will find as many "deformations" in the $\beta$-spectrum as there are mass eigenstates involved. In the simplest case involving one mass eigenstate, the endpoint energy of the $\beta$-spectrum is reduced by $m_\nu c^2$. Similarly neutrino masses manifest themselves in weak decay processes of mesons. Here the analysis of the momentum and energy distribution in the decay process provides direct information on the neutrino masses involved.

A neutrino wave function describing a weak interaction eigenstate, which is a superposition of different mass eigenstates, can lead to a complex time dependence, which manifests itself as neutrino oscillations and can lead to neutrino decay. In the following we discuss these possible manifestations and their experimental investigations.

# 2. Neutrino Masses

## 2.1 General Considerations

If we assume neutrinos to be massive particles, there arises the question how the mass eigenstates, i.e. the eigenstates of the mass projecting operator, are related to the weak eigenstates, i.e. the eigenstates of the weak interaction operator. In the most general case we cannot assume these two eigenstates to be identical. This implies that the weak interaction eigenstates can be described by an orthogonal set of corresponding mass eigenstates and mixing amplitudes if the number of weak eigenstates coincides with the number of mass eigenstates.

In various attempts it was suggested to describe all fundamental physical interactions in one single theory. The first successful unification was found for the electromagnetic interaction with the weak interaction by S.L. Glashow (1961), S. Weinberg (1967) and A. Salam (1968) and is generally known as the "standard theory of electroweak interaction". Until now there has been no experimental contradiction of this theory, which however, provides no information about neutrino masses. Neutrino masses are "artificially" set to be zero. Here neutrinos play a unique role as they are treated as purely left-handed and massless particles, in contrast to any other fermions (leptons and quarks). The whole scheme of leptons with the three known flavours is thus described in the standard theory by three left-handed doublets of charged and neutral leptons and three right-handed singlets of charged leptons,

$$\begin{pmatrix}\nu_e\\ e^-\end{pmatrix}_L \begin{pmatrix}\nu_\mu\\ \mu^-\end{pmatrix}_L \begin{pmatrix}\nu_\tau\\ \tau^-\end{pmatrix}_L \qquad \begin{pmatrix} e^-\end{pmatrix}_R \begin{pmatrix}\mu^-\end{pmatrix}_R \begin{pmatrix}\tau^-\end{pmatrix}_R$$

plus their antiparticles.

A completely symmetric description of all leptons and quarks is given with the group SO(10) as formulated by various authors [Mak62, Eli74, Bil76, Fri76, Ste80], unifying the electroweak and strong interaction.

In this theory all fermions are treated in a perfectly symmetric form as Dirac particles. These are three doublets of quarks and leptons plus their antiparticles, each being left- and right-handed.

$$\text{quarks} \qquad\qquad\qquad \text{leptons}$$

$$\begin{pmatrix}u\\ d\end{pmatrix}_{R,L} \begin{pmatrix}c\\ s\end{pmatrix}_{R,L} \begin{pmatrix}t\\ b\end{pmatrix}_{R,L} \qquad \begin{pmatrix}\nu_e\\ e^-\end{pmatrix}_{R,L} \begin{pmatrix}\nu_\mu\\ \mu^-\end{pmatrix}_{R,L} \begin{pmatrix}\nu_\tau\\ \tau^-\end{pmatrix}_{R,L}$$

ndices R, L indicate right- and left-handed particles, respectively. All particles (including the neutrinos) may be massive.

2.2 Dirac and Majorana Mass Terms

As neutrinos are electrically neutral particles, there exists, in contrast to the quarks, the possibility for neutrinos to be Dirac or Majorana particles.

The Majorana particle is defined by the identity of the particle with its own antiparticle. We therefore have to consider both possibilities formulating a theory. A mass term in the Lagrange density couples fields of different helicities.

For a Dirac field $\nu$ we get the Lagrange function for a Dirac mass term

$$L^D = m_D(\bar{\nu}_L\nu_R + \bar{\nu}_R\nu_L) = m_D\bar{\nu}\nu, \tag{1}$$

as the helicity states are given by the Weyl-Spinors

$$\nu_L = \frac{1}{2}(1-\gamma_5)\nu,$$

$$\nu_R = \frac{1}{2}(1+\gamma_5)\nu,$$

where $\nu = \nu_R + \nu_L$ is the mass eigenstate. Here $m_D$ is the Dirac mass of the neutrino field $\nu$.

From the charge conjugate fields additional combinations of right- and left-handed fields can be formed. We get for the left- and right-handed components of the charge conjugate field $\nu^C$

$$\nu_L^C := (\nu^C)_L = (\nu_R)^C,$$

$$\nu_R^C := (\nu^C)_R = (\nu_L)^C,$$

and the Majorana mass terms

$$L_A = m_A(\bar{\nu}_R^C\nu_L + \bar{\nu}_L\nu_R^C) = m_A\bar{\chi}\chi, \tag{2}$$

$$L_B = m_B(\bar{\nu}_L^C\nu_R + \bar{\nu}_R\nu_L^C) = m_B\bar{\omega}\omega, \tag{3}$$

with $\chi = \nu_L + \nu_R^C$ and $\omega = \nu_R + \nu_L^C$ being the mass eigenstates.

The mass eigenstates obey the relations

$$\chi^C = \chi$$

and

$$\omega^C = \omega.$$

For the general form of the mass term we get the Lagrange function

$$\begin{aligned} L &= m_D\bar{\nu}_L\nu_R + m_A\bar{\nu}_L\nu_R^C + m_B\bar{\nu}_L^C\nu_R + h.c \\ &= (\bar{\chi}\bar{\omega})\begin{pmatrix} m_A & \frac{m_D}{2} \\ \frac{m_D}{2} & m_B \end{pmatrix}\begin{pmatrix}\chi \\ \omega\end{pmatrix}. \end{aligned} \tag{4}$$

Diagonalizing the mass matrix leads to two Majorana mass eigenstates:

$$\eta_1 = cos\theta\chi - sin\theta\omega, \tag{5}$$

$$\eta_2 = sin\theta\chi - cos\theta\omega, \tag{6}$$

with

$$tan2\theta = \frac{m_D}{m_B - m_A}, \tag{7}$$

and masses

$$M_{1,2} = \frac{1}{2}\{(m_A + m_B) \pm \sqrt{(m_A - m_B)^2 + m_D^2}\}. \tag{8}$$

A theory which contains Dirac mass terms like in (1) does not conserve the lepton numbers $L_l(l = e, \mu, \tau ...)$ but conserves the total lepton number $L = \sum_i L_l(l = e, \mu, \tau ...)$ if $m_D$ is a mixing matrix between different flavour states. In this case processes like $\mu^+ \to e^+ + \gamma, \mu^+ \to e^+ + e^- + e^+, K^+ \to \pi^+ + \mu^\pm + e^\mp$, etc. as well as neutrino oscillations analogous to the $K_o \bar{K}_o$ oscillations in the quark sector are allowed.

Neutrinoless double $\beta$-decays

$$(A, Z) \to (A, Z+2) + e^- + e^-,$$

and processes like

$$\mu^- + (A, Z) \to e^+ + (A, Z-2),$$

as well as

$$K^+ \to \pi^- + e^+ + \mu^+$$

require Majorana masses and are neither conserving the lepton numbers $L_l(l = e, \mu.\tau ...)$ nor the total lepton number $L = \sum_l L_l$.

It may be interesting to note that we expect in the case of

$$m_A \approx -m_B$$

in (8), the neutrino to have a mass pattern which is nearly degenerate with $m_A + m_B << m_D$. For $m_A = m_B = 0$ we get a neutrino with a pure Dirac mass.

### 2.3 Neutrino Mixing

The neutrino field can be expressed in the most general form in terms of an orthogonal set of flavour eigenstates $\nu$ or mass eigenstates $\nu'$

$$\nu = \begin{pmatrix} \nu_e \\ \nu_\mu \\ \nu_\tau \\ . \\ . \end{pmatrix} \quad \nu' = \begin{pmatrix} \nu_1 \\ \nu_2 \\ \nu_3 \\ . \\ . \end{pmatrix}. \tag{9}$$

The n flavour eigenstates may be transformed into the n mass eigenstates by means of a matrix $U_{ik}$

$$\begin{pmatrix} \nu_e \\ \nu_\mu \\ \nu_\tau \\ . \\ . \end{pmatrix} = U_{ik} \begin{pmatrix} \nu_1 \\ \nu_2 \\ \nu_3 \\ . \\ . \end{pmatrix}. \tag{10}$$

If there are no flavour changing neutral currents present, the flavour eigenstates are identical with the weak eigenstates produced in decays of the weak interaction such as

$$n \rightarrow p + e^- + \bar{\nu}_e$$

or

$$\pi^+ \rightarrow \pi^o + \mu^+ + \nu_\mu.$$

The matrix $U$ is unitary and contains $\frac{1}{2}n(n-1)$ real mixing angles if CP invariance holds. In addition, there can be CP-violating phases. The number of these phases depends on whether we deal with Dirac or Majorana neutrinos. For Dirac neutrinos we get $\frac{1}{2}(n-1)(n-2)$, and for Majorana neutrinos $\frac{1}{2}n(n-1)$ phases.

In the following we discuss what consequences we expect if neutrinos acquire masses and mixing as discussed above, and how these could be tested in experiments.

## 3. Direct Neutrino Mass Experiments

### 3.1 Search for Dominant Neutrino Masses

In the search for dominant neutrino masses we assume the mixing matrix $U_{ik}$ to be approximately diagonal. This implies that to each flavour eigenstate there is one mass eigenstate coupled dominantly. The mixing to other mass eigenstates is neglected.

#### 3.1.1 Investigations of Nuclear $\beta$-Decay

If any rest masses are attributed to neutrinos emitted in nuclear $\beta$-decay, this has to be reflected in the energy balance of the $\beta$-decay reaction. In the absence of neutrino rest masses we observe the momentum spectrum of the emitted $\beta$-particle for an allowed Fermi transition

$$N_\beta \propto p_e^2 F(E_o - E)^2, \tag{11}$$

where $p_e$ is the momentum of the emitted $\beta$-particle, $F$ the Fermi function, $E_o$ the Q-value given by the mass difference $\{M_{A,Z} - (M_{A,Z\pm1} + m_e)\}$, and $E$ the energy of the $\beta$-particle; c is set to be 1. The endpoint of the spectrum is at $E = E_o$, where all the available energy of the transition is transferred to the electron, and no kinetic energy is transmitted to the emerging neutrino.

In the case of a non-vanishing neutrino mass we get

$$N_\beta \propto p_e^2 F(E_o - E)((E_o - E)^2 - m_\nu^2)^{\frac{1}{2}}. \tag{12}$$

The endpoint energy is reduced by the energy corresponding to the neutrino rest mass $m_\nu$. If we write the spectrum described in (12) in the Kurie representation $K = [N_\beta/F]^{\frac{1}{2}}$ we get for an allowed transition a straight line which for $m_\nu = 0$ cuts the abscissa at $E_{max} = E_o$. In case of $m_\nu > 0$ the spectrum is shortened and cuts the abscissa vertically at $E_{max} = E_o - m_\nu$.

Tritium $\beta$-decay is the process which has up to now been investigated most intensively with respect to a neutrino mass manifesting itself at the $\beta$-decay endpoint. The main advantage of this decay lies in its low and well determined Q value of $E_o \cong 18.59 keV$, measured by means of ion cyclotron resonance as well as with magnetic mass spectrometers [Lip85, Smi81, Nik84, Ber72], and in the relatively simple nuclear and atomic configuration of tritium and $^3He$. In fact there exists experimental evidence for an electron neutrino masses in the range

$$17\ eV < m_{\nu_e} < 40\ eV$$

found by Lubimov et al., ITEP collaboration [Lub86].

This experiment was performed with Tritium implanted in a complex organic compound (Valine). The $\beta$-momentum spectrum was analyzed with an iron- free toroidal magnetic spectrometer with 4 subsequent foci, which achieved an energy resolution of $\simeq$ 45 eV for 18 keV electrons. The main criticisms of this experiment, however, concerned the treatment of the response function of the spectrometer itself and the interactions emerging electrons could undergo in the complex valine source [Vil83, Sim84, Ber85].

The experimental spectrum as published by the ITEP group is shown in Fig.1. The results for three different sources are plotted in the Kurie representation. The solid curves represent the best fit to the data, leading to a neutrino mass in the range quoted above. The dashed lines correspond to a fit where the neutrino mass was set to be zero.

There is only one experiment, Fritschi et al., [Fri86], which has up to now achieved a comparable sensitivity to neutrino masses by investigating the endpoint of a $\beta$-spectrum. However, this experiment is only in marginal agreement with the results of the ITEP group. The spectrometer used in this experiment is similar to that of the ITEP group, with modifications to improve the energy resolution by means of a retarding electric field around the source. The energy resolution achieved was 27 eV. The major difference is found in the preparation of the $^3H$ source. They used $^3H$ implanted into carbon, which was evaporated onto an Al-foil. The group claims to have tested, experimentally and by Monte Carlo calculations, the contributions to the resolution function from interactions in the source. The experimental and calculated resolution function of the

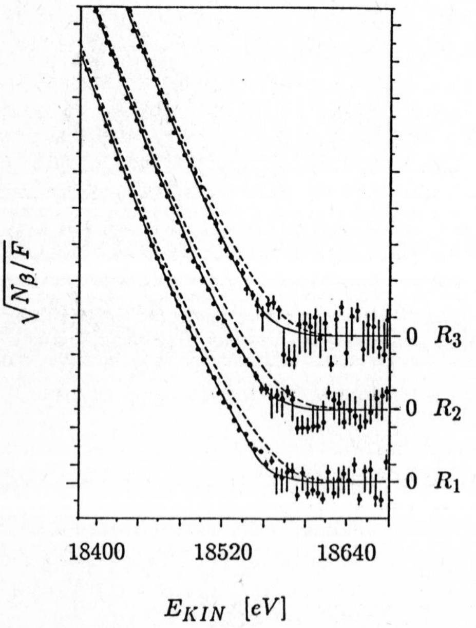


Fig.1. Kurie representation of the experimental $\beta$-spectrum as published in [Lub86]. Solid lines correspond to $17\ eV < m_{\nu_e} < 40\ eV$, dashed lines to the assumption $m_{\nu_e} = 0$

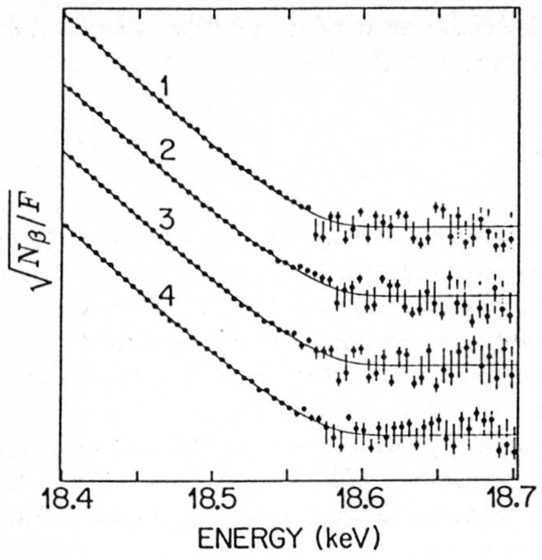


Fig.2. Kurie spectra of four different measurements at the $\beta$ endpoint of tritium from [Fri86]. The solid line corresponds to the assumption $m_{\nu_e} = 0$

spectrometer agree within 10 %. The best fit to the data corresponds to $m_\nu = 0$. Including all systematic errors, an upper limit for the neutrino mass of

$$m_{\nu_e} < 18\ eV$$

is quoted with a 90 % confidence limit (c.l.). Figure 2 shows the experimental spectra for four different measurements, together with a fit to the data (solid line) corresponding to $m_\nu = 0$. There is a large number of similar experiments already being performed or under preparation.

In particular, there is a class of experiments which try to overcome the problem of interactions in the source by using atomic tritium. An experiment at Los Alamos (USA) [Bow86] uses atomic tritium as a source contained in a 37 m long tube, differentially pumped at both ends. The emerging decay electrons are guided by a magnetic field into the magnetic spectrometer. The quoted systematic errors from the resolution and energy losses are $87(eV/c^2)^2$ and $80(eV/c^2)^2$, respectively. The main uncertainties stem from trapping of electrons in an inhomogenity of the guiding magnetic field. This problem, however, is expected to be solved in the near future. A preliminary upper limit for the neutrino mass quoted by this group is $m_\nu < 27\ eV/c^2$. There is a similar, but considerably more sophisticated, experiment under preparation at Livermore (USA), aiming for a sensitivity down to $m_\nu = 5\ eV$ to be achieved within the next year [Sto88].

### 3.1.2 Search for Muon Neutrino Masses in the Pion Decay

In close analogy to the search for electron neutrino masses in nuclear $\beta$-decay there exists the possibility to search for muon neutrino masses in the pion decay

$$\pi^+ \to \mu^+ + \nu_\mu.$$

From energy and momentum conservation (setting again c = 1) we obtain

$$m_{\nu_\mu}^2 = m_{\pi^+}^2 + m_\mu^2 - 2m_{\pi^+}\left(p_{\mu^+}^2 + m_{\mu^+}^2\right)^{\frac{1}{2}} \tag{13}$$

if the decay occurs with the $\pi^+$ being at rest. $m_i$ are the masses (energies) of the involved particles and $p_{\mu^+}$ is the muon momentum. A measurement of the muon momentum from pion decay at rest thus leads to information about the muon neutrino mass.

Evidently the sensitivity of the measurement depends on the knowledge of the pion and muon masses. The uncertainties in the parameters $m_\pi^+$, $m_{\mu^+}$ and $p_{\mu^+}$ contribute similarly to the uncertainty of the experiment. Thus it is reasonable to measure the momentum $p_{\mu^+}$ with an accuracy similar to that of the particle masses involved. The least "accurate" mass is that of the pion. Assuming the validity of the CPT- theorem it follows that

$$m_{\pi^+} = m_{\pi^-},$$

and therefore, $\Delta m_{\pi^-} = 0.0015\ MeV$ gives the systematic limit for this method. The best experimental limits obtained for $m_{\nu_\mu}$ were achieved up to now by a group at SIN in Switzerland [Dau79, Abe84].

The $\pi^+$ beam produced at the SIN proton cyclotron was stopped in a scintillator and the momentum of the emerging $\mu^+$ analyzed in a magnetic spectrometer and registered in a silicon surface barrier detector. The energy resolution of the silicon detector together with time coincidences between the silicon detector and the plastic scintillator, serving as $\pi^+$ stop, led to an effective background suppression. The stopping of $\pi^+$ particles was identified by a lower energy threshold of 3 MeV (a $\pi^+$ stop would release 3.6 MeV in the scintillator). The measured momentum $p_{\mu^+}$ obtained from a series of experiments was

$$p_{\mu^+} = 29.79139 \pm 0.00083\ MeV/c^2.$$

Together with the masses $m_{\pi^+}$ and $m_{\mu^+}$ taken from the particle data group [Par84] a mass square $m^2_{\nu_\mu}$ was derived from (13)

$$m^2_{\nu_\mu} = -0.163 \pm 0.080\ MeV^2/c^4.$$

Following the recommendations of the particle data group for the data analysis a final upper limit for the neutrino mass was achieved, being

$$m_{\nu_\mu} < 250\ keV/c^2\ at\ 90\ \%\ c.l.$$

### 3.1.3 Search for $\tau$-Neutrino Mass in the $\tau$-Decay

The search for the $\tau$-neutrino mass can be performed by investigating the $\tau$-decay. Analyzing the energy or the invariant mass of the emerging charged particles leads to information on the $\tau$-neutrino mass. The first limits on the $\tau$-neutrino mass were derived by the Delco collaboration at Spear [Bac.79]. They investigated the "quasi-$\beta$-decay" spectra

$$\tau \to e \bar{+} \nu_e + \nu_\tau$$

and

$$\tau \to \mu + \bar{\nu}_\mu + \nu_\tau.$$

The $\nu_\tau$ mass is given here by the difference between the $\tau$- momentum and the maximum momentum of the electron or muon. The limiting factor for $m_{\nu_\tau}$ is given by the experimental momentum resolution of the fastest charged particle, leading in this experiment to

$$m_{\nu_\tau} < 250\ MeV\ at\ 90\ \%\ c.l.$$

The most stringent limits on $m_{\nu_\tau}$ were achieved at the $e^+e^-$ collider DORIS at DESY in Hamburg [Schu87, Alb88]. $\tau$-leptons are produced monochromatically at $e^+e^-$ colliders of equal beam energies. Being produced in the collision

$$e^+ + e^- \rightarrow \tau^+ + \tau^-$$

each $\tau$ gets the energy $\sqrt{s}/2$, where $\sqrt{s}$ is the total energy of the electron and positron.

The $\tau$-decay channels investigated for $m_{\nu_\tau}$ determination with the ARGUS detector at DORIS are

$$\tau^\pm \rightarrow \pi^\pm + \pi^+ + \pi^- + \nu_\tau$$

and

$$\tau^\pm \rightarrow \pi^\pm + \pi^+ + \pi^+ + \pi^- + \pi^- + \nu_\tau .$$

The accelerator was operated at an energy $\sqrt{s} = (10023.4 \pm 0.3)\ MeV$. The mean energy of the $\tau$-leptons was determined with an energy of $\pm 0.15\ MeV$, whereas the gaussian width of the energy was not less than 0.5 MeV. Radiative corrections can be calculated accurately so that they do not contribute significantly to the uncertainties of the experiment. In the decay

$$\tau^\pm \rightarrow \pi^\pm + \pi^+ + \pi^- + \nu_\tau$$

the momenta $p_i$ of the pions were measured with the magnetic spectrometer of the ARGUS detector. From energy conservation and kinematics one gets for every observed decay the pion energies

$$E_i = \sqrt{p_i^2 + m_\pi^2}, \quad i = 1, 2, 3..., \tag{14}$$

the neutrino energy

$$E_\nu = \sqrt{s}/2 - E_1 - E_2 - E_3, \tag{15}$$

and the invariant mass

$$m_{3\pi}^2 = (E_1 + E_2 + E_3)^2 - (p_1 + p_2 + p_3)^2 . \tag{16}$$

The neutrinos are isotropic in the $\tau_{CM}$-system, and the $E_\nu$-spectrum is a box-spectrum in the laboratory system if the $\tau$- leptons are monochromatic. The lower edge of this box-spectrum depends on $m_\nu$ and on $m_{3\pi}$. By this method the experimental limit

$$m_{\nu_\tau} < 56\ MeV\ at\ 95\ \%\ c.l.,$$

was obtained, which depends only on kinematics and statistics.

Including all systematic uncertainties of the experiment, an upper limit of

$$m_{\nu_\tau} < 70\ MeV$$

was achieved. Figure 3 shows the neutrino energy spectrum of 1536 $\tau \rightarrow \pi\pi\pi\nu$ events used for the analysis [Alb85].

An additional investigation of $m_{\nu_\tau}$ was performed by the group, analyzing the invariant mass in the decay $\tau \rightarrow 5\pi$ [Alb88]. This method depends on the decay dynamics but not on the precise knowledge of the $\tau$-energy. Thus only the spectrometer properties and the background determines the accuracy of the experiment.

In Fig.4 one event of the type $e^+e^- \rightarrow \tau_1\tau_2$ is shown with the decays: $\tau_1 \rightarrow \nu$ plus one charged particle, $\tau_2 \rightarrow \nu$ plus five charged particles .

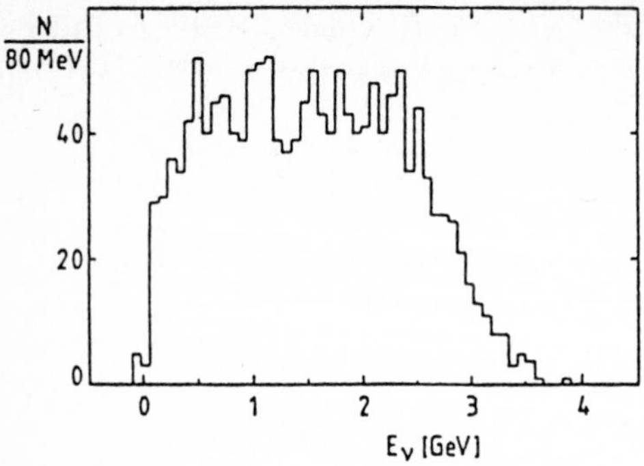


Fig.3. Neutrino energy spectrum of $\tau \rightarrow \pi\pi\pi\nu$ events from [Alb85]

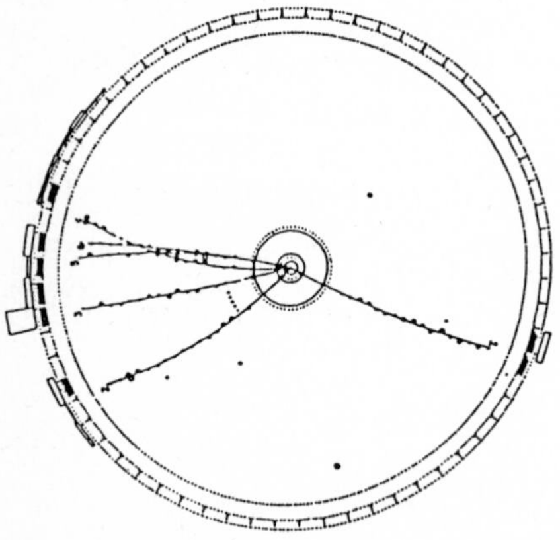

Fig.4. One event of the type $ee \rightarrow \tau\tau$ with one $\tau$ decaying into one charged particle and the other $\tau$ decaying into five charged particles [Alb88]

Events of this type, with $\tau_2 \rightarrow \nu_\tau + 5\pi$, were selected for the determination of $m_{\nu_\tau}$. After background rejection 12 events survived all cuts. Their invariant mass distribution is shown in Fig.5, using data taken between 1983 and 1986. The experimental mass resolution for the five highest-mass events is indicated by error bars. Including systematic errors, a fit to the invariant mass $m_{5\pi}$ leads to an upper limit for $m_{\nu_\tau}$ of:

$$m_{\nu_\tau} < 35 \; MeV \; at \; 95 \; \% \; c.l.$$

3.2 Neutrino Mixing in Weak Decays

In the previous chapter we discussed direct neutrino mass experiments neglecting the possibility of neutrino mixing, i.e. assuming the mixing matrix $U_{i,k}$ in (10) to be diagonal. In the following we talk about the implications of neutrino mixing in weak decays and their experimental tests.

3.2.1 Neutrino Mixing in Nuclear $\beta$-Decay

Let us assume now that the neutrino weak interaction eigenstate emitted in the nuclear $\beta$-decay has the form as described in formula (10). The consequences for the dynamics of weak decays were first discussed by E. Schrock in a series of papers [Schr81]. For simplicity we deal

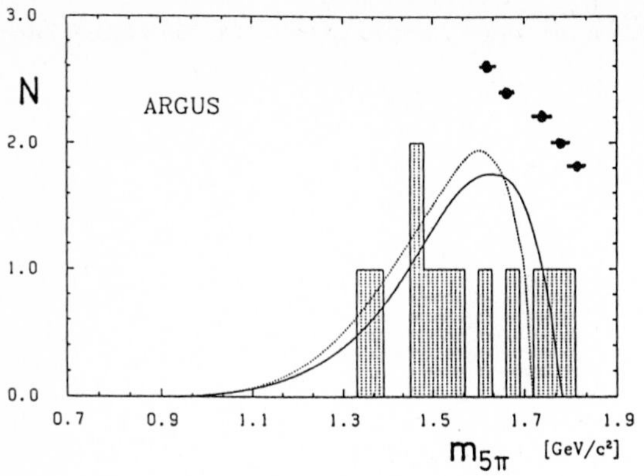


Fig.5. Distribution of the $5\pi$ invariant mass in $\tau^{\pm} \to \pi^{\pm}\pi^{+}\pi^{+}\pi^{-}\pi^{-}\nu$ decays from [Alb88]

with two flavours only. In this case the unitary matrix contains only one mixing angle [Bil78] and therefore we get for the electron neutrino

$$\nu_e = \nu_1 cos\theta + \nu_2 sin\theta. \tag{17}$$

The $\beta$-spectrum is then composed of both massive contributions to the electron neutrino [Kel80, Schr81, Sim81]:

$$N_\beta = N_\beta(\nu_1)cos^2\theta + N_\beta(\nu_2)sin^2\theta. \tag{18}$$

Under the assumption $E_o - E_\beta >> m_{\nu_1}$ (i.e. neglecting $m_{\nu_1}$) from (12) and (18), we find for the $\beta$-spectrum

$$\begin{aligned} N_\beta &\propto p_e^2 F(E_o - E)^2 cos^2\theta && for\ \ (E_o - E) < m_{\nu_2}, \\ N_\beta &\propto p_e^2 F(E_o - E)^2 cos^2\theta && \\ &\quad + p_e^2 F(E_o - E)\{(E_o - E)^2 - m_{\nu_2}^2\}^{\frac{1}{2}} sin^2\theta && for\ \ (E_o - E) > m_{\nu_2}. \end{aligned}$$

For small mixing amplitudes $sin^2\theta$ we get in the Kurie representation

$$K \propto (E_o - E)cos\theta + (sin^2\theta/2 \cdot cos\theta)\{(E_o - E)^2 - m_{\nu_2}^2\}^{\frac{1}{2}}. \tag{19}$$

Figure 6 shows schematically a Kurie spectrum expected for $\nu$-mixing with two mass eigenstates.

Experimental limits on neutrino mixing have been achieved by a variety of experiments based on these considerations. Neutrino masses from several keV to 500 keV and mixing amplitudes down to $sin^2\theta > 10^{-3}$ have been investigated [Sim81, Schr83, Alt85, Ohi85, Sim85]. In particular an evidence for neutrino mixing with $m_{\nu_2} = 17.1\ keV$ and $sin^2\theta \approx 0.03$ has been published by J.J. Simpson (1985), which, however, was not confirmed in subsequent experiments.

There exists one particularly interesting possibility to check experimental evidence for neutrino mixing in the $\beta$-decay spectrum. A given massive neutrino mixing in the $\beta$-decay initiates a kink in the $\beta$-spectrum at $E_\beta = E_o - m_{\nu_i}$ and thus should show up at different energies when $\beta$-decays with different endpoints are investigated. It is therefore rather easy to check for uncontrolled systematic errors in the individual experiment by investigating a series of different $\beta$-decays.

The first dedicated experiment in this sense was performed by studying simultaneously the $\beta^+$- and $\beta^-$-decay of $^{64}Cu$ [Schr83]. $^{64}Cu$ decays via a $\beta^+$- and $\beta^-$-emission with endpoint energies of 653 keV and 575 keV, respectively. Thus experimentally observed deformations in the $\beta$-spectra are relevant in terms of neutrino mixing only if they show up in both spectra, shifted by an energy of 78 keV in this case. In Fig.7 the results of this experiment are plotted as a function of the neutrino masses $m_\nu$ and the mixing amplitude $sin^2\theta$ as discussed in (19). The exclusion limits for the different individual $\beta$-spectra are plotted together with the final exclusion limit (solid line) from the joint analysis.

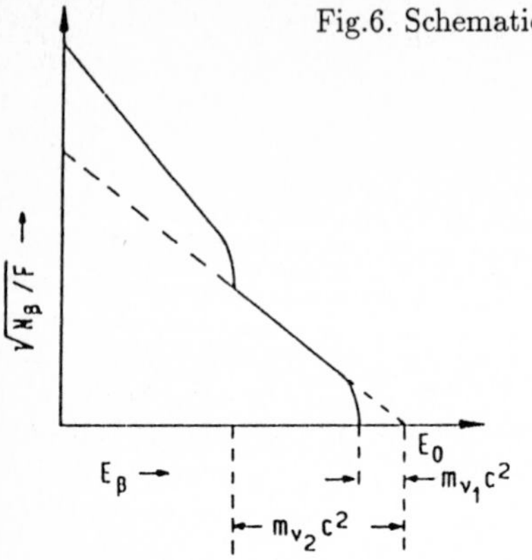


Fig.6. Schematic Kurie spectrum for $\nu$-mixing with two mass eigenstates

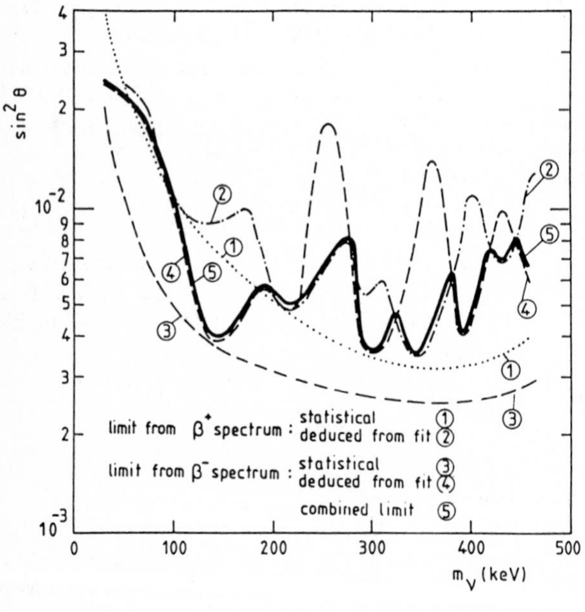


Fig.7. Exclusion plot for massive neutrino branches deduced from the $\beta^+$ and $\beta^-$ decay of $Cu^{64}$. The region above the solid line is excluded by the combined limits. For details see [Schr83]

### 3.2.2 Neutrino Mixing in Mesonic Decays

Weak decay processes of pseudoscalar mesons, such as $K$ or $\pi$, into a charged lepton and a neutrino offer the possibility of investigating two-body decay processes. This has the advantage of an expected sharp line in the momentum spectrum of the emitted charged lepton at

$$P_i = \frac{1}{2}M\left[1+\frac{m_l^4}{M^4}+\frac{m_\nu^4}{M^4}-2\left(\frac{m_l^2}{M^2}-\frac{m_\nu^2}{M^2}+\frac{m_l^2 m_\nu^2}{M^4}\right)\right]^{\frac{1}{2}}, \tag{20}$$

as a signature for a heavy neutrino emission, where $m_l$ and $m_\nu$ are the masses of the emitted charged lepton and neutrino, respectively. $M$ is the parent mass. The intensity of this extra peak $M^+ \to l^+\nu_i$ is related to the conventional peak $M^+ \to l^+\nu_l$ (l=e,$\mu$,$\tau$) via the neutrino mixing amplitude $U_{li}$ (i=1,2,3):

$$\Gamma(M^+ \to l^+\nu_i) = \rho\Gamma(M^+ \to l^+\nu_l) \mid U_{li} \mid^2, \qquad (21).$$

where $\rho$ is a kinematic factor including the phase space

$$\rho = \frac{\left\{\left(\frac{m_l}{M}\right)^2 + \left(\frac{m_\nu}{M}\right)^2 - \left(\frac{m_l}{M} - \frac{m_\nu}{M}\right)^2\right\} \cdot \left\{1 + \left(\frac{m_l}{M}\right)^4 + \left(\frac{m_\nu}{M}\right)^4 - 2\left(\frac{m_l}{M} + \frac{m_\nu}{M} + \frac{m_l m_\nu}{M^2}\right)^2\right\}^{\frac{1}{2}}}{\left(\frac{m_l}{M}\right)^2 \left(1 - \left(\frac{m_l}{M}\right)^2\right)^2}.$$

Heavy neutrinos are enhanced with respect to massless neutrinos due to the phase space.

Very stringent limits on $\nu_\mu - \nu_i$ mixing amplitudes for neutrino masses in the range $70\ MeV < m_{\nu_i} < 300\ MeV$ were achieved with an experiment investigating the decay

$$K^+ \to \mu^+ + \nu_\mu.$$

The experiment was performed at the National Laboratory for High Energy Physics (KEK) in Japan using a 550 MeV $K^+$-beam [Hay82]. The $K^+$-beam was degraded in a 7 cm Cu sheet and stopped in ten layers of plastic scintillators. Charged particles from the $K^+$-decay were momentum analyzed by a magnetic spectrograph. At the entrance and exit window of the spectrograph multiwire proportional chambers were placed to monitor the entrance and exit of charged particles. Thus by coincidence measurements between the $K^+$ stopping plastic scintillators and the proportional chambers background events could effectively be suppressed. To investigate the peak regions a momentum resolution of 0.5 % at full width half maximum was achieved for the whole spectrographic set-up.

No distinct peak was found, except for the normal one at 236 MeV/c originating from $\mu^+$ events emitted in the $K^+$-decay together with massless neutrinos. In Fig.8 the momentum spectrum of charged particles analyzed with the spectrograph is given in the upper curve. The final

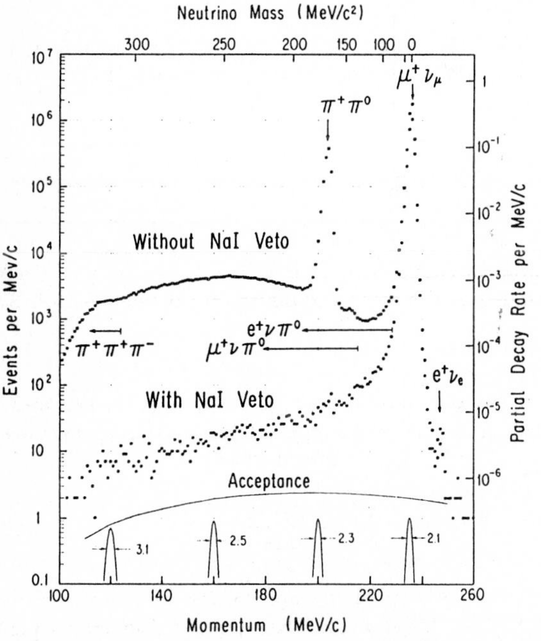


Fig.8. Momentum spectrum of charged particles from the reaction $K^+ \to \mu^+ + \nu_\mu$ analyzed with the spectrograph. For details see [Hay82] and the text

muon momentum spectrum with all background suppression applied is depicted with the lower curve. Contributions from different competing decay channels are indicated in addition, together with the relative acceptance and resolution of the system.

From the non-observation of a distinct peak in the muon momentum spectrum, an upper limit on the mixing amplitude $| U_{\mu i} |^2$ as a function of the neutrino mass $m_{\nu_i}$ was derived. This limit is depicted later in Fig.18 together with limits from other experiments. Similarly an experiment has recently been performed at the SIN proton cyclotron investigating the decay $\pi \rightarrow \mu + \nu_\mu$. With this experiment neutrino masses in the range $1\ MeV < m_{\nu_i} < 16\ MeV$ were investigated for mixing amplitudes down to $U^2_{\mu i} > 10^{-5}$ [Dau87].

## 4. Neutrino Oscillations

### 4.1 General Considerations

In the following section we discuss the implications neutrino mixing has on the time dependence of the neutrino wave function.

Using (10) for the neutrino wave function we get, e.g. for an electron neutrino eigenstate $\nu_e$ expressed in terms of mass-eigenstates $\nu_i$

$$\nu_e = \sum_i U_{ei} \nu_i. \tag{22}$$

The space-time dependence of the wave functions $\nu_i$ is given by

$$\nu_i(t) = \nu_i(0) \cdot e^{i(k_i x - \omega_i t)}, \tag{23}$$

where $k_i$ is the momentum, $x$ one space coordinate and $\omega_i$ the energy of the neutrino mass eigenstate $\nu_i$ setting $\hbar = c = 1$. For $\omega_i >> m_i$, with $m_i$ being the mass eigenvalue of $\nu_i$, we set $\omega_i \approx \omega_j = E$ and get

$$k_i x - \omega_i t \approx \frac{m_i^2}{2E} t.$$

Thus we have

$$\nu_i(t) \approx \nu_i(0) \exp\left(-i \frac{m_i^2}{2E} t\right).$$

With (22) we get

$$\nu_e(t) \approx \sum_i U_{ei} \nu_i(0) \exp\left(-i \frac{m_i^2}{2E} t\right).$$

For $t = 0$ this wave function represents the electron flavour eigenstate.

The probability to find the flavour eigenstate $\nu_\beta$ at time $t$ after $\nu_e$ was emitted is then

$$P(\nu_e \rightarrow \nu_\beta) = | \sum U_{ei} e^{-i \frac{m_i^2}{2E} t} U^*_{\beta i} |^2 . \tag{24}$$

Reducing (24) for simplicity to the case of oscillations, where only two neutrino states are involved, we get

$$U = \begin{pmatrix} cos\theta & sin\theta \\ -sin\theta & cos\theta \end{pmatrix},$$

and

$$P(\nu_e \to \nu_\beta) = \frac{1}{2} sin^2 2\theta \left\{ 1 - cos \frac{\Delta m^2}{2E} t \right\}, \tag{25}$$

with $\Delta m^2 = \mid m_1^2 - m_2^2 \mid$.

It should be mentioned that in this derivation we did not use any specific properties of the Dirac or Majorana type of neutrinos. The only presumption made was that the mass eigenvalues $m_i$ should be small enough to get coherence for the mass eigenfunctions $\nu_i$ composed in a weak decay.

Therefore, this formalism is valid for neutrino oscillations in analogy to the $K^o \bar{K}^o$ oscillation scheme (where the mass pattern is nearly degenerate $\Delta m << m_i$) as well as for neutrino flavour oscillations (where significantly different neutrino masses can be expected).

If the neutrinos transverse matter the oscillation might be altered. Due to the dependence of the interaction cross section on flavour quantum numbers for elastic neutrino scattering off electrons, electron neutrinos and muon or tau neutrinos undergo a different propagation in an electronic medium. This can in fact even lead to a resonant transition between flavour quantum numbers if appropriate electron densities are available as was first shown by S.P. Mikheyev, A. Yu. Smirnov and L. Wolfenstein. Neutrino oscillations in matter are of special interest for solar neutrinos and are discussed in a dedicated section of this book. We also refer to [Bil78, Bil87].

In an experiment we measure the time-dependence of the neutrino flavour eigenstate by means of time of flight between the neutrino source and the detector. For $m_\nu << E$ we set $v_\nu \approx c = 1$. Eventually we have to take into account the size of the neutrino source and the detector as well as the neutrino energy spectrum.

The wavelength of the neutrino oscillation can be expressed by

$$\Lambda = 4\pi \frac{E}{\Delta m^2} = 2.5 \cdot \frac{E}{\Delta m^2} [meters] \tag{26}$$

if $E$ and $\Delta m^2$ are expressed in MeV and $(eV)^2$, respectively.

We can classify the experimental search for neutrino oscillations into two groups. One is the search for the appearance of neutrinos with a given flavour quantum number from a neutrino beam with a different flavour quantum number. The other method searches for the diminution of the flux of a neutrino beam as a function of the distance between the source and the detector.

The first method has the advantage of being very sensitive to small mixing amplitudes, as it is, in principle, sufficient to detect a small number of neutrino-induced events attributed to a neutrino flavour other than that contained in the primary neutrino source.

The second method is less sensitive to mixing amplitudes being limited by the statistics and systematic uncertainties of the absolute intensity of the neutrino source, the neutrino spectrum, and the detection efficiency. However, this method is sensitive to neutrino oscillations into any neutrino species even including "sterile" neutrinos such as $\nu_R$ or $\bar{\nu}_L$ being not observed in nature, but are not excluded in several grand unified theories. In addition, this method can be sensitive to small $\Delta m^2$ by using low-energy neutrino sources.

In Fig.9 the sensitivity of neutrino oscillation experiments for different neutrino sources is schematically indicated.

## 4.2 Reactor Neutrino Oscillation Experiments

### 4.2.1 Reactor Neutrinos

The most intensive terrestrial neutrino sources are nuclear power reactors used for commercial production of electricity. A thermal power of 3 - 4 GW is typical for modern installations of this type, emitting approximately per second and GW thermal power:

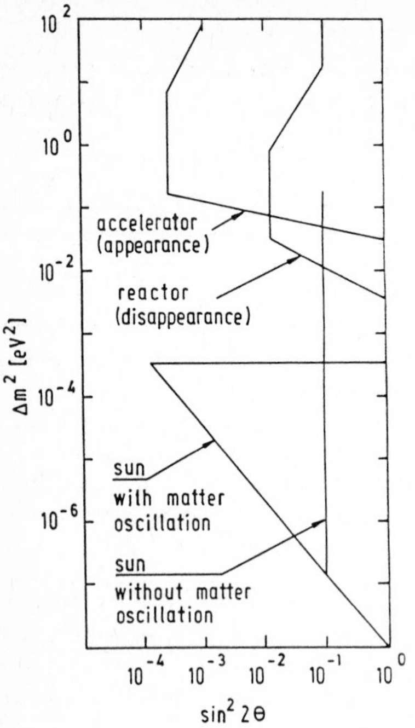


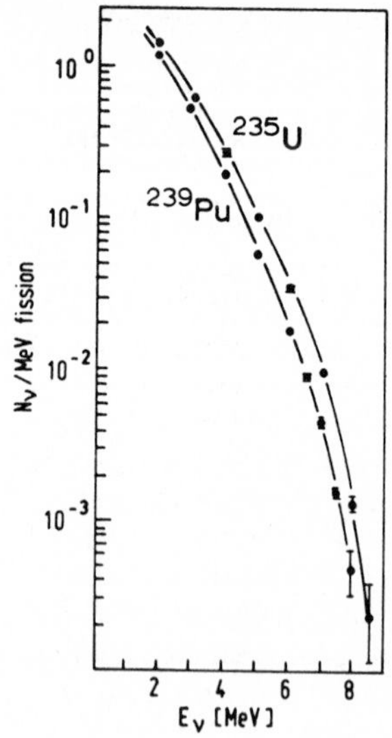


Fig.9. Sensitivity of neutrino oscillation experiments as a function of neutrino sources. Tested areas are to the right of the schematic curves

Fig.10. Experimentally obtained neutrino spectra of $^{235}U$ and $^{239}Pu$ from [Fei82, Schr85]

$$N_\nu \approx 1.6 \cdot 10^{20} \nu/s \; GW.$$

The neutrino energy spectrum stemming from nuclear $\beta$-decay of fission products ranges up to 8 MeV declining almost exponentially from 2 MeV on. There have been numerous attempts to determine the reactor neutrino spectrum on a theoretical basis using available individual $\beta$-spectra from fission products and calculating for those that are unknown. The different theoretical approaches disagreed for higher neutrino energies, by up to 50 % or more [Dav79, Avi80, Kop80, Vog81, Kla82].

There were also several attempts to determine experimentally the $\nu$-spectra by detecting the cumulated $\beta$-spectra from fission products [Fei82, Schr85, Schr86], which give the neutrino spectra from $^{235}U$, $^{239}Pu$, $^{241}Pu$ fission to an accuracy better than 6 % for the relevant energy range. The contributions of $^{238}U$ fast neutron fission and $^{241}Pu$ fission are each less than 8 % of the total neutrino production in a commercial pressurized water reactor and thus do not contribute significantly to the total uncertainty of the neutrino spectrum. Therefore commercial power reactors are not only by far the strongest but also the best determined neutrino sources being extremely pure in electron antineutrinos. Figure 10 shows the experimental neutrino spectra obtained from [Fei82] and [Schr85] for $^{235}U$ and $^{239}Pu$ fission. The predictions of [Kla82] are closest to the experimental results for $^{235}U$ and $^{239}Pu$. For the case of $^{241}Pu$ the situation is similar, the calculations of [Kla86] are here closest to the experimentally given neutrino spectra of [Schr85, Schr86].

### 4.2.2 Neutrino Oscillation Experiments at Reactors

Different experiments, searching for neutrino oscillations, were performed at various reactors. Table I lists these experiments together with the reactor sites and the distances d between the detector and the reactor core serving as neutrino source.

Table I:
Reactor Neutrino Oscillation Experiments

| Distance core-detector | Reactor | Power MW(th) | Reference |
|---|---|---|---|
| 8.75 m | ILL-Grenoble France | 57 | [Kwo81] |
| 13.6 m<br>18.3 m | Bugey France | 2800 | [Cav84] |
| 18.2 m<br>23.7 m | Savannah River / USA | 2300 | [Bau86] |
| 18.5 m<br>25.0 m | Rovno USSR | 1400 | [Afo85] |
| 37.9 m<br>45.9 m<br>64.7 m | Gösgen Switzerland | 2800 | [Zac86] |

The experiment giving the most stringent limits on the parameters $sin^2 2\theta$ and $\Delta m^2$ is the experiment performed at the Gösgen reactor in Switzerland. The reaction

$$\bar{\nu}_e + p \rightarrow n + e^+$$

was used to detect neutrinos like in most other experiments using reactor neutrinos and in particular like the experiment of C.L. Cowan and F. Reines in their first experimental proof of the existence of electron antineutrinos. For the detection of the neutron a $^3He$ multi-wire proportional chamber was used, which is very insensitive to $\gamma$-background radiation. The kinetic energy of the positron was detected in a mineral oil based scintillator, which served at the same time as neutrino target. The whole detector was constructed to be position sensitive leading to a further background suppression. By means of pulse shape discrimination, events stemming from fast neutrons (or protons) could be distinguished from $e^+$-events effectively, which allowed for a further suppression of correlated background events. The detector of the experiment at the Bugey reactor is very similar, but provides no position sensitivity and is shielded differently. In Fig.11 a schematic outline of the experimental set-up as used at the Gösgen reactor is depicted.

The experiment was performed at three different distances: 37.9 m, 45.9 m, and 64.7 m from the reactor core. At each position a total of approximately 10.000 neutrino counts were collected. Background measurements were performed during the reactor shut down intervals for the annual fuel replacement and technical servicing. The experimentally obtained $e^+$-spectra are shown in Fig.12 for the three positions together with the expected spectra using the results from [Fei82, Schr85] and calculations for $^{238}U$ and $^{241}Pu$ fission from [Vog81] (broken line). In addition, a spectrum is indicated, which results from a common fit to the data at all three distances assuming no oscillation (solid line). These two spectra (broken and solid line) agree within the experimental accuracy, indicating that no neutrino oscillation with a big amplitude ($sin^2 2\theta$) is present.

In Fig.13 the obtained areas in the $sin^2 2\theta$, $\Delta m^2$ plane for which neutrino oscillations can be excluded are presented [Zac86]. Two analyses of the experiment were performed. Analysis A is based only on the comparison of the results obtained at the three different distances from

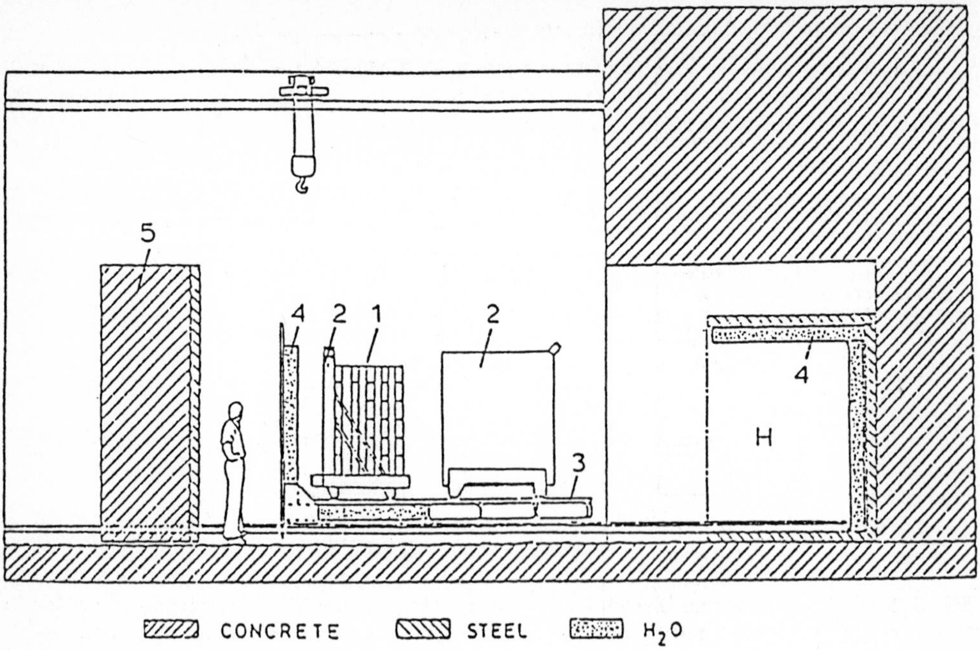


Fig.11. Experimental set-up of the Gösgen neutrino oscillation experiment. (1) central detector unit, (2) veto detector for cosmic muons, (3) rails, (4) water-tanks for n-absorption, (5) concrete door. The neutrinos enter the detector from the right side

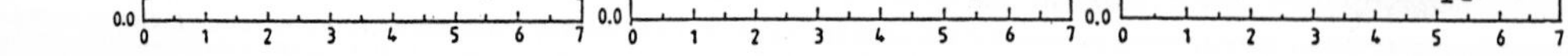

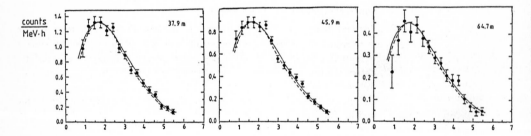

Fig.12. Experimental positron spectra from [Zac86] for three different distances from the detector to the reactor. Expected spectra for no oscillation are indicated with a broken and a solid line (see text)

the reactor, thus being essentially independent of the primary neutrino spectrum and systematic uncertainties in absolute detection efficiencies. Analysis B relies on the results of [Fei82, Schr85, Vog81] as well as on absolute efficiency measurements of the detection system. Due to the additional information used this analysis leads to much more stringent exclusion limits.

For the experiment at the Bugey reactor [Cav84] an evidence for neutrino oscillations with the parameters $\Delta m^2 = 0.2\ (eV)^2$, $sin^2 2\theta = 0.25$ was published. However, this was inconsistent with the results of [Zac86] and has recently been withdrawn [Ker88].

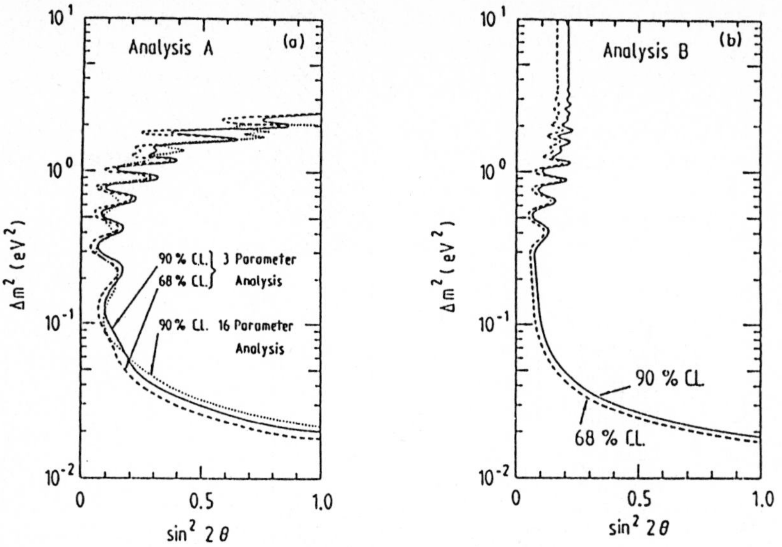


Fig.13. Exclusion limits from [Zac86] for $sin^2 2\theta$ and $\Delta m^2$ derived from an analysis relying exclusively on the information obtained in the experiments with three distances between the detector and the reactor (analysis A) and obtained from an analysis using in addition the information about the reactor neutrino spectra from [Fei82, Schr85, Vog81]. The regions on the right hand side of the curves are excluded

### 4.3 Neutrino Oscillation Experiments at Accelerators

Instead of searching for the disappearance of neutrinos from a neutrino beam with a given flavour or lepton quantum number, there exists the possibility to search for the appearance of other neutrino flavours. If the primary energy is sufficient to overcome the energy threshold for $\mu$ or $\tau$ production in the detector, the transmutation, e.g. of electron neutrinos into $\mu$ or $\tau$ neutrinos, can be identified with a high sensitivity to small mixing angles.

Numerous experiments are performed or prepared at accelerators, searching for this effect. These experiments are listed in Table II.

Table II:
Appearance experiments at accelerators

| Channel | Experiment | Limits on $\Delta m^2$ for $sin^2 2\theta = 1$ | $sin^2 2\theta$ large $\Delta m^2$ |
|---|---|---|---|
| $\nu_\mu \to \nu_e$ | COL-BNL [Bak84] | 0.6 $(eV)^2$ | $6 \cdot 10^{-3}$ |
| | BEBC/PS [Bal86] | 0.09 $(eV)^2$ | $1.3 \cdot 10^{-2}$ |
| | BNL-E734 [Ahr85] | 0.43 $(eV)^2$ | $3.4 \cdot 10^{-3}$ |
| | PS191 [Ber86] | $\Delta m^2 = 5\ eV$ | for $sin^2 2\theta = 0.03 \pm 0.01$ |
| $\bar{\nu}_\mu \to \bar{\nu}_e$ | FNAL [Tay83] | 2.4 $(eV)^2$ | $1.3 \cdot 10^{-2}$ |
| $\nu_\mu \to \nu_\tau$ | FNAL [Ush81] | 3 $(eV)^2$ | $1.3 \cdot 10^{-2}$ |
| | FNAL-E531 [Gau86] | 0.9 $(eV)^2$ | $4 \cdot 10^{-3}$ |
| $\bar{\nu}_\mu \to \bar{\nu}_\tau$ | FNAL [Asr81] | 2.2 $(eV)^2$ | $4.4 \cdot 10^{-2}$ |
| $\nu_e \to \nu_\tau$ | COL-BNL [Bak84] | 8 $(eV)^2$ | 0.6 |
| | FNAL-E531 [Gau86] | 9 $(eV)^2$ | 0.12 |

From these experiments very restrictive limits on neutrino oscillations can be set. There is one piece of evidence for neutrino oscillations quoted in [Ber86], which is going to be checked with a new experiment performed by the same group. This evidence is, however, not in good agreement with the results of [Ahr85].

To get conclusive limits on the complex possibility of three neutrino flavour mixing, an analysis was performed using all experimental data available by August 1985 [Blü85]. Using these data no significant evidence for neutrino oscillations was found. Figures 14a,b,c show the obtained limits for $\nu_e - \nu_\mu$, $\nu_e - \nu_\tau$, and $\nu_\mu - \nu_\tau$ oscillations.

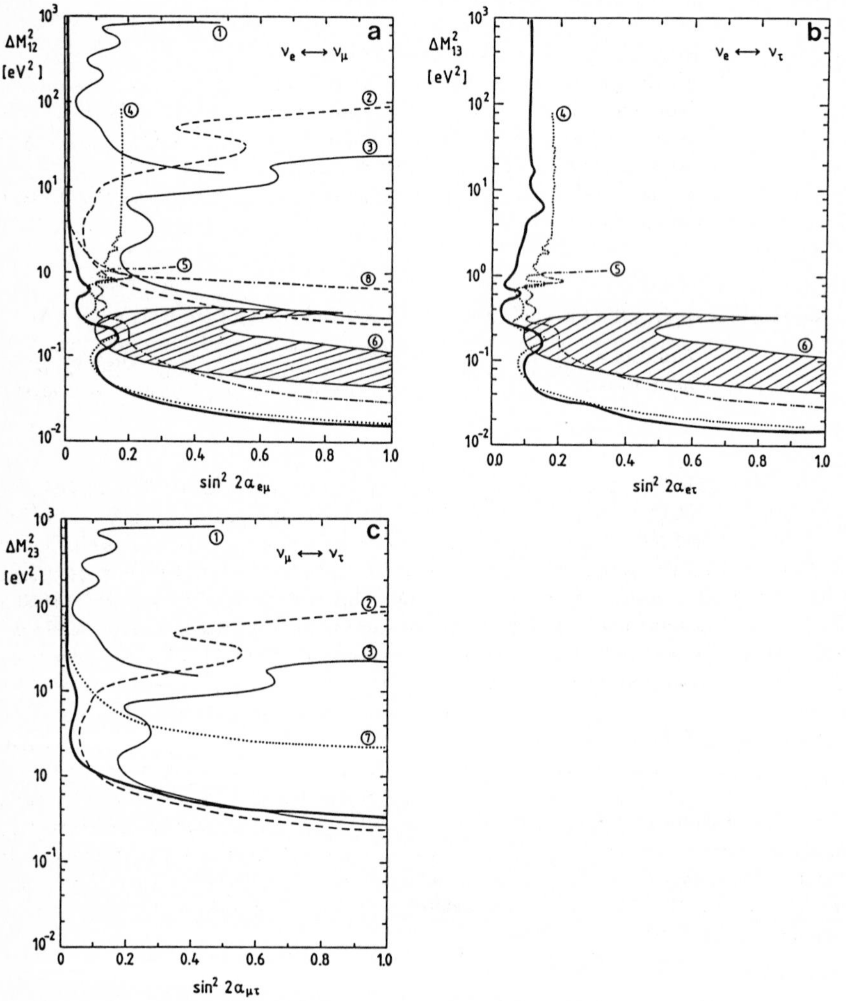


Fig.14. Limits on neutrino oscillations obtained by a common analysis of the different experimental results available for a) $\nu_e - \nu_\mu$, b) $\nu_e - \nu_\tau$, and c) $\nu_\mu - \nu_\tau$ oscillations, respectively [Blü85]. The bolder line indicates the combined limits [Blü85], excluding the regions on the right hand side

Different other experiments have been suggested to detect solar neutrinos, but none of those are accepted for funding up to now. There is one experiment, based on a water Cerenkov counter, using $D_2O$ [Aar87]. This experiment is designed to investigate the $^8B$ neutrino flux via three reactions being $\nu_e d \rightarrow ppe$; $\nu_x e \rightarrow \nu_x e$; and $\nu_x d \rightarrow \nu_x pn$, where $\nu_x$ stands for any left-handed neutrino. Thus via the combined information on charged and neutral current reactions, the neutrino oscillation hypothesis can be tested in more details. The most challenging experiment would be one, which is able to detect the spectrum of the solar neutrinos down to low energies (pp-neutrinos) in real time. Such an experiment would directly measure the relative intensities of the different neutrino components and improve our understanding of the nuclear burning in the sun; in addition it would set very sensitive limits on neutrino oscillations.

In order to perform such an experiment, the use of the reaction $\nu_e +^{115} In \rightarrow^{115} Sn^* + e^-$ was first suggested by [Rag76]. The coincident measurement of the decay $\gamma-$rays from the excited $^{115}Sn^*$, together with the kinetic energy of the emitted electron, might allow a sufficient background suppression, if adequate time and position resolution in the detector is achievable. The background suppression in this case is of special importance due to the natural $\beta$-activity of $^{115}In$. Several attempts to solve this problem are still under way [Boo85, Lun88].

More detailed theoretical aspects of oscillations of solar neutrinos, in particular in matter, are discussed in another chapter of this book.

## 4.4 Solar Neutrino Oscillations

The search for neutrino mixing and neutrino oscillations was to a large extent initiated by an experiment detecting solar neutrinos by means of the neutrino capture in $^{37}Cl$ [Dav68, Row85]

$$^{37}Cl + \nu_e \rightarrow^{37} Ar + e^- \ .$$

The reaction is identified by the subsequent decay of $^{37}Ar(t_{1/2} = 35d)$. The experiment detected only $2.1 \pm 0.3\ SNU(1\sigma)$, where $5.8 \pm 2.2\ SNU(2\sigma)$ are expected using the so-called standard solar model. One SNU stands for $1\nu$-capture per $10^{36}$ target nuclei. There is a first unpublished indication from the Kamiokande II experiment (2140 tons $H_2O$-Cerenkov detector), which seems to confirm this result by means of $\nu_e$- electron scattering events. One possible explanation for the missing solar neutrino flux is an oscillation of the emitted electron neutrinos into other neutrinos being not detected in the experiments. An alternative explanation would be a reduced electron neutrino source strength of the sun in the neutrino energy range detectable in both experiments.

The energy threshold of the $^{37}Cl$ experiment is 814 keV, the threshold for the $H_2O$ Cerenkov detector about 6 MeV. Figure 15 shows the solar neutrino spectrum according to the standard solar model [Bah82]. Only about $10^{-4}$ parts of the total neutrino flux contribute to the reaction rates mentioned above. Thus these experimental results may strongly depend on the understanding of the burning mechanism in the sun. It is crucial for a further understanding of this effect to perform an experiment, which is sensitive to the dominant part of the solar neutrino spectrum. Such an experiment (GALLEX [Ham87]) is currently under preparation, using $^{71}Ga$ as a neutrino target for the reaction

$$^{71}Ga + \nu_e \rightarrow^{71} Ge + e^-,$$

the energy threshold for this reaction is as low as 236 keV. The $^{71}Ge$ is detected via its decay ($t_{1/2} = 11.4d$) by electron capture, similarly to the $^{37}Ar$ in the $^{37}Cl$ experiment.

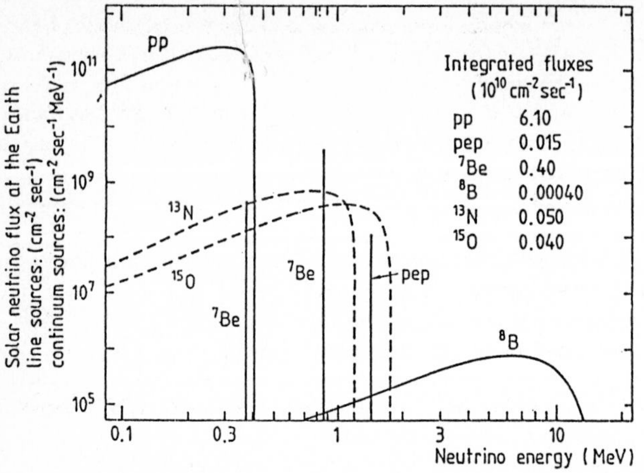


Fig.15. Solar neutrino spectrum as calculated in [Bah82]. The contributions to the neutrino flux stemming from different nuclear fusion processes in the sun are indicated

## 5. Neutrino Decay

### 5.1. General Considerations

If the neutrino wave function involves non-vanishing mass eigenstates the question arises whether radiative decays are possible. One could think of a variety of decay channels like

$$\nu \rightarrow \nu' + l_i^- + l_j^+ \qquad l_{i,j} = e, \mu, \ldots, \tag{a}$$

if the masses $m_\nu$ and $m_{\nu'}$ differ by more than $m_{l_i} + m_{l_j}$ , and

$$\begin{aligned} \nu &\rightarrow \nu' + \gamma \\ &\rightarrow \nu' + \gamma + \gamma \ , \\ &\rightarrow \nu' + Scalar \end{aligned} \tag{b}$$

where the scalar particle can be a pseudo Goldstone particle or a light Higgs particle. In addition, the "invisible" decay

$$\nu \rightarrow \nu' + \nu^* + \bar{\nu}^*,$$

where $\nu^*$ can be any neutrino-like particle, is to be expected.

For an experimental investigation the decays (a) and (b) are accessible. There are numerous arguments leading to restrictive limits for these neutrino decays.They are essentially based on cosmological and astrophysical observations. These arguments are discussed in a dedicated chapter of this book. Here we want to restrict our considerations to pure laboratory observations, which are independent of assumptions about the history of the universe and stellar evolution.

### 5.2 Neutrino Decay into Charged Leptons

A neutrino decay into charged leptons would be an inevitable consequence within the standard model if the mass difference $m_\nu - m_{\nu'}$ exceeds the mass of the emitted charged lepton pairs. The dominant part of the interaction occurs via the graph shown in Fig.16.

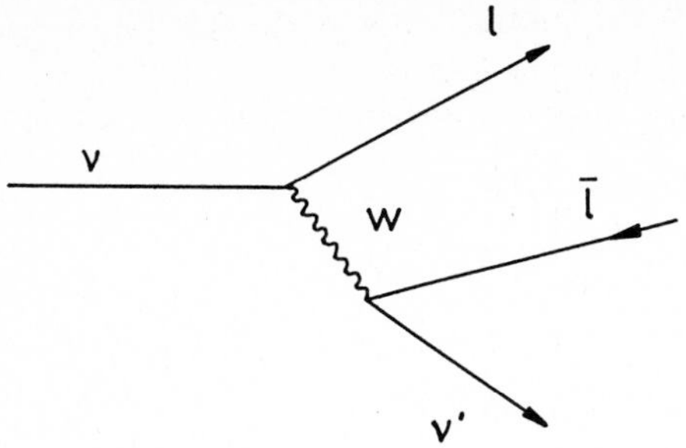


Fig.16. Heavy neutrino decay in charged $l\bar{l}$ pair

This interaction can be calculated on the basis of the standard model in analogy to the decay of the muon

$$\mu^- \rightarrow \nu_\mu + e^- + \bar{\nu}_e.$$

Assuming a decay of the neutrino into $e^+, e^-$ and $m'_\nu \rightarrow 0$ we get for the inverse lifetime of the heavy neutrino

$$\tau^{-1}(\nu \rightarrow \nu' + e^+ + e^-) = \frac{G_F^2 m_\nu^5}{192\pi^3} |U_{eH}|^2, \tag{27}$$

with $m_\nu$ in MeV and $G_F^2 = 0.206\ MeV^{-5}s^{-1}$.

Therefore, an experiment testing for these decay modes offers direct information on the neutrino mass difference $m_\nu - m_{\nu'}$ and neutrino mixing amplitudes. If the neutrino source emits electron or muon neutrinos a decay in $e^+ e^-$ pairs or heavier leptons is only possible for a sufficiently heavy neutrino mass eigenstate, mixed into the electron or muon neutrino wave function. For the electron and muon neutrinos the dominant part is proven to have a mass $m_{\nu_e} < 40\ eV$ and $m_{\nu_\mu} < 250\ KeV$, respectively. These are below the threshold for the decay mentioned above. The $\tau$-neutrino mass has an upper limit of $m_{\nu_\tau} < 35\ MeV$ and it seems that an experimental limit from direct measurements like that described in [Schu87] is probably very difficult to be obtained for masses below about 10 MeV. Therefore a search for the decay into $e^+ e^-$ pairs offers a unique possibility to investigate this mass range. This looks particularly interesting if one takes into account estimations based on cosmic $\gamma$-radiation data and cosmic nucleosynthesis. H. Harrari and Y. Niv [Har87] derive the following limits for $m_{\nu_\tau}$ using these arguments

$$m_{\nu_\tau} < 65\ eV,$$

or

$$900\ keV < m_{\nu_\tau} < 35\ MeV.$$

5.2.1 Experimental Limits on the Decay into Charged Leptons

From (27) we can directly estimate the decay rate in a detector volume V if the neutrino source emits neutrinos with which heavy neutrino masses are mixed with an amplitude $| U_{ik} |^2$. The decay rate is then

$$\frac{dN}{dt} = \frac{m_\nu}{\tau_{c.m.}} \frac{1}{cE_\nu} \frac{Vn_\nu(E_\nu)}{4\pi d^2} | U_{ik} |^2, \tag{28}$$

if the neutrinos are emitted isotropically from a source at a distance $d$ from the detector. $n_\nu(E_\nu)$ is the neutrino spectrum and $\tau_{c.m.}$ the center-of-mass neutrino lifetime.

The spectrum of the emitted $e^+ e^-$ pairs in the laboratory system is calculated from kinematics to be

$$\frac{dN}{dE} = \frac{m_\nu}{\tau_{c.m.}} \frac{V \mid U_{ik} \mid^2}{4\pi d^2} \int \frac{n(E_\nu)}{E_\nu} \frac{dn}{dE}(E, E_\nu) dE_\nu . \tag{29}$$

For the final expected counting rates we have to integrate over that part of the spectrum for which the detector is sensitive.

At high-energy accelerators different groups have searched for decays into charged leptons [Ahr87, Ber86]. The neutrino production and decay investigated in [Ahr87] is depicted in Fig.17. Traces of $\mu^-$ and $e^+$ in the detector are searched for. After subtraction of background events, stringent limits on the parameters $\mid U_{eH} \mid^2$ and $\mid U_{H\mu} \mid^2$ could be set as a function of the neutrino mass eigenvalue of $\nu_H$.

Figure 18 shows a summary of experimental limits obtained at accelerators for $\mid U_{eH} \mid^2$ and $\mid U_{H\mu} \mid^2$ up to neutrinos masses of 500 MeV [Ahr87].

As discussed previously, at nuclear power reactors neutrinos with energies up to 8 MeV are emitted. Thus equivalent tests can be performed for neutrino mixing $\mid U_{eH} \mid^2$ and neutrino decay

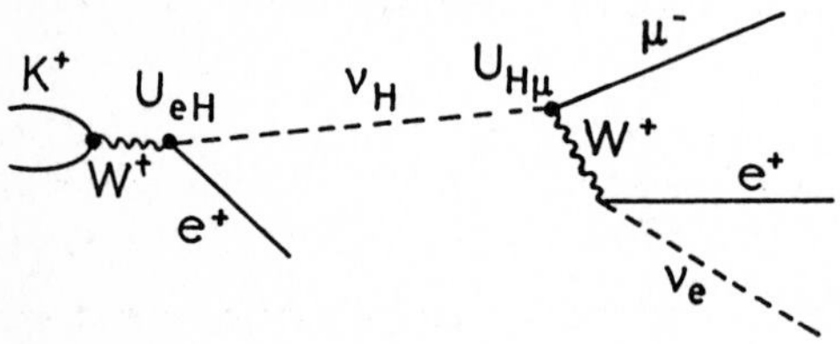


Fig.17. Heavy neutrino production and sequential decay diagram [Ahr87]

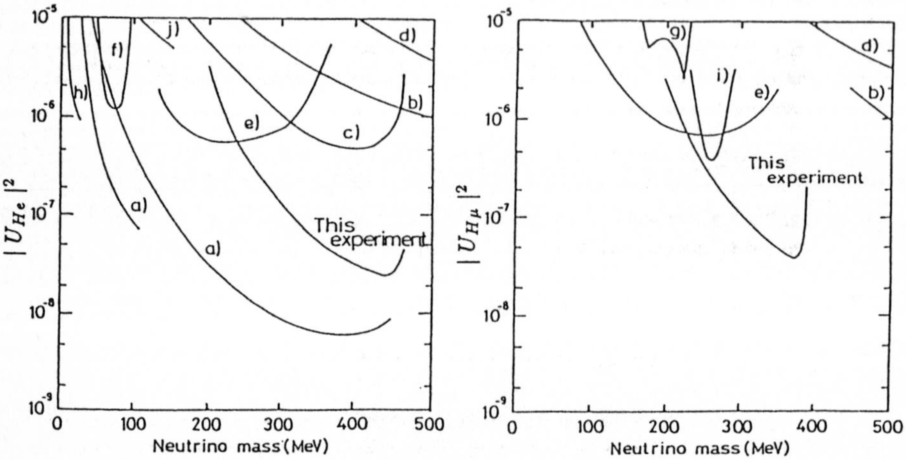


Fig.18. 90 % confidence limits of mixing matrix elements $\mid U_{He} \mid^2$ and $\mid U_{H\mu} \mid^2$ for different values of $m_{\nu_H}$, from [Ahr87] and references therein. Excluded are the areas above the curves

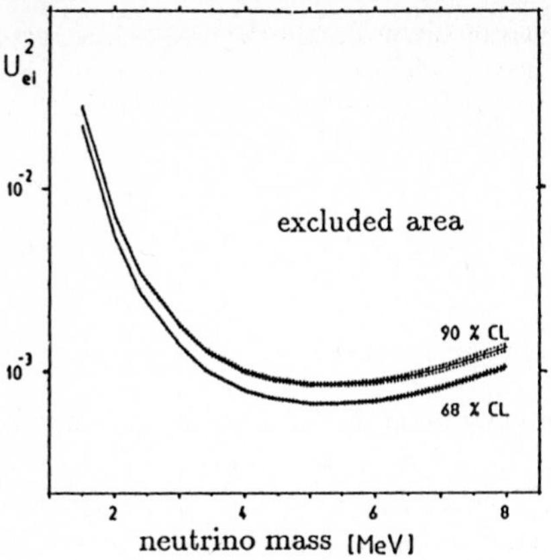


Fig.19. Exclusion limits for heavy neutrino mixing $| U_{eH} |^2$ obtained from the search for the decay of reactor neutrinos $\bar{\nu}_e$ into $e^+, e^-$ pairs [Obe87]

giving information in the range $1\ MeV < m_{\nu H} < 8\ MeV$. Such an experiment was performed at the Gösgen nuclear power reactor [Obe87]. In the experiment reactor correlated signals from $e^+e^-$ pairs were searched for in 30 liquid scintillator cells, containing in total 375 l. The limits obtained from this experiment are shown in Fig.19 for 68 % and 90 % c.l.

5.3 Neutrino Gamma Decay

If the neutrino mass is smaller than $2m_e$, then the remaining dominant visible decay mode would be

$$\nu \rightarrow \nu' + \gamma.$$

Following again the standard model the leading graphs to the decay would be those shown in Fig.20.

The calculation of this decay mode is strongly model-dependent. The standard theory predicts extremely long lifetimes.

$$\tau^{-1}(\nu \rightarrow \nu' + \gamma) = \frac{1}{8\pi} \left\{ \frac{m_\nu^2 - m_{\nu'}^2}{m_\nu} \right\}^3 (| a |^2 + | b |^2) . \tag{30}$$

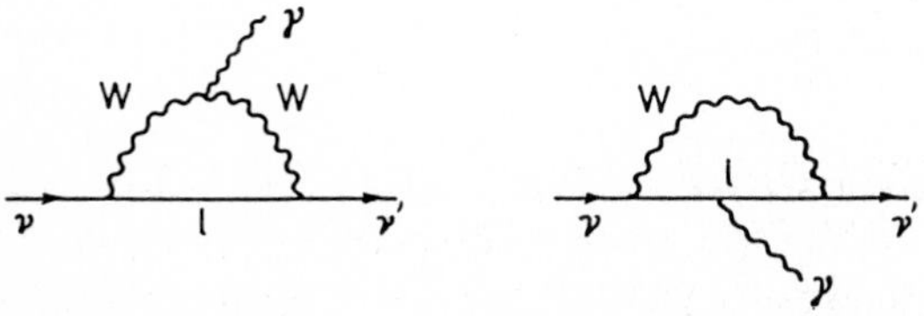


Fig.20. Graphs for the leading contributions to the $\gamma$-decay of neutrinos following the standard model

For Dirac neutrinos and $M_\nu << m_l$, where $m_l$ is the mass of the intermediate lepton l, we get

$$a_D = -\frac{eG_F}{8\sqrt{2}\pi^2}(m_\nu + m_\nu^{'}) \sum_l U_{1l}U_{2l}^* F\left(\frac{m_l}{M_W}\right)^2,$$

$$b_D = \frac{eG_F}{8\sqrt{2}\pi}(m_\nu - m_\nu^{'}) \sum_l U_{1l}U_{2l}^* F\left(\frac{m_l}{M_W}\right)^2,$$

where $U_{il}$ are mixing parameters, $F\left(\frac{m_l}{M_W}\right)^2$ is a smooth function of the lepton mass $m_l$, and $M_W$ the mass of the intermediate W-boson.

For three lepton families and $m_\nu^{'} \to 0$ we get

$$\tau_{Standard}^{-1} \approx \left\{10^{44}s\right\}^{-1} \left\{\frac{m_\nu}{eV}\right\}^5 sin^2\theta,$$

where $sin^2\theta$ is the mixing amplitude.

However, there are numerous theoretical approaches, leading to much shorter lifetimes as well [Roo87], which are shown in Fig.21.

To set experimental limits on the $\gamma$-decay of neutrinos the effect was searched for at a nuclear power reactor, in neutrinos from the sun and, recently, in neutrinos from the supernova explosion SN1987A in the Large Magellanic Cloud.

Due to angular momentum conservation the gamma emission in an assumed decay may be anisotropic in the center of mass (c.m) system. The gamma emission follows the angular distribution

$$dN\gamma_{(c.m)} = \frac{1}{2}(1 + a \cdot cos\theta)dcos\theta, \tag{31}$$

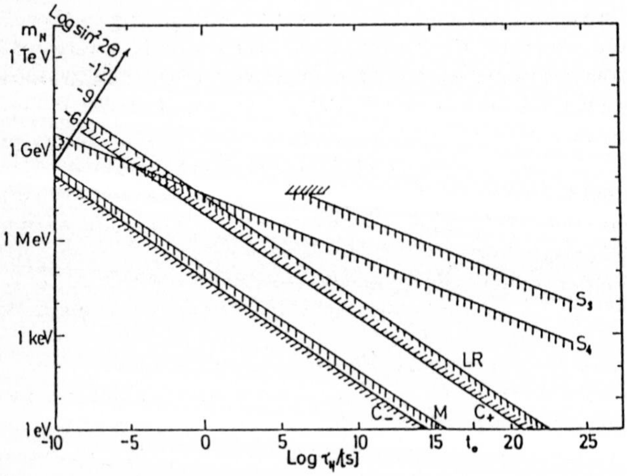


Fig.21. Theoretical limits for neutrino masses $m_\nu$ versus radiative lifetime $\tau_\nu$. $S_3$ and $S_4$ correspond to the standard model with three and four lepton families, respectively [Pal82], LR to left-right symmetric models [Pat74], M to the mirror model [Pat75] and $C_-$ and $C_+$ to a model describing neutrino excitations [Gri86]

where $a$ accounts for the polarization of $\nu$ and $\nu'$ and may have values in the range $-1 < a < +1$. Assuming an isotropic neutrino emission from the source emitting a spectrum $n_\nu(E_\nu)$, we expect at the distance $d$ the decay rate in a volume $V$ to be

$$\frac{dN_\nu}{dt} = \frac{m_\nu}{\tau_{cm}} \frac{1}{cE_\nu} \frac{Vn_\nu(E_\nu)}{4\pi d^2}. \tag{32}$$

Here $m_\nu/E_\nu$ accounts for the time dilation. In the laboratory system we get, together with (31), the photon spectrum

$$\frac{dN}{dE_\gamma} = \frac{m_\nu}{\tau_{c.m.}} \frac{V}{c4\pi d^2} \int_{E_\nu}^{\infty} \left(1 - a + \frac{2aE\gamma}{E_\nu}\right) \frac{n_\nu(E_\nu)dE_\nu}{E_\nu^2}. \tag{33}$$

An experiment [Obe87] investigating reactor neutrinos gave, in addition to the quoted limits for $\nu \to \nu' + e^+ + e^-$, limits on the decay $\nu \to \nu' + \gamma$ being analyzed in terms of (33). These results are listed in Table III for three representative values of $a$, assuming $m_{\nu'} << m_\nu$.

Table III:
Limits for radiative neutrino decay from [Obe87]

| a | $m_\nu/\tau_{c.m.}$ |
|---|---|
| -1 | $\leq 4.5 \cdot 10^{-2} eV/s$ |
| 0 | $\leq 2.6 \cdot 10^{-2} eV/s$ |
| +1 | $\leq 1.7 \cdot 10^{-2} eV/s$ |

Much more stringent limits for this decay mode can be set using solar neutrinos, presuming the solar neutrino spectrum would be known. The missing solar neutrinos according to [Row85] seem not to confirm this assumption. However, the rate measured there is only a very small part of the total solar neutrino flux, as mentioned earlier. More than 90 % of the solar neutrino spectrum stems from the pp-cycle (see Fig.15) and is largely model independent, thus representing a reasonably good base for the experiment discussed here.

The $\gamma$-flux from the direction of the sun was investigated in terms of neutrino decay, again using (33) [Raf85]. Here lifetime limits were analyzed as a function of $r$ where $r$ is the ratio between the actual center-of-mass photon energy and its possible maximal value $m_\nu/2$ obtained for $m_{\nu'} \to 0$. These results are depicted in Fig.22.

The most restrictive limits for the neutrino decay into photons can be set on the basis of data obtained with the supernova explosion SN 1987A, which are practically equivalent to a laboratory experiment.

The supernova explosion was observed on February 23, 1987, and occurred at a distance of approximately 150.000 light years from the earth. At least two detectors registered a neutrino pulse with a duration of approximately 10 s, which is attributed to the optically observed star explosion [Bio87, Hir87]. From the detected neutrino spectrum a neutrino temperature of $2\ MeV < T_{\bar{\nu}_e} < 4.5\ MeV$ [Ara87] can be evaluated on the basis of a Fermi-Dirac distribution. As the neutrino detectors (Kamiokande II and IMB) are both large water Cerenkov detectors being to more than 90 % sensitive to the reaction $\bar{\nu}_e + p \to n + e^+$, this temperature has to be attributed to electron antineutrinos.

The temperature belongs to a region within the supernova where the neutrinos are thermalized last. Using this temperature we can calculate the neutrino production of different flavours on the basis of the standard theory [Wil86] by means of the graphs shown in Fig.23.

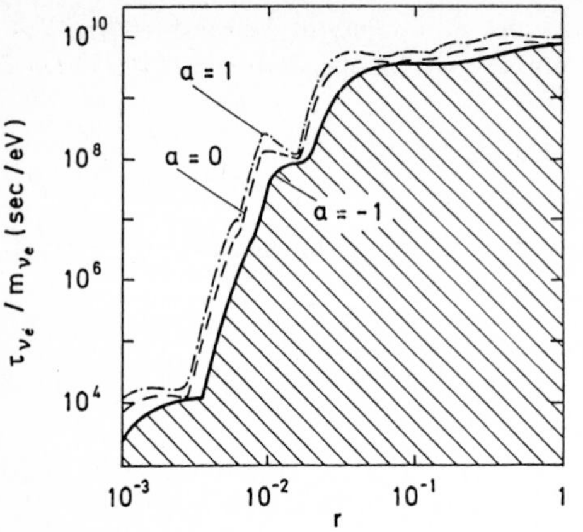


Fig.22. Possible values for the electron neutrino lifetime over mass ratio $\tau_{\nu_e}/m_{\nu_e}$ as a function of the energy fraction $r = 2 \cdot E\gamma/m_{\nu_e}$. The excluded region is the shaded area [Raf85]

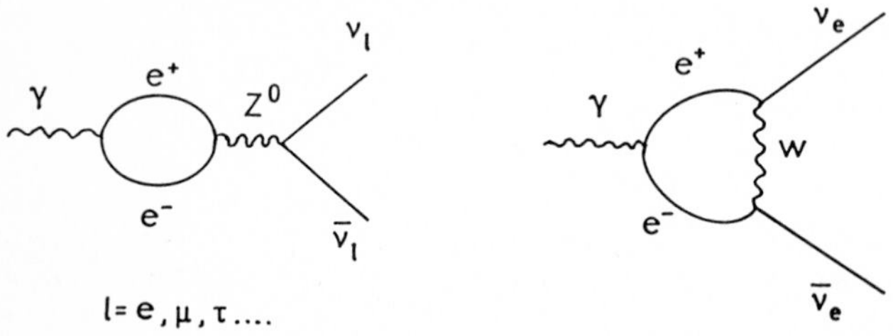


Fig.23. Graphs describing the thermal photoproduction of neutrinos

Thus with the measured rate and energy distribution of the electron antineutrino burst of SN 1987A the total neutrino spectrum including $\mu$ and $\tau$ neutrinos is determined on the basis of the standard theory, if this spectrum is not altered after the neutrino production due to later neutrino interaction with the stellar medium. For several years the Solar Maximum Mission satellite has measured the $\gamma$ -flux in orbit [For80, Ves87]. This measurement covers the direction of the supernova within the 10 s interval of the neutrino burst. No significant increase of the $\gamma$-rate in the energy range 4.1 $MeV < E_\gamma <$ 6.4 $MeV$ was detected. With these data a limit for the stability of the electron antineutrino can be deduced [Fei88] being

$$\tau/m_{\bar{\nu}_e} < 8.3 \cdot 10^{14} s/eV,$$

for $T_{\bar{\nu}_e} = 3\ MeV$ and $m_{\bar{\nu}_e} < 20\ eV$, as verified by other experiments. The limits obtained for $\nu_\mu$ and $\nu_\tau$ are given in Fig.24 as a function of the neutrino mass [Obe87, Obe87a]. In contrast to the limit for $\tau_{\nu_e}$, these limits are dependent on possible interactions the neutrinos may undergo on their way through the outer regions of the supernova.

To conclude this section, we can state that very stringent limits on neutrino decays are achieved on the basis of laboratory data. There remains only a small range for neutrino decays into $e^+e^-$-pairs with neutrino masses $1\ MeV < m_\nu < 35\ MeV$ to be investigated.

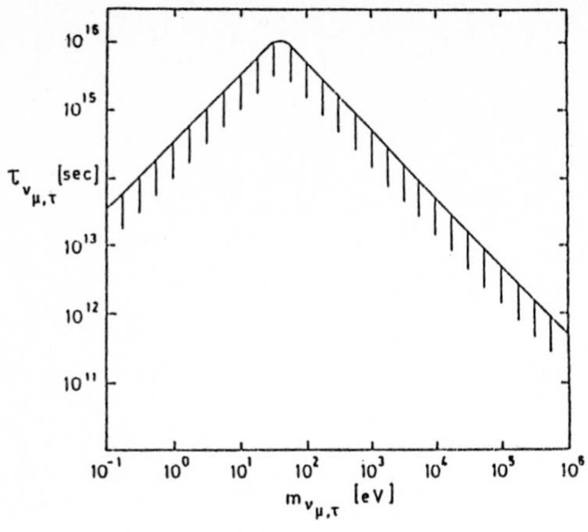


Fig.24. Lower limits on $\mu$-, $\tau$-neutrino life times as a function of $m_{\nu_{\mu,\tau}}$ up to 1 MeV. The area below the curve is excluded

## 6. Magnetic Moment of Neutrinos

In this section we want to show how new physics beyond the standard theory could be tested via the search for a magnetic moment of neutrinos. Within the standard theory, the expected magnetic moment of neutrinos is given by [Lee 77]

$$\mu_\nu = 10^{-19} \left(\frac{m_\nu}{1eV}\right) \mu_B,$$

where $m_\nu$ is the neutrino mass in eV and $\mu_B$ the Bohr magneton. Therefore, even for the present limit on the $\tau$-neutrino mass of 35 MeV it seems to be impossible to detect this value with present experimental techniques. However, for grand unification theories with left-right symmetry or theories describing neutrinos as composite particles, much larger magnetic moments [Moh80, Dom70] are allowed. In addition, there are models predicting neutrino magnetic moments even without neutrino rest masses [Fuj80].

If CPT invariance holds, then Majorana neutrinos, in contrast to Dirac neutrinos, cannot have a magnetic or electric dipole moment. Thus an experimentally verified, non-vanishing neutrino magnetic moment would directly prove that the neutrino is a Dirac particle.

There exist various experimental limits on $\mu_\nu$ as

$$|\mu_{\nu_e}| < 1.5 \cdot 10^{-10} \mu_B,$$

and

$$|\mu_{\nu_\mu}| < 1.2 \cdot 10^{-9} \mu_B$$

from accelerator data [Kyu 84], and

$$\mu_{\nu_e} < 2 \cdot 10^{-11} \mu_B$$

from cosmological arguments [Mor81]. This latter bound is based on the big bang theory and on the assumption that the synthesis of $^4He$ in the early universe was not affected by the excitation of additional neutrino helicity states due to electromagnetic interaction of neutrinos. There exists also an astrophysical bound being based on arguments considering the energy loss of degenerate dwarf stars due to neutrino pair emission. This bound is valid for neutrinos with masses smaller than 10 eV being:

$$\left[\sum_{l=e,\mu,\tau...} |\mu_{\nu_l}|^2\right]^{\frac{1}{2}} < 8.5 \cdot 10^{-4}\mu_B.$$

It would be very useful to obtain rigorous laboratory limits on $\mu_\nu$ independent on any unproven suppositions.

The existence of a magnetic moment can be determined in three ways [Cli87]:

1) By deviations from expectations of the standard model in $\nu_e$ and $\nu_\mu$ scattering off electrons at high energies.

2) By direct detection of $\nu_e, e$ scattering through the magnetic moment interaction at low energies ($E_\nu < 1MeV$).

3) By the observation of coherent interactions of neutrinos in a material (i.e. a crystal) through the magnetic moment interaction.

In addition, there exists the possibility that the solar neutrino flux could be influenced by the solar magnetic field. In fact the lack in the boron solar neutrino flux could be attributed to such an effect. It was suggested that the neutrino flux measured in the Homestake experiment could be related to the solar activity [Row85, Bar87], which could constitute evidence for a non-vanishing neutrino magnetic moment.

The cross section for magnetic moment interaction grows like $\ln(E_\nu)$ whereas the cross section for the interaction via $Z^0$ or $W$ exchange is approximately proportional to $E_\nu^2$ at neutrino energies in the range of several MeV and below. Therefore, the magnetic moment interaction may become dominant at very low neutrino energies. This could ultimately offer a unique possibility to detect neutrinos of very low energy.

## References


[Abe84] R. Abela et al.: Phys. Lett. 146B, 431 (1984)

[Afo85] A. Afoniu et al.: IETP Lett. 42, 285 (1985)

[Ahr85] L.A. Ahrens et al.: Phys. Rev. D31, 2732 (1985)

[Ahr87] L.A. Ahrens et al.: In the Proc. Telemark IV, Conference on "Neutrino Masses and Neutrino Astrophysics" Ashland, Wisconsin 1987 (World Scientific Singapore) 99,
References for Fig.19:
a) G. Bernardi et al.: Phys. Lett. 166B, 497 (1986)
b) I. Corenbosch et al.: Phys. Lett. 166B, 473 (1986)
c) F. Bergsam et al.: Phys. Lett. 128B, 361 (1983)
d) A.M. Cooper-Sarkar et al.: Phys. Lett. 160B, 207 (1985)
e) R.S. Hayano et al.: Phys. Rev. Lett. 49, 1305 (1982); T. Yamazaki: In Proc. 22nd Int. Nat. Conf. on High-Energy Physics (Leipzig), p. 262
f) D.A. Brymann et al.: Phys. Rev. Lett. 50, 1546 (1982)
g) Y. Asano et al.: Phys. Lett 104B, 84 (1981)
h) D. Toussaint, F. Wilczek: Nature 289, 777 (1981)
i) C.N. Pang et al.: Phys. Rev. D8, 1989 (1973)
j) K. Herarel et al.: Phys. Lett. 55B, 327 (1975); J. Heintze et al.: Phys. Lett. 60B, 302 (1976)

[Alb85] H. Albrecht et al. (ARGUS): Phys. Lett 163B, 404 (1985)

[Alb88] H. Albrecht et al. (ARGUS): Phys. Lett. 202B 149 (1988)

[Alt85] T. Altzitzoglou et al.: Phys. Rev. Lett. 55, 799 (1985)

[Ara87] J. Arafune, M. Fukugita: Phys. Rev. Lett. 59, 367 (1987); D.N. Schramm, XXIInd Rencontre de Moriond, March 1987

[Asr81] A.E. Asratyan et al.: Phys. Lett. 105B, 301 (1981)

[Avi80] F.T. Avignone III, Z.D. Greenwood: Phys. Rev. C22, 594 (1980)

[Bac79] W. Bacino et al. (DELCO): Phys. Rev. Lett. 42, 749 (1979)

[Bah82] J.N. Bahcall et al.: Rev. Mod. Phys. 54, 767 (1982)

[Bak84] N.J. Baker et al.: Phys. Rev. D28, 2705 (1984)

[Bal86] M. Baldo-Ceolin: XXIII Int. Conf. on High Energy Physics, Berkeley (1986)

[Bar87] R. Barbieri, G. Fiorentini: In Proc. Telemark IV Conference on "Neutrino Masses and Neutrino Astrophysics", Ashland, Wisconsin 1987 (World Scientific, Singapore)

[Bau86] N. Baumann et al.: In Proc. 12th International Conference on Neutrino Physics and Astrophysics, Sendai (Japan), June, 1986, ed. by T. Kitagaki and H. Yuta (World Scientific, Singapore)

[Ber72] K.E. Bergkvist: Nucl. Phys. B39, 317 (1972); B39, 371 (1972)

[Ber85] K.E. Bergkvist: Phys. Lett B154, 224 (1985)

[Ber86] G. Bernardi: Phys. Lett. 166B, 479 (1986)

[Bil76] S.M. Bilenky, B. Pontecorvo: Phys. Lett. 61B, 248 (1976)

[Bil78] S.M. Bilenky, B. Pontecorvo: Phys. Rep. 41, 225 (1978)

[Bil87] S.M. Bilenky, S.T. Petkov, Rev. Mod. Phys. 59, 671, (1987)

[Bio87] R.M. Bionta et al.: Phys. Rev. Lett. 58, 1494 (1987)

[Blü85] H. Blümer, K. Kleinknecht: Phys. Lett. 161B, 407 (1985)

[Boo85] N.E. Booth et al.: In Proc. of Solar Neutrinos and Neutrino Astronomy, Amer. Inst. of Phys. No. 126, 216 (1985)

[Bow86] T.J. Bowles et al.: In Proc. Int. Symposium on "Weak and Electromagnetic Interactions in Nuclei", Heidelberg, July 1986, Springer, Heidelberg, ed. by H.V. Klapdor

[Cav84] J.F. Cavaignac et al.: Phys. Lett B148, 387 (1984)

[Cli87] D.B. Cline: In Proc. Telemark IV Conference on "Neutrino Masses and Neutrino Astrophysics", Ashland, Wisconsin 1987 (World Scientific, Singapore)

[Dau79] M. Daum et al.: Phys. Rev. D20, 2692 (1979)

[Dau87] M. Daum et al.: Phys. Rev. D30, 2624 (1987)

[Dav68] R. Davis, D.S. Harmer, K.C. Hoffmann: Phys. Rev. Lett. 20, 205, (1968)

[Dav79] B.R. Davis et al.: Phys. Rev. C24, 2259 (1979)

[Dom70] G. Domogatshi, D. Nadezhirn: Yad Fiz. 12, 1233 (1970)

[Eli74] S.E. Elizier, D.A. Ross: Phys. Rev. D10, 3088 (1974)

[Fei82] F. von Feilitzsch, A.A. Hahn, K. Schreckenbach: Phys. Lett. 118B, 162 (1982)

[Fei88] F. von Feilitzsch, L. Oberauer: Phys. Lett. 200B, 580 (1988)

[For80] D.J. Forrest et al.: Solar Phys. 65, 15 (1980)

[Fri76] H. Fritzsch, P. Minkowski: Phys. Lett. 62B, 72 (1976)

[Fri86] M. Fritschi et al.: Phys. Lett. 173B, 485 (1986)

[Fuj80] K. Fujikawa, R.S. Schrock: Phys. Rev. Lett. 55, 963 (1980)

[Gau86] A. Gauthier: In Proc. XXIII Int. Conf. of High-Energy Physics, Berkeley (World Scientific, 1986, Singapore)

[Gri86] J.A. Grifols et al.: Phys. Lett. 171B, 303 (1986)

[Ham87] W. Hampel: In Proc. Workshop on Neutrino Physics, Heidelberg, Oct. 1987, Springer, Heidelberg, 1988, ed. by H.V. Klapdor and P. Povh

[Har87] H. Harrari, Y. Nir: Phys. Lett. 188B, 163 (1987)

[Hay82] R.S. Hayano et al.: Phys. Rev. Lett. 49, 1305 (1982)

[Hir87] K. Hirata et al.: Phys. Rev. Lett. 58, 1490 (1987)

[Kel80] B.H.J. Mc Kellar: Phys. Lett. 97B, 93 (1980)

[Ker88] H. de Kerret: In Proc. of the XXIII rd. Rencontres de Moriond, Neutrinos and Exotic Phenomena, Les Arc, Savoie, Jan. 1988, ed. by O. Fackler and J. Tran Thanh Van

[Kla82] H.V. Klapdor, J. Metzinger: Phys. Rev. Lett. 48, 127 (1982) and Phys. Lett. 112B, 22 (1982)

[Kla86] H.V. Klapdor: Progr. Part. Nucl. Phys. 17, 419 (1986)

[Kop80] W.I. Kopeikin: Yad. Fiz. 32, 1507 (1980)

[Kwo81] H. Kwon et al.: Phys. Rev. D24, 1097 (1981)

[Kyu84] A.V. Kyuldjiev: Nucl. Phys. B243, 387 (1984)

[Lee77] B. Lee, R.S. Schrock: Phys. Rev. D16, 1444 (1977)

[Lip85] E. Lippmaa et al.: Phys. Rev. Lett. 54, 285 (1985)

[Lub86] V.A. Lubimov: In Massive Neutrinos in Astrophysics and in Particle Physics, Proc. VIth Moriond Workshop, Tignes, Savoie, France (1986), Editions Frontières, Gif-sur-Yvette, ed. by O. Fackler and J. Tran Thanh Van

[Lun88] J.C. Lund et al.: Indium Phosphide Particle Detectors for Low Energy Solar Neutrino Spectroscopy, int. rep., Radiation Monitoring Devices, Watertown, Ma 02172, USA, (1988)

[Mak62] Z. Maki, M. Nakagawa, S. Sakata, Prog. Theor. Phys. 28, 870 (1962)

[Mik88] L.A. Mikaelyan et al: Report presented at the 13th Int. Conf. on Neutrino Physics and Astrophysics, Boston, Massachusetts, June 1988.

[Moh80] M. Mohapatra, R. Marshak: Phys. Lett. 91B, 222 (1980)

[Mor81] J. Morgan: Phys. Lett. 102B, 247 (1981)

[Nik84] N. Nikolaev et al.: JETP Lett. 39, 441 (1984)

[Obe87] L. Oberauer et al.: Phys. Lett. 198B, 113 (1987) and PHD thesis at the Technical University Munich, 1988

[Obe87a] L. Oberauer, F. von Feilitzsch: In Proc. Workshop on Neutrino Physics, Heidelberg, Oct. 1987, Springer, Heidelberg 1988, ed. by H.V. Klapdor and B. Povh

[Ohi85] T. Ohi et al.: Phys. Lett 160B, 322 (1985)

[Pal82] P.B. Pal, L. Wolfenstein: Phys. Rev. D25, 766 (1982) and references therein

[Par84] Particle Data group: Rev. Mod. Phys. 56, No. 2 Part II (1984)

[Pat74] C. Pati, A. Salam: Phys. Rev. D10, 275 (1974)
R.N. Mohapatra, J.C. Pati: Phys. Rev. D11, 599 (1975); D11, 2558 (1975)
G. Senjanovic, R.N. Mohapatra: Phys. Rev. D12, 1502 (1975)

[Pat75] J.C. Pati, A. Solam: Phys. Lett. 58B, 333 (1975)
F. Wilczek, A. Zee: Phys. Rev. D25, 553 (1983)

[Raf85] G. Raffelt: Phys. Rev. D31, 3002 (1985)

[Rei59] F. Reines, C.L. Cowan: Phys. Rev. 113, 273 (1959)

[Roo87] M. Roos: In Proc. Workshop on Neutrino Physics, and references therein, Heidelberg Oct. 1987, Springer, Heidelberg 1988, ed. by H.V. Klapdor and B. Povh

[Row85] J. Rowley, B. Cleveland, R. Davis: In AIP Conference Proc. No. 126 (Homestake 1984), ed. by M.L. Cherry, New York 1985

[Schr81] E. Schrock: Phys. Lett. 96B, 159 (1980); Phys. Rev. D24, 1275 (1981)

[Schr83] K. Schreckenbach, G. Colvin, F.v. Feilitzsch: Phys. Lett. 129B, 265 (1983)

[Schr85] K. Schreckenbach et al.: Phys. Lett. 160B, 325 (1985)

[Schr86] K. Schreckenbach et al.: Int. Symposium on Weak and Electromagnetic Interactions in Nuclei, Heidelberg, 1-5 July 1986, Springer, Heidelberg 1986, ed. by H.V. Klapdor

[Schu87] K.R. Schubert: In Proc. Workshop on Neutrino Physics, Heidelberg, Oct. 1987, Springer, Heidelberg, 1988, ed. by H.V. Klapdor and B. Povh

[Sim81] J.J. Simpson: Phys. Rev. D24, 2971 (1981)

[Sim84] J.J. Simpson: Phys. Rev. D30, 1110 (1984)

[Sim85] J.J. Simpson: Phys. Rev. Lett. 54, 1891 (1985)

[Smi81] L.G. Smith, E. Koets, A.H. Wapstra: Phys. Lett 102B, 114 (1981)

[Ste80] B. Stech: In Unification of Fundamental Particle Interactions, ed. by S. Ferrara, J. Ellis, P. van Nienwenhuizen (Plenum Press, New York, 1980) p. 23

[Sto88] W. Stöffl: In Proc. 16th INS International Symposium on Neutrino Mass and Related Topics, Tokyo 1988, to be published

[Tay83] G.N. Taylor et al.: Phys. Rev. D28, 2705 (1983)

[Ush81] N. Ushida et al.: Phys. Rev. Lett. 47, 1694 (1981)

[Ves87] W.T. Vestraud, A. Gosh, E. Chupp: International Astronomical Union Circular Nos. 4338, 4340, 4365

[Vil83] T.S. Vilov: JINR Report No. P6-83-517, Dubna (1983)

[Vog81] P. Vogel et al.: Phys. Rev. C24, 1543 (1981)

[Wil86] J.R. Wilson et al.: Ann. N.Y. Acad. Sci. 470, 267 (1986)

[Wil87] J.F. Wilkerson et al.: Phys. Rev. Lett. 58, 2023 (1987)

[Wu57] C.S. Wu et al.: Phys. Rev. 105, 1413 (1957)

[Zac86] G. Zacek et al.: Phys. Rev. D34, 2621 (1986) and references therein

# Neutrino Reactions and the Structure of the Neutral Weak Current

*K. Winter*

CERN, CH-1211 Geneva 23, Switzerland

The role of neutrino reactions in the investigation of the structure of the neutral weak current is reviewed from the discovery at CERN in 1973 to the high precision tests of the Standard Model.

## 1. Introduction

In 1973 muon-less neutrino reactions were discovered at CERN in the elastic scattering of anti-muon-neutrinos on electrons [1],

$$\bar{\nu}_\mu e \rightarrow \bar{\nu}_\mu e;$$

the neutral neutrino current $\bar{\nu}_\mu \nu_\mu$ and the neutral electron current $\bar{e}e$ couple by exchanging a neutral intermediate boson. In the reaction [2]

$$\nu_\mu N \rightarrow \nu_\mu \pi^+ \pi^- \pi^0 N,$$

muon-neutrinos scatter inelastically on nucleons and transfer part of their energy to the neutral quark currents $\bar{u}u$ and $\bar{d}d$.

Observation of an asymmetry of the cross-section of the scattering of left-handed and right-handed electrons on deuterons at SLAC in 1978 [3] demonstrated that the neutral weak $\bar{e}e$ current is violating parity. This parity violation was then confirmed by the observation that the polarization plane of a laser beam passing through bismuth vapour is rotated [4].

The existence of $\bar{\mu}\mu$ and $\bar{\tau}\tau$ neutral currents was deduced from the observation of a weak forward-backward charge asymmetry in the annihilation of electrons and positrons at the PETRA Collider at DESY [5] in 1982. The existence of other neutral quark currents, e.g. $\bar{s}s$, was deduced from the analysis of the inelasticity distribution of deep inelastic neutral current neutrino scattering on nucleons [6]. Also the neutral current $\bar{\nu}_e \nu_e$ has been observed in neutrino experiments at CERN [7].

So far only diagonal neutral currents which do not change the flavour of the particles involved have been observed . With three families of leptons and quarks, the standard theory [8] predicts the existence of six diagonal neutral lepton currents and six diagonal neutral quark currents.

The neutral current has a more complex helicity structure than the charged current. Experiments, e.g. observation of coherent $\pi^0$ production on nuclei (Chap. 3) has demonstrated the vector nature of the neutral current interaction and measurements of the inelasticity distribution in deep inelastic neutral current neutrino scattering [9] have shown that both left-handed and right-handed currents exist (Chap. 4). This observation shows directly the existence of a unified, electro-weak force. Its existence was also demonstrated by the observation of interference of amplitudes with $\gamma$ and $Z^0$ interchange, e.g. in the helicity asymmetry of electron-deuteron scattering and in the optical activity of bismuth vapour.

The structure of the current, as deduced from these experiments, implies that left-handed fermions transform as a doublet under a weak isospin rotation group and right-handed fermions transform as a singlet. The Standard Model of the electroweak gauge theory predicted this structure of neutral currents and it successfully describes a large amount of experimental data [10], which are all consistent with universal strength of the forces, $g_2$ and $g_1$, associated with the SU(2) and U(1) symmetry groups, respectively.

The fundamental quantities of the Standard Model, the coupling constants $g_1$ and $g_2$, and the masses of the weak bosons, $m_W$ and $m_Z$, are related to the angle $\Theta_W$ that describes the mixing of the two local symmetries by the relations

$$e = \frac{g_1 \, g_2}{\sqrt{g_1^2 + g_2^2}} = g_2 \sin \Theta_W,$$

$$\sin^2\Theta_W = 1 - \frac{m_W^2}{m_Z^2}. \tag{1}$$

The value of the mixing angle is not predicted by the Standard Model, while Grand Unified Theories predict a value of $\sin^2\Theta_W = 3/8$ at the unification mass.

The existence of a non-zero mixing angle defines both the structure and the strength of the neutral currents. The left-handed currents of the "up" particles of the weak isospin doublets couple with a coefficient $(1/2 - Q \sin^2\Theta_W)$, while the left-handed currents of the "down" particles of the doublets couple with a coefficient $(-1/2 - Q \sin^2\Theta_W)$. Q is the electric charge of the particle. The coupling coefficients of the right-handed currents are the same for "up" and "down" fermions, $- Q \sin^2\Theta_W$. Hence, the value of $\sin^2\Theta_W$ can be deduced from all neutral-current induced processes. In the Born approximation, one finds the values $\sin^2\Theta_W^{unc}$ obtained before 1985, which are summarized in Table 1; they are in reasonable agreement with a common value of $\sin^2\Theta_W$. The values after radiative correction, also given in Table 1, are not in significantly better agreement, owing to their relatively large errors.

TABLE 1

Values of $\sin^2\Theta_W$ obtained from various processes and the values after electroweak radiative corrections have been applied

| Experiment | $\sin^2\Theta_W^{unc}$ | Condition | $\sin^2\Theta_W^{corr}$ |
|---|---|---|---|
| Parity violation in Cs atoms [4] | 0.221 ± 0.027 | | 0.228 ± 0.027 |
| eD scattering asymmetry [3] | 0.224 ± 0.020 | $\rho = 1$ | 0.218 ± 0.020 |
| $\nu_\mu(\bar{\nu}_\mu)$e scattering [45] | 0.215 ± 0.034 | independent of $\rho$ | 0.215 ± 0.034 |
| $\nu_\mu(\bar{\nu}_\mu)$e scattering [46] | 0.209 ± 0.032 | independent of $\rho$ | 0.209 ± 0.032 |
| $\nu_\mu(\bar{\nu}_\mu)$p scattering [47] | 0.220 ± 0.016 | $M_A$ = 1.05 GeV | 0.220 ± 0.016 |
| $\nu_\mu$N deep inel. scattering [20] | 0.232 ± 0.014 | $\rho = 1$ | 0.223 ± 0.014 |
| $\nu_\mu$N deep inel. scattering [36] | 0.239 ± 0.012 | $\rho = 1$ | 0.226 ± 0.012 |

## 2. Higher Order Corrections

At a higher level of precision we expect that electroweak radiative corrections change this simple picture. The effective values of $\sin^2\Theta$ derived from the measurements will be modified differently in each process. It is therefore of great interest to make measurements with improved precision to test whether the theory can correctly predict these radiative shifts. The shifts are in close analogy to the famous Lambshift of atomic energy levels and to (g-2) of the electron and the muon. The agreement between high-precision measurements and the higher order calculations of the shifts by the theory have given us confidence that Quantum-Electro-Dynamics (QED) is a successful theory. The same tests can now be performed for the Electroweak Interaction. While the QED shifts are small, those of the electroweak interaction are large, as we shall see.

A more advanced test of the theory, at the level of the radiative corrections of order $\alpha$, can be performed by a precise determination of $\sin^2\Theta_W$ from semileptonic and leptonic neutrino scattering and from the direct measurements of $m_W$ and $m_Z$. The Fermi interaction, mediated by the weak bosons, leads to a relation between the fine-structure constant $\alpha$, the Fermi coupling constant $G_F$, the boson masses, and the weak mixing parameter $\sin^2\Theta_W$:

$$m^2_W = m^2_Z \cos^2\Theta_W = \pi\alpha/\sqrt{2}G_F \sin^2\Theta_W. \qquad (2)$$

Using $1/\alpha = 137.035963(15)$, as measured by the Josephson effect, and $G_F^{\mu} = 1.16637(2) \times 10^{-5}\ \mathrm{GeV}^{-2}$, as determined from the muon lifetime, leads in the lowest order to the relations:

$$\sin^2\Theta_W^{unc} = (37.281\ \mathrm{GeV}/m_W)^2, \qquad (3)$$

$$\sin^2 2\Theta_W^{unc} = (74.562\ \mathrm{GeV}/m_Z)^2.$$

Because of the definition of $\sin^2\Theta_W$ in (2), this relation must be radiatively corrected by using values of $\alpha$ and $G_F$ for the mass of $m_W$. The change of the value of $G_F(m^2_W)$ is negligible, whereas $\alpha$ is changed to $\alpha(m_W) = \alpha(0)/(1 - \Delta r)$, with

$$\Delta r = 0.0713 \pm 0.0013, \qquad (4)$$

and assuming $m_{Higgs} = 100$ GeV and a top quark mass of $m_t = 45$ GeV [10]. The predicted corresponding radiative shift of the W mass is about 3 GeV. The radiative correction can be experimentally determined from the measurements of $m_W$ and $m_Z$ and of $\sin^2\Theta_W$ through the relation

$$\Delta r = 1 - (\pi\alpha/m_W^2 \sqrt{2}G_F \sin^2 \Theta_W), \tag{5}$$

and the corresponding relation involving $m_Z$. A comparison with the calculated radiative shift thus tests the predictive power of the Standard Model at the level of vacuum polarization corrections. The electroweak radiative shift is approximately 50 times larger than (g - 2) of the muon, owing to the larger mass scale involved.

The weak bosons $W^{\pm}$ and $Z^0$, which mediate the charged and the neutral weak currents, were discovered at the CERN $p\bar{p}$ Collider in 1983 [11]. The measurements of $m_W$ and $m_Z$ are summarized in Table 2. The predictions of the Standard Model, using $\sin^2\Theta_W$ determined from deep inelastic neutrino scattering, are given for comparison. The successful prediction of the new mass scale of the intermediate vector bosons has been one of the triumphs of the Standard Model.

TABLE 2

The W and Z masses. The first error is statistical and the second systematic, mainly due to energy calibration.

| Group | $M_W$ | $M_Z$ | Reference |
|---|---|---|---|
| UA2 (CERN) | $80.2 \pm 0.8 \pm 1.3$ | $91.5 \pm 1.2 \pm 1.7$ | R. Ansari et al., Phys. Lett. 186B (1987) 440 |
| UA1 (CERN) | $83.5 ^{+1.1}_{-1.0} \pm 2.7$ | $93.0 \pm 1.4 \pm 3.0$ | G. Arnison et al., Phys. Lett. 166B (1986) 484 |
| UA1 + UA2 combined | $80.9 \pm 1.4$ | $92.1 \pm 1.8$ | |
| Prediction from deep inelastic $\nu$ scattering (with radiative corrections; $\sin^2\Theta_W = 0.233 \pm 0.006$) | $80.2 \pm 1.1$ | $91.6 \pm 0.9$ | |

### 3. Coherent $\pi^0$ Production by Neutral Current and Charged Current Neutrino Interactions

Is the Lorentz structure of the weak neutral current interaction of the helicity conserving vector type (V, A) or of the helicity-changing S, P or T type? Elucidating the same question for the weak charged current took nearly 25 years of experimental investigation. In neutrino-induced reactions, the answer is obscured by the so-called confusion theorem due to B. Kayser et al [12]. The process of coherent $\pi^0$ production by weak neutral-current (anti)neutrino interactions on a nucleus with mass number A probes directly the helicity structure of the interaction, owing to the intrinsic quantum numbers ($0^-$) of the neutral pion. In the limit of zero momentum transfer, the helicity of the incident and of the outgoing neutrino is depicted in Fig. 1a for V, A interaction and in Fig. 1b for S, P or T interaction. In the latter case, the cross-section must vanish because of angular momentum conservation. Hence, the $\pi^0$ spectra are distinctly different for V, A and S, P, T interactions, and experiments have shown that the interaction must be of the V, A type. Calculations of this process

$$\overset{(-)}{\nu}_\mu + A \rightarrow \overset{(-)}{\nu}_\mu + A + \pi^0 \qquad (6)$$

were carried out by Nachtmann and Lackner [13,14] and later by Rein and Sehgal [15]. Measurements at mean energies of 2 GeV at the CERN-PS were published by the Aachen-Padoua group [16] and by the Gargamelle Bubble Chamber group [17], and at a mean neutrino energy of 31 GeV and antineutrino energy of 24 GeV by the CHARM Collaboration [18]. The corresponding charged-current reaction $\bar{\nu}_\mu + A \rightarrow \mu^+ + A + \pi^-$ was studied in BEBC [19].

The coherent $\pi^0$-production is characterized by a constructive interference of the neutrino interactions on neutrons and protons within a single nucleus. The nucleus

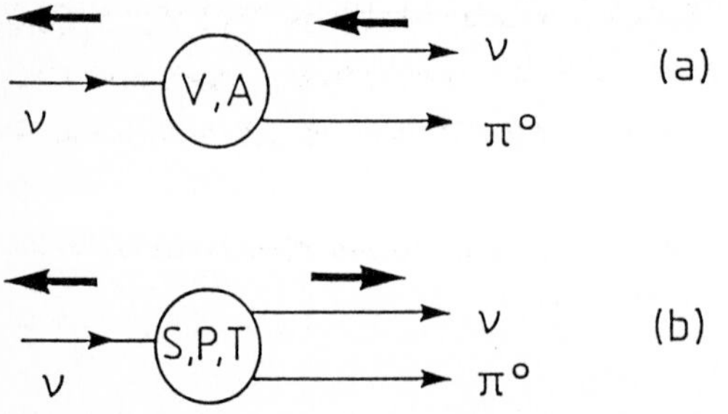


Fig. 1

Helicity of incident and outgoing neutrino (denoted by a black arrow) in the reaction $\nu_\mu A \rightarrow \nu_\mu \pi^0 A$ for (a) V, A interaction (b) S, P or T interaction in the limit of $Q^2 \rightarrow 0$

itself does not break up in this reaction and therefore a negligible amount of recoil energy is transferred to it. Coherently produced $\pi^0$ 's are emitted at small angles compared to those produced in the incoherent and resonant $\pi^0$-production; they are the main background in experiments studying elastic muon neutrino (antineutrino) scattering off electrons [20,21,22]. Coherent $\eta^0$ production probing the isoscalar axialvector part of the electroweak interaction is predicted not to exist. This prediction cannot be checked in these experiments because the detection efficiency for $\eta^0$ is negligibly small [18] or the energy is too low in the case of the CERN-PS experiments [16,17].

The predicted cross-section for coherent $\pi^0$ production is proportional to the factor

$$\sigma_{\pi^0} \propto (u_a - d_a)^2 \rho^2, \tag{7}$$

where $u_a$ and $d_a$ are the axialvector coupling constants of u and d quarks, respectively, and $\rho$ the ratio of the neutral and charged-current coupling constants. The standard model predicts $u_a = -d_a = 1/2$ and $\rho = 1$ so that a value of 1 is expected for expression (7).

In this experiment one determines the absolute value of the difference of $u_a$ and $d_a$, i.e. the isovector axialvector coupling $|\beta| = |u_a - d_a|$, by comparing the measured cross-section with the calculated value, assuming $\rho = 1$ as suggested by other measurements [23]. Because of the pure axialvector nature of the process, the cross-sections of the neutrino and antineutrino-induced reactions are expected to be equal. The measurement of $|\beta|$ allows a determination of the relative sign of $u_a$ and $d_a$. Previously this sign has been deduced from an analysis of elastic $\nu p$ scattering and $\Delta$ production [24].

The method used by the CHARM Collaboration [18] to extract the coherent $\pi^0$ candidates followed closely the procedure used in the previously published analyses of $\overset{(-)}{\nu}_\mu e$-scattering [20,21,22]. Events with narrow electromagnetic showers, initiated by a single electron or $\gamma(\pi^0)$, were selected by applying the following criteria:

1) All charged-current semileptonic events with a muon track candidate longer than 8 radiation lengths were rejected.

2) The fiducial volume was restricted to an area of $(2.3 \times 2.3)$ m$^2$ and a length corresponding to 60 marble planes extending from plane 3 to 63. For a description of the detector see ref. [25].

3) The "visible" energy was required to be larger than 6 GeV and less than 20 GeV. The lower energy cut was determined by the requirement of full trigger efficiency, whereas above 20 GeV we expected a less favourable signal to background ratio.

4) The length of the shower was required to be in the range of 84 cm to 164 cm: hadronic background is reduced by 32% and electromagnetic showers are selected with 98% efficiency.

5) The estimator which discriminated best between electromagnetic and hadronic showers was based on their different widths. The lateral shower profile was measured in the vertical and horizontal planes by the scintillator hodoscopes. The width $\Gamma$ of a Cauchy distribution fitted to the central part of the shower was calculated. A second parameter $\Delta$ using measurements from the proportional drift tubes was obtained by comparing the energy deposited in tubes further away than 6 cm from the shower axis to the total shower energy. These estimators depend only weakly on the shower energy and angle and were calibrated by measurements in a beam of pions and electrons from 5 GeV up to 50 GeV. Figure 2 shows the distributions found at 15 GeV. Events were selected if $\Gamma < 0.8$ cm and $\Delta < 0.3$. Since a $\pi^0$ beam is not available, the $\pi^0$ acceptance of these cuts has been calculated. The Electron-Gamma-Simulation program (EGS) predicts an electron acceptance of 90%, in accordance with experimental data, and an acceptance of 79% for $\pi^0$. Hadronic background is rejected by a factor larger than 100.

6) The hadronic events remaining in the sample were further rejected by the requirement of a single hit in the first tube plane after the vertex. This requirement also improves the angular resolution.

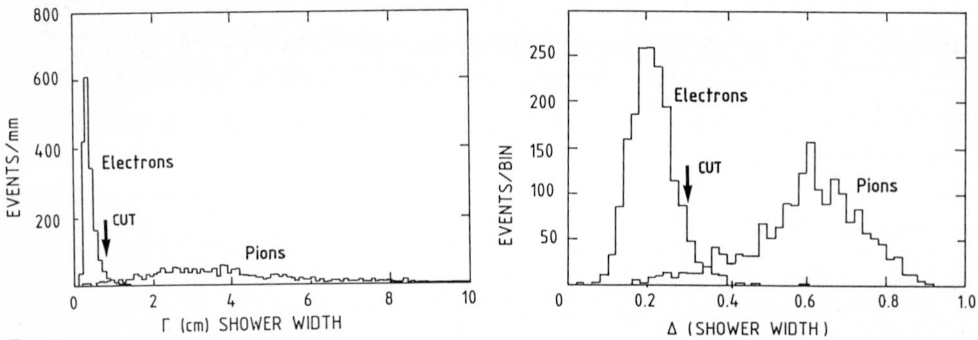


Fig. 2
Separation of electron and pion induced showers using the scintillator ($\Gamma$) and proportional drift tube ($\Delta$) dependent estimators at 15 GeV test beam energy

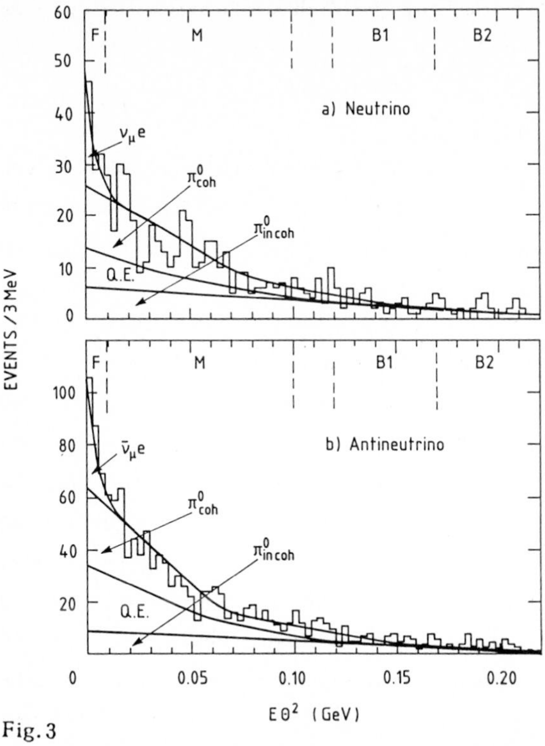


Fig. 3

$E\Theta^2$ distribution of candidates for coherent $\pi^0$ production by weak neutral-current interaction of (a) neutrinos and (b) antineutrinos

The combined selection efficiency for $\pi^0$-candidates using criteria 4, 5 and 6 is $\varepsilon_{tot}$ = (48 ± 4)%, for hadronic showers it is $\varepsilon < 1\%$.

The analysis was performed in the variable $E\Theta^2$ because the shape of the coherent signal distribution is energy independent in that representation. E denotes the measured energy of the $\pi^0$ shower and $\Theta$ its angle to the neutrino beam axis. 627 neutrino events and 1328 antineutrino events were selected with $E\Theta^2 < 220$ MeV. 90% of the coherent $\pi^0$ events are expected to have $E\Theta^2 < 100$ MeV. We estimated the background in the $E\Theta^2$ region from 100 MeV to 220 MeV and extrapolated it into the region $E\Theta^2 < 100$ MeV. We assumed 3 sources of background (see Fig. 3):

1. neutrino-electron scattering;

2. quasi-elastic charged-current events induced by the $\nu_e$ and $\bar{\nu}_e$ contamination of the beam;

3. incoherent and resonant $\pi^0$ production [15,26] and hadronic (NC) events with a large electromagnetic component.

From an analysis of the forward region ($E\Theta^2 < 9$ MeV), the main region (9MeV $< E\Theta^2 < 100$ MeV) and the background regions (120 MeV $< E\Theta^2 < 170$ MeV and 170 MeV $< E\Theta^2 < 220$ MeV) and the energy deposition observed in the first scintillator plane after the vertex ($E_{first}$), which was used to discriminate between electron-induced and $\pi^0$ induced showers, the coherent $\pi^0$ event rates were determined (see Table 3) by extrapolation.

The efficiency of detecting a coherently produced $\pi^0$ for an average marble ($CaCO_3$) nucleus in the energy window from 6 to 20 GeV was evaluated using a Monte Carlo simulation according to the theory of Rein and Sehgal [15]. The difference between the spectra of the neutrino ($\langle E_\nu \rangle = 31$ GeV) and the antineutrino beam ($\langle E_{\bar\nu} \rangle = 24$ GeV), as determined by charged-current events, was taken into account. The resulting cross-sections are given in Table 4; their statistical and systematic errors were quadratically combined. The theoretical values in Table 4 were evaluated according to ref. [15], assuming $\beta = 1$ and $\rho = 1$. Their errors are due to uncertainties in the nuclear

TABLE 3

Coherent $\pi^0$ production

Number of event candidates [18] with statistical and systematical errors

| | $N_{\gamma\pi\,raw}$ | $N_{incoh}$ | $N_{\pi\,coh}$ |
|---|---|---|---|
| $\nu$ | 383 ± 31 ± 22 | 160 ± 65 ± 48 | 223 ± 53 ± 72 |
| $\bar\nu$ | 721 ± 48 ± 50 | 239 ± 87 ± 74 | 482 ± 83 ± 99 |

TABLE 4

Summary of experimental cross-sections for coherent $\pi^0$ production per average marble nucleus in units of $10^{-40}cm^2$

| | Experiment | Theory |
|---|---|---|
| $\nu_\mu$ | $\sigma = 96 \pm 42$ | $\sigma = 74 \pm 38$ |
| $\bar\nu_\mu$ | $\sigma = 79 \pm 26$ | $\sigma = 71 \pm 34$ |

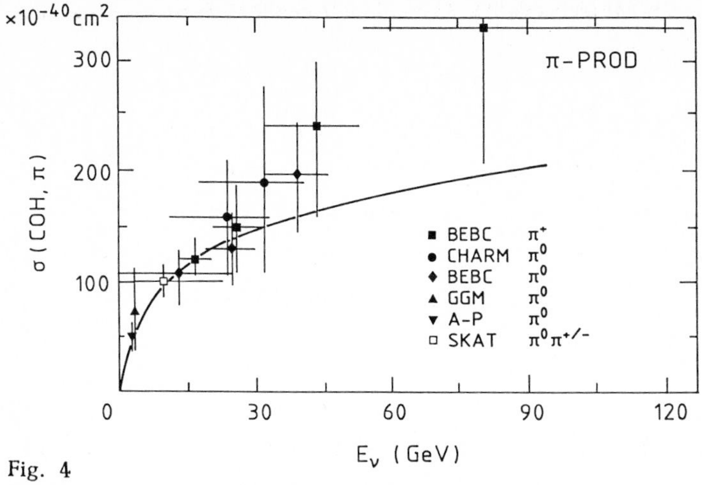


Fig. 4

Cross-sections for coherent $\pi^0$ production as a function of neutrino energy, showing a comparison of the experimental data with theoretical models. The horizontal error-bars indicate the range of neutrino-energies used for the particular points. For the CHARM results 68% of the events have a neutrino energy within the error bars.

radii. Comparing the experimental and theoretical values of Table 4, averaged for neutrinos and antineutrinos, the isovector neutral current coupling standard was determined to be

$$|\beta| = |u_a - d_a| = 1.08 \pm 0.24,$$

assuming $\rho = 1$. Statistical error (0.10) and systematic error (0.21) are quadratically combined. Figure 4 shows this result together with the theoretical predictions and results from previous experiments [16, 26, 27]. A summary of the values of $|\beta|$ published before Summer 1987 is given in Table 5. The "world average" is

$$|\beta| = 0.99 \pm 0.10, \qquad (8)$$

in close agreement with the value of 1.0 predicted by the standard model, and hence with a negative relative sign of the axialvector coupling constants of the u and d quarks.

TABLE 5

Summary of experimental determinations of $|\beta| = |u_a - d_a|$ from coherent $\pi^0$ production

| Experiment | $\lvert\beta\rvert$ |
|---|---|
| Aachen-Padova [16] | 0.93 ± 0.16 |
| CHARM [18] | 1.08 ± 0.24 |
| Skat [27] | 0.99 ± 0.20 |
| FNAL-15' [27] | 0.98 ± 0.24 |
| World average | 0.99 ± 0.10 |

## 4. Determination of the Helicity Structure of the Weak Neutral Current

Values of the coupling constants $g_L^2$ and $g_R^2$ of the left-handed and the right-handed u and d quarks have been determined from the total inclusive cross-sections of weak neutral-current (NC) and charged-current (CC) interactions induced by neutrinos and antineutrinos on isoscalar targets [28].

It should be noted that these scalar quantities do not allow a direct demonstration of the existence of a right-handed current; one needs a key to open a door. In analogy to the famous analysis of the kinematical correlation of the recoil proton in neutron beta-decay, which leads to convincing direct evidence of the left-handedness (V-A) of the weak charged-current interaction, the investigation of the differential cross-section $d\sigma/dy$ in the inelasticity y can provide the key and hence direct evidence of the conjectured contribution of a right-handed (V+A) current to the electroweak interaction. The differential cross-sections for both NC and CC events are directly related to the space-time structure of the weak-current involved. They give the admixture of V-A and V+A couplings and may indicate by a term in $y^2$ the presence of any scalar (S) or pseudoscalar (P) terms. In addition, they yield information on the antiquark content, both strange and non-strange, of the nucleon and on the $s\bar{s}$ neutral-current.

For CC muon-neutrino interactions, all the initial energy $E_\nu$ appears as detectable energy in the final state and the inelasticity value of each event is unambiguously defined by

$$y = E_h/(E_h + E_\mu) = E_h/E_\nu \, , \tag{9}$$

where $E_h$ is the hadron energy and $E_\mu$ the muon energy. For NC interactions, $E_\mu$ is replaced by $E_{\nu'}$ (the energy of the outgoing neutrino), which can obviously not be observed. One must resort to a narrow-band neutrino beam produced by a sign and momentum-selected nearly parallel beam of pions and kaons [29]. The two-body decay kinematics of the pions and the kaons provides a well-defined relationship between the neutrino energy ($E_\nu$) and the distance of the interaction vertex (R) in the detector from the beam axis, up to an ambiguity caused by the presence of neutrinos from both pion and kaon decays. The neutrino energy spectrum at a given distance R is governed by the relative production of kaons and pions, the characteristics of the magnetic focusing channel, and the geometry of the decay tunnel and the detector. These parameters are well-known and allow one to calculate by Monte Carlo methods the energy spectrum at the detector. Figure 5 shows the calculated spectrum at the CHARM detector in the 160 GeV narrow-band beam [29]. The interaction vertex of an event in the CHARM detector is measured with a typical resolution of 4 cm at 15 GeV and 3 cm at 30 GeV hadron energy [30] This accuracy is sufficient to fully exploit the energy-distance relationship of the beam. Since the neutrino energy for a given event is only known to the level of the inherent beam spread and $\pi/K$ ambiguity, an unfolding procedure is necessary to determine $d\sigma/dy$.

The approach used was to determine that form of the differential cross-section $d\sigma/dy$ which gives the best fit to the measured event distribution $d^2N/dE_h dR$ as a function of $E_h$ and the distance R of the interaction point from the beam axis in the detector. Assuming a total cross-section rising linearly with $E_\nu$ gives the relation:

$$d^2N/dE_h dR = \int F(E_\nu, R) \frac{d\sigma}{dy} \, dE_\nu \, , \tag{10}$$

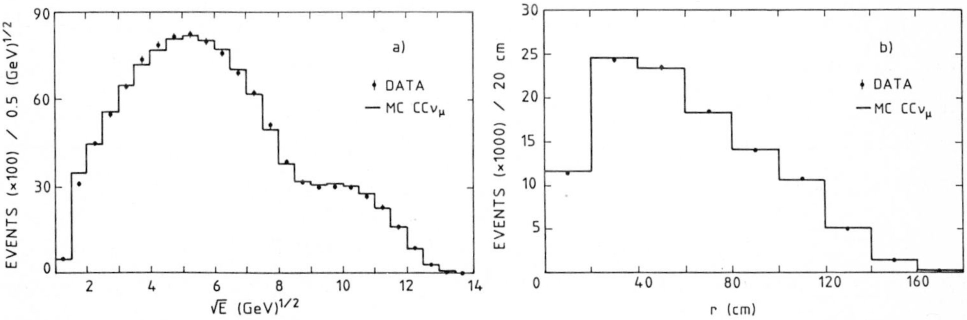


Fig. 5
The calculated muon-neutrino event energy spectrum at the CHARM detector (160 GeV narrow band beam) (a) as a function of $\sqrt{E}$ and (b) the event rate as a function of the distance r from the beam axis

where $F(E_\nu, R)$ is the calculated neutrino flux. The unfolding was performed using a parametrization of $d\sigma/dy$ as a linear combination of bell-shaped (B-spline) functions $b_i$, (y) with coefficient $a_i$:

$$d\sigma/dy = \sum_i a_i\, b_i\, (y);$$

this consisted of determining the coefficients $a_i$ in

$$d^2N/dE_h dR = \Sigma a_i \int b_i \left(\frac{E_h}{E_\nu}\right) F(E_\nu, R) dE_\nu \tag{11}$$

by a "maximum likelihood" fit to the experimental distribution, taking into account the experimental resolutions in $E_h$ and R. The integrals over the neutrino flux give the shape of the contribution to $d^2N/dE_h dR$ from each $b_i$ in y, while the coefficients $a_i$ give the relative magnitudes of these contributions [29]. The resulting unfolded differential cross-sections for both CC and NC events induced by muon-neutrinos and antineutrinos are shown in Figs. 6a and 6b. Both the CC and NC distributions have been obtained in the same fashion, the primary muon in the CC events was used for event classification only.

Electroweak radiative corrections have been applied to all distributions [31]. The verification of the validity of the method, by comparing the result for CC events with another high-precision determination of $d\sigma/dy$ using $E_\mu$ [32] to determine y will not be pursued here (see [29]).

Within the context of the quark model, and under the assumption (see Chap. 3) that the weak currents contain only V and A terms, the differential cross-sections for deep inelastic scattering on isoscalar nuclei can be written as [29]

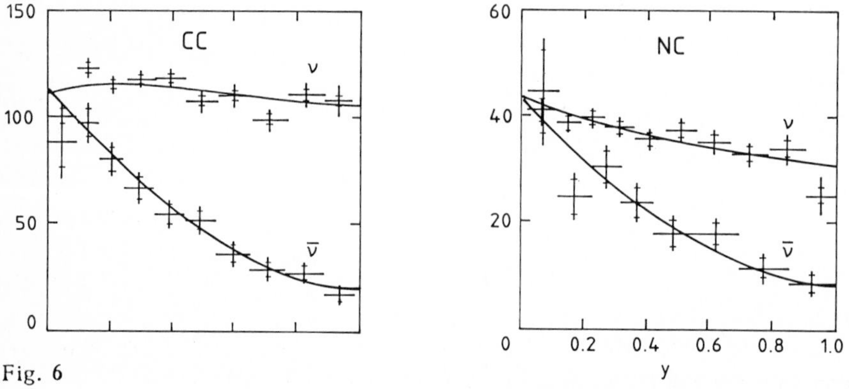


Fig. 6

The differential cross-sections $d\sigma/dy$ after resolution unfolding. The curves correspond to a two-parameter fit; (a) CC events (b) NC events

$$\frac{d\sigma}{dy}\,(\overset{(-)}{\nu} \rightarrow \overset{(-)}{\nu}) = A\Big[g^2_{L(R)}\big(Q + \bar{Q}(1-y)^2\big) + g^2_{R(L)}\big(\bar{Q} + Q(1-y)^2\big) + g^2_S\, Q_S\big(1 + (1-y)^2\big) - F_L\Big], \tag{12}$$

where A is a normalisation constant. The quark structure of the nucleon is described by $Q$, $\bar{Q}$ and $Q_S$ ; for example $Q = \int \big(u(x) + d(x)\big)dx$ is the momentum-weighted valence quark content of the nucleon. $\alpha = \bar{Q}/(Q + \bar{Q})$ is the fractional momentum-weighted sea quark content and $\beta = Q_S\,/(Q + \bar{Q})$ the fractional strange quark content. The constants $g^2_L$ and $g^2_R$ are the left-handed and right-handed couplings of the weak neutral current to "up" and "down" quarks, while $g^2_S$ is the sum of the right and left-handed couplings to strange quarks. By fitting simultaneously these expressions to the four unfolded differential cross-sections, the couplings of the weak neutral current have been determined; the resulting values are given in Table 6. The values of $\alpha$, $\beta$ and $g^2_L$, $g^2_R$ are determined by neglecting the term in $g_S$ (fit A). Note that $g^2_R$ is determined with a significance of 5.6 standard deviations. The form of the NC y-distribution in (12) precludes the possibility of varying all five parameters $\alpha$, $\beta$, $g^2_L$, $g^2_R$ and $g_S$ simultaneously in a fit. In particular, $g_S$ is strongly correlated with the value of $\beta$. In fit B the parameters $\alpha$, $\beta$, $\sin^2\Theta_W$ and $g_S$ are determined, and $g_S\,/g_d = 1.06 \pm 0.14$. Thus, one finds that the total coupling strength of the weak neutral current to the strange quark is consistent with being equal to that of the non-strange down quark, an assumption implied by the GIM mechanism [33].

TABLE 6

Results of simultaneous fits to CC and NC y distributions using an unfolding method

| Parameter | Fit A | Fit B | Fit C |
|---|---|---|---|
| $\alpha$ | 0.129 ± 0.013 | 0.129 ± 0.011 | 0.129 ± 0.011 |
| $\beta$ | 0.134 ± 0.034 | 0.118 ± 0.037 | 0.124 ± 0.030 |
| $g^2_L$ | 0.315 ± 0.004 | | |
| $g^2_R$ | 0.056 ± 0.010 | | |
| $g^2_S$ | | 0.204 ± 0.054 | |
| $\sin^2\Theta_W$ | | 0.241 ± 0.016 | 0.236 ± 0.008 |

Motivated by this consistent result, one can go on to assume that the couplings of the u, d and s quarks can all be described in terms of the Glashow-Salam-Weinberg standard model. The neutral current sector is then described by a single parameter,

$\sin^2\Theta_W$. This fit (C) gives $\sin^2\Theta_W = 0.236 \pm 0.008$, in full agreement with the result obtained from the ratio of NC and CC cross-sections for deep inelastic scattering using the same data set [23].

In the preceding sections, we first demonstrated the existence of a right-handed part in the weak-neutral current, independent of the validity of the standard model. After showing that the neutral-current coupling of the strange quark is consistent with being equal to that of the "down" quark, we have then described the neutral-current sector completely in terms of the standard model. This, of course, already assumes the presence of only V and A currents. However, if there were a scalar (S) or a pseudoscalar (P) part of the neutral-current interaction, it would contribute equally to neutrino and antineutrino interactions, and would manifest itself in $d\sigma/dy$ as a term proportional to $y^2$ with a coefficient B. Then the ratio B/A gives the relative proportions of the S or P and V, A parts. A fit to the two NC distributions gives [29] $B/A = -0.05 \pm 0.05$, implying

$$g_{sp}^2/g_{VA}^2 < 0.03 \text{ (95\% conf. level)}, \tag{13}$$

where $g_{sp}^2$ and $g_{VA}^2$ are the S, P and V, A coupling strengths, respectively. The analysis in this form disregards the possibility of a conspiracy of S, P, T terms mimicking a V, A structure in the neutral currents. This ambiguity has, however, already been resolved by the observation of coherent $\pi^0$ production (Chap. 3) and $\gamma$-$Z^0$ interference (Chap. 1).

## 5. High Precision Measurements of $\sin^2\Theta_W$

Experimental tests have now, for the first time, reached the precision required to probe the theory at the level of the radiative corrections. The present error on $\sin^2\Theta_W$ of $\pm 0.004$ corresponds to an error in $m_W$ of 0.2 GeV, whereas the total radiative shift is predicted to be 3.3 GeV.

Future improvements of the luminosity of the CERN Sp$\bar{p}$S Collider using a new antiproton collector ring (ACOL) are aimed at obtaining a data set with $10^4$ nb$^{-1}$ instead of the present integrated luminosity of $10^2$ nb$^{-1}$. These data will then give statistical errors of the masses of

$$\Delta m_W, \Delta m_Z \sim \pm 0.15 \text{ GeV}.$$

The uncertainty in the experimental mass scale will then be the dominant error. Assuming an absolute calibration of the giant calorimeters of the detectors to ± 1% one derives a limiting accuracy of

$$\Delta \sin^2 \Theta \sim \pm 0.004.$$

The scale errors cancel in the ratio of the masses, where we expect the full improvement due to the large statistics, corresponding to $\Delta\sin^2\Theta \sim \pm 0.004$.

Among the low-energy experiments, those which use neutrino scattering on nucleons have already been improved to a similar precision in $\sin^2\Theta_W$. A new experiment on neutrino-electron scattering, aiming at a higher precision (CHARM II) is presently being performed at CERN. These experiments and their limitations are described in some detail in Chaps. 6 and 7. The small theoretical error in the radiative shift requires a measurement of $m_Z$ to ± 50 MeV to be matched. This precision can be achieved at LEP.

## 6. Measurement of $\mathrm{Sin}^2\Theta_W$ in Semileptonic Neutrino Reactions

The cross-sections of neutrino scattering on isoscalar targets by the neutral and by the charged-current weak interaction are related by the following expression derived by Llewellyn-Smith [34]:

$$\frac{d^2\sigma_{NC}^{\nu,\bar{\nu}}}{dxdy} = \left(\frac{1}{2} - \sin^2\Theta + \frac{5}{9}\sin^4\Theta\right) \frac{d^2\sigma_{CC}^{\nu,\bar{\nu}}}{dxdy} + \frac{5}{9}\sin^4\Theta \frac{d^2\sigma_{CC}^{\bar{\nu},\nu}}{dxdy} . \quad (14)$$

The simplest measurements are those of the total cross-section ratios

$$R_\nu = \sigma^\nu (NC)/\sigma^\nu(CC), \; r = \sigma^{\bar{\nu}} (CC)/\sigma^\nu (CC),$$

which give the relation

$$R_\nu = \frac{1}{2} - \sin^2 \Theta + \frac{5}{9} \sin^4 \Theta + r(\frac{5}{9} \sin^4\Theta). \quad (15)$$

Equations (14) and (15) are valid if neutrino interactions with quarks and anti-quarks other than u and d can be neglected and if the Cabibbo angle is set to zero. Higher twist effects have also been neglected; they have been estimated [34] to give uncertainties smaller than $\Delta\sin^2 \Theta \sim \pm 0.005$. Of course, weak isospin symmetry is implied by (14). We know that it is broken by flavour-changing processes which have been observed to contribute to charged-current induced reactions, but not to neutral-current reactions. The energy threshold of the flavour transition s(d) + $W^+ \rightarrow$ c crosses the peak energy of the neutrino beam used for the measurement. To correct the measured cross-sections we require knowledge of the mass of the charm quark to describe the threshold behaviour. At much higher energy, for instance at HERA, a measurement of the ratio $R_\nu$ does not require a threshold correction. The corrections are applied with the help of the quark model of the nucleon. Using the best estimate of the charm quark mass, $m_c = 1.5 \pm 0.3$, introduces an uncertainty of $\Delta(\sin^2 \Theta_W) = \pm\ 0.004$. Fixing the mass at $m_c = 1.5$ GeV, the remaining theoretical uncertainty is $\Delta(\sin^2\Theta_W) = \pm\ 0.003$. The total theoretical uncertainty is therefore $\Delta(\sin^2 \Theta_W) = \pm\ 0.005$. A high precision measurement has recently been performed at CERN by the CHARM Collaboration [35] and by the CDHS Collaboration [36]. Ten years after the discovery of the neutral-current interaction by the Gargamelle team at CERN with a signal to background ratio of one to six, events can now be classified as neutral-current (NC) or charged-current (CC), by direct recognition of the muon in the fine-grain calorimeter of the CHARM detector, with less than 0.2% ambiguity. This progress is due to some important new features of this electronic detector:

- fast timing is used and events occurring upstream are vetoed, thus eliminating the so-called associated neutron background which plagued the Gargamelle experiment;

- the lateral and longitudinal dimensions of the target-calorimeter are more than 10 times larger than the interaction length of hadrons, thus giving clear signatures to neutrino interactions and muon tracks;

- detector elements of small lateral dimensions (fine-grain) and frequent segmentation of the target plates allow detection of hadron showers with high efficiency and good energy resolution $[\sigma(E_H)/E_H = 0.47/\sqrt{E_H/\mathrm{GeV}}]$ and the recognition of muons with momenta as low as 1 GeV/c;

- nearly equal response of the calorimeter ($E_e/E_H = 1.17$) to electromagnetic and hadronic showers, allowing the definition of an effectively equal energy threshold in NC and CC events which have different $\pi^0$ content.

The feasibility of precision measurement was discussed by the author at the 1982 Javea Workshop on Weak Interaction [37] and was demonstrated in detail by the CHARM Collaboration at a Physics Workshop at CERN [38].

A photograph of the CHARM detector is shown in Fig. 7. It is composed of a fine-grained calorimeter and a muon spectrometer. The calorimeter consists of 78 modules, each composed of a target plate of marble of dimensions 3m × 3m and 8cm thickness; a layer of 20 scintillation counters of 15cm width, 3m length and 3cm thickness; a plane of 128 proportional drift tubes (3cm × 3cm × 400cm) oriented at $90^0$ with respect to the scintillation counters and a plane of digital wire chambers with 1 cm wire spacing oriented parallel to the scintillators. The calorimeter was surrounded by magnetized iron frames for the detection and measurement of large angle muons. The orientation of the detector elements alternated from horizontal to vertical in successive modules. A detailed description can be found in reference [39]. Figures 8 and 9 show schematic views of a CC and NC neutrino event, respectively. Scintillation counters and proportional drift tubes which are hit by the event are shown, and the range of a 1 GeV muon is indicated. Its track can also be recognized close to the hadron shower. Charged-current (CC) events for which the primary muon cannot be identified are classified as neutral-current (NC). Some of these lost CC events have a muon with an

Fig. 7. Photograph of the CHARM detector

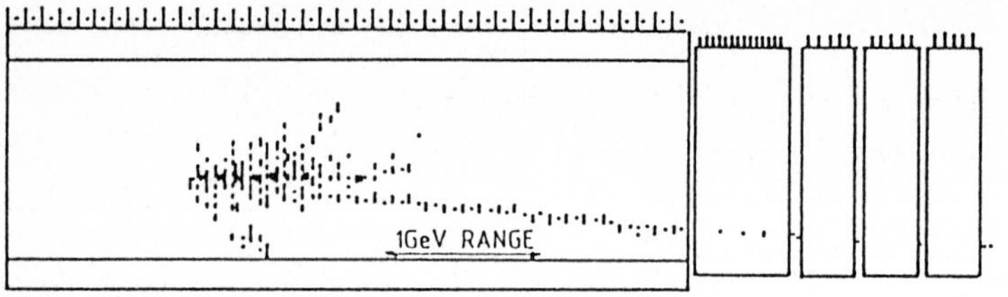


Fig. 8. Schematic view of a CC neutrino event recorded by the CHARM I detector

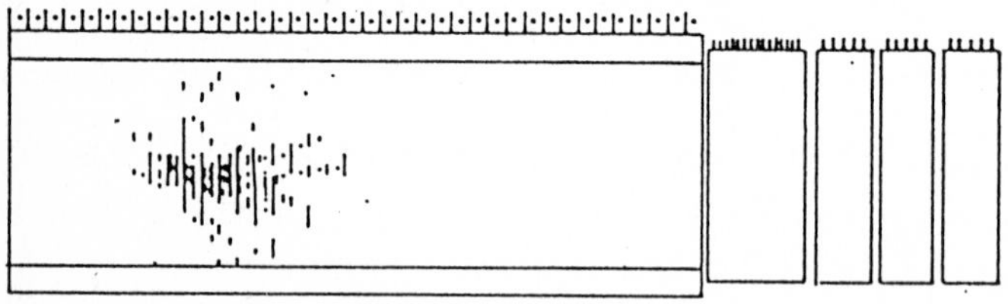

Fig. 9. Schematic view of a NC neutrino event recorded by the CHARM I detector

energy less than 1 GeV, or a muon that left the detector at the sides before depositing 1 GeV, or a muon that was obscured by the hadronic shower. A correction is required for these CC losses. As this is the largest correction required (11% of the NC events in the CHARM detector and 22% in the CDHS detector) the precision in measuring $R^{\nu}$ depends in an essential way on the reliability of estimating these losses. The uncertainty of the parent-beam momentum (± 3%) and of the muon momentum measurement by the CHARM detector can affect the correction in a systematic way. A simple and beautiful method was used to eliminate them. The correction was calculated relative to the number of events with muon momenta between 3 and 5 GeV/c. All scale errors cancel in this ratio, which was then applied by multiplying with the number of events observed in that muon momentum interval. The remaining uncertainty affecting this correction contributed an error of $\Delta R^{\nu}/R^{\nu} = \pm\ 0.32\%$. A summary of all experimental corrections is given in Table 7. A correction was applied for the small deviation from isoscalarity (N - Z) of the target material. Selecting events induced by deep inelastic scattering ($E_{hadron} >$ 4 GeV), the result of the ratio is

$$R^{\nu} = 0.3093 \pm 0.0031.$$

Table 8 shows the radiative as well as the various quark model corrections which have to be applied to determine $\sin^2\Theta_W$ from the definition of Sirlin and Marciano. The final result is

TABLE 7

Event numbers for neutrino exposure ($E_h > 4$ GeV)

| | NC | CC |
|---|---|---|
| Uncorrected data sample | 39239 ± 198 | 108472 ± 329 |
| Trigger + filter efficiency | 7 ± 4 | 0 ± 0 |
| Scan correction | 40 ± 40 | 60 ± 44 |
| Corrected raw data sample | 39286 ± 202 | 108532 ± 332 |
| WB and cosmic correction | -2310 ± 87 | -4311 ± 119 |
| - of which WB | -1998 ± 88 | -4308 ± 119 |
| - of which cosmic | -312 ± 8 | -3 ± 1 |
| Clean NBB data sample | 36976 ± 225 | 104220 ± 361 |
| Possible difference in energy cut for NC and CC | - | 0 ± 129 |
| Lost muons | -3737 ± 105 | 3735 ± 105 |
| $\pi$ and K decay | 1893 ± 50 | -1835 ± 50 |
| $K_{e3}$ CC | -1768 ± 68 | -106 ± 6 |
| $K_{e3}$ NC | -532 ± 20 | -33 ± 2 |
| Corrected event numbers | 32831 ± 283 | 105981 ± 408 |

TABLE 8

Corrections to $\sin^2\Theta_w$ ($E_h > 4$ GeV)

| Source | $\Delta\sin^2\Theta_w$ | Theoretical uncertainty |
|---|---|---|
| Muon mass | + 0.0011 | ± 0.0001 |
| $W^2$ thresholds, $F_L$ | + 0.0005 | ± 0.0005 |
| K-M mixing matrix | | ± 0.0010 |
| Strange sea for $m_c = 0$ | - 0.0074 | ± 0.0010 |
| Charm sea for $m_c = 0$ | + 0.0015 | ± 0.0010 |
| Radiative corrections | - 0.0092 | ± 0.0020 |
| Total uncertainty (fixed $m_c$) | | ± 0.0030 |
| Charm mass ($m_c = 1.5$ GeV/$c^2$) | + 0.0140 | |
| Total ($m_c = 1.5$ GeV/$c^2$) | + 0.0005 | ± 0.0030 |

$$\sin^2 \Theta_W = 0.236 + 0.012\ (m_c - 1.5) \pm 0.005\ (\text{exp}) \pm 0.003\ (\text{theor}).$$

This result, obtained by the CHARM Collaboration, is compared with other recent results from semi-leptonic neutrino scattering experiments in Table 9, assuming a charm mass of $m_c = 1.5\ \text{GeV}/c^2$. The agreement between the experiments is good, and significant in view of the fact that different experimental methods have been used, as indicated in Table 9. A combined value for $m_c = 1.5\ \text{GeV}/c^2$ is

$$\sin^2\Theta_W = 0.233 \pm 0.004 \pm 0.003.$$

Table 10 shows the averaged value of $\sin^2\Theta_W$ from the neutrino experiments and from the $m_W$ and $m_Z$ measurement, with and without the radiative corrections applied. The difference between the values from the two types of experiments shows good agreement for the corrected results, but a difference at the level of three standard deviations for the uncorrected values (the error in the difference is calculated by combining the statistical and theoretical/systematic errors quadratically). The radiative correction determined by the experiments

TABLE 9

Values of $\sin^2\Theta_W$, derived from semi-leptonic neutrino scattering experiments. The common theoretical error is ± 0.003.

| Experiment | $\sin^2\Theta_W$ | Method | Ref. |
|---|---|---|---|
| FMMF | 0.246 ± 0.016 | event-by-event | [40] |
| CCFR | 0.239 ± 0.010 | event length | [41] |
| CDHS | 0.225 ± 0.005 | event length | [36] |
| CHARM | 0.236 ± 0.005 | event-by-event | [35] |

TABLE 10

Comparison of values for $\sin^2\Theta_W$, measured in semi-leptonic neutrino scattering and from the masses of the intermediate bosons, with and without radiative correction

| Experiment | Corrected $\sin^2\Theta_W$ | Uncorrected $\sin^2\Theta_W$ |
|---|---|---|
| $\nu$N | 0.233 ± 0.004 ± 0.005 | 0.242 ± 0.004 ± 0.005 |
| $m_W$, $m_Z$ | 0.226 ± 0.008 | 0.210 ± 0.008 |
| Difference | 0.007 ± 0.010 | 0.032 ± 0.010 |

$$\Delta r \text{ (exp)} = 0.077 \pm 0.025 \text{ (exp)} \pm 0.038 \text{ (syst)}$$

agrees with the calculated value of $\Delta r$ (theor) $= 0.0713 \pm 0.0013$ assuming $m_t = 45$ GeV and $m_H = 100$ GeV [10].

Assuming therefore the validity of the radiative corrections, a precise value of the $\rho$ parameter can be obtained by combining our measured value of $R^\nu$ with the value of $\sin^2 \Theta_w$, as determined in the collider experiments from the W mass. Propagating the errors quadratically, we find that

$$\rho = 0.990 - 0.013 \ (m_c - 1.5) \pm 0.009 \pm 0.003,$$

in good agreement with the minimal Standard Model.

New physics appendages to the Standard Model or a different input value for $m_t$ would modify the prediction of $\Delta r$. We can therefore use the good agreement with experiment as a constraint. For example, if $m_t$ were 250 GeV/c$^2$, $\Delta r$ would become zero, whereas $\sin^2 \Theta_w$ hardly changes [42]. From the present experimental value of $\Delta r$, Marciano and Sirlin [43] find the constraint

$$m_t \leq 180 \text{ GeV}.$$

In the case of a heavy fourth-generation charged lepton L with a massless neutrino companion, the corresponding bound is [42]

$$m_L \leq 300 \text{ GeV}.$$

Future high-precision experiments should determine $m_w$ and $m_z$ to $\pm 0.1$ GeV and $\pm 0.02$ GeV, respectively, and, combined with the result of a determination of $\sin^2 \Theta_w$ from $\nu_\mu e$ and $\bar{\nu}_\mu e$ scattering in the CHARM II experiment to $\pm 0.005$ without theoretical uncertainty, should thus provide even better constraints (or perhaps a hint of new physics).

A final comment is concerned with the comparison of $\sin^2 \Theta_w$ predicted by Grand Unified Theories. The experimental results presented here contradict the prediction of the minimal SU(5) model [10], $\sin^2 \Theta_w = 0.214 \pm 0.004$ [SU(5)], which has already been ruled out by the experimental lower limit of proton decay [44]. It can be used to constrain the additional mass scales of non-minimal Grand Unified Theories.

For example, in the case of supersymmetric theories (SUSY), the experimental result is in good agreement with the mass constraint $M_{SUSY} \gtrsim 1$ TeV [10].

7. Measurement of $\mathrm{Sin}^2\Theta_W$ in Leptonic Neutrino Reactions

The first experimental observation of a weak neutral-current phenomenon was an event (reproduced in Fig. 10) in which a $\bar{\nu}_\mu$ scattered on an electron, found in 1972 at CERN in the Gargamelle bubble chamber [1]. Now, 15 years later, massive electronic detectors have achieved remarkable progress in this field. The CHARM Collaboration has collected about 83 events of $\nu_\mu e$ and 112 events of $\bar{\nu}_\mu e$ scattering [45]. A US-Japan Collaboration working at Brookhaven has collected 107 events of $\nu_\mu e$ and 45 events of $\bar{\nu}_\mu e$ scattering [46,47]. An event observed in the fine-grain calorimeter of the CHARM detector is reproduced in Fig. 11. Compared to the bubble chamber event in Fig. 10 it looks quite coarse. In spite of this, the events can be separated from the background and the cross-section determined.

In contrast to semileptonic neutrino scattering there are no theoretical uncertainties in extracting the weak neutral-current coupling constants of the electron, ($g_V^e$ and $g_A^e$, and $\sin^2\Theta_W$ ) from the cross-sections of neutrino electron scattering:

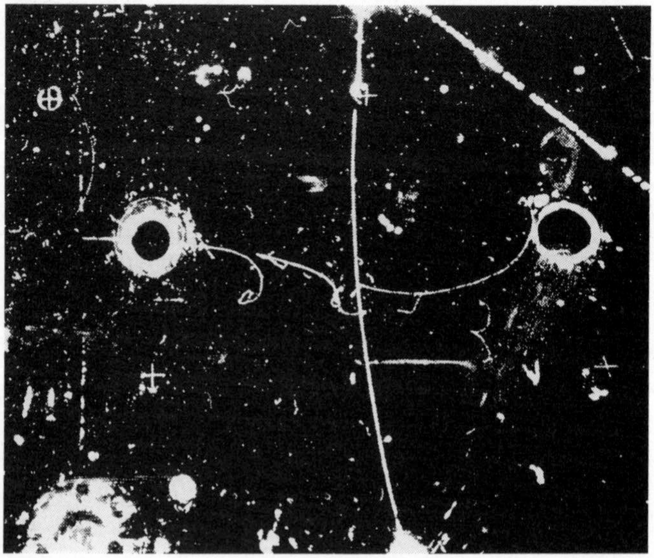

Fig. 10. First event of $\bar{\nu}_\mu e$ scattering observed in Gargamelle [1]

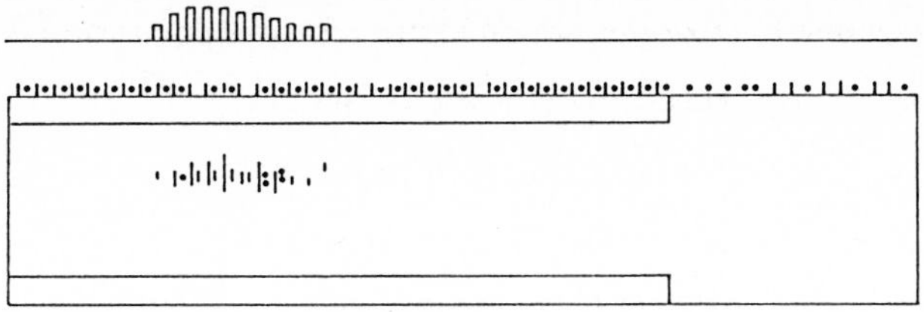

Fig. 11. Neutrino-electron scattering events observed in the CHARM I detector

$$g_V^2 + g_A^2 = (3\pi/4G_F^2\ m_e)\ (\sigma^\nu/E + \sigma^{\bar\nu}/E)$$

$$(16)$$

$$g_V^e \times g_A^e = (3\pi/4G_F^2\ m_e)\ (\sigma^\nu/E - \sigma^{\bar\nu}/E).$$

The most precise determination of the value of $\sin^2\Theta$ in the leptonic sector has been obtained by the CHARM Collaboration [45], making use of the direct relation between the ratio of $\sigma(\nu_\mu e)$ amd $\sigma(\bar\nu_\mu e)$ and $\sin^2\Theta_w$ :

$$R = \frac{\sigma(\nu_\mu e)}{\sigma(\bar\nu_\mu e)} = 3\ \frac{1 - 4\sin^2\Theta_w + (\frac{16}{3})\sin^4\Theta_w}{1 - 4\sin^2\Theta_w + 16\sin^4\Theta_w}.$$

$$(17)$$

In the vicinity of $\sin^2\Theta_w = 1/4$ this relation gives $\Delta\sin^2\Theta_w \sim 1/8\ (\Delta R/R)$ and, hence, a very precise determination of the mixing angle. The detection efficiency cancels in the ratio, many systematic uncertainties are reduced, and no absolute neutrino flux measurement is required. Radiative corrections also cancel to a large extent in the ratio and the value of $\sin^2\Theta_w$ determined by (17) is therefore practically equal to the value determined by the boson masses, as defined by Sirlin and Marciano [43]:

$$\sin^2\Theta_w = 1 - M_w^2/M_Z^2.$$

Several experimental problems have to be solved [48], namely:

1) the event rate requires a large fiducial tonnage and high selection efficiency over a wide window of recoil-electron energies;

2) the background predominantly due to quasi-elastic scattering of electron-neutrinos and to coherent $\pi^0$ production has to be reduced by a factor of about $10^3$ by precise measurements of the shower direction and efficient $e/\pi$ discrimination;

3) monitoring of the relative flux of the different beam components $\nu_\mu$, $\bar{\nu}_\mu$, $\nu_e$ and $\bar{\nu}_e$ is required to determine the cross-section ratio $\sigma(\nu_\mu e)/\sigma(\bar{\nu}_\mu e)$.

It has been demonstrated by the CHARM Collaboration [45] that these problems can be solved. In two different exposures of the fine-grain CHARM detector to the horn-focused wide-band neutrino beam of the CERN 400 GeV SPS, 83 events of $\nu_\mu e$ scattering and 112 events of $\bar{\nu}_\mu e$ scattering have been recorded. Criteria based on shower properties in low Z material (marble) were used to select these events:

i) the energy $E_F$ deposited in the first scintillator plane following the vertex was required to be $E_F < 50$ MeV, corresponding to less than 7 minimum ionizing particles;

ii) the number of wires hit in the first proportional- and in the streamer-tube planes following the vertex were required to be less than 2 or 7, respectively;

iii) the width of the shower was required to be of the order of the Molière radius to distinguish between electromagnetic and hadronic showers (see Fig. 12). Background processes were thus rejected by a factor of about $10^3$.

Measured distributions of the shower direction ($E^2\Theta^2$) for the remaining events are shown in Figs. 13 a and b. The kinematics of elastic neutrino-electron scattering limits the electron angle to

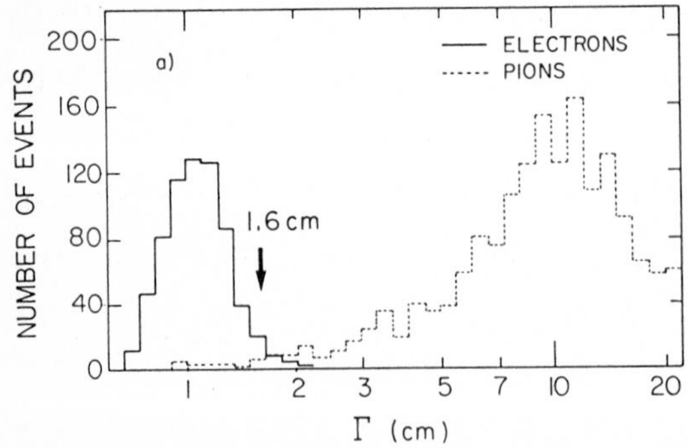


Fig. 12. Width Γ of showers induced by electrons and pions in the CHARM I detector

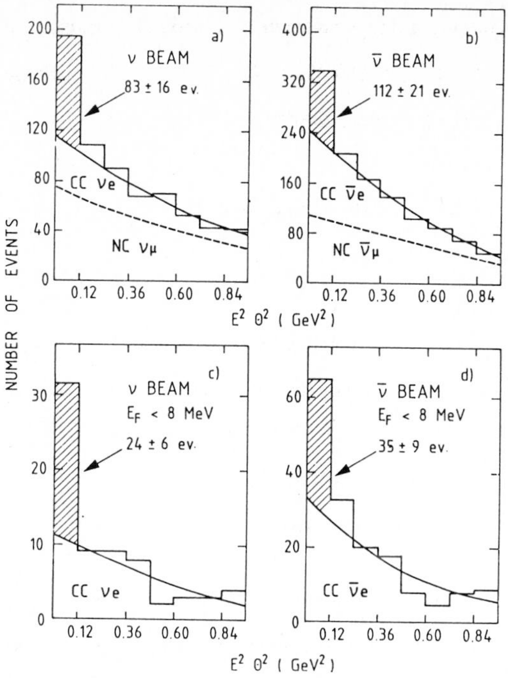


Fig. 13.
$E^2\Theta^2$ distributions for (a) neutrino, (b) antineutrino events; in (c) and (d) the additional condition $E_F < 8$ MeV is applied

$$E\Theta^2 \lesssim 2\ m_e,$$

whereas background processes due to neutrino scattering on nucleons have much broader angular distributions. According to the measured angular resolution of the detector, the bulk of the neutrino-electron scattering events (87%) is expected in the interval $E^2\Theta^2 < 0.12$ GeV$^2$ (forward region). Neutrino-electron scattering events were obtained by subtracting, in the forward region, the background measured in a reference region ($0.12 < E^2\Theta^2 < 0.54$ GeV$^2$) and extrapolated according to the hypothesis of a two-component background:

(a) quasi-elastic charged-current events on nucleons induced by the $\nu_e(\bar{\nu}_e)$ contamination (~ 2%) of the beam,

(b) neutral current events with a $\pi^0$ in the final state produced by coherent $\nu_\mu(\bar{\nu}_\mu)$ scattering on nuclei.

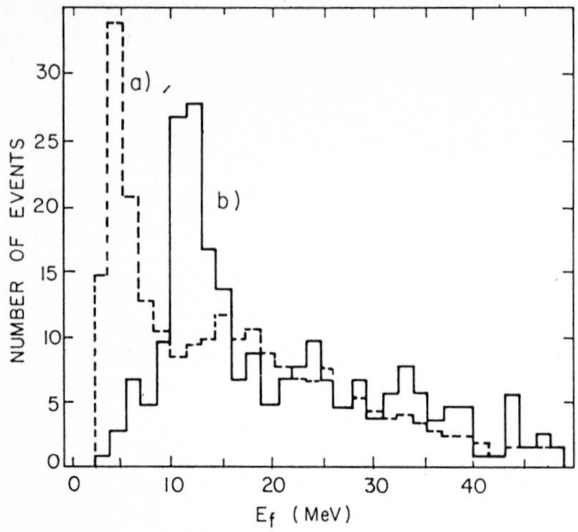


Fig. 14.
Distribution of $E_F$ (first scintillator plane following the vertex) for (a) electrons and (b) $\pi^0$

The relative amount of the two background components was evaluated from the study of the $E_F$ distribution shown in Figure 14; they are different for showers initiated by electrons, photons or neutral pions. Since the $E^2\Theta^2$ distributions of the two background components are quite similar, an error in determining the composition of the background has little effect on the number of the neutrino-electron scattering events obtained. The neutrino and antineutrino fluxes were monitored by recording events induced by quasi-elastic charged-current scattering on nucleons and the events induced by inclusive neutral and charged-current processes on nucleons. By taking the average of both flux determinations the ratio of the normalized numbers of $\nu_\mu e$ and $\bar{\nu}_\mu e$ events is found to be

$$R_{exp} = N(\nu_\mu e)/N(\bar{\nu}_\mu e) \times F = \sigma\,(\nu_\mu e)/\sigma\,(\bar{\nu}_\mu e) = 1.20\,^{+\,0.41}_{-\,0.28}\,.$$

The relation between $R_{exp}$ and $\sin^2\Theta_W$ given by (17) is shown in Fig. 15. The measured value of $R_{exp}$ agrees with the predictions of the standard model for

$$\sin^2\Theta_W = 0.211 \pm 0.035\ (\text{stat}) \pm 0.011\ (\text{syst}). \tag{18}$$

From the $E^2\Theta^2$ distributions of the events with $E_F < 8$ MeV, corresponding to a single minimum ionizing particle (see Figs. 13c and d), a ratio of $\nu_\mu(\bar{\nu}_\mu)e$ events with $E_F < 8$

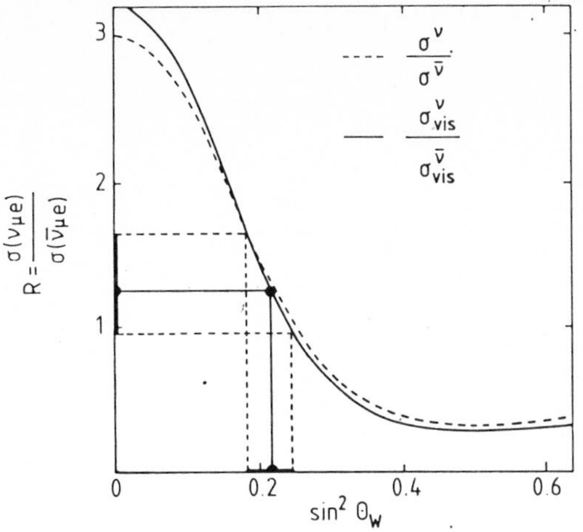


Fig. 15.

Relation between $R = \sigma(\nu_\mu e)/\sigma(\bar{\nu}_\mu e)$ and $\sin^2\Theta$ with the measured value of R and the corresponding value of $\sin^2\Theta$

MeV and with $E_F < 50$ MeV of $0.30 \pm 0.08$ was found, in very good agreement with the relative detection efficiency of $0.32 \pm 0.05$ for electrons, as measured in a calibration beam. This agreement is an experimental proof of the hypothesis that the selected events are due to neutrino-electron scattering.

Four pairs of values of the neutral-current coupling constants, $g_A^e$ and $g_V^e$, can be obtained from these measurements. The limits from other measurements in the lepton sector, those from the forward-backward asymmetry in the reaction $e^+e^- \rightarrow \ell^+\ell^-$ at PETRA and PEP [49] and from the $\bar{\nu}_e e$ scattering cross-section [50], shown in Fig. 16 select a unique solution

$$g_A^e = -0.57 \pm 0.04 \text{ (stat)} \pm 0.06 \text{ (syst)}, \qquad (19)$$

$$g_V^e = -0.06 \pm 0.07 \text{ (stat)} \pm 0.02 \text{ (syst)}.$$

A value of $g_A^e = -1/2$ is predicted by the Standard Model [8], in agreement with the experiment.

The relative strength of the neutral and charged-current coupling constants is found to be

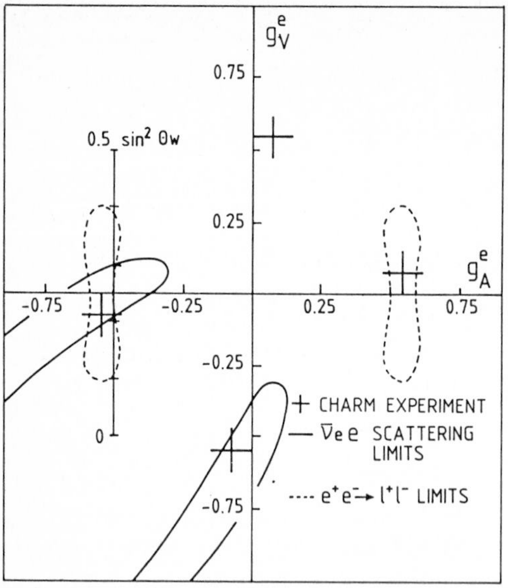


Fig. 16.
Values of $g_A{}^e$ and $g_V{}^e$, the neutral current coupling constants obtained from the measurements of $\sigma(\nu_\mu e)$ and $\sigma(\bar{\nu}_\mu e)$ by the CHARM Collaboration [49]. The limits from forward-backward asymmetry in $e^+e^- \rightarrow \ell^+\ell^-$ [45] and from $\sigma(\bar{\nu}_e e)$ measurements [50] select a unique solution.

$$\rho = 1.14 \pm 0.07 \text{ (stat)} \pm 0.12 \text{ (syst)}, \tag{20}$$

in agreement with the prediction of the Standard Model, $\rho = 1$, provided the Higgs fields form an isospin doublet. At a higher level of precision, small deviations of $\rho$ from one are expected, due to fermion doublets with large mass differences between the up and down components.

Can the present techniques used for studies of neutrino-electron scattering be sufficiently improved to match the aim of $\Delta\sin^2\Theta = \pm\ 0.005$? A new detector, dedicated to this task (CHARM II) [51], has been built and is presently taking data at CERN. The technique of low Z material calorimetry is again used together with fine-grain detection systems. The limiting accuracy of shower direction measurements is given by the Z number of the target material,

$$\sigma(\Theta) \sim \text{const. } Z/\sqrt{E}. \tag{21}$$

The mean Z of the present marble target is 13, with glass ($SiO_2$) a value of $\sim$ 11 is achieved. The accuracy depends further upon the sampling frequency (plate thickness), the grain size of the detector and the detection method.

The structure of the new, dedicated detector is sketched in Fig. 17. It consists of 420 modules of $3.7 \times 3.7\ m^2$ surface area, each composed of a 4.8 cm thick target plate (glass) and of a plane of streamer tubes with 1 cm wire spacing, read out by the wires and by crossed ($90^0$) cathode strips of 2 cm width. Using pulse height measure-

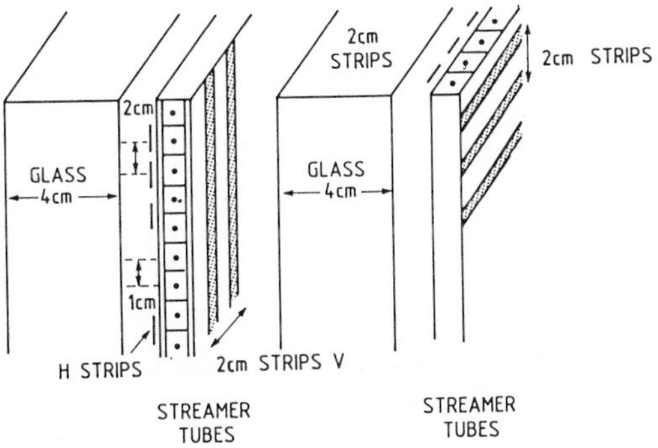


Fig. 17.
Sketch of the structure of the CHARM II detector [52]

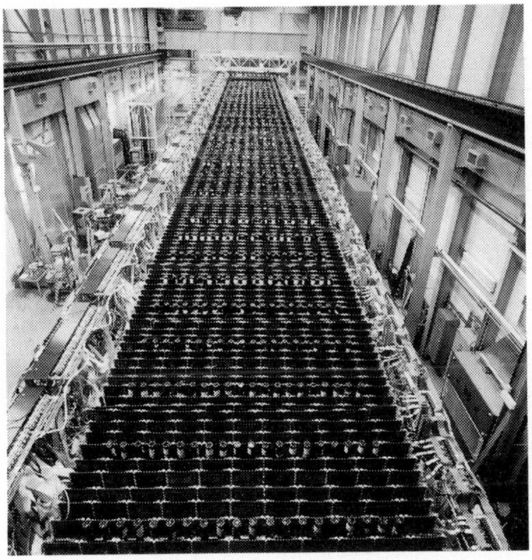

Fig. 18.
Photograph of the CHARM II detector

ments the centroid position of a track can be reconstructed with ± 3 mm accuracy from the cathode strips, whereas the wires are read out digitally to obtain unambiguous information about the track multiplicity near the vertex. A simulation of this detector by Monte Carlo methods gives an electron shower angular resolution of $\sigma(\Theta)18$ mrad/$\sqrt{E}$-/GeV, about equal to the natural angular spread of recoil electrons. A photograph of this new detector is shown in Fig. 18. Two typical events, one due to charged-current neutrino interaction (a), the other due to $\nu_\mu e \to \nu_\mu e$ scattering (b) are shown in Fig. 19. An exposure leading to 2000 $\nu_\mu e$ and 2000 $\bar{\nu}_\mu e$ events is planned.

Figure 20 shows the measurements of elastic neutrino scattering at low $Q^2$ [52]. In the forward direction, at $Q^2$ values less than 0.03 $GeV^2$ the $\nu_\mu$ and $\bar{\nu}_\mu$ distributions differ because of contributions from the reaction $\nu_\mu e \to \mu^- \nu_e$. At larger values of $Q^2$ the cross-sections differ by the V,A interference term (proportional to $Q^2/E$), which can be corrected for. The detection efficiencies and background contributions from delta resonance production differ by small amounts. A total flux ratio error of ± 2% is estimated, leading to a total error of ± 4% on R, and a corresponding experimental error of $\Delta\sin^2\Theta = \pm 0.005$. There is no theoretical uncertainty in extracting this result from (17).

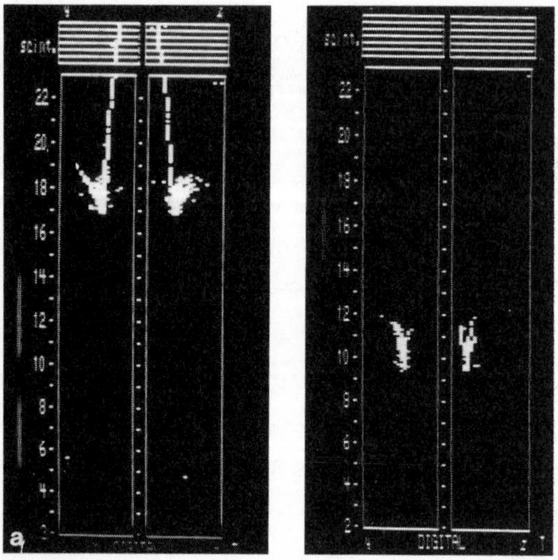

Fig. 19.
Typical neutrino events in the CHARM II detector, (a) $\nu_\mu$ charged-current interaction (b) a candidate of $\nu_\mu e \to \nu_\mu e$ scattering

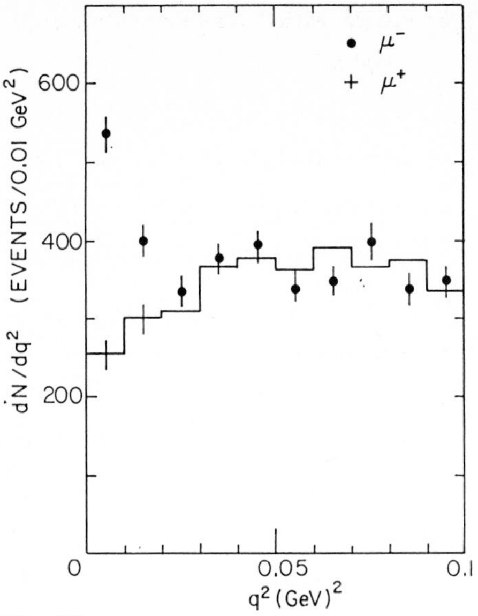


Fig. 20.

Observed $Q^2$ dependence of quasi-elastic $\nu_\mu(\mu^-)$ and $\bar{\nu}_\mu(\mu^+)$ events

## CONCLUDING REMARKS

With the recent advances in experiments determining the Lorentz structure and measuring $\sin^2\Theta_W$ in semi-leptonic and leptonic neutrino scattering, our understanding of the neutral weak current has progressed. Improvements from further measurements of $\sin^2\Theta_W$ in semi-leptonic neutrino processes are likely to be limited by theoretical uncertainties. Neutrino-electron scattering is a field that is expected to progress, following the demonstration that experiments at high energy using low-Z, fine-grain calorimeters can detect these rare events and separate them reliably from the dominant semi-leptonic reactions. Future results from the CERN $p\bar{p}$ collider on $m_W$ and $m_Z$ and from neutrino electron scattering are expected to match each other in the precision of $\sin^2\Theta_W$ and, hence, to test the theory of electroweak interaction at the level of vacuum polarisation.

# REFERENCES

[1] F.J. Hasert et al.: Phys. Lett. 46B, 121 (1973)

[2] F. J. Hasert et al.: Phys. Lett. 46B, 138 (1973)

[3] C.Y. Prescott et al.: Phys. Lett. 77B, 347 (1978); Phys. Lett. 84B, 524 (1979)

[4] M.A. Bouchiat and C.A. Piketty: J. Phys. (France) 46 (1985) 1897; S. Gilbert, N. Noecker, R. Watts, C. Wiemann: Phys. Rev. Lett. 55, 2680 (1985)

[5] See, e.g. S.L. Wu: Phys. Reports 107, 59 (1984)

[6] M. Jonker et al.: CHARM Collab., Phys. Lett. 102B, 67 (1981)

[7] J. Dorenbosch et al.: CHARM Collab., Phys. Lett. 180B, 303 (1986)

[8] S.L. Glashow: Nucl. Phys. 22, 579 (1961)
A. Salam, J. Ward: Phys. Lett. 13, 168 (1963)
S. Weinberg: Phys. Rev. Lett. 19, 1264 (1967)

[9] See ref. [6] and I. Abt: PhD Thesis, University of Hamburg (1986)

[10] U. Amaldi et al.: CERN-EP/87-93 and Phys. Rev. D (to be published)
G. Costa et al.: CERN-TH/4675-87

[11] G. Arnison et al.: UA1 Collab., Phys. Lett. 122B, 103 (1983); 126B, 398 (1983)
M. Banner et al.: UA2 Collab., Phys. Lett. 122B, 476 (1983); 129B, 130 (1983)

[12] B. Kayser et al.: Phys. Lett. 52B, 385 (1974)

[13] O. Nachtmann: Nucl. Phys. B22, 385 (1970)

[14] K.S. Lackner: Nucl. Phys. B153, 526 (1979)
and PhD Thesis, University of Heidelberg (1978), unpublished

[15] D. Rein, L.M. Seghal: Nucl. Phys. B223, 29 (1983)

[16] H. Faissner et al.: Phys. Lett. 125B, 230 (1983)

[17] E. Isiksal, D. Rein, J.G. Morfin: Phys. Rev. Lett. 52, 1096 (1984)

[18] F. Bergsma et al.: CHARM Collab., Phys. Lett. 157B, 469 (1985)

[19] P. Marage et al.: Phys. Lett. 140B, 137 (1984)

[20] M. Jonker et al.: CHARM Collab., Phys. Lett. 105B, 185 (1981)

[21] F. Bergsma et al.: CHARM Collab., Phys. Lett. 117B, 272 (1982)

[22] F. Bergsma et al.: CHARM Collab., Phys. Lett. 147B, 481 (1984)

[23] J.V. Allaby et al.: Z. f. Phys. C36, 611 (1978)

[24] P.Q. Hung, J.J. Sakurai: Ann. Rev. Part. Sci. 31, 375 (1981)

[25] A.N. Diddens et al.: CHARM Collab., Nucl. Instr. and Meth. 178, 27 (1980)

[26] P. Marage: WA59 Collab., PhD Thesis, University of Brussels (1984), unpublished

[27] J. Panman: "Electroweak Neutral Current Interactions", in Proc. of the 1987 Int. Symposium, High-Energy Lepton and Photon induced Processes, Hamburg, (to be published)

[28] See, e.g. M. Jonker et al.: CHARM Collab., Phys. Lett. 99B, 265 (1981)

[29] M. Jonker et al.: CHARM Collab., Phys. Lett. 102B, 67 (1981)
F. Bergsma et al.: CHARM Collab., Phys. Lett. 177B, 446 (1986)
I. Abt: PhD Thesis, University of Hamburg (1986)

[30] F. Bergsma et al.: CHARM Collab., Nucl. Instr. Methods 253, 203 (1987)

[31] D.Yu. Bardin, O.M. Dokuchaeva: Sov. J. Nucl. Physics 36, 282 (1982)

[32] J.G.H. de Groot et al.: Z. f. Phys. C1, 143 (1979)

[33] S.L. Glashow, J. Iliopoulos, L. Maiani: Phys. Rev. D2, 1285 (1970)

[34] C.H. Llewellyn-Smith: Nucl. Phys. B228, 205 (1983)

[35] J.V. Allaby et al.: CHARM Collab., Phys. Lett. 177B, 446 (1986);
Z. f. Phys. C36, 611 (1987)

[36] H. Abramowicz et al.: CDHS Collab., Phys. Rev. Lett. 57, 298 (1986)

[37] K. Winter: Weak Interaction Workshop, Javea (Spain), 1982 (unpublished)

[38] J. Panman: Workshop on SPS Fixed-Target Physics in the years 1984-1989, CERN 83-02, Vol. 11 p. 146

[39] A.N. Diddens et al.: CHARM Collab., Nucl. Instr. Methods 178, 27 (1980);
200, 183 (1982); 215, 361 (1983); A253, 203 (1987)

[40] D. Bogert et al.: FMMF Collab., Phys. Rev. Lett. 55, 1969 (1985)

[41] P. Reutens et al.: CCFRR Collab., Phys. Lett. 152B, 404 (1985)

[42] W.J. Marciano: in Proc. 23rd Int. Conf. on High Energy Physics, Berkeley 1986 (World Scientific, Singapore, 1987) Vol. 2 p. 299

[43] W.J. Marciano, A. Sirlin: Phys. Rev. D29, 945 (1984)

[44] M. Goldhaber: in Proc. 23rd Int. Conf. on High Energy Physics, Berkeley 1986 (World Scientific, Singapore, 1987) Vol. 1 p. 248

[45] F. Bergsma et al.: CHARM Collab., Phys. Lett. 147B, 481 (1984); Z. f. Phys. C (to be published)

[46] L.A. Ahrens et al.: Phys. Rev. Lett. 54, 18 (1985)

[47] K. Abe et al.: Phys. Rev. Lett. 56, 1107 (1986); Phys. Rev. Lett. 58, 636 (1987)

[48] K. Winter: Invited talk on $\nu_\mu e$ scattering, Europhysics Study Conference on Electro-Weak Effects at High Energy, Erice, Sicily 1983 (Plenum Press, New York, 1985) p. 141

[49] See, e.g. S.L. Wu: Rapporteur Talk, Proc. Int. Symp. on Lepton and Photon Interactions at High Energies, Hamburg 1987 (to be published)

[50] F. Reines, H.S. Gurr, H.W. Sobel: Phys. Rev. Lett. 37, 315 (1976)

[51] J.P. De Wulf et al: CHARM II Collab., Nucl. Instr. Methods A252, 443 (1986)

[52] F. Bergsma et al: CHARM Collab., Phys. Lett. 122B, 465 (1983)

# Massive Neutrinos in Gauge Theories

*P. Langacker*

Deutsches Elektronen Synchrotron, DESY,
D-2000 Hamburg 52, Fed. Rep. of Germany
Permanent address: Department of Physics,
University of Pennsylvania, Philadelphia, PA 19104, USA

The present status of several aspects of neutrino physics are summarized, including the weak interactions of neutrinos, neutrino counting, and the theoretical expectations for and experimental constraints on neutrino mass.

## 1 Introduction

Neutrinos have long been amongst the most important probes of the fundamental interactions. In the last fifteen years, in particular, neutrinos have helped establish the standard $SU_2 \times U_1$ electroweak model as correct to first approximation, have been important probes of the structure of the nucleon and of the strong interactions, and have set stringent limits on new physics beyond the standard model. Furthermore, the question of whether the neutrino has a nonzero mass is one of the most important issues in both particle physics and astrophysics: most extensions of the standard model predict a nonzero mass at some level. Masses in the 10 $eV$ range could account for the dark matter of the universe, while masses $\leq 10^{-2} eV$ could resolve the Solar neutrino problem.

In this talk I will describe several aspects of neutrino physics starting with the weak interactions of neutrinos. It will be seen that both the charged and neutral current processes are very well described by the standard model. I will then turn to the question of neutrino counting: indirect evidence leaves little doubt as to the existence of the $\tau$-neutrino, while a number of laboratory and cosmological constraints strongly suggest that the number of neutrinos (with mass $<< \frac{M_Z}{2}$ is less than 0(3-5)). Finally, I will consider the complicated subject neutrino mass: the principle theoretical models and their implications will be described, and the experimental situation will be briefly summarized.

## 2 The Weak Interactions of Neutrinos

The Glashow-Weinberg-Salam standard electroweak model [1] is based on the gauge group $SU_2 \times U_1$, with gauge couplings $g$ and $g'$ for the two factors and gauge bosons $(W^{\pm}, W^0, B)$. It incorporates the Fermi theory of the charged current weak interactions [2] and quantum electrodynamics (QED), and successfully predicted the existence and properties of a new neutral current interaction (Fig. 1). The charged and neutral current interactions are mediated by the massive gauge bosons $W^{\pm}$ and $Z$, respectively, while QED is mediated by the massless photon $A$, where

$$\begin{aligned} A &\equiv \cos\theta_W B + \sin\theta_W W^0 \\ Z &\equiv -\sin\theta_W B + \cos\theta_W W^0 \end{aligned} \tag{1}$$

*Permanent address: Department of Physics, University of Pennsylvania, Philadelphia, PA 19104.

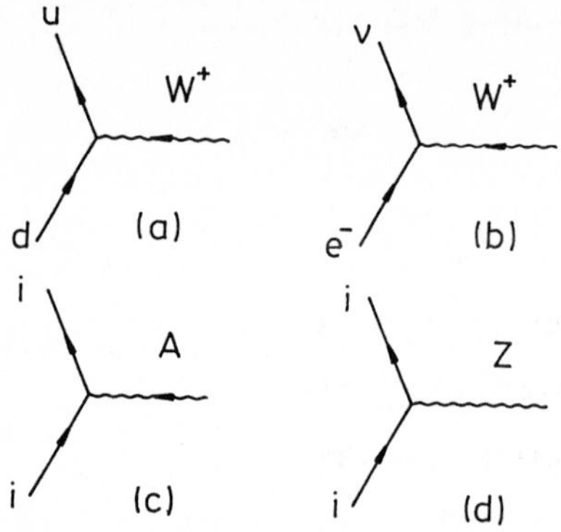


Figure 1: Charged current, QED, and neutral current interactions. The vertex factors are (a) $-\frac{ig}{2\sqrt{2}}\gamma_\mu(1+\gamma^5)V_{ud}$, (b) $-\frac{ig}{2\sqrt{2}}\gamma_\mu(1+\gamma^5)$, (c) $-ieq_i\gamma_\mu$, and (d) $-\frac{ig}{2\cos\theta_W}\gamma_\mu\left[t_{3L}(i)(1+\gamma^5)-2\sin^2\theta_W q_i\right]$. $V_{ud}$ is an element of the quark mixing matrix, $q_i$ is the electric charge of fermion $i$ (in units of $e$), and $t_{3L}(i)$, the eigenvalue of the third generator of $SU_2$, is $+\frac{1}{2}$ for $(u,\nu)$ and $-\frac{1}{2}$ for $(d,e^-)$.

In (1), $\theta_W \equiv \tan^{-1}(g'/g)$ is the weak angle. $e$, the positron electron charge, is related by

$$e = g\sin\theta_W \tag{2}$$

The $W$ and $Z$ masses are predicted in terms of $\sin^2\theta_W$, which can be determined independently from deep inelastic neutrino scattering. One has

$$\begin{aligned} M_W &= \frac{A_o}{\sin\theta_W(1-\Delta r)^{\frac{1}{2}}} \\ M_Z &= \frac{M_W}{\cos\theta_W} \end{aligned} \tag{3}$$

where $A_o = (\pi\alpha/\sqrt{2}G_F)^{1/2} = 37.281\ GeV$. $\Delta r$ is a higher order correction, mainly due to $A$, $W$, and $Z$ self-energy diagrams. It is predicted to be $0.0713 \pm 0.0013$ for top quark and Higgs boson masses of 45 $GeV$ and 100 $GeV$, respectively, while $\Delta r \to 0$ for $m_t \sim 245\ GeV$. The predictions of (3) are in striking agreement with the data from the UA1 [3] and UA2 [4] groups at CERN, and even provide a rough confirmation of the radiative corrections [5] (Table 1). The production cross sections, couplings, and angular distributions (⇒ spin) are also in agreement with expectations.

Table 1: The measured $W$ and $Z$ masses (in $GeV$), compared with the theoretical expectations [5] from deep inelastic scattering with and without radiative corrections. (The radiative corrections include $\Delta r$ from (3) as well as to the value of $\sin^2\theta_W$ extracted from experiment).

| | $M_W$ | $M_Z$ |
|---|---|---|
| UA1 + UA2 | $80.9 \pm 1.4$ | $91.9 \pm 1.8$ |
| prediction (with radiative corrections) | $80.2 \pm 1.1$ | $91.6 \pm 0.9$ |
| prediction (without radiative corrections) | $75.9 \pm 1.0$ | $87.1 \pm 0.7$ |

## 2.1 The Charged Current

The weak charged current interaction is described by the coupling

$$L = -\frac{g}{2\sqrt{2}}(J^{\mu}_{W}W^{-}_{\mu} + J^{\mu\dagger}_{W}W^{+}_{\mu}) \tag{4}$$

between the massive $W^{\pm}$ bosons and the charged fermion current $J^{\mu}_{W}$, given (for massless neutrinos), by

$$J^{\mu\dagger}_{W} = (\bar{u}\ \bar{c}\ \bar{t})\ \gamma^{\mu}(1+\gamma^5)V \begin{pmatrix} d \\ s \\ b \end{pmatrix} + (\bar{\nu}_e\ \bar{\nu}_\mu\ \bar{\nu}_\tau)\ \gamma^{\mu}(1+\gamma^5) \begin{pmatrix} e^- \\ \mu^- \\ \tau^- \end{pmatrix}. \tag{5}$$

The weak charged current is purely $V - A$, which means that it involves only the left-chiral $(1+\gamma^5)$ projections of the quark and lepton fields [6]. In (5),

$$V \equiv \begin{pmatrix} V_{ud} & V_{us} & V_{ub} \\ V_{cd} & V_{cs} & V_{cb} \\ V_{td} & V_{ts} & V_{tb} \end{pmatrix} \tag{6}$$

is the unitary Cabibbo-Kobayashi-Maskawa [7] (CKM) quark mixing matrix, which is due to the mismatch between the weak interactions and the quark mass matrix. $V_{ij}$ describes the relative amplitude for the transition $d_j \to u_i$. Experimentally,

$$V \simeq \begin{pmatrix} \cos\theta_c & \sin\theta_c & 0 \\ -\sin\theta_c & \cos\theta_c & 0 \\ 0 & 0 & 1 \end{pmatrix} + O(\theta_c^2), \tag{7}$$

where $\sin\theta_c \simeq 0.23$ is the sine of the Cabibbo angle. For massless neutrinos there is no analogue of $V$ in the leptonic current. Since there are no mass terms to define the neutrino flavours one can simply define $\nu_e$ as the state produced in weak transitions involving the electron, etc. For momenta small compared to $M_W$, weak charged current processes can be described by an effective four-fermion interaction (Fig. 2) $-L^{CC}_{eff} = \frac{G_F}{\sqrt{2}} J^{\mu\dagger}_{W} J_{W\mu}$, where the fermi constant $G_F$ is given (to lowest order) by

$$G_F = \frac{\sqrt{2}g^2}{8M_W^2} = 1.16637\times10^{-5}GeV^{-2}. \tag{8}$$

The numerical value is determined from muon decay.

The standard model predictions for the weak charged current have been extensively tested in a variety of processes [8]. In particular, there have been many precise tests in the purely leptonic

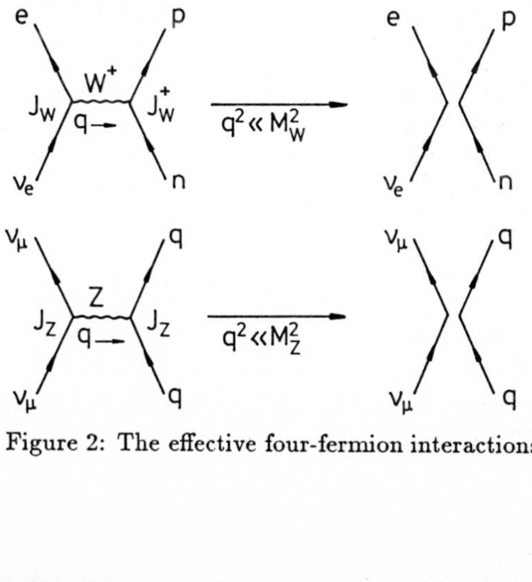


Figure 2: The effective four-fermion interactions for $\nu_e n \to e^- p$ and $\nu_\mu q \to \nu_\mu q$.

sector (which is free from any uncertainties from the strong interactions), including $\mu$ and $\tau$ decay and $\nu_\mu e \to \mu^- \nu_e$ scattering. In a recent model-independent analysis of muon decay and inverse decay data, Fetscher, Gerber, and Johnson[9] have considered the most general local derivative-free four-fermion interaction for muon decay, assuming only Lorentz invariance, separately conserved electron and muon lepton numbers [10], and massless neutrinos. They found that the data uniquely require $V - A$ couplings for the leptonic interactions (Fig. 3 and Table 2). The other invariants, involving $V + A$ as well as scalar, pseudoscalar, and tensor operators are all required to be small, with stringent limits on the coefficients of all operators except the scalar interaction involving left-chiral $e$ and $\mu$.

One way of seeing to what extent pure $V - A$ is required is provided by a series of measurements of polarized $\mu^+$ decay asymmetries at TRIUMF [11]. They find that the mass of $W_R$, a hypothetical gauge boson coupling to right-chiral $(V + A)$ current [12] in $\mu$ decay, must exceed 400 $GeV$, in contrast to the ordinary $W$ (coupling to $V - A$) mass of $80.9 \pm 1.4$ $GeV$. The same results can be used [9] to infer that $1 - |h_{\nu_\mu}| < 0.0032$, where $h_{\nu_\mu}$ is the helicity of $\nu_\mu$ produced in $\pi_{\mu 2}$ decays. This is in striking agreement with the $V - A$ prediction [13] of $h_{\nu_\mu} = -1$.

Similarly, $L^{CC}_{eff}$ has been extensively tested in a variety of semi-leptonic decay processes, such as $\beta$, hyperon, $\pi$, $K$, $c$, and $b$ decays. The results are in impressive agreement with the predictions of the standard model. In particular, the $V - A$ nature of the charged current interaction and the relative strength of the various weak processes, as predicted in (5), are quantitatively confirmed. For example, from $\mu$, $\beta$, $K$, hyperon, and $b$ decays one can extract the CKM matrix elements $|V_{ud}|$, $|V_{us}|$, and $|V_{ub}|$. One finds [14]

Table 2: Limits on the branching ratios $Q^\gamma_{\epsilon\mu}$ for muon decay via scalar ($\gamma = S$), vector ($\gamma = V$), and tensor ($\gamma = T$) interactions, from Fetscher et al.[9] $\epsilon$ and $\mu$ are the chiralities of the $e$ and $\mu$, respectively. The $Q^\gamma_{\epsilon\mu}$ are related to the couplings in Fig. 3 by $Q^\gamma_{\epsilon\mu} = \lambda_\gamma |g^\gamma_{\epsilon\mu}|^2$, where $\lambda_S = \frac{1}{4}$, $\lambda_V = 1$, and $\lambda_T = 3$.

| Quantity | Limit (90% $c.l.$) |
|---|---|
| $Q^S_{RR} + Q^V_{RR}$ | $< 0.002$ |
| $Q^S_{LR} + Q^V_{LR} + Q^T_{LR}$ | $< 0.008$ |
| $Q^S_{RL} + Q^V_{RL} + Q^T_{RL}$ | $< 0.04$ |
| $Q^S_{LL} + Q^V_{LL}$ | $> 0.95$ |
| $Q^S_{LL}$ | $< 0.21$ |
| $Q^V_{LL}$ | $> 0.79$ |

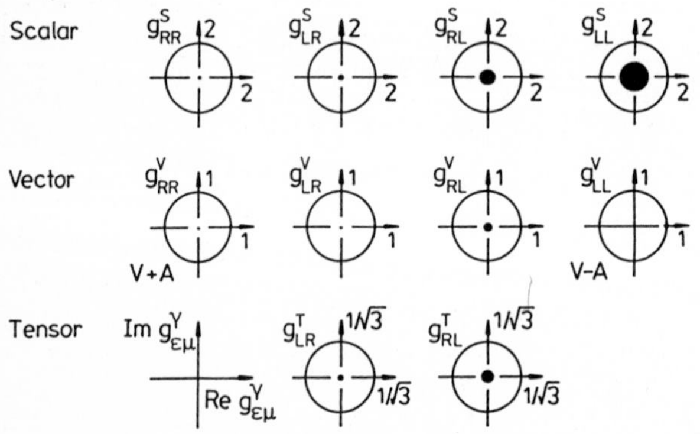


Figure 3: Values of scalar, vector, and tensor interactions in muon decay, as determined by Fetscher et al [9]. The subscripts refer to the chiralities of the $e$ and $\mu$, respectively.

$$|V_{ud}^2| + |V_{us}^2| + |V_{ub}^2| = 0.9979 \pm 0.0021, \tag{9}$$

in remarkable agreement with the expectation of unity from universality [15] (i.e. from the unitarity of $V$).

The leptonic and semi-leptonic data combined leave little room for any deviation from the standard model in charged current processes. In particular, we can be certain that neutrinos are produced by (almost) canonical $V - A$ interactions in weak decays.

The semi-leptonic charged current interaction has also been extensively tested [16] in neutrino scattering processes such as quasi-elastic $\nu_\mu n \to e^- p$ and deep inelastic $\nu_\mu N \to \mu^- X$. These processes are more useful as probes of the hadron than of the neutrino. They have been very useful in testing the QCD-improved proton model and in measuring the relative amount of $u$ and $d$ quarks, antiquarks, and strange quarks in the proton, as well as in determining the CKM elements $V_{cd}$ and $V_{cs}$. $L_{eff}^{CC}$ has also been qualitatively tested in $|\Delta S| = 1$ nonleptonic decays and $\Delta S = 0$ parity violating interference effects, but in these cases hadronic uncertainties obscure the interpretation of the experiments. Higher order weak effects have been semi-quantitatively tested in the $K_L - K_S$ mass difference, the CP-violating parameters $\epsilon$ and $\epsilon'$ observed in $K$ decays [17,18], and, recently, in the $B_d^0 \leftrightarrow \bar{B}_d^0$ oscillations observed by the ARGUS collaboration [19] at DESY.

## 2.2 The Neutral Current

The weak neutral current interaction is

$$L = -\frac{g}{2\cos\theta_W} J_Z^\mu Z_\mu, \tag{10}$$

where

$$\begin{aligned} J_Z^\mu &= \tfrac{1}{2}\bar{u}\gamma^\mu(1+\gamma^5)u - \tfrac{1}{2}\bar{d}\gamma^\mu(1+\gamma^5)d \\ &+ \tfrac{1}{2}\bar{\nu}\gamma^\mu(1+\gamma^5)\nu - \tfrac{1}{2}\bar{e}\gamma^\mu(1+\gamma^5)e \\ &- 2\sin^2\theta_W J_{EM}^\mu \end{aligned} \tag{11}$$

(+ heavy fermion terms), and

$$J_{EM}^\mu = \sum_i q_i \; \bar{\psi}_i\gamma^\mu\psi_i = \frac{2}{3}\bar{u}\gamma^\mu u - \frac{1}{3}\bar{d}\gamma^\mu d - \bar{e}\gamma^\mu e + \cdots \tag{12}$$

is the (purely vector) electromagnetic current. $Z$ couples to both left and right-chiral fermions, but with different strength. For low momenta compared to $M_Z$, (10) implies the effective four-fermion interaction

$$-L_{eff}^{NC} = \frac{G_F}{\sqrt{2}} J_Z^\mu J_{Z\mu}, \tag{13}$$

The neutral current interaction has been observed and quantitatively tested in a wide variety of weak processes, including deep inelastic $\overset{(-)}{\nu}_\mu N$ scattering from isoscalar and proton targets, elastic $\overset{(-)}{\nu}_\mu p$ scattering, coherent $\nu N \to \nu\pi^0 N$ scattering, elastic $\overset{(-)}{\nu}_i e$ ($i = e, \mu$) scattering, and $e^+e^- \to$ hadrons. In addition, weak-electromagnetic interference has been studied in polarized $eD$ and $\mu C$ scattering, atomic parity violation, and forward-backward asymmetries in $e^+e^- \to e^+e^-$, $\mu^+\mu^-$, $\tau^+\tau^-$, $c\bar{c}$, and $b\bar{b}$. All processes are in excellent agreement with the standard model predictions, as can be seen in Fig. 4 and Table 3. Combined with the $W$ and $Z$ masses the standard model is quantitatively confirmed over an enormous momentum range, $10^{-6}\ GeV^2 < |Q^2| < 10^4\ GeV^2$. It is almost certainly correct to first approximation.

Let us now examine the neutral current interactions of neutrinos in more detail. It is convenient to write the terms in $-L_{eff}^{NC}$ relevant to $\nu$-hadron processes in a form that is valid in an arbitrary gauge theory (assuming massless left-handed neutrinos). One has

$$-L^{\nu H} = \frac{G_F}{\sqrt{2}}\bar{\nu}\gamma^\mu(1+\gamma^5)\nu \left\{\sum_i [\epsilon_L(i)\,\bar{q}_i\gamma_\mu(1+\gamma^5)q_i + \epsilon_R(i)\,\bar{q}_i\gamma_\mu(1-\gamma^5)q_i]\right\}, \tag{14}$$

where in the standard model [20]

$$\begin{aligned}
\epsilon_L(u) &= \frac{1}{2} - \frac{2}{3}\sin^2\theta_W \\
\epsilon_L(d) &= -\frac{1}{2} + \frac{1}{3}\sin^2\theta_W \\
\epsilon_R(u) &= -\frac{2}{3}\sin^2\theta_W \\
\epsilon_R(d) &= +\frac{1}{3}\sin^2\theta_W
\end{aligned} \tag{15}$$

It is also convenient to define the variables

$$\begin{aligned}
g_L^2 &\equiv \epsilon_L(u)^2 + \epsilon_L(d)^2 \simeq \frac{1}{2} - \sin^2\theta_W + \frac{5}{9}\sin^4\theta_W \\
g_R^2 &\equiv \epsilon_R(u)^2 + \epsilon_R(d)^2 \simeq \frac{5}{9}\sin^4\theta_W,
\end{aligned} \tag{16}$$

and

$$\theta_i \equiv \tan^{-1}(\epsilon_i(u)/\epsilon_i(d)),\ i = L \text{ or } R \tag{17}$$

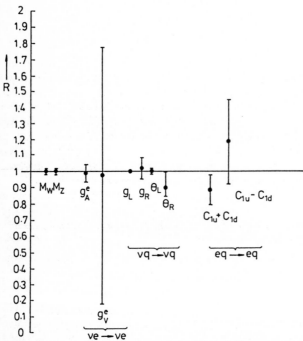

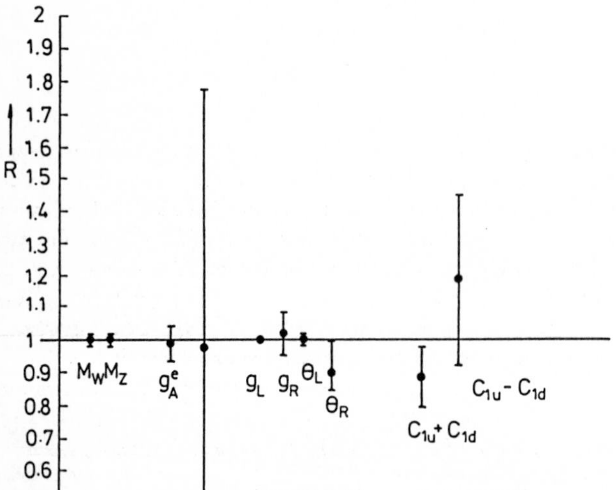


Figure 4: Experimental values of the $W$ and $Z$ masses and the neutral current couplings, relative to the standard model predictions for the global best fit value $\sin^2\theta_W = 0.230$ (the value of $g_L$ should be regarded as the major determinant of $\sin^2\theta_W$ rather than a prediction). $C_{1i}, i = u, d$ are the coefficients in $-L_{eff}^{NC}$ of the parity-violating $eq$ interaction $\frac{G_F}{\sqrt{2}}\bar{e}\gamma_\mu\gamma^5 e\bar{q}_i\gamma^\mu q_i$. The other quantities are defined in the text. The error bars on $g_V^e$ are large only because the predicted value (-0.045) is so small.

Table 3: Values of the model independent neutral current parameters, compared with the standard model prediction for $\sin^2\theta_W = 0.230$. Correlations are not given for the neutrino-hadron couplings because of the non-Gaussian $\chi^2$ distributions. However, the neutrino-hadron constraints are accurately represented by the ranges of the variable $g_i^2$ and $\theta_i$, $i = L, R$, which are very weakly correlated.

| Quantity | Experimental Value | Standard Model Prediction | Correlation | |
|---|---|---|---|---|
| $\epsilon_L(u)$ | $0.339 \pm .017$ | 0.345 | | |
| $\epsilon_L(d)$ | $-0.429 \pm .014$ | $-0.427$ | | |
| $\epsilon_R(u)$ | $-0.172 \pm .014$ | $-0.152$ | | |
| $\epsilon_R(d)$ | $-0.011^{+.081}_{-.057}$ | 0.076 | | |
| $g_L^2$ | $0.2996 \pm 0.0044$ | 0.301 | | |
| $g_R^2$ | $0.0298 \pm 0.0038$ | 0.029 | | |
| $\theta_L$ | $2.47 \pm 0.04$ | 2.46 | | |
| $\theta_R$ | $4.65^{+0.48}_{-0.32}$ | 5.18 | | |
| $g_A^e$ | $-0.498 \pm .027$ | $-0.503$ | | $-0.08$ |
| $g_V^e$ | $-0.044 \pm .036$ | $-0.045$ | | |
| $C_{1u}$ | $-0.249 \pm 0.071$ | $-0.191$ | $-0.98$ | $-0.88$ |
| $C_{1d}$ | $0.381 \pm 0.064$ | 0.340 | | 0.88 |
| $C_{2u} - \frac{1}{2}C_{2d}$ | $0.19 \pm 0.37$ | $-0.039$ | | |

At present the most precise determinations of $\sin^2\theta_W$ are from deep inelastic neutrino scattering from (approximately) isoscalar targets. The ratio $R_\nu \equiv \sigma_{\nu N}^{NC}/\sigma_{\nu N}^{CC}$ of neutral to charged current cross sections has been measured to 1% accuracy by the CDHS [21] and CHARM [22] collaborations, so it is important to obtain theoretical expressions for $R_\nu$ and $R_{\bar\nu} \equiv \sigma_{\bar\nu N}^{NC}/\sigma_{\bar\nu N}^{CC}$ (as functions of $\sin^2\theta_W$) to comparable accuracy. Fortunately, most of the uncertainties concerning the strong interactions (as well as neutrino spectra) cancel in the ratio. For neutral current parameters in the vicinity of the standard model $\simeq 90\%$ of $R_\nu$ can be predicted from isospin alone [23]. The remaining 10% (from such effects as quark mixing and the $s$ sea) is strongly constrained by independent measurements involving deep inelastic $e$, $\mu$, and charged-current $\nu$ scattering, including dimuon production, and can be estimated to the necessary (10%) accuracy.

A simple zeroth order approximation (ignoring quark mixing, the $s$ and $c$ sea, and certain tiny higher twist effects) is

$$\begin{aligned} R_\nu &= g_L^2 + g_R^2 r \\ R_{\bar\nu} &= g_L^2 + \frac{g_R^2}{r}, \end{aligned} \tag{18}$$

where $r \equiv \sigma_{\bar\nu N}^{CC}/\sigma_{\nu N}^{CC}$ is the ratio of $\bar\nu$ and $\nu$ charged current cross sections, which can be measured directly. (In the simple parton model, ignoring hadron energy cuts, $r \simeq (\frac{1}{3} + \epsilon)/(1 + \frac{1}{3}\epsilon)$, where $\epsilon \sim 0.125$ is the ratio of the fraction of the nucleon's s momentum carried by antiquarks to that carried by quarks. i.e. $\epsilon \equiv (\bar U + \bar D)/(U + D)$, where $U \equiv \int_o^1 xu(x)dx$ is the first moment of the $u$ quark distribution.) In practice, (18) must be corrected for quark mixing, the $s$ and $c$ seas, $c$ quark threshold effects (which mainly affect $\sigma^{CC}$ - these turn out to be the largest theoretical uncertainty), non-isoscalar target effects, $W - Z$ propagator differences, and radiative corrections (which lower the extracted value of $\sin^2\theta_W$ by $\sim 0.009$.). Details of the neutrino spectra, experimental cuts, $x$

and $Q^2$ dependence of structure functions, and longitudinal structure functions enter only at the level of these corrections and therefore lead to very small uncertainties. Altogether, the theoretical uncertainty is $\Delta\sin^2\theta_W \sim \pm 0.005$, which would be very hard to improve in the future.

There are also a number of measurements [24] of deep inelastic $\overset{(-)}{\nu}_\mu$ scattering from non-isoscalar targets, which are useful for determining the isospin structure of the neutral current interaction. [25] The most recent result (from BEBC [26]) determines the ratio of neutral to charged current cross sections to around 7% accuracy for both $\nu_\mu$ and $\bar{\nu}_\mu$.

The differential cross sections for elastic $\overset{(-)}{\nu}_\mu p \to \overset{(-)}{\nu}_\mu p$ scattering have been precisely measured in the BNL E734 experiment [27]. Four groups [24] have measured the cross section for coherent $\nu N \to \nu\pi^0 N$, for which the hadronic matrix elements can be estimated fairly reliably [28] using PCAC.

From these results [29] the neutrino-hadron couplings can be determined uniquely and (for the left-handed couplings) precisely. The extracted couplings, shown in Fig. 5 and Table 3, are in impressive agreement with the standard model predictions.

Similarly, for an arbitrary gauge theory with massless left-handed neutrinos, the four-fermion interaction for $\overset{(-)}{\nu}_\mu e$ scattering is

$$-L^{\nu e} = \frac{G_F}{\sqrt{2}} \bar{\nu}_\mu \gamma^\mu (1+\gamma^5)\nu_\mu \, \bar{e}\gamma_\mu(g_V^e + g_A^e\gamma^5)e \tag{19}$$

(for $\overset{(-)}{\nu}_e e$ the charged current contribution must be included). In the standard model

$$\begin{aligned} g_V^e &= -\tfrac{1}{2} + 2\sin^2\theta_W \\ g_A^e &= -\tfrac{1}{2}, \end{aligned} \tag{20}$$

up to radiative corrections.

The laboratory cross section for $\overset{(-)}{\nu}_\mu e \to \overset{(-)}{\nu}_\mu e$ elastic scattering is

$$\frac{d\sigma_{\nu_\mu,\bar{\nu}_\mu}}{dy} = \frac{G_F^2 m_e E_\nu}{2\pi}\left[(g_V^e \pm g_A^e)^2 + (g_V^e \mp g_A^e)^2(1-y)^2 - (g_V^{e2} - g_A^{e2})\frac{y m_e}{E_\nu}\right], \tag{21}$$

where the upper (lower) sign refers to $\nu_\mu(\bar{\nu}_\mu)$, $y \equiv T_e/E_\nu$ (which runs from 0 to $(1+\frac{m_e}{2E_\nu})^{-1}$) is the ratio of the kinetic energy of the recoil electron to the incident $\overset{(-)}{\nu}$ energy, and $G_F^2 m_e/2\pi = 4.31 \times 10^{-42}\ cm^2/GeV$. For $E_\nu \gg m_e$ this yields a total cross section

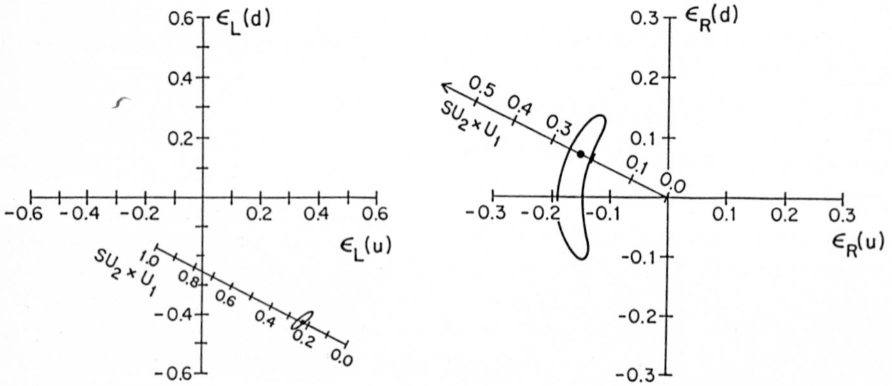


Figure 5: Allowed regions at 90% *c.l.* for the (weak) model independent $\nu q$ parameters $\epsilon_i(u)$ and $\epsilon_i(d)$, $i = L$ or $R$ and the predictions of the standard model as a function of $\sin^2\theta_W$.

$$\sigma = \frac{G_F^2 m_e E_\nu}{2\pi}\left[(g_V^e \pm g_A^e)^2 + \frac{1}{3}(g_V^e \mp g_A^e)^2\right]$$
$$\sim \frac{G_F^2 m_e E_\nu}{2\pi}\begin{cases} 1 - 4\sin^2\theta_W + \frac{16}{3}\sin^4\theta_W, & \nu_\mu e \\ \frac{1}{3} - \frac{4}{3}\sin^2\theta_W + \frac{16}{3}\sin^4\theta_W, & \bar\nu_\mu e \end{cases} \tag{22}$$

The most accurate leptonic measurements [30,31] of $\sin^2\theta_W$ are from the ratio $R \equiv \sigma_{\nu_\mu e}/\sigma_{\bar\nu_\mu e}$, in which many of the systematic uncertainties cancel. Radiative corrections, which are small compared to the precision of present experiments, increase the extracted $\sin^2\theta_W$ by $\simeq 0.002$.

The $\bar\nu_e e$ cross section was measured a decade ago at the Savannah River reactor [32], while $\nu_e e \to \nu_e e$ has been measured recently at Los Alamos [33]. These are not nearly so precise as the $\overset{(-)}{\nu}_\mu e$ measurements, but are interesting because they involve both neutral and charged current contributions. (The cross sections for $\overset{(-)}{\nu}_e e$ may be obtained from (21) by replacing $g_{V,A}^e$ by $g_{V,A}^{e\prime} \equiv g_{V,A}^e + 1$, where the 1 is due to the charged current.)

In fact, the Los Alamos result strongly supports destructive interference ($g_A^e < 0$) between the two amplitudes and rules out constructive interference ($g_A^e > 0$).

The results of the various reactions [5] for the $\nu e$ couplings are shown in Fig. 6. The $\bar\nu_\mu e$ data alone allow four solutions (which differ by $g_i^e \leftrightarrow -g_i^e$ and $g_V^e \leftrightarrow g_A^e$). The reactor $\bar\nu_e e$ results eliminate C, while the Los Alamos $\nu_e e$ experiment eliminates solutions C and D. The remaining two solutions (axial dominant (A) and vector dominant (B)) are consistent with all $\nu e$ data. However, solution (B) is eliminated by the $e^+e^- \to \mu^+\mu^-$ forward-backward asymmetry under the (now very reasonable) assumption that the neutral current is dominated by the exchange of a single $Z$. The remaining solution (A) is in excellent agreement with the standard model prediction, as can be seen in Table 3.

The $\nu$ - hadron and $\nu e$ interactions are therefore uniquely determined and are consistent with the standard model within uncertainties. Similar statements hold for the $e$-hadron and $e^+e^-$ couplings [5]. Having established the standard model couplings as correct to first approximation, the neutral current and boson mass results can be used to test the standard model more stringently and to set limits on possible new physics.

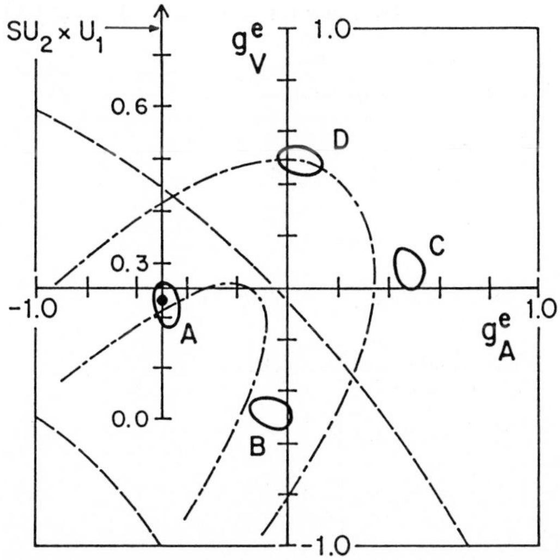


Figure 6: Allowed regions (90% c.l.) for the $\nu e$ parameters $g_V^e$ and $g_A^e$, for $\bar\nu_\mu e$ (solid lines), reactor $\bar\nu_e e$ (dot-dash), and $\nu_e e$ (dash).

The values of $\sin^2\theta_W$ and, equivalently, $M_Z$ (using (3)) determined from various processes are shown in Table 4 and Fig. 7. They are in impressive agreement with each other, reconfirming the quantitative success of the standard model. The best fit to all data yields [34] $\sin^2\theta_W = 0.230 \pm 0.0048$ and $M_Z = 92.0 \pm 0.7\ GeV$, where the errors include full statistical, systematic, and theoretical uncertainties.

As can be seen in Fig. 7 consistency of the various $\sin^2\theta_W$ values (especially those obtained from deep inelastic $\nu N$ and the $W$, $Z$ masses) depends sensitively on the top quark mass, which enters the radiative corrections. In fact, one can use these results to set an upper limit [5] $m_t < 200\ GeV$ (90% *c.l.*), with similar limits applying to the splitting between the masses of possible fourth generation fermions. Similarly, the deep inelastic neutrino data can be combined with the $W$ and $Z$ masses to determine $\Delta r$ in (3). One finds [5] $\Delta r = 0.077 \pm 0.037$, in excellent agreement with the value $0.0713 \pm 0.0013$ predicted for $m_t = 45\ GeV$ and $M_H = 100\ GeV$, and providing a rough test of the theory at the level of radiative corrections (see also Table 1).

The best fit value of $\sin^2\theta_W \equiv 1 - \frac{M_W^2}{M_Z^2}$ corresponds to the modified minimal subtraction value [35]

$$\sin^2\hat{\theta}_W(M_W) = 0.228 \pm 0.0044 \tag{23}$$

This is larger by $\simeq 2.5\ \sigma$ than the prediction $0.214^{+0.003}_{-0.004}$ of minimal $SU_5$ (for $\Lambda^{(4)}_{\overline{MS}} = 150^{+150}_{-75}\ MeV$) and other "great desert" models. Similar conclusions hold for all values of $m_t$ and $M_H$, as can be seen in Fig. 8. Of course, the simplest grand unified theories (GUTs) have been excluded for some time by the nonobservation of proton decay [36], but the additional evidence is welcome, especially since variations on the simplest GUTs can yield much longer lifetimes.

The fact that the $\sin^2\hat{\theta}_W(M_W)$ value in (23) is close to but not identical with the $SU_5$ prediction can be taken as a hint that the basic ideas of GUTs may be roughly correct, but that there is additional structure in the desert. For example, (23) is closer to (but still somewhat below)

Table 4: Determination of $\sin^2\theta_W$ and $M_Z$ (in $GeV$) from various reactions. The central values of all fits assume $m_t = 45\ GeV$ and $M_H = 100\ GeV$ in the radiative corrections. Where two errors are shown the first is experimental and the second (in square brackets) is theoretical, computed assuming 3 fermion families, $m_t < 100\ GeV$, and $M_H < 1\ TeV$. In the other cases the theoretical and experimental uncertainties are combined.

| Reaction | $\sin^2\theta_W$ | $M_Z$ |
|---|---|---|
| Deep inelastic (isoscalar) | $0.233 \pm .003 \pm [.005]$ | $91.6 \pm 0.4 \pm [0.8]$ |
| $\overset{(-)}{\nu}_\mu p \to \overset{(-)}{\nu}_\mu p$ | $0.210 \pm .033$ | $95.0 \pm 5.2$ |
| $\overset{(-)}{\nu}_\mu e \to \overset{(-)}{\nu}_\mu e$ | $0.223 \pm .018 \pm [.002]$ | $93.0 \pm 2.7$ |
| $W, Z$ | $0.228 \pm .007 \pm [.002]$ | $92.3 \pm 1.1$ |
| Atomic parity violation | $0.209 \pm .018 \pm [.014]$ | $95.1 \pm 3.9$ |
| SLAC eD | $0.221 \pm .015 \pm [.013]$ | $93.3 \pm 2.7$ |
| $\mu C$ | $0.25 \pm .08$ | $89.6 \pm 9.7$ |
| All data | $0.230 \pm 0.0048$ | $92.0 \pm 0.7$ |

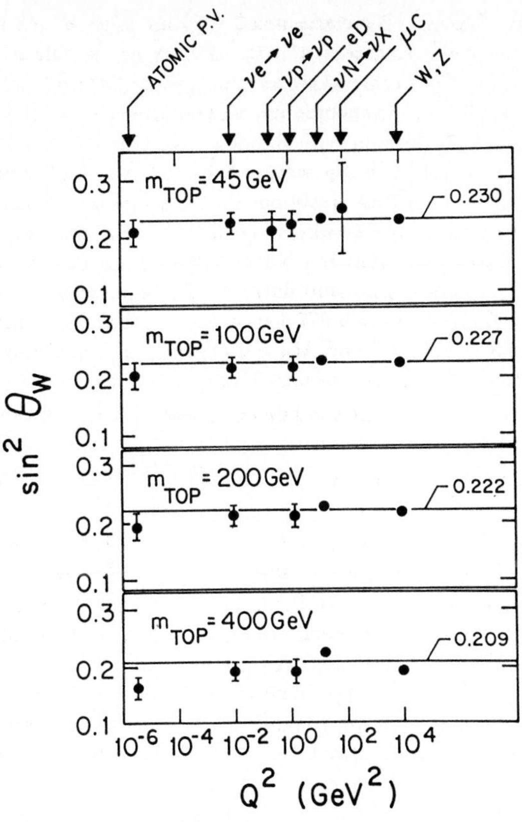


Figure 7: (a) $\sin^2\theta_W$ for various reactions as a function of the typical $Q^2$, determined for $m_t = 45$ *GeV*. The best fit line $\sin^2\theta_W = 0.230$ is also shown. (b-d) $\sin^2\theta_W$ values determined for $m_t = 100$, 200, and 400 *GeV*.

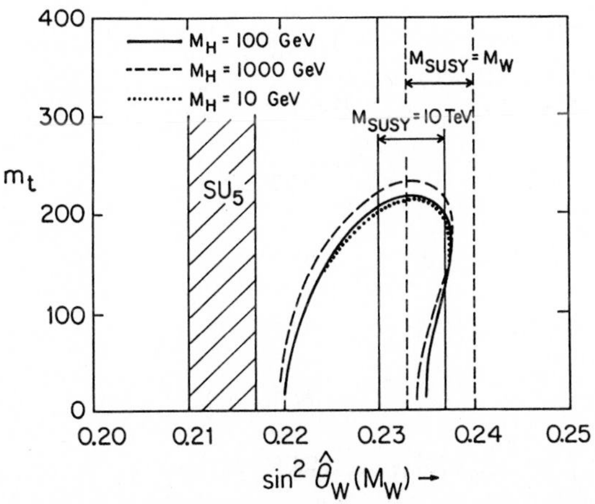


Figure 8: Allowed regions (90% *c.l.*) in $\sin^2\hat{\theta}_W(M_W)$and $m_t$ for fixed values of $M_H$. Also shown are the predictions of ordinary and supersymmetric GUTs, assuming no new thresholds between $M_W$ or $M_{SUSY}$ and the unification scale.

the prediction of the simplest supersymmetric GUTs. (Typically $0.237^{+0.003}_{-0.004}$ for $M_{SUSY} \sim M_W$, decreasing by $\sim 0.003$ for $M_{SUSY} \sim 10\ TeV$). The agreement is better for larger $m_t$ (Fig. 8). Similarly, $SO_{10}$ models [36] with three stages of symmetry breaking can be compatible with (23).

The neutral current data can be used to place rather stringent constraints on certain deviations from the standard model, such as the existence of Higgs triplets with significant vacuum expectation values [5], or the mixing between ordinary and exotic fermions [10]. The $e^+e^- \to b\bar{b}$ forward-backward asymmetry [37] excludes all topless models not involving exotic quarks. Many extensions of the standard model predict the existence of additional $Z$ bosons [5], which could conceivably be light enough to be experimentally relevant. Some limits on the masses $M_2$ and mixing angle $\theta$ between the new and ordinary $Z$ are shown for a class of $E_6$ models in Fig. 9. These neutral current limits are somewhat more stringent [38] than limits from direct searches $\bar{p}p \to Z_2 + X$, $Z_2 \to l^+l^-$ at the $Sp\bar{p}S$ except for a small region in $\beta$ near the $Z_\eta$. Nevertheless, the limits (typically $120 - 300\ GeV$) are still relatively weak. In constrast, there is a non-rigorous but plausible lower limit [39] from the $K_L - K_S$ mass difference of several $TeV$ on the mass of the new charged bosons in many $SU_{2L} \times SU_{2R} \times U_1'$ models. This situation will presumably change in the near future: for example, the FNAL $\bar{p}p$ collider should be sensitive to bosons up to around $400\ GeV$ and the SSC would be sensitive up to several $TeV$.

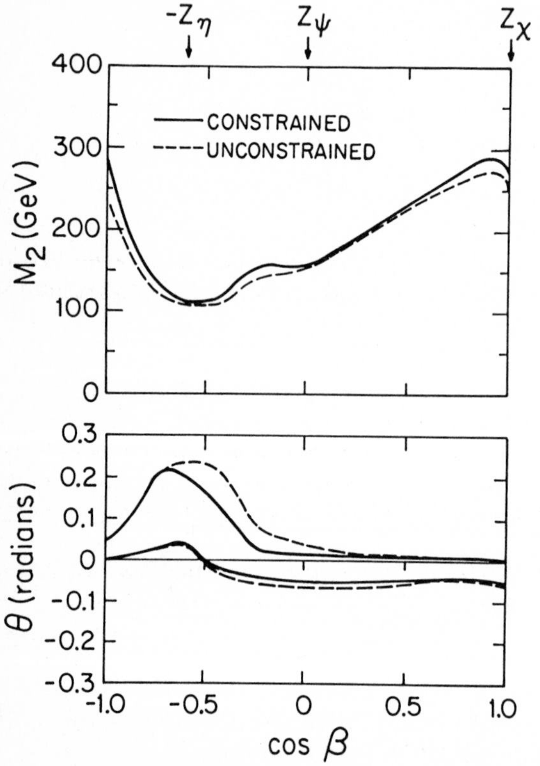


Figure 9: Lower limits on $M_2$ and allowed $\theta$ range (both at 90% $c.l.$) for an $E_6$ boson $Z(\beta) = \cos\beta\ Z_\chi + \sin\beta\ Z_\psi$, where $Z_\chi$ and $Z_\psi$ refer to the breaking patterns $SO_{10} \to SU_5 \times U_{1\chi}$ and $E_6 \to SO_{10} \times U_{1\psi}$, respectively, and $Z_\eta = -Z(\pi - \tan^{-1}\sqrt{\frac{5}{3}})$ occurs in many superstring models. Constrained and unconstrained refer to whether or not it is assumed that $SU_2$ breaking is due to Higgs doublets only.

# 3 Neutrino Counting

Table 5: Limits on the number $N_\nu$ of neutrino flavors and the mass ranges to which they apply. The laboratory limits are at 90% *c.l.*

| $N_\nu$ | mass range | source | reference |
|---|---|---|---|
| $N_\nu \geq 2$ | – | direct | |
| $N_\nu \geq 3$ | – | $\tau$ properties | |
| $N_\nu \leq 4$ | $m_\nu < 1\ MeV$ | nucleosynthesis | [40] |
| $N_\nu \leq 6$ | $m_\nu < O(MeV)$ | SN1987A energetics | [41] |
| $N_\nu \leq \begin{cases} 7.5\ (\text{ASP}) \\ 4.9\ (\text{combined}) \end{cases}$ | $m_\nu < O(5\ GeV)$ | $e^+e^- \not\to \gamma\nu\bar{\nu}$ | [42] |
| $N_\nu \leq \begin{cases} 5,\ m_t < 40\ GeV \\ 3,\ m_t > 50\ GeV \end{cases}$ | $m_\nu < O(40\ GeV)$ | $R$ | [43] |

Constraints on the number of neutrino flavors are listed in Table 5.

There is direct laboratory proof for the existence of only two neutrinos, $\nu_e$ and $\nu_\mu$. However, indirect evidence leaves little doubt as to the separate existence of the $\nu_\tau$. If there were no $\nu_\tau$ then, up to mixing effects, the $\tau_L^-$ would have to be a singlet under $SU_2$ transformations. Including mixing, the two left-handed lepton doublets and one charged singlet would be

$$\begin{pmatrix} \nu_1 \\ U_{1i}e_i^- \end{pmatrix}_L \begin{pmatrix} \nu_2 \\ U_{2i}e_i^- \end{pmatrix}_L U_{3i}e_{iL}^- \qquad (24)$$

where $(e_1, e_2, e_3)_L \equiv (e, \mu, \tau)_L$ and $U$ is a unitary matrix. However, one knows that the $\mu$ and $e$ weak interactions are canonical - there is little room for mixing with an $SU_2$ singlet. From $\mu$, $\beta$, $K$, and hyperon decays and the $W$ mass one can show [10]

$$|U_{13}|, |U_{23}| < 0.05 \qquad (25)$$

(this is confirmed by the absence of $\tau^- \to \mu^-\mu^-\mu^+$ decays). On the other hand, the $\tau$ lifetime [44] $\tau_\tau = (3.07 \pm 0.09) \times 10^{-13}$ sec, which agrees at least roughly with the value $(2.87 \pm 0.06) \times 10^{-13}$sec expected if $\nu_\tau$ exists, implies

$$|U_{13}|^2 + |U_{23}|^2 = 0.94 \pm 0.04, \qquad (26)$$

in clear conflict with (25). An independent argument is that $A^\tau$, the axial vector coupling of $\tau$ in the weak neutral current, is determined from the $e^+e^-$ forward-backward asymmetry to be $A^\tau = -0.46 \pm 0.05$. This is in agreement with the value $-\frac{1}{2}$ expected if the $\tau_L^-$ is in an $SU_2$ doublet with its own partner ($\nu_\tau$), and disagrees with the value (zero) expected if $\tau_L^-$ and $\tau_R^-$ are in $SU_2$ singlets. [45] Hence, the $\nu_\tau$ almost certainly exists, but it would nevertheless be desirable to observe it directly.

There are several upper limits on the number of neutrinos with normal weak interactions. An upper limit of $N_\nu \leq 4$ neutrino flavors with masses $\leq 1\ MeV$ is determined by nucleosynthesis [40] (the abundance primordial $^4He$). Extra neutrino flavors [46] would cause the universe to expand faster, causing the $\nu_e n \leftrightarrow e^- p$ reactions to freeze out earlier (when there are more neutrons), leading to too much $^4He$.

Limits can also be set from the cross section for $e^+e^- \to \gamma\nu\bar{\nu}$ (with only the photon observed), which effectively sums the number of neutrinos. The ASP experiment at PEP obtained [42] $N_\nu < 7.5$. Combined with cross section limits from MAC and CELLO this implies $N_\nu < 4.9$ (90% *c.l.*) sensitive to masses less than several $GeV$. Finally, the $Z$ width increases by 170 $MeV$ for each new neutrino with mass $\leq 40\ GeV$. Indirect limits on $\Gamma_Z$ already exist from the ratio

$$R = \frac{\sigma_{\bar{p}p \to W} B_{W \to l\nu}}{\sigma_{\bar{p}p \to Z} B_{Z \to l^+l^-}} = \frac{\sigma_{\bar{p}p \to W}}{\sigma_{\bar{p}p \to Z}} \frac{\Gamma_{W \to l\nu}}{\Gamma_{Z \to l^+l^-}} \frac{\Gamma_Z}{\Gamma_W}. \qquad (27)$$

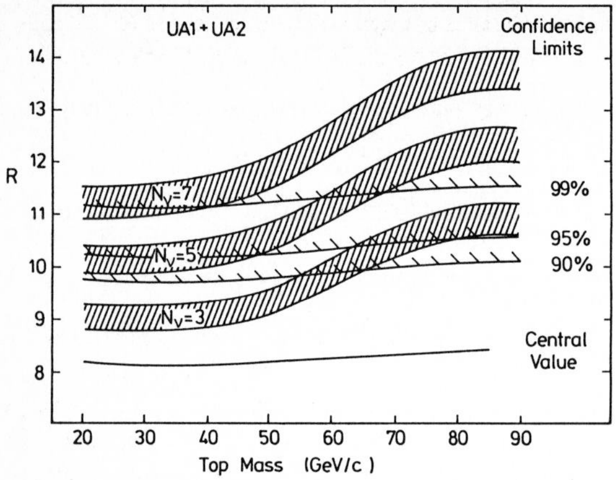


Figure 10: The value of $R$ (27) as a function of $N_\nu$ and $m_t$, and the experimental results form UA1 and UA2.

Using the measured $R$ and theoretical values for the cross section ratio and leptonic widths one determines $\Gamma_Z/\Gamma_W$, which is sensitive to both $N_\nu$ and the $t$ quark mass. Recent estimates [43] typically yield $N_\nu \leq 5$ for $m_t \leq 40\ GeV$ and $N_\nu \leq 3$ for $m_t \geq 50\ GeV$, (the larger $m_t$ range is favored by $B-\bar{B}$ oscillations [19] and the non-observation [47] of the $t$ by UA1), and incidentally suggest the upper limit $m_t \leq 65\ GeV$. These limits are suggestive but should be viewed with caution. As can be seen in Fig. 10 the bounds essentially disappear if one increases the uncertainty in either $R$ itself or the cross section ratio.

Future direct measurements of $\Gamma_Z$ at SLC and LEP should ultimately yield a precision of $\Delta\Gamma_Z \simeq 35\ MeV$, which is equivalent to an uncertainty [48] of $\Delta N_\nu \sim 0.2$. It should be possible to obtain an independent measurement of $\Gamma_{Z\rightarrow\nu\bar{\nu}}$ accurate to $\simeq 50\ MeV$ by measuring $e^+e^-\rightarrow\gamma Z\rightarrow\gamma\nu\bar{\nu}$ above the $Z$ pole.

# 4 Neutrino Mass

In the minimal $SU_2 \times U_1$ model the neutrinos are predicted to be massless. However, extensions of the standard model involving new $SU_2$-singlet neutral fermions (the right-handed neutrino partners needed for Dirac mass terms) or new Higgs representations (to generate Majorana masses) allow non-zero masses. [49] In fact, most extensions of the standard model (e.g. most grand unified theories other than $SU_5$) involve one or both of these mechanisms. Furthermore, non-zero masses could have important implications for the missing Solar neutrinos and/or the missing (dark) matter of the universe.

## 4.1 Weyl, Majorana, and Dirac Neutrinos.

For the weak interactions it is convenient to deal with Weyl two-component spinors $\psi_L$ or $\psi_R$, each of which represents two physical degrees of freedom. The field $\psi_L$ can annihilate a left-handed ($L$) particle or create a right-handed ($R$) antiparticle, while $\psi_L^\dagger$ annihilates a $L$-particle or creates an $R$-antiparticle. For a $\psi_R$ field the roles of $L$ and $R$ are reversed. An ordinary four-component Dirac field $\psi$ can be written as the sum $\psi = \psi_L + \psi_R$ of two Weyl fields, where $\psi_L$ and $\psi_R$ are just the chiral projections

$$\psi_{L,R} = P_{L,R}\psi, \tag{28}$$

with $P_{L,R} = (1 \pm \gamma_5)/2$.

Alternatively, one can consider Weyl fermions that do not have distinct partners of the opposite chirality. We will see below that such spinors correspond to particles that are either massless or carry no conserved quantum numbers.

In the free field limit a Weyl field $\psi_L$ can be written as

$$\psi_L(x) = \sum_{\vec{p}} \left[ b_L(\vec{p}) u_L(\vec{p}) e^{-ip\cdot x} + d_R^\dagger(\vec{p}) v_R(\vec{p}) e^{+ip\cdot x} \right], \tag{29}$$

where $\sum_{\vec{p}}$ represents $\int d^3\vec{p}/\sqrt{(2\pi)^3 2E}$. In(29), $b_L$ and $d_R$ are annihilation operators for $L$ particles and $R$-antiparticles, respectively, and $u_L$ and $v_R$ are the corresponding (4-component) spinors satisfying $P_L u_L = u_L$, $P_L v_R = v_R$, $P_R u_L = P_R v_R = 0$. For a $\psi_R$ spinor one simply interchanges $L$ and $R$. Equation (29) differs from an ordinary (Dirac) free field in that there is no sum over spin.

It is apparent from (29) that each left-handed (right-handed) particle is necessarily associated with a right-handed (left-handed) antiparticle. The right-handed antiparticle [50] field $\psi_R^c$ is not independent of $\psi_L$, but is closely related to $\psi_L^\dagger$. One has

$$\psi_R^c = C\bar{\psi}_L^T, \tag{30}$$

where $C$ is the charge conjugation matrix, defined by $C\gamma_\mu C^{-1} = -\gamma_\mu^T$. Similarly, for a $R$-Weyl spinor, $\psi_L^c = C\bar{\psi}_R^T$. In the special case that $\psi_L$ is the chiral projection $P_L\psi$ of a Dirac field $\psi$, $\psi_R^c$ is just the $R$-projection $P_R\psi^c$ of the antiparticle field $\psi^c = C\bar{\psi}^T$.

If $\psi_R$ and $\psi_R^c$ both exist, they have the opposite values for all additive quantum numbers. Since the quarks and charged leptons carry conserved quantum numbers (e.g. color and electric charge), they must be Dirac fields - i.e. $\psi_R$ and $\psi_R^c$ must be distinct. The only quantum number associated with the neutrinos is lepton number, however, and it is conceivable that that is violated in nature. As we will see, that will allow for two very different possibilities for neutrino mass.

The known neutrinos of the first family are the left-handed electron neutrino $\nu_{eL}$ and its CP partner, the right-handed "antineutrino" $\nu_{eR}^c = C\bar{\nu}_{eL}^T$. These are associated with the $e_L^-$ and $e_R^+$, respectively, in ordinary charged current weak interactions.

Mass terms always take left- and right-handed fields into each other. If one introduces a new field $N_R$ (distinct from $\nu_R^c$) and its CP conjugate $N_L^c = C\bar{N}_R^T$ into the theory, then one can write a Dirac (lepton number conserving) mass term

$$-L_{Dirac} = m_D \bar{\nu}_L N_R + h.c., \tag{31}$$

which connects $N_R$ and $\nu_L$. In this case $\nu_L$, $N_R$, $N_L^c$ and $\nu_R^c$ form a four component Dirac particle - i.e. one can define $\nu \equiv \nu_L + N_R$, $\nu^c = N_L^c + \nu_R^c = C\bar{\nu}^T$, so that

$$-L_{Dirac} = m_D \bar{\nu}\nu. \tag{32}$$

Clearly lepton number is conserved in this case, because there is no transition between $\nu$ and $\nu^c$. In the free field limit the Dirac neutrino field $\nu$ has the canonical expression

$$\nu_{Dirac}(x) = \sum_{\vec{p}} \sum_{S=L,R} \left[ b_S(\vec{p}) u_S(\vec{p}) e^{-ip\cdot x} + d_S^\dagger(\vec{p}) v_S(\vec{p}) e^{+ip\cdot x} \right], \tag{33}$$

Usually, the $N_R$ is an $SU_2 \times U_1$ singlet, with $m_D$ generated by an ordinary Higgs doublet, and $L = L_e + L_\mu + L_\tau$ is conserved in the three family generalization. This possibility is most similar to the way in which masses are generated for the other fermions ($e^-$, $u$, $d$, etc.) in the standard model, but it is difficult to understand why $m_{\nu_e}$ is so small in this case.

Another possibility [51] is that $N_R$ is a known doublet neutrino, such as $\nu^c_{\tau R}$. This is a variation on the Konopinski-Mahmoud model. [52] Then $\nu_{eL}$, $\nu^c_{\tau R}$, $\nu_{\tau L}$ and $\nu^c_{eR}$ can be combined to form a Dirac neutrino with $L_e - L_\tau$ conserved.

For the generalization of (31) to $F$ fermion families one has

$$- L_{Dirac} = \bar{n}^0_L m_D N^0_R + h.c., \tag{34}$$

where $m_D$ is an arbitrary [53] $F \times F$ mass matrix, and $n^0_L$ and $N^0_R$ are $F$-component vectors; thus $n^0_L = (n^0_{1L}\, n^0_{2L} \cdots n^0_{FL})^T$, where $n^0_{iL}$ are the "weak eigenstate" neutrinos - i.e. $n^0_{iL}$ is associated with $e^-_{iL}$ in weak transitions. The weak eigenstate neutrinos are related to the neutrinos $n_{iL}$, $N_{iR}$ of definite mass by unitary transformations

$$\begin{aligned} n^0_L &= V_L n_L \\ N^0_R &= V_R N_R. \end{aligned} \tag{35}$$

$V_L$ and $V_R$ are $F \times F$ unitary matrices, determined by

$$V^\dagger_L m_D V_R = m_d = \mathrm{diag}(m_1\, m_2 \cdots m_F) \tag{36}$$

where $m_d$ is the diagonal matrix of physical neutrino masses. $V_L$ and $V_R$ can be determined by

$$V^\dagger_L m_D m^\dagger_D V_L = V^\dagger_R m^\dagger_D m_D V_R = m^2_d \tag{37}$$

($m_D m^\dagger_D$ and $m^\dagger_D m_D$ are Hermitian). In general $V_L$ and $V_R$ are unrelated. If there are no degeneracies then $V_L$ and $V_R$ are determined uniquely by (37) up to diagonal phase matrices; i.e. if $V_{L,R}$ satisfy (37) then so do $V_{L,R} K_{L,R}$, where $K_{L,R}$ are diagonal phase matrices associated with the unobservable phases of the $n_{iL}$ and $N_{jR}$ fields. Usually one chooses $K_L$ to put $V_L$ into a simple conventional form. Then $K_R$ is determined by the requirement that $m_d$ be real.

$V_L$ modifies the leptonic weak charged current in (5) to [54]

$$J^{\mu\dagger}_W = (\bar{\nu}_e\, \bar{\nu}_\mu\, \bar{\nu}_\tau)\; V^\dagger_L \gamma^\mu (1 + \gamma^5) \begin{pmatrix} e^- \\ \mu^- \\ \tau^- \end{pmatrix} \tag{38}$$

so that $V^\dagger_L$ is just the analogue of the CKM quark mixing matrix. It describes the relative strengths [55] of the weak transition between the various charged leptons and neutrinos of definite mass.

In a Majorana (lepton number violating) mass term one avoids the need for a new fermion field by coupling the $\nu_L$ to its CP conjugate $\nu^c_R$:

$$\begin{aligned} -L_M &= \tfrac{1}{2} m \bar{\nu}_L \nu^c_R + h.c. \\ &= \tfrac{1}{2} m \bar{\nu}_L C \bar{\nu}^T_L + h.c. \end{aligned} \tag{39}$$

$L_M$ can be thought of as creating or annihilating two neutrinos, and violates lepton number by $\Delta L = \pm 2$. $\nu_L$ and $\nu^c_R$ can be combined to form a two component Majorana neutrino $\nu = \nu_L + \nu^c_R$, so that $-L_M = \frac{1}{2} m \bar{\nu} \nu$. From (30) we see that $\nu = C\bar{\nu}^T$, i.e. a Majorana neutrino is its own antiparticle. In the free field limit $\nu$ is just

$$\nu(x) = \sum_{\vec{p}} \sum_{S=L,R} \left[ b_S(\vec{p}) u_S(\vec{p}) e^{-ip\cdot x} + b^\dagger_S(\vec{p}) v_S(\vec{p}) e^{+ip\cdot x} \right], \tag{40}$$

i.e. it has the same form as for a free Dirac field (cf (33)) except that there is no distinction between $b$ and $d$ annihilation operators.

The Majorana mass $m$ in (39) can be generated by the vacuum expectation value (VEV) of a new Higgs triplet [56] or as a higher order effective operator. Majorana masses are popular amongst theorists because they are so different from quark and lepton masses, and there is therefore the possibility of explaining why $m_{\nu_e}$ is so small (if it is non-zero).

For $F$ fermion families, the Majorana mass term is

$$-L_M = \tfrac{1}{2}\bar{n}_L^0 M n_R^{0c} + h.c. \tag{41}$$

where $M$ is an $F \times F$ Majorana mass matrix and $n_L^0$ and $n_R^{0c}$ are $F$ component vectors: i.e. $n_L^0 = (n_{1L}^0 \ldots n_{FL}^0)^T$, $n_R^{0c} = (n_{1R}^{0c} \ldots n_{FR}^{0c})^T$, where $n_{iL}^0$ and $n_{iR}^{0c}$ are weak eigenstate neutrinos and "antineutrinos", related by

$$n_{iR}^{0c} = C\bar{n}_{iL}^{0T} \tag{42}$$

From (42) one can prove the identity

$$\bar{n}_{iL}^0 n_{jR}^{0c} = \bar{n}_{jL}^0 n_{iR}^{0c}, \tag{43}$$

from which it follows that the Majorana mass matrix $M$ must be symmetric: $M = M^T$. Proceeding in analogy to the Dirac case, one can relate the $n_{iL}^0$ and $n_{jR}^{0c}$ to mass eigenstate neutrino fields by

$$\begin{aligned} n_L^0 &= U_L n_L \\ n_R^{oc} &= U_R n_R^c, \end{aligned} \tag{44}$$

where $U_L$ and $U_R$ are $F \times F$ unitary matrices chosen so that

$$U_L^\dagger M U_R = M_d = \mathrm{diag}(m_1\ m_2\ \ldots m_F), \tag{45}$$

where $M_d$ is a diagonal matrix of Majorana mass eigenvalues. Unlike the Dirac case (for which $m_D$ was an arbitrary matrix and $V_L$ and $V_R$ unrelated), the symmetry of $M$ implies a relation between $U_L$ and $U_R$, viz

$$U_L = U_R^* K^\dagger, \tag{46}$$

where $K$ is unitary and symmetric. That is, just as in the Dirac case, $U_L$ is determined from

$$U_L^\dagger M M^\dagger U_L = M_d^2 \tag{47}$$

to be of the form $U_L = \hat{U}_L K_L$, where $K_L$ is a matrix of phases that can be chosen for convenience. $U_R$ is then determined from (46), where $K$ is chosen so that $M_d$ is real and positive. If there are no degeneracies then $K$ is just a matrix of phases. [57] One can always pick $K_L$ such that $K = I$, but it is not always convenient to do so.

In terms of the mass eigenstates, (41) reduces to

$$\begin{aligned} -L_M &= \tfrac{1}{2}\sum_{i=1}^{F} m_i \bar{n}_{iL} n_{iR}^c + h.c. \\ &= \tfrac{1}{2}\sum_{i=1}^{F} m_i \bar{n}_i n_i, \end{aligned} \tag{48}$$

where $n_i = n_{iL} + n_{iR}^c$ is the $i^{th}$ Majorana mass eigenstate. [58] Written in terms of the $n_{iL}$, the weak charged current assumes a form analogous to (38), with $U_L^\dagger$ replacing $V_L^\dagger$ to describe the leptonic mixing. [59]

There are several physical distinctions between Dirac and Majorana neutrinos. If the $\nu_e$ is Majorana, for example, one could have the sequence $\pi^+ \to e^+\nu_e$ followed by $\nu_e p \to e^+ n$. The combined process violates lepton number by two units and is allowed for Majorana but not Dirac neutrinos. Similarly, a hypothetical heavy neutrino $N$ would undergo the decays $N \to e^+ q_1 \bar{q}_2$ and $N \to e^- \bar{q}_1 q_2$ with equal rates if it is Majorana, while for a Dirac particle one would have $N \to e^- \bar{q}_1 q_2$, $N^c \to e^+ q_1 \bar{q}_2$ only [60]. There are differences due to Fermi statistics in the production of $\nu\nu$ (Majorana) or $\nu\nu^c$ (Dirac) pairs near threshold [61], and finally Majorana neutrinos cannot have electromagnetic form factors, such as magnetic moments [62].

It is important to keep in mind, however, that these distinctions must all disappear in the limit that the neutrino mass can be neglected. For $m_\nu \to 0$ the $\nu_R$ component of a Dirac neutrino decouples, and both Majorana and Dirac neutrinos reduce to Weyl two-component neutrinos - there is no difference between them. [63] In particular, lepton number conservation is reestablished smoothly as $m_\nu \to 0$ for a Majorana neutrino, because in that limit helicity - which is conserved up to corrections of order $m_\nu/E_\nu$ - plays the role of an approximate lepton number. For example, the $\nu_e$ produced in $\pi^+ \to e^+\nu_e$ has $h_\nu = -1$ up to corrections of order $(m_\nu/E_\nu)^2$ ( in rate), while the reaction $\nu_e p \to e^+ n$ has a cross section that is suppressed by $(m_\nu/E_\nu)^2$ for the wrong (negative) helicity.

In many models Dirac and Majorana mass terms are both present. For one doublet neutrino $\nu_L^0$ (with $\nu_R^{0c} = C\bar{\nu}_L^{0T}$) and one new singlet $N_R^0$ (with $N_L^{0c} = C\bar{N}_R^{0T}$), for example, one could have the general mass term

$$-L = \frac{1}{2}\begin{pmatrix}\bar{\nu}_L^0 & \bar{N}_L^{0c}\end{pmatrix}\begin{pmatrix} m_t & m_D \\ m_D^T & m_S \end{pmatrix}\begin{pmatrix}\nu_R^{0c} \\ N_R^0\end{pmatrix} + h.c., \tag{49}$$

where $m_D = m_D^T$ is a Dirac mass generated by a Higgs doublet (analogous to (31)), $m_t$ is a Majorana mass for $\nu_L^0$ generated by a Higgs triplet or effective interaction (cf. (39)), and $m_S$ is a Majorana mass for $N_R^0$, generated by a Higgs singlet or bare mass. Similarly, for $F$ families (49) still holds provided one interprets $\nu_L^0$ , $N_L^{0c}$ , $\nu_R^{0c}$ , and $N_R^0$ as $F$ component vectors, and $m_t$, $m_D$, and $m_S$ as $F \times F$ matrices (with $m_t = m_t^T$, $m_S = m_S^T$). Then, (49) becomes simply

$$-L = \frac{1}{2}\bar{n}_L^0 M n_R^{oc} + h.c., \tag{50}$$

where $n_L^0 \equiv (\nu_L^0,\ N_L^{0c})^T$ and $n_R^{0c} \equiv (\nu_R^{0c},\ N_R^0)^T$ are $2F$ component vectors and

$$M = \begin{pmatrix} m_t & m_D \\ m_D^T & m_S \end{pmatrix} \tag{51}$$

is a symmetric $2F \times 2F$ Majorana mass matrix. Equation (50) can be diagonalized in exact analogy with (41-48), yielding finally

$$-L = \frac{1}{2}\sum_{i=1}^{2F} m_i \bar{n}_{iL} n_{iR}^c + h.c. \tag{52}$$

i.e. there are in general $2F$ Majorana neutrinos, related to $n_L^0$, $n_R^{0c}$ by unitary transformations similar to (44). Unlike the pure Majorana case, however, there is now mixing between particles with different weak interaction properties (e.g. $n_{iL} = (U_L^\dagger)_{ij} n_{jL}^0$ is a mixture of $SU_2$ doublets and singlets), which can have important consequences for neutrino oscillations [64] and decays.

It is instructive to see how the Dirac case ($m_t = m_S = 0$) emerges as a limiting case of (49). For a single family one has

$$M = m_D\begin{pmatrix} 0 & 1 \\ 1 & 0 \end{pmatrix}. \tag{53}$$

Since $M$ is Hermitian (for $m_D$ real) one can diagonalize it by a unitary transformation $U_L$. One finds

$$U_L^\dagger M U_L = m_D\begin{pmatrix} 1 & 0 \\ 0 & -1 \end{pmatrix}, \tag{54}$$

with $U_L = \frac{1}{\sqrt{2}}\begin{pmatrix} 1 & 0 \\ 0 & -1 \end{pmatrix}$; i.e. the mass eigenstates are

$$\begin{aligned} n_{1L} &= \frac{1}{\sqrt{2}}(\nu_L^0 + N_L^{0c}) \\ n_{2L} &= \frac{1}{\sqrt{2}}(\nu_L^0 - N_L^{0c}) \\ n'_{1R} &= \frac{1}{\sqrt{2}}(\nu_R^{0c} + N_R) \\ n'_{2R} &= \frac{1}{\sqrt{2}}(\nu_R^{0c} - N_R). \end{aligned} \tag{55}$$

The negative mass eigenvalue in (54) can be removed by redefining [65] the right-handed fields $n_{1R} = n'_{1R}$, $n_{2R} = -n'_{2R}$. This is nothing more than taking

$$U_L^\dagger M U_R = m_d = m_D \begin{pmatrix} 1 & 0 \\ 0 & 1 \end{pmatrix}, \tag{56}$$

where $U_R$ is given by (46) with $K = \begin{pmatrix} 1 & 0 \\ 0 & -1 \end{pmatrix}$. Finally, the two Majorana states $n_1 = n_{1L} + n_{1R}^c$ and $n_2 = n_{2L} + n_{2R}^c$ are degenerate. We can therefore reexpress $L$ in the new basis

$$\begin{aligned} \nu &\equiv \frac{1}{\sqrt{2}}(n_1 + n_2) = \nu_L^0 + N_R^0 \\ \nu^c &\equiv \frac{1}{\sqrt{2}}(n_1 - n_2) = N_L^{0c} + \nu_R^{0c}, \end{aligned} \tag{57}$$

yielding

$$\begin{aligned} -L &= \tfrac{1}{2} m_D (\bar{n}_{1L} n_{1R}^c + \bar{n}_{2L} n_{2R}^c) + h.c. \\ &= m_D \bar{\nu}_L^0 N_R^0 + h.c. \\ &= m_D \bar{\nu} \nu. \end{aligned} \tag{58}$$

This is just a standard Dirac mass term, with a conserved lepton number (i.e. no transition between $\nu$ and $\nu^c$). A Dirac neutrino is therefore nothing but a pair of degenerate two-component Majorana neutrinos ($n_1$ and $n_2$), combined to form a 4-component neutrino with a conserved lepton number.

Similarly, the Dirac limit for $F$ families ($m_t = m_S = 0$ in (51)), can be obtained by choosing

$$U_L = \frac{1}{\sqrt{2}} \begin{pmatrix} V_L & V_L \\ V_R^* & -V_R^* \end{pmatrix} \tag{59}$$

and $K = \begin{pmatrix} I & 0 \\ 0 & -I \end{pmatrix}$, where $V_L$ and $V_R$ are the $F \times F$ unitary matrices that diagonalize $m_D$ (in (36). One then obtains

$$U_L^\dagger \begin{pmatrix} 0 & m_D \\ m_D^T & 0 \end{pmatrix} U_R = \begin{pmatrix} m_d & 0 \\ 0 & m_d \end{pmatrix}, \tag{60}$$

so that one obtains $F$ pairs of degenerate Majorana neutrinos, which can be combined into $F$ Dirac neutrinos.

One sometimes refers to a pseudo-Dirac neutrino, which is just a Dirac neutrino to which is added as small lepton number-violating perturbation. For example, for $F = 1$ one could modify the Dirac mass in (53) to

$$M = \begin{pmatrix} \epsilon & m_D \\ m_D & 0 \end{pmatrix}, \tag{61}$$

with $\epsilon << m_D$. One then finds two Majorana mass eigenstates $n_\pm$, with

$$\begin{aligned} n_{+L} &= n_{1L} + \frac{\epsilon}{4} n_{2L} \\ n_{-L} &= -\frac{\epsilon}{4} n_{1L} + n_{2L}, \end{aligned} \tag{62}$$

($n_{1L}$ and $N_{2L}$ are defined in (55)), with masses $m_D \pm \frac{\epsilon}{2}$.

Other important special cases of (51) are considered below.

## 4.2 Models of Neutrino Mass

There are many models for neutrino mass [49], all of which have good and bad features. The major classes of models are listed in Table 6, along with the most natural scales for the neutrino masses and for $\langle m_{\nu_e} \rangle$, an effective mass relevant to neutrinoless double $\beta$ decay.

Table 6: Models of neutrino mass, along with their most natural scales for the light neutrino masses.

| Model | $m_{\nu_e}$ | $\langle m_{\nu_e} \rangle$ | $m_{\nu_\mu}$ | $m_{\nu_\tau}$ |
|---|---|---|---|---|
| Dirac | $1 - 10\ MeV$ | 0 | $100\ MeV - 1\ GeV$ | $1 - 100\ GeV$ |
| pure Majorana [56] (Higgs triplet) | arbitrary | $m_{\nu_e}$ | arbitrary | arbitrary |
| GUT seesaw [66,67] ($M \sim 10^{14}\ GeV$) | $10^{-11}\ eV$ | $m_{\nu_e}$ | $10^{-6}\ eV$ | $10^{-3}\ eV$ |
| intermediate seesaw [68] ($M \sim 10^{9}\ GeV$) | $10^{-7}\ eV$ | $m_{\nu_e}$ | $10^{-2}\ eV$ | $10\ eV$ |
| $SU_{2L} \times SU_{2R} \times U_1$ seesaw [69] ($M \sim 1\ TeV$) | $10^{-1}\ eV$ | $m_{\nu_e}$ | $10\ KeV$ | $1\ MeV$ |
| light seesaw [70] ($M \ll 1\ GeV$) | $1 - 10\ MeV$ | $\ll m_{\nu_e}$ | - | - |
| charged Higgs [71] | $< 1\ eV$ | $\ll m_{\nu_e}$ | - | - |

Dirac neutrinos are exactly like other fermions. They involve a conserved total lepton number (though the individual $L_e$, $L_\mu$, and $L_\tau$ lepton numbers are violated by mixing in general) and therefore do not lead to neutrinoless double beta decay. The problem with Dirac neutrinos is that it is hard to understand why the neutrinos are so much lighter than the other fermions. In the standard model Dirac mass are generated by the vacuum expectation value (VEV) $v = \sqrt{2}\langle \varphi^0 \rangle \simeq (\sqrt{2}G_F)^{-1/2} \simeq 246\ GeV$ of the neutral component of a doublet [72] of Higgs scalar fields. One has

$$m_D = h_\nu v, \tag{63}$$

where $h_\nu$ is the Yukawa coupling

$$L = -\sqrt{2} h_\nu (\bar{\nu}_L\ \bar{e}_L) \begin{pmatrix} \varphi^0 \\ \varphi^- \end{pmatrix} N_R + h.c. \tag{64}$$

of the neutrino to $\varphi^0$.

A $\nu_e$ mass in the 20 $eV$ range would require an anomalously small Yukawa coupling $h_{\nu_e} \leq 10^{-10}$. Moreover, $h_{\nu_e}$ would have to be smaller by $m_{\nu_e}/m_e \leq 10^{-4}$ than the analogous Yukawa coupling for the electron. Of course, we do not understand the masses of the other fermions either (or why they range over at least five orders of magnitude), so it is hard to totally exclude the possibility that $h_{\nu_e}$ is simply small. Nevertheless, the possibility seems sufficiently ugly that it is hard to take seriously unless some mechanism (other than fine-tuning) for the smallness is proposed.

One possibility is that $h_\nu$ is actually zero to lowest order (tree level) due to some new symmetry, and that $h_\nu$ is only generated as a higher order correction (i.e. so that $m_\nu/m_e$ is some power of $\alpha$.) This is a very attractive possibility, but no particularly compelling models to implement it have emerged. The idea has recently been resurrected in some superstring inspired models [73], which have difficulty incorporating the seesaw type ideas described below.

Majorana mass terms for the ordinary $SU_2$-doublet neutrinos involve a transition from $\nu_R^c$ ($t_3 = -\frac{1}{2}$) into $\nu_L$ ($t_3 = +\frac{1}{2}$), and therefore must be generated by an operator transforming as a triplet under weak $SU_2$.

The simplest possibility is the Gelmini-Roncadelli model [56], in which one introduces a triplet of Higgs fields $\vec{\varphi}_t = (\varphi_t^0,\ \varphi_t^-,\ \varphi_t^{--})$ into the theory. The Yukawa coupling

$$
\begin{aligned}
L &= \tfrac{1}{2} h_t (\bar{\nu}_L\ \bar{e}_L) \vec{\tau} \cdot \vec{\varphi}_t \begin{pmatrix} e_R^c \\ -\nu_R^c \end{pmatrix} \\
&= \tfrac{1}{2} h_t (\bar{\nu}_L\ \bar{e}_L) \begin{pmatrix} \varphi_t^- & \sqrt{2}\varphi_t^0 \\ \sqrt{2}\varphi_t^{--} & -\varphi_t^- \end{pmatrix} \begin{pmatrix} e_R^c \\ -\nu_R^c \end{pmatrix}
\end{aligned}
\tag{65}
$$

then generates a Majorana mass

$$
m_t = h_t v_t \tag{66}
$$

for the $\nu$, where $v_t = \sqrt{2}\langle\varphi_t^0\rangle$ is the VEV of the Higgs triplet. Since both $h_t$ and $v_t$ are unknown the neutrino mass is unrelated to the other fermions and can in principle be arbitrarily small, at least at tree level.

However, small $m_{\nu_e}$ is not explained in such models - it is merely parametrized and in fact is almost as problematic as a Dirac mass. The weak neutral current (and $W$ and $Z$ masses) require [5] $v_t \leq 0.08v \sim 20\ GeV$. For $v_t$ close to this limit one requires $h_t \leq 10^{-9}$, i.e. almost as bad a fine-tuning as the Dirac case. For $v_t \ll v$ one can tolerate more reasonable values for $h_t$, but then it is difficult to understand the large hierarchy in vacuum expectation values. One generally expects all non-zero VEV's to be comparable in magnitude unless fine-tunings are performed on the parameters in the Higgs potential. Even if one does this, higher order corrections are likely to upset the hierarchy. [74]

The VEV $\langle\varphi_t^0\rangle \neq 0$ necessarily violates lepton number conservation by two units (the Yukawa coupling in (65) does not by itself violate $L$ because $\varphi_t$ can be regarded as carrying two units of $L$). If the rest of the Lagrangian conserves $L$ then lepton-number is spontaneously broken, and there will be an associated massless Goldstone boson, the triplet-Majoron. (This is the version of the model that is usually considered [56].) In this case limits based on stellar energy loss (carried off by Majorons) require [75] $v_t \leq 2 - 10\ KeV$. Implications of the Majoron for neutrino decay and annihilation, cosmology, and neutrinoless double beta decay will be mentioned below.

It is also possible introduce other couplings into the Higgs triplet model which explicitly break lepton number conservation, such as a cubic interaction between $\vec{\varphi}_t$ and two Higgs doublets. (This violates $L$ since $\vec{\varphi}_t$ was assigned $L = 2$ to make (65) invariant). In that case all of the new scalar particles associated with $\varphi_t$ become massive - i.e. there is no Majoron.

Another mechanism for introducing a Majorana mass is to consider the induced interaction (Fig. 11).

$$
L_{eff} = \tfrac{1}{2}\frac{C}{M}(\bar{\nu}_L\ \bar{e}_L)\vec{\tau}\begin{pmatrix} e_R^c \\ -\nu_R^c \end{pmatrix} \cdot (\varphi^-\ \ -\varphi^0)\vec{\tau}\begin{pmatrix} \varphi^0 \\ \varphi^- \end{pmatrix} \tag{67}
$$

between two leptons and two Higgs doublets. The Higgs fields in (67) are arranged to transform as an $SU_2$ triplet, so $L_{eff}$ is $SU_2 \times U_1$ invariant; however, $L_{eff}$ is non-renormalizable, as is evidenced by the dimensional coupling $C/M$, where $M$ is a mass. $L_{eff}$ cannot therefore be an elementary

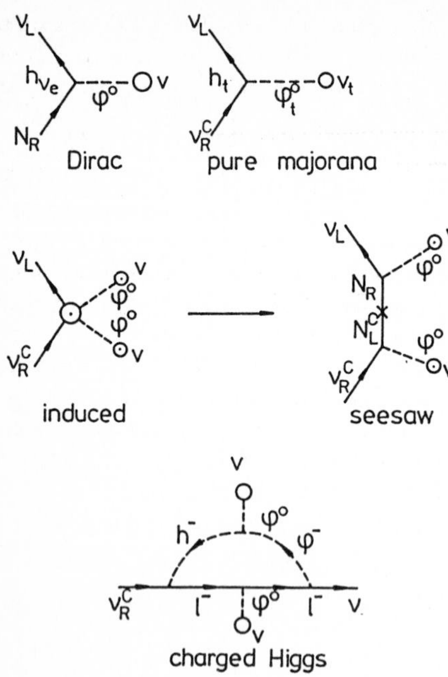


Figure 11: Dirac, pure Majorana, induced, and charged Higgs generated neutrino masses.

coupling, but it could be an effective four-particle interaction induced [76] by new physics at some large mass scale $M$ (just as the four-fermion weak interaction is a nonrenormalizable effective interaction that is really generated by $W$ and $Z$ exchange). When $\varphi^0$ is replaced by its vacuum expectation value, (67) yields an effective Majorana mass $m \sim Cv^2/M$, which is naturally small for $M \gg v$. For example, if (67) were somehow induced by quantum gravity one would expect $M \sim 10^{19}$ $GeV$ (the Planck scale). Then for $C \sim 1$ one would have $m_\nu \sim 10^{-5}$ $eV$.

The most popular realisation of this idea is the seesaw model, [66] in which the underlying physics is the exchange of a very heavy $SU_2$-singlet Majorana neutrino $N_R^0$, as indicated in Fig. 11. The seesaw model for one family is a special case of the general mass matrix in (49), in which $m_D$ is a typical Dirac mass (typically assumed to be comparable to $m_u$ or $m_e$ for the first family) connecting $\nu_L^0$ to a new $SU_2$-singlet $N_R^0$ and $m_S \gg m_D$ is a Majorana mass for $N_R^0$, presumably comparable to some new (large) physics scale. One typically assumes that $m_t = 0$ in the seesaw model, i.e. that there is not a Higgs triplet as well. [77] In that case, (49) yields two Majorana mass eigenstates $n_1$ and $n_2$ with

$$\begin{aligned} \nu_L^0 &= n_{1L}\cos\theta + n_{2L}\sin\theta \\ N_L^{0c} &= -n_{1L}\sin\theta + n_{2L}\cos\theta \\ \nu_R^{0c} &= -(n_{1R}^c\cos\theta + n_{2R}^c\sin\theta) \\ N_R^0 &= -n_{1R}^c\sin\theta + n_{2R}^c\cos\theta. \end{aligned} \tag{68}$$

The physical masses [78] are

$$\begin{aligned} m_1 &\simeq \frac{m_D^2}{m_S} \ll m_D \\ m_2 &\simeq m_S \end{aligned} \tag{69}$$

and the mixing angle is

$$\tan\theta = \left(\frac{m_1}{m_2}\right)^{1/2} \simeq \frac{m_D}{m_S} \ll 1. \tag{70}$$

Hence, one naturally obtains one very light neutrino, which is mainly the ordinary $SU_2$ doublet $(\nu_L^0, \nu_R^{0c})$, and one very heavy neutrino, which is mainly the singlet $(N_L^{0c}, N_R^0)$.

If one does allow $m_t \neq 0$ (but $\ll m_S$) then there are still two Majorana neutrinos with masses $|a - \frac{m_D^2}{m_S}|$ and $m_S$, respectively, while $\theta \sim m_D/m_S \ll 1$ still holds. (The minus sign in $\nu_R^{0c}$ is removed if $a - \frac{m_D^2}{m_S}$ is positive). In this case, however, one loses the natural explanation of why $m_1$ is so small, unless $m_t$ is itself induced by the underlying physics and is of the same order as $m_D^2/m_S$.

The seesaw model is easily generalized to $F$ families. One then has the general $2F \times 2F$ Majorana mass matrix in (51). Assuming that the eigenvalues of $m_S$ are all much larger than any of the components of $m_D$ or $m_t$ (if it is non-zero) one can calculate the eigenvalues and mixing matrices to leading order in $m_S^{-1}$. One finds that there are $F$ light Majorana neutrinos (consisting of the $F$ doublets $(\nu_L^0, \nu_R^{0c})$, up to corrections of order $m_D m_S^{-1}$ and $F$ heavy Majorana neutrinos (consisting of the singlets $(N_L^{0c}, N_R^0)$, to $O(m_D m_S^{-1})$). That is, one can write

$$\begin{aligned} \begin{pmatrix} \nu_L^0 \\ N_L^{0c} \end{pmatrix} &= U_L \begin{pmatrix} n_{lL} \\ n_{hL} \end{pmatrix} \\ \begin{pmatrix} \nu_R^{0c} \\ N_R^0 \end{pmatrix} &= U_R \begin{pmatrix} n_{lR}^c \\ n_{hR}^c \end{pmatrix}, \end{aligned} \tag{71}$$

where $n_{lL}$ and $n_{hL}$ are $F$ component vectors of light and heavy Majorana mass eigenstates, respectively, and similarly for $n_{lR}^c, n_{hR}^c$. As usual, $U_L$ and $U_R$ are $2F \times 2F$ unitary matrices which diagonalize $M$ in (51), viz

$$U_L^\dagger \begin{pmatrix} m_t & m_D \\ m_D^T & m_S \end{pmatrix} U_R = m_d = \begin{pmatrix} m_l & 0 \\ 0 & m_h \end{pmatrix}, \tag{72}$$

where $m_l$ and $m_h$ are diagonal $F \times F$ matrices of the $F$ light and $F$ heavy eigenvalues, respectively. To leading order in $m_S^{-1}$ one can write $U_L^\dagger$ and $U_R$ in block diagonal form

$$U_L^\dagger = K U_R^T = \begin{pmatrix} K_1 & 0 \\ 0 & K_2 \end{pmatrix} \begin{pmatrix} A^T & -A^T m_D m_S^{-1} \\ D^T m_S^{-1\dagger} m_D^\dagger & D^T \end{pmatrix}, \tag{73}$$

where $A^T$ and $D^T$ are unitary (to leading order) $F \times F$ matrices defined by

$$\begin{aligned} m_l &= K_1 A^T (m_t - m_D m_S^{-1} m_D^T) A \\ m_h &= K_2 D^T m_S D \end{aligned} \tag{74}$$

i.e. the mass matrix for the light neutrinos is $m_t - m_D m_S^{-1} m_D^T$, which is diagonalized by $A$, while that for the heavy neutrinos is $m_S$, diagonalized by $D$. $K_1$ and $K_2$ are diagonal phase matrices which ensure that $m_l$ and $m_h$ are real and positive. We see from (71-74) that indeed there are $F$ heavy states with masses of $O(m_S)$, and in the simplest case $m_t = 0$ there are $F$ states which are naturally very light $(O(m_D^2 m_S^{-1}))$. (For $m_t \neq 0$ one must separately assume $m_t$ is small). Furthermore, the mixing between the light and heavy sectors is very small (of $O(m_D m_S^{-1})$), while the matrices $A$ and $D$, which describe mixings within the two sectors, are in general arbitrary.

There are several classes of seesaw models [66], depending on the scale of $m_S$. In simple grand unified models one assumes that the scale is a typical GUT unification scale of around $10^{14}$ $GeV$. In many such models (e.g. $SO_{10}$) one has that the neutrino Dirac mass matrix $m_D$ is the same as $m_u/k$ where $m_u$ is the $u$-quark mass matrix and $k \simeq 4.7$ represents the running of the Yukawa couplings between the GUT scale and low energies. If one makes the somewhat ad-hoc assumption that the matrix $m_S$ is just $M_X I$, where $M_X \sim 10^{14}\ GeV$ is the unification scale and $I$ is the identity matrix, one has (for $m_t = 0$) the light eigenvalues

$$m_{\nu_i} \sim \frac{m_{u_i}^2}{M_X k^2} \tag{75}$$

$\sim 10^{-11}\ eV,\ 10^{-6}\ eV,\ 10^{-3}\ eV$, i.e. the neutrino masses are naturally expected to be extremely tiny, and to scale like the squares of the $u$, $c$, and $t$ quark masses. (Equation (75) was computed for $m_{top} \sim 50\ GeV$). Several caveats are in order: the assumption of $m_S \sim M_X I$ was quite arbitrary. One could easily imagine that the eigenvalues of $m_S$ are smaller than $M_X$ due to small Yukawa coupling couplings (increasing $m_{\nu_i}$). Also, they need not be the same. For example, if the $m_S$ eigenvalues followed the same family hierarchy as the ordinary fermions (i.e. $m_{S_i} \propto m_{u_i}$) then one would have $m_{\nu_i}$ scaling as $m_{u_i}$ rather than $m_{u_i}^2$. (A similar linear hierarchy ensues in some variant GUTs in which $m_S$ is zero at tree level but is generated at higher orders [79,36]. Of course, more complicated patterns for $m_S$ and $m_D$ (in (74)) are also possible. Furthermore, in many cases loop corrections to the (GUT) Higgs potential may induce [77] VEV's for Higgs representations that can yield a non-zero triplet terms $m_t$ in (72). These are most likely to affect the smallest masses (e.g. $m_{\nu_e}$). Equation (75) should therefore be regarded only as a typical order of magnitude.

If one does assume that $m_S = M_X I$, however, then $m_u^2/M_X$ is diagonalized by the same transformations that diagonalize $m_u$. Since one also has equal electron and $d$-quark mass matrices (i.e. $m_e = m_d/k$) in most simple GUTs the final result is that flavor mixing in the lepton sector (analogous to (38)) is described by the same mixing matrix as the CKM quark mixing matrix. This result continues to hold [67] approximately for a far wider class of $m_S$ than does the simple mass prediction in (75).

Lower mass scales for $m_S$ imply larger values for the light neutrino masses (and generally less predictive power for $m_D$). Several authors [68] have suggested that the heavy Majorana scale could be the intermediate range $10^8 - 10^{12}\ GeV$ associated with invisible axions. For $m_{D_i} \sim m_{e_i}$ and $m_S \sim 10^9\ GeV I$, for example, one obtains the values $\sim 10^{-7}\ eV,\ 10^{-2}\ eV,\ 10\ eV$ for $m_{\nu_e}, m_{\nu_\mu}, m_{\nu_\tau}$, respectively.

If $m_S$ is in the several $TeV$ range (as expected in some left-right symmetric [80] $SU_{2L} \times SU_{2R} \times U_1$, models [69], for example) one typically expects (for $m_{D_i} \sim m_{e_i}, m_S \propto I$) $m_{\nu_e}, m_{\nu_\mu}, m_{\nu_\tau}$ to have relatively large values $10^{-1}\ eV$, $10\ KeV$, and $1\ MeV$, respectively. As we will see, such models run into severe cosmological difficulties unless the mass hierarchy is somehow modified or a fast decay mechanism is found for the $\nu_\mu$ and $\nu_\tau$. Of course, one could also have $m_S$ much smaller than the $SU_{2L} \times SU_{2R} \times U_1$ scale (e.g. in the $10\ GeV - 100\ GeV$ range), with corresponding larger masses for the light neutrinos. Similar statements apply to models with extra $Z$ bosons in the $100\ GeV - 10\ TeV$ range, which usually also have heavy Majorana neutrinos.

Finally, one can consider light seesaw models, in which typically $m_S \ll 1\ GeV$. Such models are very artificial and abandon the principal advantages of the seesaw, because both $m_D$ and $m_S$ must be taken unnaturally small to obtain an acceptable $\nu_e$ mass. Their only virtue is that they yield strongly suppressed neutrinoless double beta decay rates, even though the neutrinos are Majorana.

Seesaw models were first introduced in GUT type models in which lepton number is explicitly violated by the gauge interactions. One can also consider non-gauge seesaw models [81] in which lepton number is spontaneously broken by the VEV of the Higgs field which generates $m_S$. Such models imply the existence of a massless Goldstone boson, the singlet-Majoron. [82] Unlike the triplet-Majoron in the Gelmini-Roncadelli model,[56] which can couple strongly to the ordinary neutrinos (coupling $\sim h_t$), the singlet-Majoron effectively decouples from ordinary particles. That is, it couples strongly to the heavy neutrino, with a coupling of order $m_D m_S^{-1}$ to off-diagonal $n_l n_h$ vertices, and with strength $(m_d m_S^{-1})^2$ to light neutrinos.

It is difficult to implement the seesaw model in most superstring inspired models, because there is no Higgs field available to generate a large $m_S$. It has been suggested [83] that $m_S$ could be generated by a higher order effective operator, but such model may run into serious cosmological problems [84].

There have also been variant seesaw models constructed [85] in which the light neutrinos occur in degenerate pairs which can be combined from Dirac neutrinos with a conserved $L$.

Finally, I mention the charged Higgs models [71], in which small Majorana masses are generated by loop diagrams involving new charged Higgs bosons with explicit $L$-violating couplings (Fig. 11). Viable versions often lead to pseudo-Dirac neutrinos. The approximately conserved lepton number is typically $L_e - L_\mu + L_\tau$, for example, rather than $L$. The actual mass scale depends on unknown Yukawa couplings and masses.

## 4.3 Experimental Constraints

There are a number of excellent reviews [49] of the experimental status of neutrino mass. My major purpose in this section is to comment on the implications of the various theoretical models for the different types of experiments.

## 4.4 Kinematic Tests

Direct kinematic limits on the masses of the $\nu_e$, $\nu_\mu$, and $\nu_\tau$ are given in Table 7. The ITEP group [86] has long claimed evidence for a non-zero $\nu_e$ mass in the 20 $eV$ range from tritium $\beta$ decay, but this has not been confirmed by other groups, and in fact the Zurich-SIN measurement is on the verge of conflicting with the ITEP result. In addition the neutrinos from supernova 1987A observed by the Kamiokande [93] and IMB [94] experiments place upper limits in the 20 $eV$ range on the $\nu_e$ mass (otherwise the arrival times of the detected neutrinos would be spread out more than is observed), but it is hard to make this limit precise because it depends on the details of the neutrino emission [90].

A 20 $eV$ neutrino mass is just in the range that would be most interesting cosmologically, so clearly it is essentially to resolve the situation. Hopefully, the current and next generation of tritium $\beta$ decay experiments will be sensitive down to a few $eV$, but it is doubtful whether experiments of this type will ever be able to probe to much lower scales. As can be seen in Table 6, none of the models really predict $m_{\nu_e}$ in the 20 $eV$ range (the $SU_{2L} \times SU_{2R} \times U_1$ models come closest), but most can accomodate masses in this range by fine-tuning parameters.

As can be seen in Table 7, the direct kinematic limits on $m_{\nu_\mu}$ (from $\pi_{\mu 2}$ decay) and on $m_{\nu_\tau}$ (from $\tau \rightarrow \nu_\tau + 5\pi$) are relatively weak. The experiments are extremely difficult (the mass scales being probed are very much smaller than the energies released in the decays), so it is unlikely that these measurements will improve by much more than a factor of two.

Table 7: Kinematic limits/values on neutrino masses.

| | |
|---|---|
| $17\ eV < m_{\nu_e} < 40\ eV$ | ITEP [86] |
| $m_{\nu_e} < 18\ eV$ | Zurich [87] |
| $m_{\nu_e} < 27\ eV$ | LANL [88] |
| $m_{\nu_e} < 32\ eV$ | INS-Tokyo [89] |
| $m_{\nu_e} < O(20\ eV)$ | SN1987A [90] |
| $m_{\nu_\mu} < 0.25\ MeV$ | SIN [91] |
| $m_{\nu_\tau} < 50\ MeV$ | ARGUS [92] |

## 4.5 Heavy Neutrinos

There are many limits [49,95] on possible small admixtures of heavy neutrino states in the $\nu_e$ o $\nu_\mu$, including universality tests in nuclear $\beta$ decay, searches for secondary peaks or distortions c the lepton spectra in $\beta, \pi$, and $K$ decay, searches for the decay products of heavy neutrinos (e.g $\nu_h \to \nu_e e^+ e^-$) produced in beam dumps, $e^+e^-$ annihilation, or neutrino scattering [96]. The limits on the mass $m_i$ versus mixing angle $U_{ai}$, $a = e$ or $\mu$, where

$$\nu_a^0 = \sum_i U_{ai} \nu_i \tag{76}$$

are shown in Fig. 12.

It is seen that the constraints on $|U_{ai}|^2$ are quite impressive, especially for $m_i$ in the range $10\ MeV - 10\ GeV$, where they are comparable to the expectations in (70) of a seesaw model with $m_1 \sim 10\ eV$ and $m_2 = m_i$. The lower part of this range corresponds to the masses expected in the "light-seesaw" model (Table 6), while the $1\ GeV - 1\ TeV$ range is consistent with $SU_{2L} \times SU_{2R} \times U_1$ models. [69]

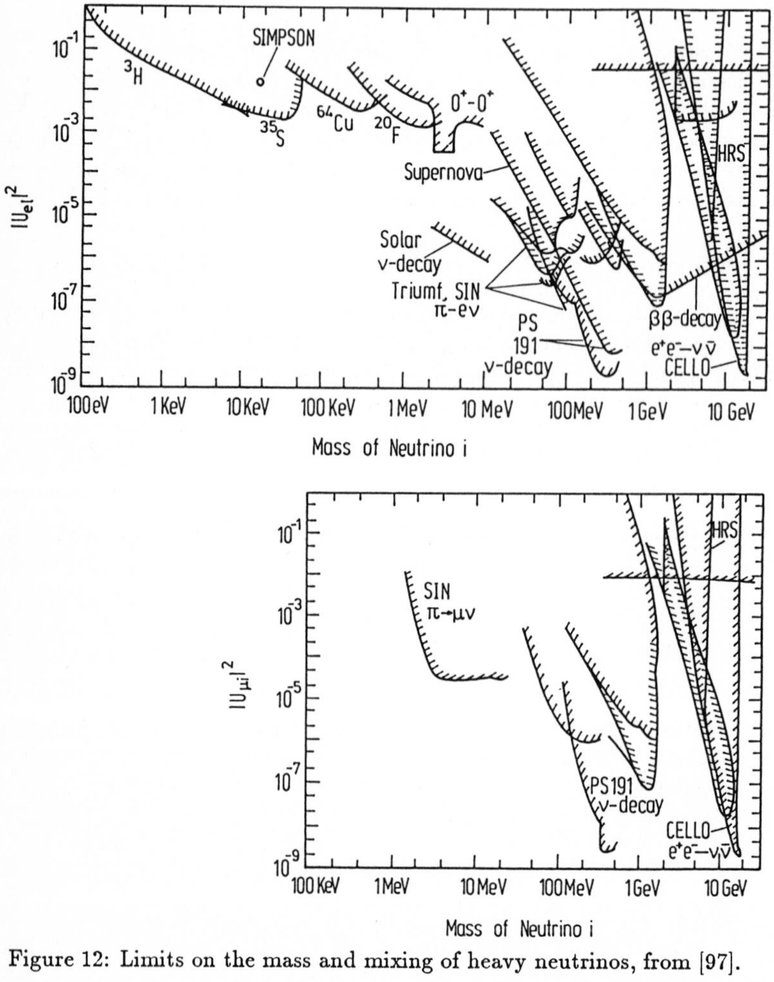


Figure 12: Limits on the mass and mixing of heavy neutrinos, from [97].

Also, most models with extra $Z$ bosons in the 100 $GeV - 1\ TeV$ range predict [98] the existence of heavy $SU_2$-singlet Majorana or Dirac neutrinos [95,99,100]. The extra $Z$'s typically couple to these new neutrinos and other exotic fermions much more strongly than to the ordinary fermions. Future hadron colliders should therefore be able to extend the search for heavy neutrinos via

$$\begin{aligned} \overset{(-)}{p}\, p &\rightarrow Z' \rightarrow N N^c \\ \overset{(-)}{p}\, p &\rightarrow W_R \rightarrow N l \end{aligned} \tag{77}$$

into the several hundred $GeV$ range. The subsequent decays of the $N$'s should be a superb probe of underlying physics. In models with just an extra $Z$, for example, the $N$ is expected to decay due to mixing with the light neutrinos. The $N$ can then decay [95,99] via virtual $W$ or $Z$ exchange [101] into such modes as $3\nu_1$, $\nu_1 l^+ l^-$, $\nu_1 q\bar{q}$, and $\nu_1 e^+ \mu^-$. On the other hand, in $SU_{2L} \times SU_{2R} \times U_1$ models the $N$ will generally decay via virtual $W_R$ exchange [100], and for the lightest $N$ the decay should usually be into $l^\pm q\bar{q}$. Moreover, the decay modes should easily establish whether the heavy neutrino is Majorana or Dirac, because in the former case the decays $N \rightarrow l^+ \bar{q} q$ and $N \rightarrow l^- q\bar{q}$ would be equally likely [60] (though with different angular distributions).

It is of course also possible that a heavy neutrino could simply be a massive $4^{th}$ generation neutrino.

As has already been mentioned, heavy neutrinos in the $GeV - TeV$ range are likely to give too large $m_{\nu_\mu}$ and $m_{\nu_\tau}$ unless the typical seesaw hierarchy $m_{\nu_i} \propto m_{l_i}^n$ or $m_{u_i}^n$, $n = 1$ or 2, for the light neutrinos is avoided or new physics is invoked to ensure fast decays or annihilations for the $\nu_\mu$ and $\nu_\tau$. On the other hand, if such new physics is present some of the limits in Fig. 12 (those based on decays) may no longer be valid, because in many cases the heavy neutrinos will decay rapidly into unobservable channels (e.g. $\nu_h \rightarrow \nu_l +$ Majoron) before reaching the detector.

## 4.6 Neutrino Oscillations

Neutrino oscillations are a beautiful example of a common quantum phenomenon: viz that if one starts at time $t = 0$ in a state that is not an energy eigenstate [102] then at later times it can oscillate into another (orthogonal) state. For example, suppose that the $\nu_e^0$ and a second neutrino $\nu_a^0$ (e.g. $\nu_a^0 = \nu_\mu^0$ or $\nu_\tau^0$) are mixtures of two mass eigenstates $\nu_1$ and $\nu_2$ with mixing angle $\theta$:

$$\begin{aligned} \nu_e^0 &= \cos\theta\, \nu_1 + \sin\theta\, \nu_2 \\ \nu_a^0 &= -\sin\theta\, \nu_1 + \cos\theta\, \nu_2 \end{aligned} \tag{78}$$

If at time $t = 0$ the weak eigenstate $\nu_e^0$ is produced (e.g. in the process $\pi^+ \rightarrow \pi^0 e^+ \nu_e^0$) then at time $t$ it will have evolved into the state

$$\begin{aligned} \nu_e^0(t) &= \cos\theta\, \nu_1 e^{-iE_1 t} + \sin\theta\, \nu_2 e^{-iE_2 t} \\ &\simeq \cos\theta\, \nu_1 e^{\frac{-im_1^2 t}{2p}} + \sin\theta\, \nu_2 e^{\frac{-im_2^2 t}{2p}}. \end{aligned} \tag{79}$$

In the second form I have assumed relativistic neutrinos $E_i = \sqrt{p^2 + m_i^2} \sim p + m_i^2/2p$ with definite momentum [103] $p >> m_i$, and have neglected an irrelevant overall phase $\exp(-ipt)$. The state $\nu_e^0(t)$ has a non-trivial overlap with $\nu_a^0$. After traveling a distance $L \sim t$, there will be a probability

$$\begin{aligned} P(\nu_e \rightarrow \nu_a) &= |\langle \nu_a^0 \mid \nu_e^0(t) \rangle|^2 \\ &= \sin^2 2\theta \sin^2\left(\frac{\Delta m^2 L}{4p}\right) \\ &= \sin^2 2\theta \sin^2\left(\frac{1.27 \Delta m^2 (eV^2) L(m)}{p(MeV)}\right) \end{aligned} \tag{80}$$

that the state will have evolved into $\nu_a^0$ (as can be observed in the process $\nu_a N \rightarrow e_a N'$, for example), and a probability

$$P(\nu_e \rightarrow \nu_e) = 1 - P(\nu_e \rightarrow \nu_a) \tag{81}$$

that the state will remain a $\nu_e^0$. In (80), $\Delta m^2 = m_1^2 - m_2^2$, and the last form is valid for $\Delta m^2$ in $eV^2$, $L$ in $m$, and $p$ in $MeV$. It is seen that the $\nu_e \rightarrow \nu_a$ probability depends on both the mixing angle $\theta$ and on $\Delta m^2 L/p$. For moderate values of the latter quantity the probability oscillates as a function of $L$ and $p$, while for very large values the oscillations are averaged by a finite-sized detector or non-monochromatic source, (the second factor in (80) averages to 1/2). It is easy to generalize [49] (80) to the case that the initial neutrino is a mixture of more than two mass eigenstates, as in (76). One obtains

$$P(\nu_e \rightarrow \nu_a) = \sum_i |U_{ei} U_{ai}^*|^2 + Re \sum_{i \neq j} U_{ei} U_{ai}^* U_{ej}^* U_{aj} e^{\frac{-i(m_i^2 - m_j^2)L}{2p}} \tag{82}$$

Neutrino oscillations can be searched for in (a) appearance experiments, in which one looks for the interactions of $\nu_a$ in a detector, and (b) disappearance experiments, in which one looks for a reduced $\nu_e$ flux. In both cases one can compare the observed counting rate with the expectation from known backgrounds (appearance) or from the expected flux (disappearance) as determined, for example, by measuring the electron spectrum from $n \rightarrow p e^- \bar{\nu}_e$ in reactor $\bar{\nu}_e$ oscillation experiments. A much cleaner technique is to search for actual oscillations in the appearance or disappearance probabilities as a function of $L$ or $p$, such as by using two detectors at different distances form the source.

There are many limits on neutrino oscillations from accelerator experiments [49] (e.g. counter and emulsion experiments and beam dumps, searching for $\nu_\mu \rightarrow \nu_e$, $\nu_\mu \rightarrow \nu_\tau$ and $\nu_e \rightarrow \nu_\tau$, as well as $\nu_\mu$ disappearance), and reactors [49] ($\bar{\nu}_e$ disappearance), as well as on the oscillations of $\nu_\mu$ produced in cosmic ray interactions in the atmosphere [104]. (Implications for the Solar neutrino problem are discussed below). The results of these searches [105] are summarized in Fig. 13. The Bugey reactor experiment [106] reports a positive signal for $\bar{\nu}_e$ disappearance, but their results are contradicted by the Gösgen experiment [107]. Similarly, the CERN PS-191 counter experiment [108] reports an excess of $\nu_e$ events in a $\nu_\mu$ beam, but their signal is in conflict with several other $\nu_\mu \rightarrow \nu_e$ experiments [97]. Clearly, a clarification of the situation is essential .

From Fig. 13 it is clear that there are stringent limits on neutrino mixings for $|\Delta m^2|$ above $\simeq 1\ eV^2$. This should be contrasted with the suggested value $m_{\nu_e} \sim 17 - 40\ eV$ by the ITEP experiment [86]. If the ITEP result is correct then most likely the $\nu_e$ could not have any significant mixing with other neutrinos (the alternative possibility, that the $\nu_e$ is almost degenerate with another neutrino flavor so that $|\Delta m^2| << m_{\nu_e}^2$, seems rather contrived but cannot be excluded). A comparison of Fig. 13 with the expectations of various models (Table 6) suggests that $\nu_\mu \rightarrow \nu_\tau$ oscillations may be the most optimistic possibility for the future. Many of the seesaw-type models predict that the lepton mixing angles are roughly correlated with the corresponding quark mixing angles. This would suggest $\sin^2 2\theta \sim 10^{-4}$, $10^{-2}$, $10^{-1}$ for $\nu_e \leftrightarrow \nu_\tau$, $\nu_\mu \leftrightarrow \nu_\tau$, and $\nu_e \leftrightarrow \nu_\mu$, respectively.

Oscillations between ordinary $SU_2$ doublet neutrinos ($\nu_e^0, \nu_\mu^0, \nu_\tau^0$, and possible fourth family $\nu'$s), known as first class or flavor oscillations, occur for pure Dirac and pure Majorana neutrinos, as well as in the multi-family seesaw models. In models involving both Dirac and Majorana mass terms of comparable magnitude, however, there can be additional light neutrinos, and the mass eigenstates can have significant admixtures of both $SU_2$ doublets and singlets. In this case second class oscillations [64] can occur, in which the ordinary neutrinos oscillate into $SU_2$ singlets with negligible interactions. These "sterile" neutrinos are essentially undetectable, so second class oscillations can be observed [109] only in disappearance experiments. Of course, first and second class oscillations can occur simultaneously. For three families, for example, there could be oscillations between six Majorana neutrinos (3 doublets and 3 singlets).

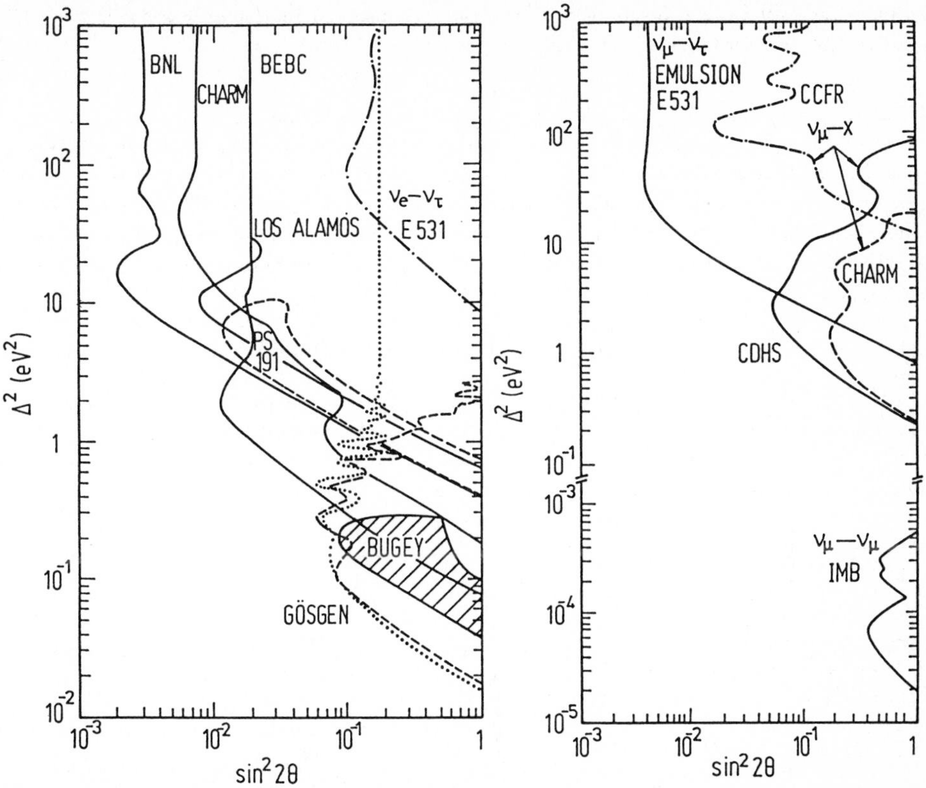


Figure 13: 90% *c.l.* limits on neutrino oscillations, from [97]. (a) $\nu_\mu \to \nu_e$ (BNL, CHARM, BEBC, Los Alamos, PS-191), $\nu_e \to \nu_\tau$ (E531), and $\bar{\nu}_e \to \nu_X$ (Bugey, Gösgen). (b) $\nu_\mu \to \nu_\tau$, $\nu_X$, $\nu_\mu$. The Bugey [106] and PS-191 [108] regions are allowed by positive results. The other contours are exclusion plots (the regions to the right are excluded).

Yet another possibility [110] are models in which the ordinary neutrinos have small mixings with heavy neutrinos. In that case the neutrinos actually produced in weak processes are the projections of the weak eigenstates onto the subspace of light or massless neutrinos. It can easily occur that the projections of the $\nu_e^0$ and $\nu_\mu^0$, for example, are not orthogonal. The result is that a $\nu_\mu^0$ could produce an $e^-$ in a subsequent reaction. Such a non-orthogonality would mimic the effects of oscillation appearance experiments, even if the masses of the light neutrinos are zero or negligible.

## 4.7 Cosmology

There are many limits on neutrino mass and decays from cosmology [111]. Ordinary light or massless neutrinos would have been produced by such weak processes as $e^+e^- \leftrightarrow \nu\nu^c$ in the early universe. As long as the weak reaction rate [112]

$$\Gamma_{weak} \sim \langle \sigma v \rangle n_T \sim G_F^2 T^5 \tag{83}$$

($\langle \sigma v \rangle \sim G_F^2 T^2$ is the thermally averaged cross section times relative velocity, and $n_T \sim T^3$ is the density of target particles, where $T$ is the temperature) was large compared to the expansion rate

$H \sim T^2/m_p$ (where $m_p = G_N^{-1/2} \sim 10^{19}\ GeV$ is the Planck scale) the number of neutrinos stayed in equilibrium. However, as soon as $T$ dropped below the temperature

$$T_D \sim (G_F^2 m_P)^{-1/3} \simeq 3\ MeV \tag{84}$$

for which $\Gamma_{weak} \sim H$, the weak rate became negligible and the neutrinos decoupled, i.e. effectively stopped interacting. According to most models these neutrinos should remain in the present universe, undisturbed from the first second of the big bang except for a redshifting of their momenta by the expansion of the universe. They are analogous to the $2.7°K$ microwave radiation (which decoupled later). If the neutrino masses are much less than $1\ eV$ there should be $\simeq 50$ neutrinos/$cm^3$ of each type ($\nu_{eL}, \nu_{eR}^c$, etc) with momenta characterized by a thermal spectrum with temperature $\simeq 1.9°K$ ($10^{-4}\ eV$). Despite the large number of neutrinos ($\simeq 10^{10}$ per baryon) they are essentially impossible to detect [113] - [115] because their cross section $\sim G_F^2 E_\nu^2 \sim 10^{-62} cm^2$ is so low. [116]

The major cosmological bound is based on the energy density of the present universe. There are predicted to be so many relic neutrinos that even for a small mass in the $10\ eV$ range they would be important. Limits on the energy density imply

$$\sum_i m_{\nu_i} < 40\ eV \tag{85}$$

where the sum extends over the light, stable (at least compared to the age of the universe) doublet neutrinos. Conversely, a neutrino with mass in this range would dominate the energy density and could account for the dark (missing) matter in galaxies and clusters [117]. In particular, for the ITEP value $m_{\nu_e} \sim (17 - 40)\ eV$, the $\nu_e$ would be an ideal candidate for the dark matter, but one would probably then have to find a mechanism to explain why the $\nu_e$ is the heaviest neutrino.

Similarly, the energy density associated with light or massless neutrinos for $T \sim T_D$ affects nucleosynthesis and leads to the limit $N_\nu \leq 4$ (section III).

There are also a variety of constraints on unstable neutrinos. An ordinary doublet mass eigenstate neutrino $\nu_2$ (with $m_{\nu_2} > m_{\nu_1}$) is expected to decay into

$$\begin{aligned} \nu_2 &\rightarrow \nu_1\gamma, \quad (m_{\nu_2} < 2m_e) \\ \nu_2 &\rightarrow \nu_1 e^+ e^-, \quad (2m_e < m_{\nu_2} < m_\mu + m_e). \end{aligned} \tag{86}$$

The first decay occurs at one loop, while the second occurs at tree level. Both decays are very slow for small $m_{\nu_2}$ and the decay products are detectable. There are a large variety of cosmological and astrophysical constraints [118] on $m_{\nu_2}$ and $\tau_{\nu_2}$ from the present energy density, the growth of galaxies, the distortion of the $2.7°K$ background radiation, the non-observation of the decay photons, supernovae, and nucleosynthesis and breakup. For reasonable mixing angles these limits exclude the range $40\ eV - (20 - 40)\ MeV$ for ordinary neutrinos [119] decaying according to (86). Combined with laboratory limits this implies [118,120] that the $\nu_\mu$ and $\nu_\tau$ (i.e. their dominant mass eigenstate components) should be lighter than $40\ eV$. In particular, this poses serious problems for the $TeV$ scale seesaw model.

Most of the cosmological limits can be evaded if new physics is invoked to allow fast and invisible (except for the relativistic energy of the decay products) decays or annihilation for the heavy neutrinos. One possibility is the decay $\nu_2 \rightarrow 3\nu_1$. However, the rate for this mode from off-diagonal $Z$ couplings [121] is too slow, while models in which the couplings of a Higgs triplet [122] (present in $SU_{2L} \times SU_{2R} \times U_1$) are arranged to allow a fast decay generally run into problems [123] with $\mu \rightarrow 3e$.

More promising are models in which $\nu_2 \rightarrow \nu_1 G$, where $G$ is a Goldstone boson [124]-[127] associated with a spontaneously broken global symmetry. Likely examples are the case that $G$ is a familon [124] (a Goldstone boson associated with a broken family symmetry) or a triplet-Majoron [125]. In fact, for triplet-Majorons one expects the annihilation process $\nu\nu \rightarrow MM$ (which begins when $T$ drops below $v_t$) to have removed any relic neutrinos from the present universe [56]. In familon models some care must be taken to avoid unacceptably large flavor changing neutral current effects. The decay $\nu_2 \to \nu_1 M$ is too slow in the simpler versions of the singlet-Majoron model [126] to avoid cosmological problems.

The role of spontaneous $L$ violation in Majoron models in reducing possible initial large lepton asymmetries to cosmologically interesting values at the time of nucleosynthesis is discussed in [128].

## 4.8 Double Beta Decay

Another important source of information on the $\nu_e$ mass (if it is Majorana) is neutrinoless double beta decay ($\beta\beta_{0\nu}$).

First consider the lepton-number conserving two-neutrino ($\beta\beta_{2\nu}$) process $(Z, N) \leftrightarrow (Z+2, N-2)e^-e^-\nu_e^c\nu_e^c$, which can be thought of as two ordinary beta decays occurring in the same nucleus (Fig. 14). In the context of neutrino mass this process is mainly of interest as a calibration of the calculated nuclear matrix elements that are needed for the neutrinoless case. There has long been a two order of magnitude discrepancy between the predicted rates [129], e.g. for $^{130}Te \to {}^{130}Xe$, and indirect measurements by geochemical techniques [130]. Within the last year, however, this discrepancy has gone away. The geochemical measurements were confirmed by the first laboratory observation of double beta decay (at Irvine [131].) In addition, several groups [129] have found that previously neglected ground state correlation effects could suppress the matrix element by the required order of magnitude. Furthermore, there is no analogous uncertainty in the $\beta\beta_{0\nu}$ case.

The neutrinoless double beta decay process $(Z, N) \to (Z+2, N-2)e^-e^-$, which violates lepton number by two units, can proceed through the second diagram [132] in Fig. 14. In the absence of mixing the quantity $\langle m_{\nu_e}\rangle$, the effective Majorana neutrino mass, is

$$\langle m_{\nu_e}\rangle = \begin{cases} 0, & \text{Dirac neutrino} \\ m_{\nu_e}, & \text{unmixed Majorana neutrino} \end{cases} \tag{87}$$

Although the matrix element is proportional to $\langle m_{\nu_e}\rangle$, which is necessarily very small, $\beta\beta_{0\nu}$ has an enormous advantage in phase space over $\beta\beta_{2\nu}$ and could be observable for $\langle m_{\nu_e}\rangle$ in the $eV$ range. Of course, the sum of the electron energies should be a sharp peak in $\beta\beta_{0\nu}$ (and a continuum for $\beta\beta_{2\nu}$), so the principal difficulty is controlling the background. [133,49] Currently, the most sensitive experiments are for $^{76}Ge \to {}^{76}Se\ e^-e^-$. No evidence for $\beta\beta_{0\nu}$ has been observed, [134] and the lower limit on the lifetime is [97] $\tau_{1/2} > 9 \times 10^{23}\ yr$ (68% $c.l.$). According to several calculations of the nuclear matrix elements [135] this implies $\langle m_{\nu_e}\rangle \leq 1\ eV$. However, a recent estimate by Engel et al. [136] yielded a much weaker limit $\langle m_{\nu_e}\rangle \leq 11\ eV$, so caution is advisable.

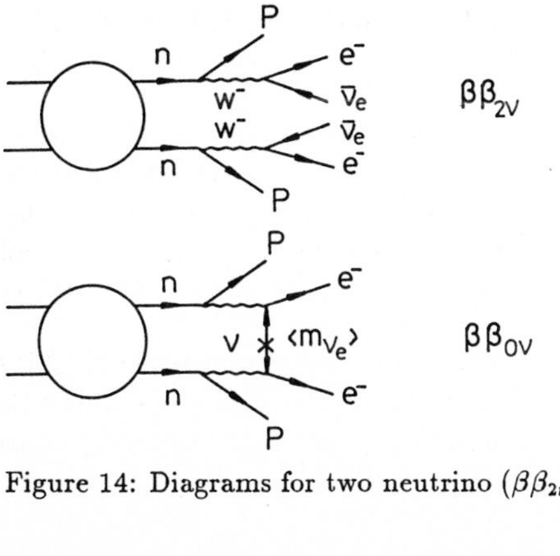


Figure 14: Diagrams for two neutrino ($\beta\beta_{2\nu}$) and neutrinoless ($\beta\beta_{0\nu}$) double beta decay.

Even the largest value $\langle m_{\nu_e} \rangle \leq 11\ eV$ is smaller than the range $m_{\nu_e} \sim (17-40)\ eV$ suggested by the ITEP experiment. If the latter is correct the simplest possibility is that the $\nu_e$ is Dirac. Another possibility [137] is that the $\nu_e^0$ is a mixture of Majorana mass eigenstate neutrinos, as in (44). Then, $\langle m_{\nu_e} \rangle$ becomes

$$\langle m_{\nu_e} \rangle = \sum_i m_i U_{Lei}^2 \xi_i F(m_i, A), \tag{88}$$

where $m_i \geq 0$ is the physical mass of the $i^{th}$ mass eigenstate, $U_{Lei}$ is the mixing matrix element ($\nu_{eL}^0 = \sum U_{Lei} \nu_{iL}$) and $\xi_i = \pm 1$ is the CP parity of $\nu_{iL}$. $\xi_i$ is just $K_{ii}$ in (46), and a negative value $\xi_i = -1$ means simply that the eigenvalue of $M$ in (41) was negative before choosing $K$ to redefine $\nu_R^c$. In (88), $F(m_i, A)$ is a nucleus dependent propagator correction, [138] defined by

$$F(m_i, A) = \frac{\langle e^{-m_i r}/r \rangle}{\langle 1/r \rangle}. \tag{89}$$

It is $\sim 1$ for $m_i \ll 10\ MeV$. For $m_i \gg 10\ MeV$, $F(m_i, A) \ll 1$ (it falls as $m_i^{-2}$) and allows the possibility [139] of $A$ dependence of $\langle m_{\nu_e} \rangle$.

Because of the possibility of negative contributions to $\langle m_{\nu_e} \rangle$ it is conceivable that there are cancellations so that $\langle m_{\nu_e} \rangle$ is much smaller than the mass of the dominant Majorana component of $\nu_e$ (e.g. $m_1 \sim (17-40)\ eV$). Such a cancellation is actually not so contrived as it might first appear. If all of the $m_i$ are small enough that $F(m_i, A) = 1$ then from (45) $\langle m_{\nu_e} \rangle$ is just the $M_{ee}$ component of the original Majorana mass matrix in (41). As we have seen, $M_{ee}$ must be generated by a Higgs triplet and vanishes in many models. In fact, the light seesaw model of Table 6 automatically leads to $\langle m_{\nu_e} \rangle = 0$ for sufficiently small $m_i$. For two neutrinos, for example, $\langle m_{\nu_e} \rangle = m_1 \cos^2\theta - m_2 \sin^2\theta$, which vanishes by (69) and (70). However, the light seesaw model was devised just in order to give $\langle m_{\nu_e} \rangle = 0$. For seesaw models with more natural scales $m_2 \gg 10\ MeV$ one has that $F(m_i, A) \ll 1$ and $U_{e1} \sim 1$, so that $\langle m_{\nu_e} \rangle \sim m_{\nu_e}$. In most Majorana models, therefore, one expects $\langle m_{\nu_e} \rangle \sim m_{\nu_e}$ unless fine-tuned deviations from the seesaw formula are invoked.

Whether or not the cancellation of the terms in (88) is natural, one can consider whether it is phenomenologically viable. For two neutrinos, for example, the conditions $m_1 \sim 20\ eV$, and $\langle m_{\nu_e} \rangle \ll m_1$ imply

$$\tan^2\theta = \frac{m_1}{m_2 F(m_2, A)}, \tag{90}$$

where $m_1 \leq m_2$, $\tan^2\theta \leq 1$ since the ITEP experiment presumably measures the dominant component of $\nu_e$. However, the reactor oscillation limits in Fig. 13 allow only two possibilities. One is that $m_1 \simeq m_2$, $0 \sim 45^0$. In that case $\nu_1$ and $\nu_2$ can be combined to form a Dirac neutrino (or pseudo-Dirac if the degeneracy is not exact), possibly with a non-canonical lepton number (such as $L_e - L_\mu + L_\tau$) conserved. Alternatively, one can have $m_2 \geq 450\ eV$. However, the various laboratory and cosmological limits exclude [70] almost all values of $m_2$ except for small windows around 40 $MeV$ and 2 $GeV$. Hence, if the ITEP results turn out to be correct they would almost certainly imply either (a) the $\nu_e$ is Dirac, or (b) there is new physics (such as a Majoron) that evades the cosmological bounds.

There are additional contributions to neutrinoless double beta decay in $SU_{2L} \times SU_{2R} \times U_1$ models [140]. Typically, such models contain additional charged $W_R^\pm$ bosons which couple to right-handed currents $\bar{e}_R \gamma^\mu N_R$, where $N_R$ is a heavy Majorana neutrino. The exchange of a $N_R$ (rather than a $\nu_L$ in Fig. 14) yields a new contribution $M_N F(M_N, A)(M_{W_L}/M_{W_R})^4$ to $\langle m_{\nu_e} \rangle$, which sets non-trivial constraints [69] on $M_N$ and $M_{W_R}$. Furthermore, mixed contributions involving one ordinary left-handed current $\bar{e}_L \gamma^\mu \nu_L$ and one right-handed current $\bar{e}_R \gamma^\mu N_R$ can yield contributions to to $0^+ \to 2^+$ decay amplitudes that are not directly proportional to a neutrino mass [141] However, the relevant amplitudes are of order [49,140]

$$\left(\frac{M_{W_L}}{M_{W_R}}\right)^2 \theta, \ \zeta\theta, \tag{91}$$

where $\theta$ is a light-heavy neutrino mixing angle and $\zeta$ is the $W_L - W_R$ mixing angle. One typically expects $(M_{W_L}/M_{W_R})^2$ and $\zeta$ to be less than $10^{-3}$. Since we expect $\theta \sim m_D/M_{W_R} \leq 10^{-4} - 10^{-5}$ in a typical $TeV$-seesaw, the expected values for the quantities in (91) are smaller than the experimental limits (of $\sim 10^{-6}$).

One typically has $\langle m_{\nu_e}\rangle \ll m_{\nu_e}$ for the charged Higgs models [71] because the antisymmetry of the relevant Yukawa coupling forces $M_{ee}$ to vanish.

## 4.9 The Solar Neutrino Problem

For some years the event rate in the $^{37}Cl \rightarrow {}^{37}Ar$ Solar neutrino experiment [142] ($2.0 \pm 0.3\ SNU$ [143]) has been considerably below the prediction [144] $5.8 \pm 2.2\ SNU$ of the standard Solar model. The discrepancy has recently been confirmed by the Kamiokande group which reports [145] an upper limit on the $\nu_e$ flux (from $\nu_e e$ elastic scattering) that is less than half the expected event rate. One explanation for the discrepancy is the existence of vacuum oscillations of the $\nu_e$ into other neutrinos. These could be important for neutrino mass-squared differences [146] $\Delta m^2 \equiv m_1^2 - m_2^2$ as small as $\Delta m^2 \sim (10^{-11} - 10^{-10})\ eV^2$, but only if the mixing angles are large.

Another possibility [147] is that the $\nu_e$ is a Dirac particle with a magnetic moment in the range $\mu_{\nu_e} \sim (0.6 - 10) \times 10^{-10} \mu_B$. The $\nu_e$ spin could then process in the Solar magnetic field into a sterile right-handed $\nu_e$, thus reducing the observed flux by a factor $\simeq 2$. The necessary value of $\mu_{\nu_e}$ is barely consistent with laboratory limits [148] but is probably excluded by astrophysical constraints from nucleosynthesis and stellar cooling [149] (Table 8). The worst objection, however, is that the necessary $\mu_{\nu_e}$ is unnaturally high. In the standard model with a Dirac mass one expects [150]

$$\mu_\nu = \frac{3G_F m_\nu m_e}{4\pi^2\sqrt{2}} \mu_B \sim 3\times10^{-19} \left(\frac{m_\nu}{1eV}\right) \mu_B \tag{92}$$

which is many orders of magnitude too small. Non-standard models [151] can yield larger $\mu_\nu$, but to obtain a sufficiently large value appears highly contrived.

Other canonical explanations involve non-standard Solar models. The existing experiments are mainly sensitive to the relatively high energy (from $0.81\ MeV$ up to $14\ MeV$) neutrinos from $^8B$ decay. The flux of these $^8B$ neutrinos depends very sensitively on the temperature of the Solar core and could be changed significantly by modifications of the standard Solar model. Recently, there has been much attention to the possibility that weakly interacting massive particles (WIMPs),

Table 8: Limits on the neutrino magnetic moments. A value $\mu_{\nu_e} \sim (0.6 - 10) \times 10^{-10} \mu_B$ would be needed to resolve the Solar $\nu$ problem.

| | |
|---|---|
| laboratory [148] | $\mu_{\bar\nu_e} < 1.5\times10^{-10}\mu_B$ |
| | $\mu_{\nu_\mu} < 9.5\times10^{-10}\mu_B$ |
| Stellar cooling [149] ($\gamma\rightarrow\nu\bar\nu$) | $\mu_\nu < 0.8\times10^{-11}\mu_B$ |
| Nucleosynthesis [149] ($\nu_R$ produced by spin precession) | $\mu_\nu < 0.5\times10^{-10}\mu_B$ |
| Standard model [150] (Dirac mass) | $\mu_\nu \sim 3\times10^{-19}\left(\frac{m_\nu}{1\ eV}\right)\mu_B$ |

which could form the dark matter, could carry energy out of the Solar core and lower the central temperature slightly. [111] Less exotic modifications of the standard model are also possible.

A $^{71}Ga \rightarrow {}^{71}Ge$ experiment could distinguish the nonstandard Solar model from the first two possibilities. Most of the expected $^{71}Ga$ event rate is from the low energy $pp$ neutrinos, the flux of which can be inferred from the over-all Solar luminosity and is relatively insensitive to the temperature of the Solar core. The predicted $^{71}Ga$ event rate of $\simeq 107$ $SNU$ can be reduced at most to around 78 $SNU$ in most non-standard Solar models [144,152]. The traditional view has been that a flux lower than this would imply large vacuum oscillations, which would reduce the $^{71}Ga$ rate by a factor comparable to the $^{37}Cl$ event rate reduction for most oscillation parameters (e.g. to around 40 $SNU$).

Yet another possibility, i.e. that neutrinos decay between the Sun and the Earth, is all but excluded by the survival of neutrinos from supernova 1987A, except in some two-component models with large mixing angles. [153]

Recently, Mikheyev and Smirnov [154] have proposed an elegant new solution to the Solar neutrino problem, in which even tiny vacuum mixing angles can be amplified by the coherent interactions of $\nu_e$ with matter.

Considering $\nu_e \leftrightarrow \nu_\mu$ oscillations for definiteness, the vacuum oscillation equation in (79) can be described in terms of the weak basis states $|\nu_e\rangle$ and $|\nu_\mu\rangle$ by

$$|\nu(t)\rangle = \nu_e(t)|\nu_e\rangle + \nu_\mu(t)|\nu_\mu\rangle, \tag{93}$$

where the coefficients satisfy the Schrödinger-like equation

$$i\frac{d}{dt}\begin{pmatrix} \nu_e(t) \\ \nu_\mu(t) \end{pmatrix} = M_0 \begin{pmatrix} \nu_e(t) \\ \nu_\mu(t) \end{pmatrix}, \tag{94}$$

with

$$M_0 = \begin{pmatrix} \frac{\Delta m^2}{2p}\cos 2\theta & \frac{-\Delta m^2}{4p}\sin 2\theta \\ \frac{-\Delta m^2}{4p}\sin 2\theta & 0 \end{pmatrix}, \tag{95}$$

where an irrelevant term proportional to the identity (which only affects the overall phase) has been dropped. Wolfenstein pointed out [155] that in the presence of matter, $M_0$ is replaced by the $M'$, where

$$M' = M_0 + \begin{pmatrix} \sqrt{2}G_F n_e & 0 \\ 0 & 0 \end{pmatrix} \tag{96}$$

and $n_e$ is the density of electrons. The new term [155] - [157] is the effect of the coherent forward scattering amplitude for $\nu_e e^- \rightarrow \nu_e e^-$ via the charged current. The effects of neutral current scattering from $e^-$, $p$, and $n$ have been neglected because they are the same for $\nu_e$ and $\nu_\mu$ and only contribute to the overall phase. For $\Delta m^2 < 0$ (i.e. $m_{\nu_e} < m_{\nu_\mu}$ [157]) there is a critical density [158] $n_e^{crit} = -\Delta m^2 \cos 2\theta/(2\sqrt{2}G_F p)$ for which the diagonal elements of $M'$ are equal (i.e. zero). At that density a resonance occurs, i.e. even a tiny off diagonal mixing term leads to large mixing effects.

In particular, if $n_e$ in the Sun varies slowly an adiabatic approximation applies [154,159]. $\nu_e'$s produced in the core of the Sun (where $n_e > n_e^{crit}$) correspond to the larger mass eigenstate $\nu_2$ of $M'$ (Fig. 15). Outside the Sun, on the other hand, the higher energy state $\nu_2$ corresponds to $\nu_\mu$ for $\Delta m^2 < 0$. Hence, if the variation of $n_e$ with the distance from the center of the Sun is sufficiently slow, the initial $\nu_e$ will be adiabatically converted to $\nu_\mu$ as they pass through the resonance.

A number of authors [152],[154],[159] - [161] have analyzed the implications of the Mikheyev-Smirnov-Wolfenstein (MSW) effect for the Solar neutrinos quantitatively. It is found that there are three classes of parameters which can explain the reduction of $^8B$ neutrinos observed in the $^{37}Cl$ experiment. These roughly form the sides of a triangle, as is illustrated schematically in Fig. 16. For solution (a) corresponding to $|\Delta m^2| \sim 5 \times 10^{-5}$ $eV^2$, $\sin^2 2\theta \geq 4 \times 10^{-4}$, the adiabatic

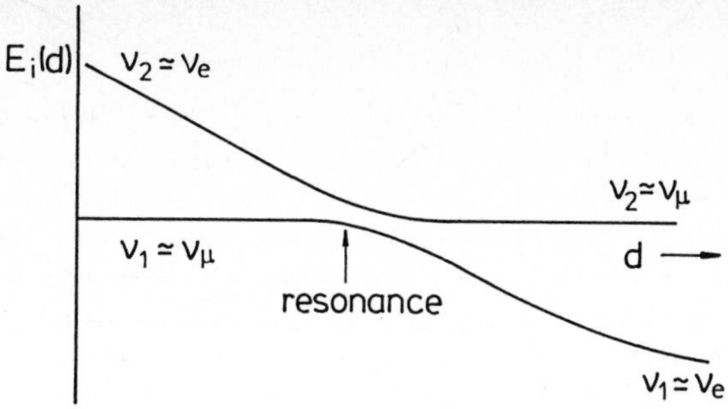


Figure 15: The energy eigenvalues of $M'$ as a function of $d$, the distance from the center of the Sun.

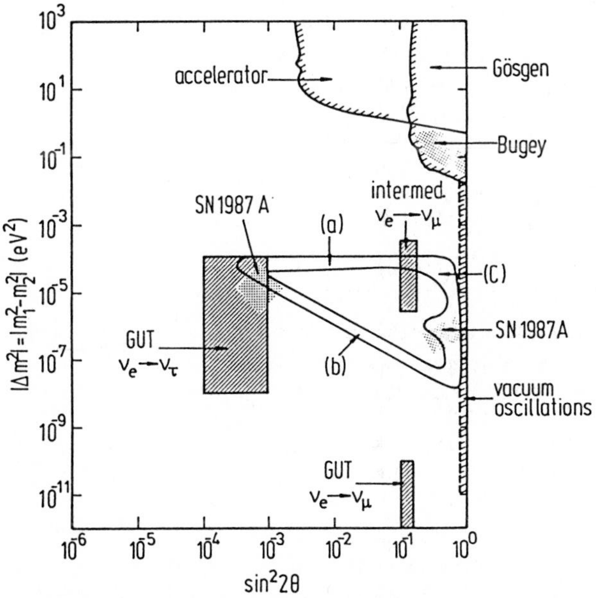


Figure 16: A schematic view of the regions in the $\Delta m^2$ - $\sin^2 2\theta$ plane which can explain the Solar neutrino problem via the Mikheyev-Smirnov-Wolfenstein (MSW) effect.

hypothesis is valid and $\simeq 100\%$ conversion occurs. However, only the high energy $^8B$ neutrinos actually encounter a resonance layer (the central density is too low for the low energy neutrinos) and are converted. For this parameter range one expects little reduction in the counting rate for the gallium experiment (i.e. the effect is similar to non-standard Solar models in that respect).

For solution (b), extending down to $|\Delta m^2| \sim 10^{-8}\ eV^2$ one has [160] $|\Delta m^2| \sin^2 2\theta \sim 10^{-7.5}\ eV^2$. Here the adiabatic approximation starts to break down. All neutrino energies are affected, but the conversion probability is less than unity. For these solutions one expects a significant reduction in the gallium counting rate, similar to large vacuum oscillations or magnetic moments. Solution (c), corresponding to large vacuum mixing angles, is an extension of the vacuum oscillation solution. In the middle of region (c) one expects a large day-night asymmetry in the $\nu_e$ counting rate due to MSW regeneration in the Earth at night [152].

The MSW solution is very elegant, but it severely complicates the task of sorting out which if any of the proposed solutions to the Solar $\nu$ problem is correct. It will take an ambitious program of experiments [152,162] to clarify the matter.

The first two events from SN1987A observed by the Kamiokande experiment [93] point back towards the supernova. They are consistent with $\nu_e$ from the initial neutronization burst, scattering via $\nu_e e^- \rightarrow \nu_e e^-$. However, they could also be $\bar{\nu}_e$ from the subsequent thermal burst, scattering via $\bar{\nu}_e p \rightarrow e^+ n$, which has a much larger cross section (and which produces an isotropic distribution of positrons). If they are indeed $\nu_e$ they are problematic for the MSW mechanism because one expects $\nu_e \rightarrow \nu_\mu$ conversions on the way out of the supernova. However, there are two parameter regions (shown in Fig. 16), which would still be consistent [163], corresponding respectively to incomplete conversion in the supernova and reconversion in the Earth. Unfortunately there is no way to determine whether the two events are really $\nu_e$'s.

The MSW mechanism is consistent with the expectation of GUT [67] and intermediate scale [68] seesaws. As is illustrated in Fig. 16 the predictions of the GUT seesaw are consistent with $\nu_e \leftrightarrow \nu_\tau$ conversions in the Sun. In this case, the mass ranges are too small to ever see any direct laboratory effects of neutrino mass. The intermediate scale seesaw could account for the Solar $\nu$ problem via $\nu_e \leftrightarrow \nu_\mu$ conversions. In that case, $\nu_\mu \leftrightarrow \nu_\tau$ oscillations could well be observable in the laboratory.

It is also possible that the Solar $\nu$ problem could be explained by small neutrino masses, but that neutrino appearance experiments might nevertheless yield positive signals due to non-orthogonal neutrino states (induced by mixings between very light and very heavy neutrinos. [110])

# 5 Summary

- The predictions of the standard $SU_2 \times U_1$ model for the $W$ and $Z$, the charged current, and the neutral current interaction are qualitatively confirmed. In particular, the charged and neutral current interactions of the neutrino are correctly described by the standard model to excellent precision. Furthermore, neutrino interactions are superb probes of the strong interactions and of possible new physics.

- Indirect arguments indicate that the $\nu_\tau$ must exist. Nucleosynthesis constraints imply that there are no more than 4 neutrinos with masses $\leq 1\ MeV$. $e^+e^-$ and $\bar{p}p$ constraints imply $\leq (3-5)$ neutrinos with masses up to $\sim 40\ GeV$.

- The question of whether the neutrinos have mass is vital for both particle physics and cosmology. However, there is at present no compelling evidence for neutrino mass. The ITEP result $m_{\nu_e} \sim (17-40)\ eV$ has not been confirmed by other experiments and is on the verge of being excluded. Although there are two positive indications of neutrino oscillations (with different parameters), these are contradicted by other experiments. The negative results suggest $m_{\nu_i} \simeq O(1\ eV)$ unless there are very small mixings or degeneracies. There are also stringent limits on incoherent mixing with heavy neutrinos.

- The MSW solution to the Solar neutrino problem would work for $\nu_\mu$ or $\mu_\tau$ in the $10^{-2}\ eV$ range, consistent with intermediate mass or GUT seesaws.

- The nonobservation of neutrinoless double beta decay implies $\langle m_{\nu_e} \rangle < 1 - 11\ eV$. If the ITEP result is correct this would most likely imply a Dirac neutrino or new physics to evade cosmological bounds.

- A $\nu$ mass in the 5 - 40 $eV$ range would dominate the energy density of the universe and would be an excellent candidate for the dark matter, though other mechanisms would have to be invoked to explain the initial formation of galaxies. Conversely, the light stable neutrinos must be lighter than $\simeq 40\ eV$. A variety of astrophysical, laboratory, and cosmological bounds exclude unstable neutrinos up to $\sim 40\ MeV$ (unless new physics is invoked for fast, invisible decays or annihilations), implying that $m_{\nu_\mu}$, $m_{\nu_\tau} < 40\ eV$.

Acknowledgement

It is a pleasure to thank the Alexander von Humboldt-Stiftung, the DESY theoretical group, the Max Planck Inst. für Kernphysik, and the U.S. Department of Energy grant DE-AC02-ER0-3071 for support during the preparation of this paper.

# References

[1] S. Weinberg, Phys. Rev. Lett. 19, 1264 (1967); A. Salam in *Elementary Particle Theory*, ed. N. Svartholm (Almquist and Wiksells, Stockholm, 1969) p. 367; S.L. Glashow, J. Iliopoulos, and L. Maiani, Phys. Rev. D2, 1285 (1970).

[2] As modified to include parity violation, quarks, flavor mixing, and the intermediate vector boson.

[3] UA1: G. Arnison et al, Phys. Lett. 166B, 484 (1986).

[4] UA2: R. Ansari et al., Phys. Lett. 186B, 440 (1987).

[5] The results given here are from U. Amaldi, A. Böhm, L.S. Durkin, P. Langacker, A.K. Mann, W.J. Marciano, A. Sirlin, and H.H. Williams, Phys. Rev. D36, 1385 (1987). Very similar conclusions are reached in an analysis by G. Costa, J. Ellis, G.L. Fogli, D.V. Nanopoulos, and F. Zwirner, CERN preprint CERN-TH.4675/87.

[6] Left ($L$) and right ($R$) chiral $(1 \pm \gamma^5)$ are equivalent to negative and positive helicity, respectively, for relativistic fermions, up to corrections of order $m/E$.

[7] For a recent review, see A. Sirlin, *1987 Int. Symp. on Lepton and Photon Interactions at High Energies*, Hamburg, July 1987.

[8] For reviews, see G. Barbiellini and C. Santoni, Riv. Nuo. Cim. 9(2), 1 (1986); E.D. Commins and P.H. Bucksbaum, *Weak interactions of leptons and quarks*, (Cambridge Univ. Press, Cambridge, 1983).

[9] W. Fetscher, H.-J. Gerber, and K.F. Johnson, Phys. Lett. 177B, 102 (1986); H.-J. Gerber, *Int. Europhysics Conference on High Energy Physics*, Uppsala (Sweden), June, 1987; W. Fetscher, $12^{th}$ *Int. Conf. of Neutrino Physics and Astrophysics*, Sendai, Japan, June 1986.

[10] For a generalization, see P. Langacker and D. London, to be published.

[11] A. Jodidio et al., Phys. Rev. D34, 1967 (1986); see also I. Beltrami et al., Phys. Lett. B194, 326 (1987).

[12] The limit on $W_R$ assumes that the neutrinos coupled to $e_R^-$ and $\mu_R^-$ in the $V + A$ current are sufficiently light to be produced in $\mu$ decay, and does not apply if the $V + A$ current involves heavy majorana neutrinos.

[13] It is known independently that $h_{\nu_\mu} < 0$.

[14] A. Sirlin, Phys. Rev. D35, 3423 (1987), and [7].

[15] The result (9) also tests the radiative corrections to the standard model, which can only be consistently calculated within a gauge theory (they would be infinite in a 4-Fermi or intermediate vector boson theory). Without the radiative corrections, the right hand side of (9) would be 1.036, in apparent violation of unitarity [14].

[16] See, for example, F. Sciulli, *1985 Int. Symposium on Lepton and Photon Interactions at High Energies*, eds. M. Konuma and K. Takahashi (Nissha, Kyoto, 1986) p.8.

[17] I. Manelli et al., to be published.

[18] CP violation is incorporated in the standard model by a phase in the (small) components of $V$.

[19] H. Albrecht et al., Phys. Lett. 192B, 245 (1987).

[20] The expressions in (15) are modified slightly by radiative corrections. See [5].

[21] CDHS: H. Abramowicz et al., Phys. Rev. Lett. 57, 298 (1986).

[22] CHARM: J.V. Allaby et al., Phys. Lett. 177B, 446 (1986).

[23] C.H. Llewellyn Smith, Nucl. Phys. B228, 205 (1983).

[24] For a review, see [5].

[25] It is crucial to use the ratio of valence quark moments $D_v/U_v = 0.39 \pm 0.06$ determined from charged current scattering, [5,16] rather than the naive ratio 0.5.

[26] BBCIMOU: G.F. Jones et al., Phys. Lett. 178B, 329 (1986).

[27] E734: L.A. Ahrens et al., Phys. Rev. D35, 785 (1987).

[28] D. Rein and L.M. Sehgal, Nucl. Phys. B223, 29 (1983).

[29] Other reactions, such as inclusive and exclusive incoherent pion production and $\bar{\nu}_e D \to \bar{\nu}_e np$ are in qualitative agreement with the standard model, but are not used in the analysis because of large uncertainties in the hadronic matrix elements.

[30] CHARM: F. Bergsma et al., Phys. Lett. 147B, 481 (1984).

[31] BNL E734: L.A. Ahrens et al., Phys. Rev. Lett. 54, 18 (1985).

[32] Savannah River: F. Reines et al., Phys. Rev. Lett. 37, 315 (1976).

[33] LANL ILM: R.C. Allen et al., Phys. Rev. Lett. 55, 2401 (1985).

[34] The radiative corrections are computed using the Sirlin definition [24] $\sin^2\theta_W \equiv 1 - \frac{M_W^2}{M_Z^2}$ of the renormalized weak angle.

[35] Here the value is for fixed $m_t = 45\ GeV$, $M_H = 100\ GeV$, to facilitate comparison with $SU_5$.

[36] For a recent review, see P. Langacker, *Weak and Electromagnetic Interactions in Nuclei*, ed. H.V. Klapdor (Springer-Verlag, Berlin, 1986) p. 879; Phys. Rep. 72, 185 (1981).

[37] R. Marshall, Rutherford preprint RAL-87-031.

[38] V. Barger et al., Phys. Rev. D35, 2893 (1987).

[39] G. Beall, M. Bander, and A. Soni, Phys. Rev. Lett. 48, 848 (1982).

[40] G. Steigman et al., Phys. Lett.176B, 35 (1986).

[41] J. Ellis and K. Olive, Phys. Lett. 193B, 525 (1987).

[42] For reviews, see D.L. Burke, *Proc. of the Theoretical Advanced Study Institute in Particle Physics*, Santa Cruz, July 1986, SLAC-PUB-4284; N.G. Deshpande, *Neutrino Physics and Neutrino Astrophysics*, ed. V. Barger et al., (World, Singapore, 1987) p. 78.

[43] UA1: C. Albajar et al., Phys. Lett. 198B, 271 (1987); A.D. Martin, R.G. Roberts, and W.J. Stirling, Phys. Lett. 189B, 220 (1987); F. Halzen, C.S. Kim, and S. Willenbrock, MAD/PH/342 (1987).

[44] H.-J. Gerber, [9]; K. Mursula and F. Scheck, Nucl. Phys. B253, 189 (1985).

[45] $A^\tau$ also rules out the possibility that $\tau_R^-$ is assigned to an $SU_2$ doublet with $\nu_{\tau R}$ ($A^\tau = 0$ or $\frac{1}{2}$, depending on whether $\tau_L$ is in a doublet or singlet, respectively). These models are also excluded by the energy spectrum in leptonic $\tau$ decays [44].

[46] The nucleosynthesis constraint applies to all light neutrino degrees of freedom that were present in significant numbers at $T \sim 1\ MeV$. It does not apply to the right handed ($SU_2$ singlet) partners of the ordinary neutrinos that must be introduced if they are to have small Dirac masses, unless new interactions (e.g. new charged or neutral gauge bosons) are introduced as well. Otherwise, the singlet neutrinos have essentially no interactions and would not have been produced in significant numbers in the early universe. See G. Steigman, K.A. Olive, and D. Schramm, Phys. Rev. Lett. 43, 239 (1979); Nucl. Phys. B180, 497 (1981); J. Ellis et al., Phys. Lett. 167B, 457 (1986).

[47] UA1: C. Albajar et al., to be published.

[48] It is important to obtain an accuracy on $\Delta N_\nu$ of much less than one, because other neutral weakly interacting particles can contribute fractional values to the apparent value of $N_\nu$. For example, supersymmetric scalar neutrinos contribute $\Gamma_Z \sim 85\ MeV$, equivalent to $\frac{1}{2}$ unit of $N_\nu$.

[49] There are several recent reviews of neutrino mass. See, for example, F. Boehm and P. Vogel, *Physics of Massive Neutrinos*, (Cambridge Univ. Press, Cambridge, 1987); J.D. Vergados, Phys. Rev. (reports) 133, 1 (1986); S.M. Bilenky and S.T. Petcov, Rev. Mod. Phys. 59, 671 (1987); R. Eichler, *1987 Int. Symp. on Lepton and Photon Interactions at High Energies.*

[50] Under a CP transformation $\psi_L(\vec{x},t) \to \gamma^0 \psi_R^c(-\vec{x},t)$, so $\psi_L$ and $\psi_R^c$ are essentially CP conjugates.

[51] L. Wolfenstein, Nucl. Phys. B186, 147 (1981).

[52] E.S. Konopinski and M. Mahmoud, Phys. Rev. 92, 1045 (1953).

[53] Fermion mass matrices need not be Hermitian. One could generalize still further and introduce $K(\neq F)$ $N_R^0$ fields, in which case $m_D$ would be $F \times K$ dimensional.

[54] We have chosen a basis for the charged leptons so that the analogue of $V_L$ for the electrons is the identity matrix.

[55] $V_L^\dagger$ contains $F^2$ real parameters, of which $F(F-1)/2$ are angles, analogous to the Cabibbo angle, which describe transitions between families and lead to such effects as neutrino oscillations. $2F-1$ parameters are unobservable phases, which can be removed by appropriate choices of phases for the $n_{iL}$ and $e_{iL}^-$ fields (corresponding phases must be chosen for the $n_{iR}$ and $e_{iR}$ to ensure real masses). The remaining $(F-1)(F-2)/2$ parameters are phases which could lead to CP violation in the leptonic sector.

[56] G.B. Gelmini and M. Roncadelli, Phys. Lett. 99B, 411 (1981); H. Georgi et al., Nucl. Phys. B193, 297 (1983).

[57] More generally, $K$ is block diagonal, the blocks being unitary symmetric matrices in the subspaces of degenerate eigenvalues.

[58] One has $n_{iR}^c = K_{ji} C \bar{n}_{jL}^T$, so that $n_i$ is self-conjugate under the generalized C transformation $n_i \to K_{ji} C \bar{n}_j^T$.

[59] The counting of angles in $U_L^\dagger$ is the same as for the Dirac case. However, there are an additional $F-1$ CP violating phases associated with the fact that $U_L$ and $U_R$ are related by (46). For example, if one chooses $K_L$ so that $K = I$, then the phases of the $n_{iL}$ are fixed and cannot be redefined to remove phases from $U_L$. See S.M. Bilenky, J. Hosek, and S.T. Petcov, Phys. Lett. 94B, 495 (1980); M. Doi et al., Phys. lett. 102B, 323 (1981).

[60] W.Y. Keung and G. Senjanovic, Phys. Rev. Lett. 50, 1427 (1983); J.F. Gunion and B. Kayser, *Proc. 1984 Summer Study on the SSC*, eds R. Donaldson and J.G. Morfin (APS, 1984) p. 153.

[61] H. Goldberg, Phys. Rev. Lett. 50, 1419 (1983); L. M. Krauss, Phys. Lett. 128B, 37 (1983), Nucl. Phys. B227, 556 (1983).

[62] For a review, see B. Kayser, Comm. Nucl. Part. Phys. 14, 69 (1985).

[63] New interactions could in principle distinguish the two cases even for $m_\nu = 0$, depending on whether they conserved lepton number.

[64] S.M. Bilenky and B. Pontecorvo, Lett. Nuovo Cimento 17, 569 (1976). The implications of the existence of Dirac and Majorana neutrino mass terms were also studied in: V. Barger et al., Phys. Rev. Lett. 45, 692 (1976); S.M. Bilenky, J. Hosek and S.T. Petcov, Phys. Lett. 94B, 495 (1980); J. Schechter and J.W. F. Valle, Phys. Rev. D22, 2227 (1980); T.P. Cheng and L.F. Li, Phys. Rev. D22, 2860 (1980); T. Yanagida and M. Yoshimura, Progr. Th. Phys. 64, 1870 (1980).

[65] The relative sign of $n_{1R}$ and $n_{2R}$ will reemerge, however, in neutrinoless double beta decay, where it will be seen to be observable.

[66] M. Gell-Mann, P. Ramond, and R. Slansky, in *Supergravity*, eds. F. van Nieuwenhuizen and D. Freedman, (North Holland, Amsterdam, 1979) p. 315; T. Yanagida, Prog. Th. Physics B135, 66 (1978).

[67] P. Langacker, S. T. Petcov, G. Steigman, and S. Toshev, Nucl. Phys. B282, 589 (1987).

[68] R.N. Mohapatra and G. Senjanovic, Z. Phys. C17, 53 (1983); Q. Shafi and F.W. Stecker, Phys. Rev. Lett. 53, 1292 (1984); K. Kang and M. Shin, Nucl. Phys. B287, 687 (1987); K. Kang and A. Pantziris, Phys. Lett. 193B, 467 (1987); P. Langacker, R.D. Peccei, and T. Yanagida, Mod. Phys. Lett. A1, 541 (1986).

[69] R.N. Mohapatra and G. Senjanovic, Phys. Rev. D23, 165 (1981). For constraints on the heavy neutrino mass, see R.N. Mohapatra; Phys. Rev. D34, 909 (1986).

[70] P. Langacker, B. Sathiapalan, and G. Steigman, Nucl.Phys. B266, 669 (1986).

[71] A. Zee, Phys. Lett. 93B, 389 (1980); 161B, 141 (1985); Nucl. Phys. B264, 99 (1986); K. Tamvakis and J. Vergados, Phys. Lett. 155B, 373 (1985); G.K. Leontaris and K. Tamvakis, Phys. Lett. 191B, 421 (1987); M. Fukugita and T. Yanagida, Phys. Rev. Lett. 58, 1807 (1987).

[72] Since $\nu_L$ transforms according to a doublet representation of $SU_2$ and $N_R$ is a singlet, the Higgs field coupling $\nu_L$ to $N_R$ must be an $SU_2$ doublet.

[73] A. Masiero, D.V. Nanopoulos, and A.Sanda, Phys. Rev. Lett. 57, 663 (1986).

[74] Unlike similar hierarchy problems in comparing the weak and unification scale, we know that there is no supersymmetry to control the radiative corrections.

[75] D.S.P. Dearborn, D.N. Schramm, and G. Steigman, Phys. Rev. Lett. 56, 26 (1986); H.Y. Cheng, Phys. Rev. D36, 1649 (1987); and references therein.

[76] R. Barbieri, J. Ellis, and M.K. Gaillard, Phys. Lett. 90B, 249 (1980).

[77] In many models which implement the seesaw mechanism Higgs triplet fields are also present. Even if these have zero VEV at tree level they may obtain VEV's due to higher order corrections to the Higgs potential. See M. Magg and C. Wetterich, Phys. Lett. 94B, 61 (1980); D. Chang and R.N. Mohapatra, Phys. Rev. D32, 1248 (1985); J. Bijnens and C. Wetterich, DESY 87-042 (1987).

[78] A simple diagonalization of $\begin{pmatrix} 0 & m_D \\ m_D & 0 \end{pmatrix}$ yields eigenvalues $-m_D^2/m_S$ and $m_S$. The sign of the smaller eigenvalue is reversed by choosing $K = \begin{pmatrix} -1 & 0 \\ 0 & 1 \end{pmatrix}$, which accounts for the minus sign in $\nu_R^{0c}$.

[79] E. Witten, Phys. Lett. 91B, 81 (1980).

[80] Some versions of these models [77] are expected to have induced values for $m_t$ in (72).

[81] Y. Chikashige, R.N. Mohapatra, and R.D. Peccei, Phys. Lett. 98B, 265 (1981).

[82] In many of the intermediate mass seesaw models [68] the singlet-Majoron and the invisible axion are the same particle.

[83] S. Nandi and U. Sarkar, Phys. Rev. Lett. 56, 564 (1986); J. Derendinger, L. Ibanez, and H.P. Nilles, Nucl. Phys. B267, 365 (1986).

[84] J. Ellis et al., Phys. Lett. 188B, 415 (1987).

[85] M. Roncadelli and D. Wyler, Phys. Lett. 133B, 325 (1983); A.S. Joshipura et al., Phys. Lett. 156B, 353 (1985); P. Roy et al, Phys. Rev. D30, 2385 (1984); O. Shanker, Nucl. Phys. B250, 351 (1985); J. Oliensis and C.H. Albright, Phys. Lett. 160B, 121 (1985).

[86] S. Boris et al., Phys. Rev. Lett. 58, 2019 (1987).

[87] M. Fritschi et al., Phys. Lett. 173B, 485 (1986).

[88] J.F. Wilkerson et al, Phys. Rev. Lett. 58, 2023 (1987).

[89] H. Kawakami et al., *'86 Massive Neutrinos in Physics and Astrophysics*, ed. O. Fackler and J. Tran Than Van.

[90] See, for example, D. Schramm, *1987 Int. Symp. on Lepton and Photon Interactions at High Energies*; H. Meyer, $X^{th}$ *Workshop on Particles and Nuclei*, Heidelberg, Oct. 1987; E.W. Kolb et al., Phys. Rev. D35, 3598 (1987).

[91] R. Abela et al., Phys. Lett. B146, 431 (1984).

[92] ARGUS collaboration, quoted by R. Eichler, [49].

[93] K. Hirakata et al., Phys. Rev. Lett. 58, 1490 (1987).

[94] R.M. Bionta et al., Phys. Rev. Lett. 58, 1494 (1987).

[95] For reviews and a general discussion, see M. Gronau, C.N. Leung, and J.L. Rosner, Phys. Rev. D29, 2539 (1984); F.J. Gilman and S.H. Rhie, Phys. Rev. D32, 324 (1985). F.J. Gilman, Comm. Nucl. Part. Phys. 16, 231 (1986); R.E. Shrock, Phys. Rev. D24, 1232, 1275 (1981).

[96] CCFRR: S.R. Mishra et al., Phys. Rev. Lett. 59, 1397 (1987).

[97] R. Eichler, [49].

[98] For references, see [5].

[99] M.J. Duncan and P. Langacker, Nucl. Phys. B277, 285 (1986).

[100] B. Kayser and P. Langacker, to be published.

[101] There will be off-diagonal $\bar{\nu}_1\nu_2 Z$ vertices because of the mixing between $SU_2$ doublets and singlets.

[102] A similar phenomenon occurs for a system of coupled classical oscillators.

[103] For a discussion of quantum mechanical subtleties, such as the uncertainty in $p$, see F. Boehm and P. Vogel, [49].

[104] IMB, J.M. Lo Secco et al., Phys. Lett. 184B, 305 (1987).

[105] The limits in Fig. 13 are obtained by analyzing the data in terms of the formula in (80) for oscillations between two types of neutrino only. Slightly weaker limits are obtained if one allows mixings between 3 neutrino flavors $(\nu_e, \nu_\mu, \nu_\tau)$. See K. Kleinknecht, Comm. Nucl. Part. Phys. 16, 267 (1986).

[106] J.F. Cavaignac et al., Phys. Lett. 148B, 387 (1984).

[107] G. Zacek et al., Phys. Rev. D34, 2621 (1986).

[108] G. Bernardi et al., Phys. Lett. 181B, 173 (1986).

[109] Another manifestation of the underlying physics would be off-diagonal neutral currents, so that a heavy neutrino could decay into three light neutrinos.

[110] D. Wyler and L. Wolfenstein, Nucl. Phys. B218, 205 (1983); B.W. Lee and R.E. Shrock, Phys. Rev. D16, 1444 (1977); P. Langacker and D. London, to be published.

[111] For a review, see G. Gelmini, this volume.

[112] Both $\Gamma_{weak}$ and $H$ have additional factors involving the number of types of relativistic particles at temperature $T$.

[113] P. Langacker, J.P. Leveille, and J. Sheiman, Phys. Rev. D27, 1228 (1983).

[114] N. Cabibbo and L. Maiani, Phys. lett. 114B, 115 (1982).

[115] For a recent review, see B. Müller, *Proc.* $X^{th}$ *Workshop on Particles and Nuclei: Neutrino Physics*, Heidelberg, Oct. 1987.

[116] For masses in the 10 $eV$ range the $\nu's$ would be clustered in the galaxy and their cross sections would be increased by $10^8$, but direct detection still appears virtually impossible.

[117] Massive neutrinos could easily account for the scales of galactic clusters. However, it is hard to see how such structures could have fragmented into galaxies in the time available, unless another ingredient, such as cosmic strings, provided the seed.

[118] For a recent review, see M. Roos, $X^{th}$ *Workshop on Particles and Nuclei*; M. Turner, AIP Conf. Proc. 72, 266 (1980).

[119] Most of these limits apply to light $SU_2$-singlet neutrinos as well, because in many cases their small mixing with ordinary doublet neutrinos is sufficient to produce them prolifically in the early universe. See [70].

[120] S. Sarker and A.M. Cooper, Phys. Lett. B148, 347 (1984); L. Krauss, Phys. Rev. Lett. 53, 1976 (1984); R. Cowsik, Phys. Lett. 151B, 52 (1985); H. Harari and Y. Nir, Phys. Lett. 188B, 163 (1987), Nucl. Phys. B292, 251 (1987).

[121] Y. Hosotani, Nucl. Phys. B191, 411 (1981); M. Davis et al., AP.J. 250, 423 (1981); P. Hut and S. White, Nature 310, 637 (1984); P. Binetruy et al., Phys. Lett. 134B, 147 (1984); A.A. Natale, Phys. Lett. 141B, 323 (1984).

[122] M. Roncadelli and G. Senjanovic, Phys. Lett. 107B, 59 (1983).

[123] P.B. Pal, Nucl. Phys. B227, 237 (1983); H. Harari and Y. Nir, [120].

[124] D.B. Reiss, Phys. Lett. 115B, 217 (1982); F. Wilczek, Phys. Rev. Lett. 49, 1549 (1982); G. Gelmini, S. Nussinov, and T. Yanagida, Nucl. Phys. B219, 31 (1983).

[125] See [56].

[126] [81] and J. Schechter and J.W.F. Valle, Phys. Rev. D25, 774 (1982); G.B. Gelmini and J.W.F. Valle, Phys. Lett. 142B, 181 (1984); G.B. Gelmini et al., Phys. Lett. 146B, 311 (1984); A. Kumar and R. N. Mohapatra, Phys. Lett. 150B, 191 (1985).

[127] For recent reviews, see [75].

[128] P. Langacker, G. Segre, and S. Soni, Phys. Rev. D26, 3425 (1982).

[129] For a recent review, see A. Faessler, $X^{th}$ *Workshop on Particles and Nuclei.*

[130] T. Kirsten et al., Phys. Rev. Lett. 50, 474 (1983), Z. Phys. C16, 189 (1983).

[131] M.K. Moe, $X^{th}$ *Workshop on Particles and Nuclei.*

[132] Neutrinoless double beta decay could conceivably occur through other mechanisms, such as Higgs exchange, or could be accompanied by the emission of a triplet-Majoron. [134] See G.K. Leontaris and J.D. Vergados, CERN-TH.4835/1987. J.D. Vergados, [49].

[133] F.T. Avignone, this volume.

[134] The PNL/USC group [133] has reported evidence for $^{76}Ge \rightarrow \ ^{76}Se\ e^-e^- +$ Majoron, but their results are in contradiction with other experiments.

[135] W.C. Haxton and G.J. Stephenson, Prog. Part. Nucl. Phys. 12, 409 (1984); T. Tomoda et al., Nucl. Phys. A452, 591 (1986); K. Grotz and H.V. Klapdor, Nucl. Phys. A460, 395 (1986); T. Tomoda and A. Faessler, to be published.

[136] J. Engel, P. Vogel, and M.R. Zirnbauer, to be published.

[137] L. Wolfenstein, Phys. Lett. 107B, 77 (1981); M. Doi et al., Phys. Lett. 102B, 323 (1981).

[138] See W.C. Haxton and G.J. Stephenson, [135], J.D. Vergados, [49].

[139] A. Halprin, S.T. Petcov, and S.P. Rosen, Phys. Lett. 125B, 335 (1983); C.N. Leung and S.T. Petcov, Phys. Lett. 145B, 416 (1984).

[140] M. Doi, T. Kotani, and E. Takasugi, Prog. Th. Phys. (Supp) 83, 1 (1985).

[141] However, in a gauge theory some neutrino must be massive. Otherwise the diagrams cancel. See B. Kayser, 7th *Moriond Workshop on Searches for New and Exotic Phenomena*, Les Arcs, Jan. 1987.

[142] J.K. Rowley, B.T. Cleveland and R. Davis, AIP Conference Proc. 126, 1 (1985).

[143] A Solar neutrino unit ($SNU$) is defined as $10^{-36}$ interactions per target nucleus per second.

[144] J.N. Bahcall et al., Rev. Mod. Phys. 54, 767 (1982); J.N. Bahcall, AIP Conference Proc. 126, 60 (1985).

[145] H. Hirata et al, to be published.

[146] For recent discussions, see V. Barger et al., Phys. Rev. D24, 538, (1981); S.L. Glashow and L.M. Krauss, Phys. Lett. 190B, 199 (1987); L. Krauss and F. Wilczek, Phys.Rev. Lett. 55, 122 (1985).

[147] M.B. Voloshin and M.I Vysotsky, ITEP 86-1(1986); L.B. Okun et al., Zh. Eksp. Teor. Fiz. 91, 754 (1986) (JETP 64, 446 (1986)). For earlier references, see [67].

[148] A.V. Kyuldjiev, Nucl. Phys. B243, 387 (1987); K. Abe et al., Phys. Rev. Lett. 58, 636 (1987).

[149] For a recent discussion, see M. Fukujita and S. Yazaki, Kyoto preprint. RIFP-709.

[150] B.W. Lee and R. Shrock, Phys. Rev. D16, 1444 (1977); K. Fujikawa and R. Shrock, Phys. Rev. Lett. 45, 963 (1980).

[151] M.J. Duncan et al., Phys. Rev. Lett. 191B, 304 (1987); J. Liu, Phys. Rev. D35, 3447 (1987); M. Fukugita and T. Yanagida, [71].

[152] For a review, see W. Hampel, this volume.

[153] F.A. Frieman, H.E. Haber, and K. Freese, Phys. Lett. 200B, 115 (1988).

[154] S.P. Mikheyev and A. Yu Smirnov, Yad Fiz. 42, 1441 (1985) (Sov. Jour. Nucl. Phys. 42, 1441 (1985)); Nuo. Cim. 9C, 17 (1986).

[155] L. Wolfenstein, Phys. Rev. D17, 2369 (1968); D20, 2634 (1979).

[156] V. Barger et al., Phys. Rev. D22, 2718 (1980).

[157] A crucial sign error was corrected in [113].

[158] The scattering amplitude in $M'$ changes sign for antineutrinos. Hence, resonance can occur for either $\nu_e$ or $\bar{\nu}_e$, depending on the sign of $\Delta m^2$.

[159] H.A. Bethe, Phys. Rev. Lett. 56, 1305 (1986).

[160] S.P. Rosen and S.M. Gelb, Phys. Rev. D34, 969 (1986).

[161] P. Langacker, S.T. Petcov, G. Steigman, and S. Toshev, Nucl. Phys. B282, 589 (1987).

[162] E.W. Beier et al., 1986 Snowmass Workshop, Pennsylvania UPR-0140E.

[163] J. Arafune et al., Phys. Rev. Lett. 59, 1864 (1987); P.O. Lagage et al., Phys. Lett. 193B, 127 (1987); L. Wolfenstein, Phys. Lett. 194B, 197 (1987); D. Nötzgold, Phys. Lett. 196B, 315 (1987); S. P. Rosen, Los Alamos LA-UR-87-1296.

# Neutrinos in Left-Right Symmetric, SO(10) and Superstring Inspired Models

*R.N. Mohapatra* *

Department of Physics and Astronomy, University of Maryland, College Park, MD 20742, USA

We review the properties of neutrinos such as masses, decays and magnetic moments in the left-right symmetric as well as superstring inspired models.

## 1. Introduction

Since Pauli postulated the neutrino more than fifty years ago, it has played a key role in the development of weak interaction theory as well as our understanding of the evolution of the universe to its present form. We know that, neutrinos are unique in that they are nearly massless, (or perhaps massless?) neutral spin half particles that participate only in weak and gravitational interactions. Furthermore, leaving aside the proton, neutron, electron and the photon, the only other kind of matter that fills the universe as abundantly as radiation is the neutrino: the neutrino to photon ratio in the universe today is about one-third. Besides, it also plays a major role in the energy loss from stellar objects starting from the main sequence stars (such as the sun) to the exploding supernovae (such as the SN 1987a). Our own sun bombards us (on earth) with nearly sixty billion neutrinos per $cm^2$ per sec. We practically "swim" in the ocean of neutrinos. Yet other than its spin, some of its interactions and the fact that there are 3 kinds of neutrinos we really know very little else about is properties. In the absence of a full picture (as, say, we have for the electron or the proton), the discussion of the properties of the neutrinos involve clever combinations of results from null experiments, theoretical model building and often, sheer intuition and guess work. In this review, I discuss the properties of neutrinos that can be inferred from some popular models of elementary particle physics available today. The two classes of model we will discuss are the left-right symmetric models[1,2] and superstring models.[3] Both of these models are extensions of physics beyond that predicted by the standard model of Glashow, Weinberg and Salam[4] and have been subjects of extensive investigation in the recent past.

This paper is organized as follows: in sec. II, we outline the phenomenological (and cosmological) constraints on massive neutrinos and their life times; in sec. III, we briefly introduce the various possibilities for neutrino masses such as Dirac, Majorana and pseudo-Dirac masses and their connection to $B - L$ symmetry; in sec. IV, we begin the discussion of left-right symmetric models and phenomenological constraints on various parameters of the model; sec. V is devoted to the discussion of Dirac vrs. Majorana masses in left-right

*Work supported by a grant from the National Science Foundation.

symmetric models as well as the $\nu$-decays in both cases; in sec. VI we discuss the embedding of these models in SO(10) grandunified theories; in sec. VII, we discuss neutrino masses in superstring models and the various possibilities, that exist for neutrino masses in these models.

## 2. Observational Constraints on Massive Neutrinos:

There are several kinds of observational constraints on the properties of massive neutrinos: those implied by (a) cosmological observations; (b) solar neutrino puzzle; (c) supernova, SN1987a and (d) those implied by accelerator and reactor experiments attempting to detect neutrino stability. Below, we briefly summarize the first three cases.

**(a) Cosmological Constraints:** The cosmological constraints date back to the pioneering works of Cowsik and Mclelland[5] and Gershtein and Zeldovich[6] who pointed out that, there exists a neutrino sea all around us, with a temperature of $1.9^o\ K \simeq 1.6 \times 10^{-4}$ ev. Usual considerations of Fermi-Dirac statistics then leads to the conclusion that the present number density of each neutrino species is given by $n_\nu \simeq \frac{3}{11} n_\gamma \simeq 109/cm^3$. If neutrinos have mass $m_{\nu_i}$ and are stable, then their contribution to the mass density of the universe must be smaller than the critical mass density of the universe leading to an upper limit on the sum of the masses of the stable neutrinos:

$$\left(m_{\nu_e} + m_{\nu_\mu} + m_{\nu_\tau} + \cdots\right) n_\nu \leq 10^4\, h_o^2\, ev$$

where $h_o$ denotes the uncertainty in the Hubble constant (i.e. $H_o^{-1} = 10^{10}/h_o$ years) (or the age of the universe) and we have $.4 \leq h_o \leq 1$, with smaller $h_o$ corresponding to an older universe. For $h_o \approx .5$ to 1, we get,

$$\Sigma\, m_{\nu_i} \leq 25 \text{ to } 100\ ev\ . \tag{2.1}$$

Later on it was pointed out[7] that for stable heavier neutrinos, there is a lower bound of about 2 GeV. The situation is, however, very different for unstable neutrinos as was first pointed out by Dicus, Kolb and Teplitz.[8] The reason of course is that, if a neutrino species decays in the course of the evolution of the universe, it no longer contributes to mass density at present; the decay products, however, carry the information about the mass in their energy but energy, unlike the mass gets redshifted with the cosmological evolution and if the decay happens early enough, the energies could be sufficiently redshifted to be harmless. In this way, one gets a constrained joint bound on mass and lifetime as follows:

$$m_{\nu_H} \left(\frac{\tau_{\nu_H}}{\tau_U}\right)^{1/2} \leq 25\ ev \text{ to } 100\ ev\ . \tag{2.2}$$

There are three principal decay modes of heavier neutrinos that one may consider to avoid the bound in eqn. (2.1):

a) $\nu_H \to \nu_e \gamma$: Cosmology implies the following bound on this decay mode. Unless the half-life is less than[8a] $10^4$ sec., the decay photons will disintegrate the deuterium and affect the isotropy of the observed black body radiation. In gauge models, these decay times are too long to be useful in avoiding the cosmological bounds. (See later). Therefore, much attention has been focussed in the following two decay modes:

b) $\nu_H \rightarrow \nu_e + \chi$, where $\chi$ is the Majoron,[9] which represents the Goldstone boson corresponding to spontaneous breaking of global $B - L$ quantum number. These decay modes are known[10] to lead to life-times much shorter than the photonic decay mode.

c) $\nu_H \rightarrow 3\nu_e$: This mode, in certain models such as the left-right symmetric ones, is important for consistency of neutrino masses with cosmology but often requires fine tuning of parameters in gauge models[11] since by an $SU(2)_L$ rotation, $\nu_\mu \rightarrow 3\nu_e$ gets related to $\mu \rightarrow 3e$ decay, which is known to have a tiny upper bound on its branching ratio ( $\lesssim 10^{-11}$). In some models with Majorana neutrinos $\nu_H \rightarrow 3\nu_e$ can be induced via flavor changing neutral current coupling $Z$, but these effects are small.

It is important to point out that arbitrary neutrino mass can be generated by including the right-handed neutrino in the standard model; but the decay modes in cases (b) and (c) involve new physics beyond this and are therefore extremely important. For instance, evidence for neutrino mass in the forbidden range ($10^{-7}$ to 2 GeV) combined with cosmological constraints will imply the existence of the massless Majoron or an ultralight neutral Higgs boson (in the Gev range).[12]

**(b) Constraints Implied by Solar Neutrino Puzzle:** Another class of constraints on the properties of the neutrino can be inferred from recent discussions of the solar neutrino puzzle. To present those constraints, we briefly remind the reader about the essential aspects of solar neutrino puzzle. It is well-known that sun shines because of hydrogen burning to helium in the hot core of the sun ($T_{core} \sim 1.5 \times 10^7\,{}^\circ K$, $E \sim 1.3$ kev). In this process, two electron neutrinos are emitted for every 26.5 Mev of radiation emitted. From the total luminosity of the sun, one can conclude that there is a flux of roughly $10^{10}$ $\nu_e/cm^2$ sec. on earth. These neutrinos cover the entire energy range from zero to 14 Mev, with most of them concentrated in the lower energy range.

The experiment of Davis[13] exposes $C_2Cl_4$ to this neutrino beam, and is sensitive to the higher energy Boron $\nu_e$'s, which cause the reaction $\nu_e + {}^{37}Cl \rightarrow {}^{37}Ar + e^-$. According to extensive theoretical calculation of Bahcall et al[14] using the standard solar model, the Davis experiment should see 7.9 ± 2.4 SNU's (1 SNU = $10^{-36}$ captures/atom sec.) On the other hand, only 2.1 ± 0.4 SNU's are detected in the Davis experiment. This discrepancy between theory and observation is known as the solar neutrino puzzle.

It may be that the solar core where the high energy Boron neutrinos are produced via the reaction ${}^8B \rightarrow {}^8B_e^* + e^+ + \nu_e$ has a temperature ($T_c$) lower than is assumed in the solar model due to some yet unknown mechanism. A 15% reduction in $T_c$ would be enough to reconcile data with theoretical predictions.

A more interesting possibility from the point of view of particle physics is that the discrepancy has to do with the properties of the neutrino. If we accept this point of view, an immediate conclusion is that the neutrinos must be massive. Furthermore, they must have one of the following properties such that after their production in the central core, most of them oscillate into another paticle which has no observable interaction with chlorine. The possible candidates for such particles in known particle physics models are: $\nu_\mu$, $\nu_\tau$, right-handed neutrino $\nu_R$ or singlet neutrino $\nu_s$. Depending on which particle we assume, we obtain a certain constraint on the observed left-handed neutrino properties, summarized below:

**(i) vacuum oscillation:**

If $\nu_e$, $\nu_\mu$ and $\nu_\tau$ mix among themselves, a fraction of $\nu_e$ will change their flavor in transit from sun to the earth. This probability is given by (for the case of two flavors)

$$|\langle \nu_e(t) \mid \nu_e(o)\rangle|^2 = 1 - \frac{1}{2} sin^2 2\theta \left(1 - cos \frac{\Delta m^2 t}{2k}\right) \tag{2.3}$$

Using $t = L/c$, ($L$ is the Earth-Sun distance) one can immediately conclude that a sixty percent reduction in the neutrino signal corresponds to $\Delta m^2 \simeq 10^{-10}\ ev^2$. If we do not assume any finetuning of parameters, this implies that, $m_{\nu_e} << m_{\nu_\mu} \simeq 10^{-5}$ eV, for maximal mixing ($\theta = \frac{\pi}{4}$). This kind of neutrino spectrum is not accessible to laboratory experiments.

**(ii) Vacuum oscillation into sterile neutrinos:**

An interesting variation of the above idea is that neutrino is a pseudo-Dirac particle (see sec. 3), with a Majorana mass mixing $\delta m \simeq 10^{-11}$ eV. In such a case, the $\nu_{eL}$ will oscillate into its Dirac partner in transit from sun to the earth if $m_{\nu_e}\delta m \simeq 10^{-10}\ eV^2$ or for $m_{\nu_e} \simeq 10$ ev.[16] If the Dirac partner happens to be sterile with respect to weak interaction, it will escape detection. This scheme has two advantages over case (i); first, here the neutrino mass being in the eV range is accessible to in the laboratory experiments; secondly, the maximal mixing (i.e. $\theta = 45°$) arises in a natural manner.[16]

**(iii) Neutrino oscillation in matter:**

This solution was proposed by Mikheyev and Smirnov[17] using a mechanism proposed by Wolfenstein.[17] According to this proposal, neutrino scattering in a medium changes the form of the neutrino mass matrix. As a result, neutrino flavor oscillation can occur for a much wider range of mass differences for smaller values of laboratory mixing angles between $\nu_e$ and $\nu_{\mu,\tau}$. Detailed analysis has yielded[18] the range of masses and mixings that can solve the solar neutrino puzzles to be the following:

$$\Delta m^2_{ei} \simeq 3 \times 10^{-5} - 10^{-4}\ ev^2$$

and

$$sin^2 2\theta_{ei} \geq 10^{-3}\ . \tag{2.4}$$

Again, we assume no finetuning, this would imply that, $m_{\nu_e} << m_{\nu_\mu} \simeq 10^{-2}$ ev or $m_{\nu_e} << m_{\nu_e}$ $\simeq 10^{-2}$ ev. As is clear, this spectrum (to be called MSW spectrum in this article) is different from the above two cases.

**(iv) $\nu_e$-Decay:** If the electron neutrino decays on its way to the earth with an appropriate life time that would also provide a resolution of the solar neutrino puzzle. This possibility was originally proposed by Bahcall et al;[19] but at that time no plausible particle physics model was known for such decay. With the introduction of the idea of the Majoron,[9] this possibility was revived again, since a model where the decay of type $\nu_e \to \nu_{\mu,\tau}$ + Majoron occurs could now be constructed.[20] A solution of the solar neutrino puzzle requires that

$$\gamma \tau_{\nu_e} \simeq 500 \ sec. \tag{2.5}$$

where $\gamma$ is the relativistic time dilation factor.

**(v) Magnetic Moment of the Neutrino:**

A quite fascinating resolution of the solar neutrino puzzle with several interesting and testable predictions is the suggestion[21,22] that the neutrino is a Dirac particle with a non-vanishing magnetic moment $\mu_\nu$. In such a case in the magnetic field of about $10^4$ Gauss in the convective zone of the sun, the left-handed neutrino emitted from the solar core undergoes precession thereby becoming a right-handed neutrino which has no visible weak interactions. A value of $\mu_\nu \approx (3-1) \times 10^{-10}\ \mu_B$ (where $\mu_B = e\hbar/2m_e c$) is required to flip the helicity of 60% of the emitted neutrinos. This value of the magnetic moment is close to the upper bound, obtained from considerations of energy loss from stars[23] as well as nucleosynthesis,[24] and was therefore acceptable at the time it was proposed. This model predicts that the intensity of solar neutrinos observed on earth should be correlated with the 11 year sun-spot cycle and that, there should be a half yearly variation. However, as we will see below, such a large value of $\mu_\nu$ may not be consistent with observations from supernova.

**(c) SN1987a Constraints:**

Let us now turn to the information gained from the observations of neutrino signals in the underground detectors at Kamioka mine in Japan[25] and by the IMB group in Ohio[26] from the supernova 1987a. As is well known, neutrino emission is one of the primary mechanisms for energy loss in stars. In the case of a dying star, whose iron core collapses leading to a neutron star or a black hole, the neutrino emission accounts practically for 99% of the emitted energy, $Q$. Since the core contracts mainly through gravitational attraction, this energy $Q$ represents the binding energy of the core. Using the information that core mass $M_c \simeq (1.4-2)M_O$ and core radius $R_c \simeq 10$ Km., the binding energy $B \cdot E \equiv Q$ is

$$Q = \frac{G_N M_C^2}{R_C} \simeq (2-4) \times 10^{53}\ .ergs \tag{2.6}$$

In the standard model of quark-lepton interactions, the neutrinos carry this energy to the earth in the form of $\nu_e$, $\bar{\nu}_e$, $\nu_\mu$, $\bar{\nu}_\mu$, $\nu_\tau$ and $\bar{\nu}_\tau$. These neutrinos interact with water Cherenkov detector in the earth to produce electrons (via $\nu_e e^- \to \nu_e e^-$ reaction) or positrons via $\bar{\nu}_e p \to e^+ n$ reactions. From the observed $e^\pm$ energy and angular distribution one can infer the energy of the incident $\nu_e$ and $\bar{\nu}_e$. From the references 25 and 26, we conclude that perhaps all the events are due to $\bar{\nu}_e$ interactions with incident $\bar{\nu}_e$ energy $E_{\bar{\nu}_e} \simeq$ 10-20 Mev. From this, one can infer that almost all the energy released in gravitational collapse of the core is emitted in the form of neutrinos. It is, therefore, to be expected that any new property of the neutrino not implied by the standard model will be constrained by these observations. We summarize some of these constraints below.

**(i) Constraints on $m_{\nu_e}$:** For an anti-neutrino, $\bar{\nu}_e$ of mass $m_{\bar{\nu}_e}$ and energy $E_{\bar{\nu}_e}$, its transit time is given by

$$t \simeq \frac{D}{c}\left[1+\frac{1}{2}\frac{m_{\bar{\nu}_e}^2}{E_{\bar{\nu}_e}^2}\right]. \tag{2.7}$$

Since most of the neutrino arrive (at Kamioka) in a bunch of about 8 within a time space of 2 sec., and typical core collapse time is of the order of a second or two, the $\bar{\nu}_e$-mass must be constrained[27] to be less than 10-30 eV.

**(ii) $\tau_{\nu_e}$:** From the mere fact that the electron neutrinos arrive on earth after transversing 50 Kpc (which is about $1.7 \times 10^5$ light years), one easily concludes that,

$$\gamma\tau_{\bar{\nu}_e} \geq 1.7 \times 10^5 \ yrs. \tag{2.8}$$

By CPT theorem, $\nu_e$ must have the same bound on it lifetime, thereby ruling out neutrino decay as a solution of the solar neutrino puzzle.

**(iii) Magnetic moment of the neutrino $\mu_{\nu_e}$:**

It has recently been shown[28] that SN1987a observations severely constrain $\mu_{\nu_e}$. The reason for this is that for large values of $\mu_{\nu_e}$, the degenerate $\nu_{eL}$'s in the supernova core can flip helicity to $\nu_{eR}$ via the photon exchange reaction $\nu_L e^- \to \nu_R e^-$. But since $\nu_R$ interctions are very weak, the meanfree path of $\nu_R$'s produced exceeds the core radius (unless there is a neutral right-handed gauge boson $Z_R$ of mass $M_{Z_R} \leq 300$ Gev). The emitted $\nu_R$'s, therefore, have high energy (i.e. $E_{\nu_R} \approx$ 100-200 Mev) and they escape the core without further interaction, thus providing an additional mechanism of energy loss from the supernova. Demanding that $Q_{\nu_R} < Q$, we obtain[28]

$$\mu_{\nu_e} \leq 10^{-12}\,\mu_B \tag{2.9}$$

It further turns out that in the Galactic magnetic fields of $10^{-6}$ Gauss, the emitted high energy $\nu_R$'s can flip helicity to become $\nu_L$ before they reach the earth, providing possibilities of high energy neutrino signals ($E_{\nu_e} \geq 100$ Mev) in the underground detectors. Again absence of any such signals implies a stronger bound

$$\mu_{\nu_e} \leq 10^{-13}\,\mu_B\,. \tag{2.10}$$

This rules out neutrino magnetic moment as an explanation of the solar neutrino puzzle, (unless, of course, there is a $Z_R$ of mass less than 300 Gev in which case, the emitted $\nu_R$'s get trapped in the supernova core).

**(iv) Constraints on neutrino radiative decays and neutrino mixings::** Finally, we turn to the constraints on radiative neutrino decays and neutrino flavor mixing from supernova observations. If the heavier neutrinos $\nu_\mu$ and $\nu_\tau$ decayed into $\nu_e$ + photon, photons would be observable; however, lack of any signal in the kev and Mev range[29,30] has been shown to imply rather stringent constraints on allowed values of neutrino mixings.

Laboratory experiments also have led to constraints on neutrino masses as follows:[31] $m_{\nu_e} \leq 18$ ev, $m_{\nu_\mu} \leq .25$ Mev and $m_{\nu_\tau} \leq 70$ Mev. There exist constraints on the mixing angles from various accelerator experiments, which we do not include in this article, since we will not be using it in the subsequent sections.

## 3. Majorana, Dirac or Pseudo-Dirac Neutrino Masses:

Unlike a boson or an electrically charged fermion, a neutral spin half object such as the neutrino can have several kinds of masses. To understand this, we consider 2-component Weyl spinor $\nu$ to denote the chiral spin 1/2 neutrino emitted in the process of $\beta$-decay. It transforms as a $\left(\frac{1}{2}, 0\right)$ representation of the Lorentz group and admits a Lorentz invariant mass term such as:

$$\mathcal{L}^{(1)}_{mass} = i\, m\, \nu^T \sigma_2 \nu + \text{hermitean conjugate} . \tag{3.1}$$

If we assume that the neutrino carries a global quantum number such as lepton number (under which $\nu \rightarrow e^{i\alpha}\,\nu$), then $\mathcal{L}^{(1)}_{mass}$ violates lepton number by two units. In this case, the neutrino is said to have a Majorana mass and is called a Majorana particle. If the only spin half neutral fermion (with properties similar to the neutrino) around is $\nu$, then no other mass term is possible for the neutrino. So, the neutrino will either remain massless or become a Majorana particle. On the other hand, if there is another spin 1/2 neutral lepton $N$ present, then, one can envisage a generalized form for the mass term:

$$\mathcal{L}^{(2)}_{mass} = i \left(\nu^T\, N^T\right) \begin{pmatrix} m_{LL} & m_{LR} \\ m_{LR} & m_{RR} \end{pmatrix} \sigma_2 \begin{pmatrix} \nu \\ N \end{pmatrix} \tag{3.2}$$

(We will denote this $2 \times 2$ mass matrix in eqn. (3.2) by $M$). If all elements of $M$ are not-vanishing, again no global symmetry will be respected by $\mathcal{L}^{(2)}_{mass}$; therefore, in this case, the neutrino will also be a Majorana particle and lepton number will be violated by two units. One can, however, have special situations:

a) $m_{LL} = m_{RR} = 0$: In this case, one can define the lepton number $U(1)$ symmetry as follows: $\nu \rightarrow e^{i\alpha}\,\nu$ and $N \rightarrow e^{-i\alpha}\,N$. $\mathcal{L}^{(2)}_{mass}$ will then respect lepton number. The $\nu$ and $N$ combine to form a four-component Dirac neutrino. In the two component notation, one has the following eigenstates:

$$\nu_\pm = \frac{\nu \pm N}{\sqrt{2}} \tag{3.3}$$

with mass eigenvalues $\pm m_{LR}$; making the mass positive changes $\nu_-$ to $i\nu_-$. In the four component language, we write,

$$\psi = \begin{pmatrix} \nu \\ i\sigma_2\, N^* \end{pmatrix} . \tag{3.4}$$

Then,

$$\psi^c = C\bar{\psi}^T = -\gamma_2\psi^\star = \begin{pmatrix} N \\ i\sigma_2\nu^\star \end{pmatrix} . \tag{3.5}$$

We then define:

$$\psi_\pm = \frac{1}{\sqrt{2}}\left(\psi \pm \psi^c\right) = \begin{pmatrix} \nu_\pm \\ i\sigma_2\nu_\pm^\star \end{pmatrix} . \tag{3.5}$$

From (3.5), it is obvious that, $\psi_+$ and $\psi_-$ are even and odd under charge conjugation respectively.

b) $m_{LR} >> m_{LL} = m_{RR}$: In this case, the neutrino is called a pseudo-Dirac particle.[32] In this case there is a slight mixing between the CP-odd and CP-even eigenstates $\psi_\pm$; although the neutrino is a predominantly Dirac particle, it has a very slight admixture of Majorana character. This situation is of particular interest in the study of evolution of a chiral left-handed neutrino state, such as that emitted in beta decay. When $m_{LL}$ or $m_{RR}$ are nonzero (and small, $m_{LL}$ , $m_{RR} << m_{LR}$), the eigenstates of $M$ are still $\nu_\pm$; but their masses are split by an amount $\delta m$, i.e.

$$m_\pm = \mp m_{LR} + m_{LL} . \tag{3.7}$$

The left-handed neutrino oscillates into the extra sterile lepton $N$, with a probability given by

$$P_{\nu N}(t) \simeq sin^2 \frac{1}{2}\left(\frac{m_{LL}}{R}\right)\left(\frac{m_{LR}}{E}\right) t . \tag{3.8}$$

As discussed earlier if $m_{LL} \times m_{LR} \simeq 10^{-10}$ eV$^2$, this can provide an alternative solution to the solar neutrino problem.[16]

c) $m_{LL} = 0$; $m_{LR} << m_{RR}$: In this case, the neutrino is a Majorana particle. But unlike cases (a) and (b) where the two neutrino states are nearly degenerate, in this case, there are two neutrinos, one heavy ($m_H \simeq m_{RR}$) and one light ($m_L \simeq m_{LR}^2/m_{RR}$). This is well known as the "see-saw" mechanism[33] and provides a "naturally" small value for the neutrino mass without any finetuning of parameters. The eigenstates are also nearly pure, i.e.

$$\nu_{Light} \simeq \nu + \left(\frac{m_{LR}}{m_{RR}}\right) \cdot N$$

$$\nu_{Heavy} \simeq N - \left(\frac{m_{LR}}{m_{RR}}\right) \nu . \tag{3.9}$$

Before closing this section, we wish to reemphasize the point that Majorana mass terms require the breaking of Lepton number, $L$. Since in general gauge theories the only gauge-anomaly free combination of quantum numbers is $B-L$, we will talk about breaking of $B-L$ symmetry. In the context of gauge theories, there are three possible ways to achieve this breaking:

(i) Explicit breaking of $B-L$ symmetry, where Lagrangian contains terms that break $B-L$ symmetry.

(ii) Spontaneous breaking of local $B-L$ symmetry[2]

(iii) Spontaneous breaking of global $B-L$ symmetry.[9]

In sec. 4-7 of the article, we will focus on the local $B-L$ symmetry models (case ii) and its connection to neutrino mass. In sec. 8, we return to case (iii), where a Goldstone boson (the Majoron[9]) appears in the theory coupled to the neutrino. This particle leads to many interesting phenomenological and cosmological implications for neutrino decay and neutrino scattering. For instance, heavier neutrinos can now decay via the emission of Majoron such as $\nu_H \to \nu_e + \chi$; there can also be neutrino annihilation such as $\nu\nu \to \chi\chi$. These new processes effect cosmological constraints on the neutrino masses. In the laboratories, one can have new modes for neutrinoless double $\beta$-decay such as $(N,Z) \to (N-2, Z+2) + 2e^- + \chi$. We discuss these questions in sec. 8.[17]

## 4. Brief Description of Left-Right Symmetric Models:

In the standard model of Glashow, Weinberg and Salam, only one helicity component is introduced into the theory to fit observations in muon and beta decay. Since the model has exact $B-L$ symmetry, it follows from the discussions of sec. 3 that neutrinos must be massless. In this model, the left-handed nature of weak interactions is also built in by hand, because of which no fundamental understanding of the origin of parity violation emerges. In order to provide an understanding of parity violation, the left-right symmetric theories of weak interactions were developed in a series of papers[1] in 1974-75. In these models, the basic interactions are invariant under parity transformations: the observed near maximal parity violation at low energies is then understood to be a consequence of spontaneous symmetry breaking. Before proceeding to the discussion of neutrino masses in this model, we present the basic ideas of the model and review the constraints on the new parameters of the model such as the masses of the right-handed gauge bosons, right-handed neutrino masses, etc.

### Motivation and description of the Model

There is a rather compelling reason why left-right symmetry ought to be a fundamental and natural symmetry of the quark-lepton world. It is well-known that weak interactions are symmetric between quarks and leptons. We can use this symmetry to write a unified formula for electric charge in terms of weak isospin and $B-L$ quantum number, in a manner analogous to the Gell-Mann-Nishijima formula for the hadronic world, i.e.

$$Q = I_{3W} + \frac{B-L}{2} \,. \tag{4.1}$$

Since at low energies electroweak processes involve only the left-handed weak isospin, we must rewrite the eqn. (4.1) as

$$Q = I_{3L} + I_{3R} + \frac{B-L}{2} \,. \tag{4.2}$$

It is then natural to assume that Nature realizes the full right-handed weak isospin in the same way as it realizes the left-handed one. This then leads to the left-right symmetric model of weak interactions, based on the gauge group $SU(2)_L \times SU(2)_R \times U(1)_{B-L}$.[1] The quarks and leptons are assigned to the gauge group as follows: (denote $Q \equiv (u,d)$ and $\psi \equiv (\nu, e^-)$)

$$Q_L : (2,1,1/3)\,; \quad Q_R : (1,2,1/3)\,; \quad \psi_L : (2,1,-1)\,; \quad \psi_R : (1,2,-1)\,. \tag{4.3}$$

(where the numbers within the bracket represent the transformation property under the gauge group). It is clear that gauge interactions are symmetric between left and right-handed fermions; therefore, prior to spontaneous breaking of gauge symmetry, weak interactions like strong, electromagnetic and gravitational interactions conserve parity. This can be seen explicitly, by writing the gauge interactions

$$\mathcal{L}_{WK} = \frac{g}{2}\left[\vec{J}_{\mu,L}\cdot\vec{W}_L^{\mu} + \vec{J}_{\mu,R}\cdot\vec{W}_R^{\mu}\right] + g'V_\mu^{B-L}\cdot B^\mu\,. \tag{4.4}$$

The $\vec{W}_L$, $\vec{W}_R$ and $B$ are the gauge bosons corresponding to $SU(2)_L$, $SU(2)_L$ and $U(1)_{B-L}$ gauge symmetries. As long as the $W_L$ and $W_R$ have the same mass, the weak interactions conserve parity. The observed parity violation at low energies is then attributed to vacuum being parity asymmetric, which manifests itself in $m_{W_R} >> m_{W_L}$. The symmetry breaking pattern responsible for the parity violation observed at low energies can be obtained using the following generic pattern for the Higgs multiplets: (we drop the reference to color gauge group from now on) $\phi$: $(2,2,0)$, $H_L\,(a,1,b) + H_R\,(1,a,b)$ where $a$ and $b$ will be fixed later. The first stage of the symmetry breaking is implemented by the neutral component of the right-handed multiplet $H_R$ acquiring a vacuum expectation value (v.e.v.) i.e., $\langle H_R^{\circ}\rangle = v_R$. This reduces the electroweak symmetry to $SU(2)_L \times U(1)_Y$, which is subsequently broken by $\phi$ acquiring a v.e.v.

$$\langle\phi\rangle = \begin{pmatrix} \kappa & 0 \\ 0 & \kappa' e^{i\alpha} \end{pmatrix}\,. \tag{4.5}$$

The second stage of the symmetry breaking can induce[18] a v.e.v. for $\langle H_L^{\circ}\rangle = v_L \simeq \gamma\,\kappa^2/v_R$ assuming $\kappa' >> \kappa$; the quarks and leptons acquire a mass at the second stage, as do the $W_L$ and $Z$-bosons. The form of the low energy charged current interaction in this model is as follows:

$$H_{WK}^{c.c} = \frac{G_F}{\sqrt{2}}\left[\left(\cos^2\zeta + \eta\sin^2\zeta\right) J_{\mu_L}^{+}\, J_L^{\mu^-} + \left(\eta\cos^2\zeta + \sin^2\zeta\right) J_{\mu_R}^{+}\, J_R^{\mu^-}\right.$$
$$\left. + e^{i\alpha}\cos\zeta\,\sin\zeta\,(1-\eta)\, J_{\mu_L}^{+}\, J_R^{\mu^-} + \text{h.c.}\right] \tag{4.6}$$

where $\eta = \left(\frac{m_{W_L}}{m_{W_R}}\right)^2$ and $\zeta$ is the mixing angle between the charged $W$-bosons and is given in terms of symmetry breaing parameters as $\zeta \simeq \left(\kappa\kappa'/m_{W_R}^2\right)$. The currents $J_{\mu L,R}$ are given as follows:

$$J_{\mu L}^{+} = \overline{P}\,\gamma_\mu\,(1+\gamma_5)\,U_L N + \overline{E}^{o}\,\gamma_\mu\,(1+\gamma_5)\,V_L E^- \tag{4.7}$$

where $P$, $N$, $E^o$ and $E^-$ denote the up quark, down quark, neutrino and electron column vector involving all generations and $U_L$ and $V_L$ denote the charged current mixing matrices for the quark and lepton sector. The corresponding right-handed currents are obtained by replacing $L$ by $R$ and $\gamma_5$ by $-\gamma_5$. We would now like to confront the Hamiltonian in eqn.

(4.6) with experiment; and since weak processes involving leptons are less complicated by strong interactions, we would like to study purely leptonic and semileptonic processes to gain information and $\eta$ and $\zeta$. For this purpose, we need to know the nature of the neutrino and the structure of the weak right-handed current, specifically the quark and lepton mixing matrix in the right handed sector. As we saw in sec. 3, in a theory with two spin half neutral leptons, many possibilities exist for the neutrino masses. We must therefore study the experimental limits on $W_R$, $Z_R$ and $\nu_R$ for various cases separately.

**Limits on the Masses of the $Z_2$, $W_R$ and the Right-Handed Neutrino:**

To search for the signatures of new physics associated with the right-handed symmetry of weak interactions, we must determine to what extent known low energy physics restricts the mass of $Z_2$, $W_R$ and $\nu_R$. The most model independent limit arises on the mass of the $Z_2$-boson. The form of the neutral current interaction of $Z_1$ and $Z_2$ can be written in these models in the form[35,36]

$$\mathcal{L}_{WK}^{N.C.} = \frac{g}{\cos\theta_W}\left[Z_{1\mu}\left\{J_L^\mu + \eta_Z J^\mu\right\} + \frac{1}{(\cos 2\theta_W)^{1/2}} Z_{2\mu} J^\mu\right] \tag{4.8}$$

where $\eta_Z = (M_{Z_1}/M_{Z_2})^2$ and $J = \sin^2\theta_W J_L + \cos^2\theta_W J_R$; $J_{L,R} = \left(I_{3L,R} - Q\sin^2\theta_W\right)$. We, then, see that the effects of right-handed bosons vanish as $M_{Z_2} \to \infty$ and we obtain the neutral current interaction of the standard model. Typical precision of neutral current experiments is about 10%, which allows one to have a rather light $Z_2$; a recent analysis[36] yields $M_{Z_2} \geq 275$ Gev for these models.

As far as the charged $W_R$-boson is concerned, we consider both the light Dirac and Majorana cases. And, we will further assume that the left and right handed mixing matrices are equal. This is the case of manifest left-right symmetry[37] which emerges for the simplest choice of Higgs multiplets $\phi$. In this case, if the neutrino is a Dirac particle with $\nu_L$ and $\nu_R$ forming this particle or is a Majorana particle such that $m_{\nu_R} \leq 10$ Mev, the most model independent bounds on $m_{W_R}$ and $\zeta$ come from the analysis of a recent TRIUMF muon decay experiment.[38] In this experiment the endpoint spectrum of the positrons moving along a direction opposite to the spin of a stopped 100% polarized $\mu^+$ is measured. In pure $V-A$ theory, this spectrum vanishes at the end point, thus any non-vanishing would be an indication of the presence of $V+A$ currents. The quantity measured this way, to be called $R$ is given in terms of the muon decay parameters $\delta$, $\xi$ and $\rho$ as

$$R = 1 - \frac{\delta\xi}{\rho} P_\mu \tag{4.9}$$

where in terms of the $\eta$ and $\zeta$ defined after eqn. (4.6) we get

$$R = 4\eta^2 + 2\zeta^2 + 4\eta\zeta\,. \tag{4.10}$$

The experimental result for $1-R = .99863 \pm .00046$ (stat) $\pm .00075$ (syst.). This implies that $m_{W_R} \geq 432$ Gev for arbitrary $\zeta$ and $m_{W_R} \geq 514$ Gev for $\zeta = 0$; for arbitrary $m_{W_R}$, -0.050 $\leq$

$\zeta \leq .035$ and $|\zeta| \leq .035$ for $m_{W_R} \to \infty$. These analyses have recently been extended[39] to the case of non-manifest left-right symmetry,[40] where bounds on $m_{W_R}$ become weaker.

Turning now to the case of the heavy right-handed neutrino (i.e. $m_{\nu_R}$ in the Gev range), the only bound comes from non-leptonic decays such as $K$-decays[44] and $K_L - K_S$ mass difference[42] ($\Delta M_K$), the latter case leading to the most stringent bound. The reason for this bound is that calculation of the one loop $\Delta S = 2$ effective Hamiltonian arising from the $W_L - W_R$ exchange leads to an enhanced contribution to $\Delta M_K$ as follows

$$\Delta m_K = \Delta m_{K_{LL}} \left[1 - 430\,\eta\right] . \tag{4.11}$$

Thus we see that unless $\eta \leq 2.3 \times 10^{-3}$ or $m_{W_R} \geq 1.6$ Tev, eqn. (4.11) will give the wrong sign and contradict observations. This is the most stringent bound on $m_{W_R}$.

Coming now to the bound on the mixing parameter $\zeta$, we have already seen that for Dirac neutrinos muon decay experiments provide a bound $\zeta < 3.5\%$. For heavy right-handed neutrinos, the most model independent bound comes from the study[43] of $y$-distribution in deep inelastic scattering of antineutrinos off nuclei. In the absence of $V + A$ current, the valence quarks lead to $(1-y)^2$ type distribution, whereas the left-right mixing leads to flat $y$-distribution proportional to $\zeta$. Analysis of existing experiments[44] lead to $\zeta < .1$. Other more model dependent bounds can be obtained by considering deviations from current algebra relations in non-leptonic decays, ($\zeta \leq 4 \times 10^{-3}$) as well as from beta decays[45] of nucleus combined with unitarity of the $KM$-matrix ($\zeta < .005$). Finally, we note that with the simplest Higgs structure (one $\phi$ and $\Delta_{L,R}$), one has the following theoretical upperbound[46] on the mixing parameter $\zeta \leq \left(\frac{m_{W_L}}{m_{W_R}}\right)^2$. If we use the bound from the $K_L - K_S$ mass difference, it implies $\zeta \leq 2 \times 10^{-3}$, which is the most stringent, though model dependent bound on this parameter.

**Bound on the mass of the Majorana Neutrino $\nu_R$:**

As mentioned earlier, the most natural way to understand small neutrino mass is to assume that the right-handed neutrino is a heavy Majorana particle. While the smallness of $m_\nu$ implies that $m_{\nu_R}$ is in the tens of Gev range, we would like to discuss the limits on $m_{\nu_R}$ from phenomenological constraints. An interesting limit that correlates $(m_{\nu_R})_{min}$ with $m_{W_R}$ arises from present experimental limits[47] on neutrinoless double $\beta$-decay.[48] The point is that a heavy right-handed neutrino contributes to $(\beta\beta)_{ov}$ decay via the exchange of $W_R$ boson with an amplitude proportional to $G_F^2\, \eta^2\, m_{\nu_R}\, \langle e^{-m_{\nu_R}^{r}/r} \rangle_{nuc.}$ and this contribution adds incoherently with the usual light neutrino contribution. Using the available nuclear matrix element calculations,[49] we obtain the bound shown in fig. 1. An upper limit on $m_{\nu_R} \leq m_{W_R}$ arises from considerations of vacuum stability, which is the right-hand line in fig. 1. Also, we infer from Fig. 1 that, combination of vacuum stability and $(\beta\beta)_{ou}$ decay lead to a lower bound on $M_{W_R}$ of 800 Gev.

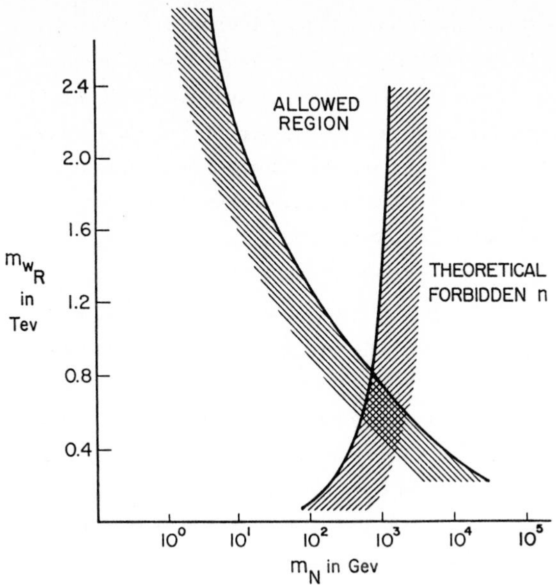


Fig. 1. Correlated bounds on $m_{N_R}$ and $m_{W_R}$ from experiments on neutrinoless double beta decay (left arm) and considerations of vacuum stability. The shaded region is forbidden.

## 5. Dirac vs. Majorana Neutrino in Left-Right Models:

Due to left-right symmetry, $\nu_L$ is accompanied by $\nu_R$ leading to a rich set of possibilities for the neutrino masses. The detailed nature, however, depends on the pattern of symmetry breaking which must lead to the standard model at low energies. We see that a crucial role is played by the Higgs boson that is responsible for breaking the $SU(2)_R$ symmetry, i.e. the nature of the Higgs multiplets $H_L$ and $H_R$. Two possible choices of $H_L$ and $H_R$ can be considered: the first one which we will call the canonical choice was introduced in ref. 35 (case i) and it leads to Majorana neutrinos and provides an understanding of the smallness of the neutrino mass being related to the suppression of $V+A$ currents. The second one we mention here leads to light Dirac neutrinos (case ii) at the price of expanding the model to include singlet leptons.[50]

Case (i): Here, $H$ are chosen to be triplets under the weak gauge group and are denoted by $\vec{\Delta}_L(3,1,2)+\vec{\Delta}_R(1,3,2)$. They couple to leptons only as follows:

$$\mathcal{L}'_Y = h\left\{\psi_L^T\, c^{-1}\,\tau_2\vec{\tau}\cdot\vec{\Delta}_L\psi_L + \psi_R^T\, C^{-1}\,\tau_2\vec{\tau}\cdot\vec{\Delta}_R\psi_R\right\} + \text{h.c.} \tag{5.1}$$

Choosing $\langle\Delta_R^\circ\rangle = v_R$ and assuming that the neutrino Dirac mass coming from the $SU(2)_L$ breaking is $m_e$, we obtain the following $2\times 2$ mass matrix for $(\nu, N)'$

$$\begin{array}{c} \\ \nu \\ N \end{array}\begin{array}{c} \begin{array}{cc} \nu \qquad & \qquad N \end{array} \\ \begin{pmatrix} h\gamma\,\kappa^2/v_R & m_e \\ m_e & h\,V_R \end{pmatrix} \end{array}. \tag{5.2}$$

This mass matrix provides a qualitative explanation for the small neutrino mass, which comes out to be $m_\nu \simeq (h\gamma\kappa^2 - m_e^2)/v_R$; according to this formula, $m_\nu \to 0$ as $v_R \to \infty$ in the absence of right-handed weak interactions; it is, therefore, aesthetically pleasing that if neutrino has a small mass, it is not just a freak accident of nature but is related to a new intrinsic property of weak interactions being ultimately parity conserving. Turning this question around, we could say that once mass of the neutrino is known, that would provide a rough idea about the mass scale for right-handed interactions. There is, however, one technical difficulty with eqn. (5.2); to get acceptable values for neutrino mass (i.e. ev or less), we will have to tune the parameter $\gamma \simeq 10^{-6}$ or so. It turns out that such a small value arises automatically,[51] if parity symmetry is broken at a scale $M_P$ higher than the breaking scale of $SU(2)_R$ gauge symmetry.[52] In fact, in this case, we find, $v_L \simeq \frac{\kappa^2 v_R}{M_P^2}$ which is of order $10^{-10}$ or so for $v_R$ in the Tev range and $M_P \simeq 10^9$ Gev. This has the impact that at $\mu \simeq m_{W_R}$, $g_L \neq g_R$ and the multiplet $\Delta_L$ decouples from low energies. Alternatively, it is worth pointing out if $\gamma$ is set to zero at the tree level, then it receives contributions at the one loop level only from Yukawa couplings, which are much smaller than one ($\sim 10^{-2}$ or so). So if we assume that such one loop graphs are cut off in momentum at $\mu \simeq M_{Planck}$, we would expect $\gamma \simeq \frac{h^4}{16\pi^2} \ell n\,(M_P/m_W) \lesssim 10^{-8}$ or so, which may justify the finetuning adopted earlier.

Turning to eqn. (4.4), we note that the right-handed leptonic current involves the right-handed Majorana neutrino. As a result, whether the right-handed leptonic currents manifest themselves in low energy decay processes depends on the mass, $m_{N_R} \simeq h v_R$, which for natural values of coupling parameters is expected to be in the tens to hundreds of Gev range. We will derive constraints on this mass from neutrinoless double beta decay.

A further point worth discussing in connection with the mass matrix in eqn. (5.2) is that if the scale $v_R$ is high, then for arbitrary $\gamma$, we get $m_{\nu_i} \simeq h\gamma \cdot \kappa^2/v_R$. We find $m_{\nu_i} \approx 10$ eV requires for $h \simeq 10^{-3} - 10^{-4}$, $\gamma \simeq 10^{-1}$, $v_R \simeq 10^8 - 10^9$ Gev. An important characteristic of this point of view is that all light neutrinos can be nearly degenerate in mass, which could manifest themselves in substantial neutrino oscillations.

The mass eigenstates for neutrinos in this case (ignoring mixings between generations) is given by (setting $\gamma \simeq 0$)

$$\begin{aligned} \nu_e &\simeq \nu + \xi N \\ N_e &\simeq N - \xi \nu \end{aligned} \tag{5.3}$$

where

$$\xi \simeq (m_{\nu_e}/m_{N_e})^{1/2} .$$

Due to this admixture of the right-handed neutrinos, the GIM cancellation in this sector does not occur; as a result, one obtains flavor changing neutral current decays of $Z$ such as $Z \to \nu_e \bar{\nu}_\mu$. This, then, opens up a channel for heavier neutrinos to decay via $Z$-exchange, e.g. $\nu_\mu \to \nu_e \bar{\nu}_e \nu_e$. This decay amplitude is, however, small in practice (see below).

Turning to the neutrino masses, we find (in the simplest approximation), the following mass formula for the neutrinos, i.e.

$$m_{\nu_i} \simeq m_{e_i}^2/m_{N_i} \tag{5.4}$$

where $i$ denotes the generation and $\ell_i$ denotes the charged lepton. Assuming $m_{N_i}$ to be generation independent, we find that, $m_{\nu_e} : m_{\nu_\mu} : m_{\nu_\tau} = m_e^2 : m_\mu^2 : m_\tau^2$. This relation was first proposed in ref. 18. If we assume, $m_{\nu_e} \simeq 1$ to 5 ev, we obtain from the above scaling law, $m_{\nu_\mu} \simeq 40$ - 200 kev and $m_{\nu_\tau} \simeq 10$ - 50 mev. Now, recalling our discussion in sec. 2, we see that, for stable neutrinos, these values are in the forbidden range. Therefore, for the model to be acceptable, $\nu_\mu$ and $\nu_e$ must decay with typical life times of about $\tau_{\nu_\mu} \simeq 10^{12}$ sec. or less. The photonic decay mode, which occurs via a one loop graph[54] is bigger than $10^{16}$ sec. for $m_{\nu_\mu} \simeq 200$ kev. This is clearly much too long for our purpose. Let us, therefore, turn to the .$3\nu_e$ decay mode of $\nu_\mu$. This decay mode can arise in two ways in left-right symmetric models: one, via the flavor changing $Z$-couplings to neutrinos and two, via the exchange of neutral Higgs boson $\Delta_L^o$. The first mechanism leads to an effective $\nu_\mu \to 3\nu_e$ interaction with effective couplings strength $\sim G_F \left(m_{\nu_\mu}/m_{N_e}\right)^{1/2}$; since $G_F \left(m_{\nu_\mu}/m_\mu\right)^{1/2} \simeq 10^{-8}$ Gev$^{-2}$. This leads to a lifetime of order

$$\tau_{\nu_\mu} \geq 10^{17} \ sec. . \tag{5.5}$$

In the second mechanism, suggested in ref. 11 the strength of the four neutrino interaction is of order $\sim \left(h_{e\mu}\, h_{ee}/m^2_{\Delta_L^o}\right)$, where $m_{\Delta_L^o}$ is the mass of the neutral Higgs boson, $h_{\mu e}$ and $h_{ee}$ are the Yukawa couplings of $\Delta_L$ with leptons. Since $m_{\Delta_L^o}$ is unknown, it can be arbitrary. To see how big it is, we note that, to satisfy eqn. (2.2), we find the following constraint:[12]

$$h_{e\mu}\, h_{ee}/M^2_{\Delta_L^o} \geq 6 \times 10^{-6}\, Gev^{-2} \left(\frac{m_{\nu_\mu}}{40\, kev}\right)^{-3/2} . \tag{5.6}$$

By an $SU(2)_L$-rotation, the $\Delta_L^{++}$ gives rise to the highly suppressed decay mode $\mu \to 3e$, leading to the constraint that,

$$h_{\mu e}\, h_{ee}/m^2_{\Delta_L^{++}} \leq 10^{-10}\, Gev^{-2} . \tag{5.7}$$

Since $\Delta_L^{++}$ and $\Delta_L^o$ are members of the same $SU(2)_L$ multiplet, their mass difference can at most be 250 GeV. This, together with eqn. (5.6) and (5.7) implies that,[12] $m_{\Delta_L^o} \simeq$ few Gev. This can therefore be tested once the $Z$-width is measured since $B\,(Z \to \Delta_L^o\, \Delta_L^{o*})\,/B\,(Z \to \nu\bar{\nu})$ $\simeq 2$.

Finally, we turn to the Majoron decay mode of $\nu_\mu$ and $\nu_\tau$ in left-right symmetric models.[55] If the left-right symmetric model is extended by the inclusion of the left-right symmetric doublets $\chi_L\left(\frac{1}{2}, 0, 1\right)$ and $\chi_R\left(0, \frac{1}{2}, 1\right)$, then, the model can admit an extra global symmetry, (similar to the $B - L$ quantum number). Spontaneous breaking of this symmetry leads to the existence of a mass less Goldstone boson $\chi$. The Goldstone boson couples to neutrinos and allows for the decay mode $\nu_\mu \to \nu_e \chi$. The typical strength of this decay mode can be big enough[54] to avoid the constraints from cosmological mass density.

**Case (ii)** Now, we turn to the Dirac neutrinos in left-right symmetric models. As mentioned earlier, this requires the introduction of new singlet fermions $\theta_1$ and $\theta_2$ and new set of Higgs bosons $\chi_L\left(\frac{1}{2},0,1\right) \oplus \chi_R\left(0,\frac{1}{2},1\right)$. The existence of these multiplets enables one to write the following Yukawa coupling:

$$\mathcal{L}_Y = f\left(\bar{\psi}_L\,\chi_L\,\theta_1 + \bar{\psi}_R\,\chi_R\,\theta_2\right) + \mu\bar{\theta}_1\theta_2 + h.c.\,. \tag{5.8}$$

The first point we wish to note about eqn. (5.8) is that the parameter $\mu$ does not receive infinite contributions from higher loops and is, therefore radiatively stable. The significance of this observation is that if we fix $\mu$ to be $\simeq 1-10$ Mev it will not change in higher orders. Minimization of the potential leads to $\langle\chi_R\rangle \simeq v_R\,, \langle\chi_L\rangle \simeq \frac{\mu'\kappa}{V_R} \approx 10^{-2}\,\mu'$ where we choose $\mu' \approx \mu$. (This is because as $\mu \to 0\,, \mu' \to 0$). Then, $\langle\chi_L\rangle \simeq 10^{-4}-10^{-5}$ Gev. In the subsequent discussion, we will ignore this. We then obtain the following $4\times 4$ mass matrix:

$$\begin{array}{c} \\ \nu \\ \theta_1 \\ N \\ \theta_2 \end{array}\begin{array}{c} \begin{array}{cccc} \nu & \theta_1 & N & \theta_2 \end{array} \\ \begin{pmatrix} 0 & 0 & m_D & 0 \\ 0 & 0 & 0 & \mu \\ m_D & 0 & 0 & f\,V_R \\ 0 & \mu & f\,V_R & 0 \end{pmatrix} \end{array}\,. \tag{5.9}$$

This matrix leads to two Dirac neutrinos, one heavy with mass $\sim f\,v_R$ and another light, with mass, $m_\nu \sim m_D\mu/f\,v_R$. This light four component spinor has the correct weak interaction properties to be identified as the neutrino.

Coming to the mass spectrum, we find that for $f \simeq 1$ and $v_R \sim$ Tev, $m_\nu \simeq 1$ eV. To extend this model to include th higher generations, we include 3 pairs of $(\theta_1\,,\,\theta_2)$. If we assume $f$ and $\mu$ to be generation independent, then neutrino masses scale linearly with charged lepton masses. Thus, if $m_{\nu_e} \sim 10$ eV, we would have $m_{\nu_\mu} \sim 2$ kev and $m_{\nu_\tau} \sim 40$ kev. We again see the need for $\nu_\mu$ and $\nu_\tau$ to decay to avoid cosmological mass constraints.

Let us discuss the magnetic moment of neutrino. It is well-known that a Dirac neutrino can have a magnetic moment, $\mu_\nu$. In fact in the case conventional left-right models with Dirac neutrinos, $\mu_\nu$ has been calculated.[55] The interest in this parameter is due to recent observations by Okun, Voloshin and Vysotski[56] that the observed deficit in the number of solar neutrinos reaching the earth can be explained if $\mu_\nu \simeq 10^{-11}\,\mu_e$. In conventional left-right[1] symmetric models with no extra singlet neutrinos, the principal diagram contributing to $\mu_\nu$ is shown in fig. 2 and has the magnitude

$$\mu_\nu \simeq \frac{\alpha\,m_e}{8\pi\,sin^2\,\theta_W\,m_{W_L}^2}\,\xi \simeq 8\times 10^{-11}\,\xi\cdot\mu_e\,. \tag{5.10}$$

If we take the phenomenological upperbounds of $10^{-2}$ on $\xi$, $\mu_\nu \simeq 10^{-12}\,\mu_e$, which is smaller than that required for solar neutrino puzzle. Note that in a model like this, the smallness of the neutrino mass is put in by hand. If, however, we consider a model like the one proposed in this section, (where small $m_\nu$ arises without finetuning of parameters), we estimate

$$\mu_\nu \simeq \frac{\alpha}{8\pi\,sin^2\,\theta_W}\cdot\left(\frac{m_e}{m_W^2}\right)\left(\frac{m_D}{m_{W_R}}\right)\cdot\xi\,. \tag{5.11}$$

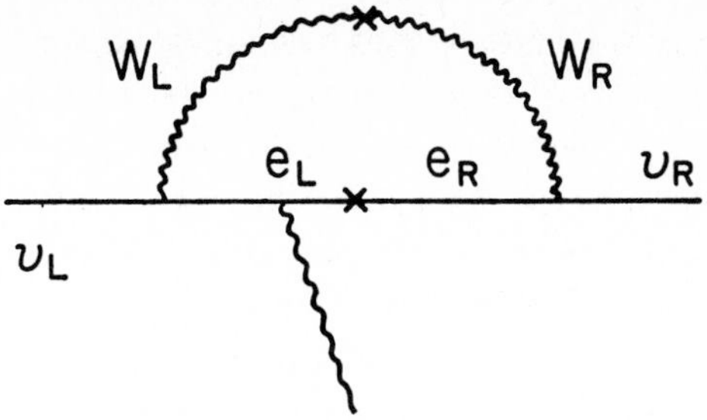


Fig. 2. The diagram contributing to $\nu_H \to \nu_e \gamma$ decay as well as the magnetic moment of the neutrino in left-right symmetric models.

Due to the additional suppression factor $(m_D/m_{W_R})$, which is of order $10^{-5}$, implying that, $\mu_\nu \simeq 10^{-17}\,\mu_e$, making it completely irrelevant for any discussion of solar neutrino puzzle.

Finally, we wish to note that one can construct a left-right symmetric model by addition of weak iso-singlet quarks and leptons such that the Dirac mass of the neutrino vanishes at the tree level and arises at the one-loop level via $W_L - W_R$ mixing.[40] In this case, the Dirac neutrino mass is given by:

$$m_{\nu_i} = \left(\frac{\alpha}{4\pi\, sin^2\,\theta_W}\right)^2 \left(\frac{m_b m_t}{m^2_{W_R}}\right) \cdot m_{\ell_i}\,. \tag{5.12}$$

For $m_{W_R} \simeq 10$ Tev (say), this will lead to $m_{\nu_\mu} \simeq 5 \times 10^{-4}$ ev, (and $m_{\nu_e} \simeq 5 \times 10^{-6}$ eV) which is in the range of interest for the MSW solution to the solar neutrino puzzle.

## 6. $SO(10)$ Embedding:

In this section, we discuss the $SO(10)$ embedding of the left-right symmetric models discussed in the previous section and any new possibilities for neutrino masses, that may arise in these models. It is of course well known that, the simplest grandunification model that leads to left-right symmetric models at low energies is the $SO(10)$ model. The fermions (both quarks and leptons) belong to the 16-dimensional spinor representation of $SO(10)$. The symmetry breaking pattern of interest to us is the following one:

$$\begin{array}{c} SO(10) \longrightarrow G \longrightarrow SU(3)_C \times SU(2)_L \times SU(2)_R \times U(1)_{B-L} \\ \downarrow \\ G_{std.} \end{array}$$

We will first consider the scenario that leads to Majorana neutrinos: this is achieved by including the Higgs multiplets, which transform as $\{54\}$ $(S)$, $\{45\}$ $(A)$, $\{126\}$ $(\Sigma)$, and $\{10\}$ $(H)$ dimensional representations. (The symbols within the parenthesis denote those representations in what follows). The $\{54\}$-dimensional representation breaks $SO(10)$ down to $SU(4)_C \times SU(2)_L \times SU(2)_R$, whereas $\{45\}$ breaks this group further down to $SU(3)_C \times SU(2)_L \times SU(2)_R \times U(1)_{B-L}$ without the discrete $D$-parity symmetry that relates the left and right-handed gauge couplings. The {126}-dimensional representation contains the $\Delta_L$ and $\Delta_R$ defined earlier and therefore serves to break not only $SU(2)_R$ symmetry but also gives a heavy mass to the right-handed neutrinos. Due to the difference between the scales $M_P$ of $D$-parity breaking and $M_R$ of $SU(2)_R$-breaking, the $\Delta_L^o$ v.e.v. is suppressed by an additional factor of $(v_R/M_P)^2$ with respect to $\kappa^2/v_R$. This gives the real see-saw form to the neutrino mass matrix.[59]

Having outlined the overall scenario, we mention two interesting specific cases that emerge in this model: in one case, the $D$-parity breaking scenario enables the $sin^2\,\theta_W$ constraints to be satisfied even with a low $M_{W_R}$ (in the Tev range) leading neutrino mass spectrum of ev-kev-Mev type as discussed in sec. 5; in the second case, if the $B-L$ breaking scale (which coincides with $SU(2)_R$ scale in the cases we consider) is identified with the Peccei-Quinn scale,[58] then a unique $SO(10)$ model emerges, that simultaneously provides a solution to solar neutrino problem via the MSW mechanism.

In order to comment on these two possibilities, we first explain the meaning of $D$-parity. When $SO(10)$ breaks down to one of its maximal subgroups, $SO(4)\times SO(6)$, there is an additional discrete symmetry which can also remain unbroken. We call this discrete symmetry as $D$-parity. Under $D$-parity, $q\rightarrow q^C$ and $\vec{W}_L\rightarrow\vec{W}_R$; so as far as the quark sector is concerned it acts as charge conjugation or parity; but its effect on Higgs bosons is more complicated; therefore we prefer to call it $D$-parity rather than $C$ or $P$. Its practical impact is to maintain the equality of gauge couplings constant of the $SU(2)_L$ and $SU(2)_R$ groups. Until the work of ref. 52, it was not realized that $SU(2)_R$ and $D$-parity could be broken separately. If $SU(2)_R$ and $D$-parity were broken together, the observed value of $sin^2\,\theta_W$ could only be consistent with $SO(10)$ model if the value of $m_{W_R}$ is rather high ($\sim 10^{10}-10^{12}$ Gev). On the other hand, if $D$-parity was broken "strongly" (i.e. if $g_L/g_R$ differed very much from 1 at the scale $m_{W_R}$), one could have $m_{W_R}\simeq$ from TeV's without spoiling the agreement with $sin^2\,\theta_W$. Since in the neutrino mass matrix, $m_{W_R}$ determines the mass of the right-handed neutrino, $m_{RR}$, we would get an ev-Kev-Mev type neutrino spectrum. The question, then, is: do we have a built-in mechanism to achieve sufficiently fast neutrino decay? This is now more subtle due to the existence of $D$-parity breaking; since the mass of the entire $\vec{\Delta}_L$ triplet is now lifted to the $D$-parity breaking scale $M_P$. Thus, $\nu_\mu\rightarrow 3\nu_e$ decay modes are suppressed. On the other hand, one can have the Majoron decay modes present by including Higgs multiplets {16}, if we forbid the $\{16\}\cdot\{16\}\cdot\{126\}$ Higgs coupling a là ref 54. In such a case, a consistent $SO(10)$ model with $m_{\nu_e}$ in the ev range could be constructed.

Let us now turn to the second scenario for the neutrino mass suggested in ref. 58. The theoretical motivation for the model is the observation that while the vectorial $B-L$ symmetry is anomaly-free and is, in fact, a part of $SO(10)$ symmetry, the axial $B-L$ is plagued with anomalies arising from the QCD gluon contributions and is the source of the well-known strong CP-problem. The suggestion of Peccei and Quinn was of course to make axial $B-L$ into a continuous global symmetry of the gauge model (operating both on quarks and Higgs bosons) and use it to convert $\theta$ into symmetry parameter, in which case $\theta=0$ is as physical as $\theta\neq 0$. The breaking of this symmetry leads to axion and the presently consistent axion models[59] seem to require that $10^9$ Gev $\leq M_{PQ}\leq 10^{12}$ GeV. Since $PQ$ symmetry and vectorial gauge

$B-L$ symmetry as so closely related, it was suggested in ref. 58 that perhaps they may be broken at the same scale. In fact, $SO(10)$ models, as mentioned before, are required (without or with a mild $D$-parity breaking) to have $M_{B-L} \simeq 10^{12}$ Gev.

This has profound implications as far as neutrino mass is concerned. The neutrino masses are now given by:

$$m_{\nu_i} \simeq m_{u_i}^2 / 9\, M_{BL} \tag{6.1}$$

where $m_{u_i}$ is the up-quark mass of the corresponding generation and with $M_{BL} \simeq 10^{13}$ Gev, we get, $m_{\nu_e} \simeq 10^{-8}$ ev; $m_{\nu_\mu} \simeq 10^{-3}$ ev, $m_{\nu_\tau} \simeq 4$ ev. We then see that $m_{\nu_\mu}$ has mass in the right range to provide a solution to the solar neutrino problem a là MSW. It is worth pointing out that this model is rather unique and leads to a proton lifetime of $2.8 \times 10^{34}$ years.[58,60] This could therefore be tested in the next generation of proton decay experiments.

Finally, it should also be noted that the left-right symmetric model for the Dirac neutrino also has a simple extension to $SO(10)$, where we include a single extra singlet fermion $\theta$ and replace the $\{126\}$ Higgs by a $\{16\}$-dimensional Higgs boson, which, however, does not have a coupling of type $\{16\}\{16\}\{10\}$. This leads to the following mass matrix for $(\nu, N, \theta)$:

$$\begin{array}{c} \\ \nu \\ N \\ \theta \end{array} \begin{array}{c} \begin{array}{ccc} \nu & N & \theta \end{array} \\ \begin{pmatrix} 0 & m_D & 0 \\ m_D & 0 & V_R \\ 0 & V_R & 0 \end{pmatrix} \end{array} . \tag{6.2}$$

This leads to a massless neutrino. On the other hand, if a mass term for $\theta$ $\left(\mu\theta^T C^{-1} \theta\right)$ is added, it leads to a light Majorana neutrino $\nu_e$, with $m_{\nu_e} \simeq \frac{m_D^2 \mu}{V_R^2}$, which is ultralight. This last case is similar to solution proposed by the author[61] to solve the neutrino mass problem in superstring models.

## 7. Neutrinos in Superstring Inspired Models:

In this section, we consider the properties of neutrinos in the superstring model. To make this discussion useful, we first give a miniature review of the gross features of superstring inspired models. These models imply a supersymmetric $E_6$-grandunified model with matter fields belonging to $\{27\}$-dimensional representations of the $E_6$-group and with Higgs fields also transforming as $\{27\} + \{\overline{27}\}$ fields under $E_6$. There are additional $E_6$-singlet fields. In particular note the absence of any $E_6$ representation such as $\{351\}$, which contains the $\{126\}$-dimensional representation of $SO(10)$ that was essential to the see-saw mechanism for small neutrino masses. Therefore, the nature of neutrino masses are likely to be more complicated in superstring models. To facilitate this discussion, we first write down the decomposition of the $\{27\}$-dim. representation of $E_6$ under $SO(10)$ as well as $SU(3)_c \times SU(3)_L \times SU(3)_R$.

$$E_6 \longrightarrow SO(10) \times U(1)_X \;:$$

$$\{27\} = \{16\}_1 \oplus \{10\}_{-2} \oplus \{1\}_4$$

$$\text{Particle Content}: \quad \{16\}; \quad \begin{pmatrix} u_1 & u_2 & u_3 & \nu \\ d_1 & d_2 & d_3 & e^- \end{pmatrix}_{L+R}$$

$$\begin{array}{ll} \{10\} : & g\,,\, g^c \text{ singlet quark} \\ & \begin{pmatrix} H_u^o & H_u^+ \\ H_d^- & H_d^o \end{pmatrix} \\ \{1\} & n_o \end{array} \tag{7.1}$$

$$E_6 \supset SU(3)_c \times SU(3)_L \times SU(3)_R :$$

$$\{27\} \supset (3,3,1) \oplus (\bar{3},1,\bar{3}) \oplus (1,\bar{3},3)$$

$$(3,3,1) : \begin{pmatrix} u \\ d \\ g \end{pmatrix} \; ; \; (\bar{3},1,\bar{3}) : \begin{pmatrix} u^c \\ d^c \\ g^c \end{pmatrix}$$

$$(1,\bar{3},3) : \begin{pmatrix} H_u^o & H_u^+ & e^+ \\ H_d^- & H_d^o & \nu^c \\ e^- & \nu & n_o \end{pmatrix} \equiv L \tag{7.2}$$

We see from eqn. (7.1) and (7.2) that aside from the known fields, there are eleven extra fermions out of which three are neutral. In total, each generation have 5 neutral leptons. Add to this the neutral gauginos (4 colorless ones) and many Higgsinos plus $E_6$ singlet fermions, which makes a total of more than 20 neutral fermions. All these in general can mix making any understanding of the nature of the neutrino next to impossible in general. We will therefore simplify the picture by first ignoring generation mixing and secondly by assuming that there exist intermediate mass scales, which decouple the neutral leptons $H_u^o - H_d^o$. We also further assume that, in these models, $B - L$ symmetry scale is either an intermediate scale such as $10^{11} - 10^{12}$ Gev or a Tev. In this simplified approximation, four mechanisms have been proposed to understand neutrino masses, which we briefly summarize below.

### (i) *R*-parity violation Mechanism:[61]

In this mechanism, it is assumed that the low energy gauge group is given by $SU(3)_c \times SU(2)_L \times U(1)_{I_{3R}} \times U(1)_{B-L}$ and that the superpartner of $\nu^c$ (the anti right-handed neutrino) acquires a v.e.v. of order $V_R$ to break $I_{3R}$ and $B-L$. Because of this, the gaugino corresponding to the broken generator (denoted by $\lambda_{Y'}$) couples to the $\nu^c$ and leads to the following mass matrix gaugino-neutrino mass matrix:

$$\begin{array}{c} \\ \nu \\ \nu^c \\ \lambda_{Y'} \end{array} \begin{array}{c} \begin{array}{ccc} \nu & \nu^c & \lambda_{Y'} \end{array} \\ \begin{pmatrix} 0 & m_D & 0 \\ m_D & 0 & g\,v_R \\ 0 & g\,v_R & \mu \end{pmatrix} \end{array} . \tag{7.3}$$

Here $\mu$ is the Majorana mass for $\lambda_Y$-gaugino that can arise in the two-loop order. This matrix can be diagonalized leading to a pseudo-Dirac 4-component heavy lepton ($m \simeq g\, v_R$) and a light 2-component Majorana neutrino

$$\nu_e \simeq \nu + \frac{m_D}{g\, v_R}\, \nu^c \tag{7.4}$$

with mass

$$m_{\nu_e} \simeq \frac{m_D^2 \cdot \mu}{(g\, v_R)^2}\, . \tag{7.5}$$

Usually, in supergravity models, $\mu \simeq 100$ Gev or so; assuming $(g\, v_R) \simeq 1$ Tev, we get $m_{\nu_e} \simeq 10^{-1}$ ev. If, on the other hand, we choose $v_R \simeq 10^{12}$ Gev, $m_{\nu_e} \simeq 10^{-9}$ ev. Thus, smallness of neutrino mass is understood without any unnatural fine tuning of parameters. It has been pointed out[45] that, in these models, the $H_u^o$-Higgsino mixes with the $\nu$ with mixing proportional to $h_\nu^D\, v_R$; this modified eqn. (7.5) and $m_\nu$ has a component which grows with $v_R$. Thus, if we want to understand the small $m_\nu$, we must have an upperbound on $v_R$ of a few Tev. A limitation of this method so that it works only for one generation.

### (ii) $E_6$-Singlet Mechanism:

Use of $E_6$-singlets ($S$) for studying the neutrino mass problem was hinted at by Witten[63] and was analyzed in detail in ref. 64. here, one has to rely on possible symmetries that may arise in superstring models on specific Calabi-Yau spaces to restrict the Yukawa couplings in such a way that we have couplings of the following type in the superpotential: using the notation of (7.2),

$$W = W_o + \lambda_1\, L L L_H + \lambda_2\, L \bar{L}_H\, S \tag{7.6}$$

where subscript $H$ stands for the Higgs field. If we then give v.e.v. to $\tilde{\nu}_H^c$, and $(H_u^o)_{Higgs}$, then, for each generation we get the following type of $3 \times 3$ mass matrix:

$$\begin{array}{c} \\ \nu \\ \nu^c \\ S \end{array} \begin{array}{c} \begin{array}{ccc} \nu & \nu^c & S \end{array} \\ \begin{pmatrix} 0 & m_D & 0 \\ m_D & 0 & \lambda_2\, v_R \\ 0 & \lambda_2\, v_R & 0 \end{pmatrix} \end{array} . \tag{7.7}$$

This, as discussed in sec. 6, leads to a massless neutrino and a massive Dirac neutrino. If we include a possible Majorana mass term for the $S$-fermion of order of 100 Gev, the mass matrix looks like that in eqn. (7.3) with similar values for $m_\nu$. Note that these values fall outside the range needed to solve the solar neutrino puzzle for $v_R \simeq 1$ Tev or $10^{11}$ Gev.

### (iii) Higher Dimensional Operators:

It was proposed in ref. 65 that one may solve the neutrino mass problem in superstring models with two intermediate mass scales corresponding to $\langle \tilde{\nu}_H^c \rangle \approx \langle \tilde{n}_{o_H} \rangle \approx 10^{11} - 10^{12}$ Gev. In this case, a higher dimension term in the superpotential of the form $L \bar{L}_H \cdot \bar{L}_H / M_{Pl}$ leads to an effective see-saw-like mechanism with right-handed neutrino mass $m_{N_R} \simeq \langle\tilde{\nu}^c\rangle^2/M_{Pl} \simeq 10^4 - 10^6$ Gev. Using the standard see-saw formula, we get $m_{\nu_e} \simeq m_e^2/m_{N_R} \simeq 10^{-2}$ ev $- 10^{-4}$ ev $m_{\nu_\mu} \simeq 400$ ev $-$ 4 ev and $m_{\nu_\tau} \simeq 160$ Kev $-$ 1.6 Kev. Since in the Superstring model, there is no known way for $\nu_\mu$ and $\nu_\tau$ to decay, this scenario will be ruled out by cosmology. However, there exist arguments from proton decay that $\langle\nu_H^c\rangle \simeq \langle n_{oH}\rangle$ may be of order $10^{14}$ Gev, leading to $m_{N_R} \simeq 10^{10}$ Gev in which case, $m_{\nu_e} \simeq 10^{-8}$ ev, $m_{\nu_\mu} \simeq 4 \times 10^{-4}$ ev and $m_{\nu_\tau} \simeq .16$ ev. This kind of a spectrum is cosmologically acceptable and with slight variation of coupling parameters could be useful for the solution of solar neutrino puzzle via the MSW mechanism.

### (iv) One Loop Induced Neutrino Mass:[66]

It has also been proposed that existence of some accidental symmetries may force the tree level Dirac mass of the neutrino to vanish. As a result, the dominant contribution to neutrino masses can arise at the one loop level from diagrams of the type shown in fig. 3. Here also the neutrino masses are expected to be small and could be of interest for discussion of the MSW effect.

Whereas in the above discussion, we have kept the different mechanisms (i), (ii) and (iii) separate, one may keep all these mechanisms together as has been recently attempted.[67] The above conclusions remain unaffected in this general analysis.

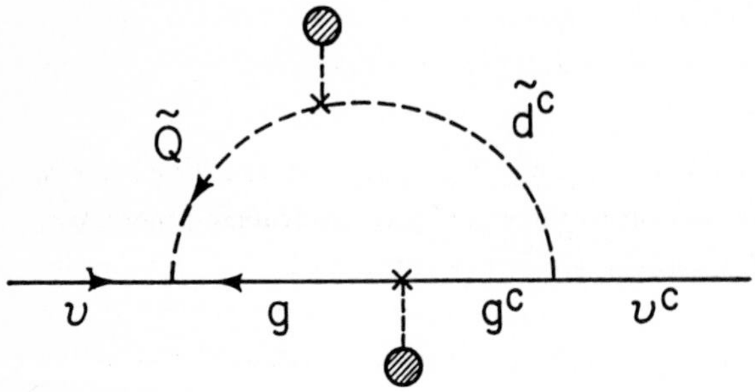


Fig. 3. One loop graph for neutrino Dirac mass in some versions of superstring inspired models.

## 8. Spontaneous Breaking of Global B-L Symmetry and the Majoron:

As we saw earlier in this article, if the neutrino is a Majorana particle, the low enegy theory must break $B-L$ symmetry. A novel possibility first suggested by Chikashige, Mohapatra and Peccei[9] is that the Majorana character of the neutrino may arise from spontaneous breaking of a global $B-L$ symmetry. This leads, by Nambu-Goldstone theorem to the existence of a massless scalar boson coupled to leptons, called the Majoron[68] in ref. 9. While apriori existence of a massless boson would appear to be disastrous since, it would lead to $1/r$ type long range forces, it was pointed out in ref. 9 that as a consequence of shift invariance associated with Goldstone bosons, their diagonal couplings in the non-relativistic limit are always spin dependent, i.e. $[(\vec{\sigma}_1 \cdot \vec{\nabla}_1)(\vec{\sigma}_2 \cdot \vec{\nabla}_2)]\,\frac{1}{R}$ type and are not therefore easily observable. A concrete realization of this idea was proposed in reference 9 by a minimal extension of the standard model that includes the right handed neutrino and a lepton number carrying neutral singlet Higgs boson. Let us start by describing this model, which is based on the standard model gauge group $SU(2)_L \times U(1)_Y$ with usual particle assignments: Quarks: $Q_L \equiv (\frac{1}{2}, \frac{1}{3})$; $u_R \equiv (0, \frac{4}{3})$, $d_R \equiv (0, -\frac{2}{3})$; leptons: $\psi_L \equiv (\frac{1}{2}, -1)$, $e_R \equiv (0, -2)$; the Higgs field: $\varphi \equiv (\frac{1}{2}, +1)$. To this we add the right handed neutirno $\nu_R = (0,0)$ (one for each generation) and a neutral singlet Higgs boson $\Delta^\circ \equiv (0,0)$. We choose the Yukawa couplings and the Higgs potential so that the model has a global lepton-number symmetry with $\Delta^\circ$ having $L = 2$, i.e. in our lagrangian we include the couplings $\nu_R^T c^{-1} \nu_R \Delta^\circ$, $\bar{\psi}_L \varphi \nu_R$ and forbid the scalar self-coupling $\varphi^+ \varphi \Delta$, (which is otherwise allowed by the gauge symmetry). The Higgs potential is then so chosen as to lead to a minimum where the following Higgs fields have vacuum expectation values:

$$\langle \varphi \rangle = \begin{pmatrix} 0 \\ v/\sqrt{2} \end{pmatrix} \quad \text{and} \quad \langle \Delta^\circ \rangle = v_{BL}\,. \tag{8.1}$$

The global lepton number (or more precisely $B - L$) symmetry is then broken spontaneously. The associated Goldstone boson (denoted by $\chi$) is given by $\chi = I_m \Delta^\circ$ and couples in the non-diagonal basis to $\nu_R$'s. Furthermore, the neutrino mass matrix (per generation) takes the following see-saw form,[9]

$$M = \begin{pmatrix} 0 & h_\ell v/\sqrt{2} \\ h_\ell v/\sqrt{2} & h^M v_{BL} \end{pmatrix} \tag{8.2}$$

leading to the light neutrino masses given by the formula $m_\nu \simeq \frac{(h_\ell v)^2}{h^M V_{BL}}$; if we approximate $h_\ell$ to be the same as the Yukawa coupling of the charged lepton of the corresponding generation, then $m_{\nu_e} \simeq m_e^2/2h^M v_{BL}$ and for $h^M \simeq 1$ and $v_{BL} \simeq$ few Tevs, we get neutrino masses below the laboratory bounds. Shortly after the proposal of the idea of the Majoron, it was also pointed out[69] that, $\nu_{\mu,\tau}$ can decay to $\nu_e + \chi$ making an eV-Kev-Mev type spectrum for neutrinos consistent with cosmological constraints. More refined analysis of this question carried out in the context of both $SU(2)_L \times U(1)_Y$[70] as well as $SU(2)_L \times SU(2)_R \times U(1)_{B-L}$ models[54] have confirmed this belief. Thus, an immediate advantage of the concept of Majoron is that an eV-keV-Mev type neutrino spectrum is theoretcally and cosmologically consistent. Turning this question around, any evidence for an eV-Kev-Mev type neutrino spectrum will also, at the same time, be an evidence for the existence of the Majoron.

It is worth emphasizing at this point that Majoron couples[9] to $\nu_L$ at the tree level in the diagonal basis with a coupling strength $\approx (m_{\nu_L}/v_{BL})$ which is expected to be of order $\sim 10^{-11} - 10^{-12}$. This leads to $\nu\nu \to \chi\chi$ annihilation cross section of order of $10^{-66}$ cm$^2$ for $E_\nu \simeq$ few Mev, which is utterly negligible. The coupling of the Majoron to charged fermions $u, d, e$ arises at the one loop level via the exchange of $W$ or $Z$ bosons and has the magnitude $\approx \frac{G_F}{\pi} m_f m_\nu$, where $f = u, d, e$. This is of order $\approx 10^{-16}$ or so. As a result, Majoron emission contributes a negligible amount to the energy loss of red giants as well as other stars. We see from this that the singlet Majoron is highly invisible. As we see below, there exist modification of this idea introduced and analyzed by Gelmini and Roncadelli[71] and Georgi, Glashow and Nussinov,[72] following the work of ref. 9, which leads to a number of testable laboratory prediction of the Majoron idea.

The triplet Majoron model[71,72] extends the standard model by introducing a Higgs triplet $\vec{\Delta}_L$ with $Y = 2$ and lepton number $L = 2$.

$$\vec{\tau} \cdot \vec{\Delta}_L = \begin{pmatrix} \Delta_L^+ & \sqrt{2}\,\Delta_L^{++} \\ \sqrt{2}\,\Delta_L^o & \Delta_L^+ \end{pmatrix} . \tag{8.3}$$

The introduction of this multiplet allows for the existence of the coupling $h\,\psi_L^T c^{-1} \tau_2 \vec{\tau} \cdot \vec{\Delta}_L \psi_L$. This Yukawa coupling includes a direct coupling of type $\nu_L^T\, c^{-1}\, \nu_L \Delta_L^o$ in contrast with the singlet Majoron. The global $B - L$ symmetry of the model is spontaneously broken by a vacuum expectation value of $\langle \Delta_L^o \rangle = v_T$, which then leads to a neutrino Majorana mass $m_\nu$ $= h_m\, v_L$. Furthermore, since the Majoron field in this case is given by

$$\chi = \frac{\sqrt{2}\, v_T Im\varphi^o - v\, Im\, \Delta^o}{\sqrt{v^2 + 2v_T^2}} \tag{8.4}$$

the left-handed neutrinos are rather strongly coupled to the Majoron, leading to testable predictions in laboratory process such as $\mu \to e\chi\chi$, $\pi \to e\nu\chi$, $(N, Z) \to (N-2, Z+2) + 2e^-$ $+ \chi$, etc.

Let us now discuss the constraints on $v_L$ and coupling $h_m$ from observations. The neutral current interaction in general $SU(2)_L \times U(1)$ theories, includes a parameter $\rho$, whose tree level value in the standard model is given by unity in agreement with experiment to within one percent. In the presence of the Higgs triplet that acquire non-zero v.e.v., $v_T$, we get,

$$\rho \simeq 1 - 4\,\frac{v_T^2}{v^2} . \tag{8.5}$$

Neutral current observations therefore imply that $v_T \leq 25$ GeV. However, further constraints on $v_T$ arise from the triplet Majoron (TM) coupling to $u$, $d$, $e$, which is given by

$$f_{uu\chi} \approx \frac{g\, m_u}{m_W} \times \left(\frac{v_T}{v}\right) < 10^{-12} \tag{8.6}$$

and leads to the bound $v_T \leq 10$ kev.[73] Laboratory bounds on the electron neutrino mass, then, implies, $h_{ee} \lesssim 10^{-3}$. Bounds of the same order of magnitude, also emerge from the study of $(\beta\beta)_{o\nu}$ decay[34] (see fig. 4) and studies of $\pi \to e\nu\chi$,[74] $\mu \to e\chi\chi$.[75] A precise measurement of the $Z$-width at LEP will be a crucial test of the triplet Majoron model since its contribution is equivalent to that of two extra neutrino species. An immediate theoretical question then arises as to what is responsible for the small symmetry breaking scale. One can speculate that this may be related to the existence of very high scale in the theory. It is, however, not easy to construct such a model.

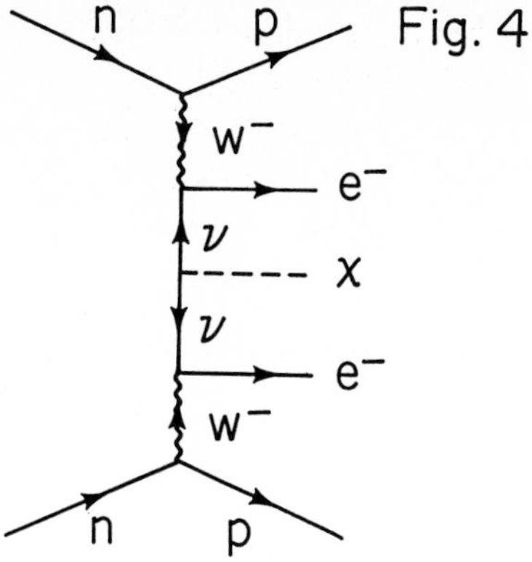


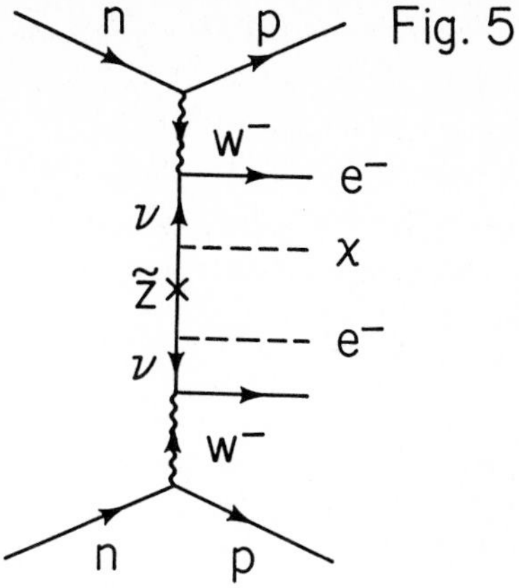


Fig. 4. The Feynman diagram for single Majoron emission in neutrinoless double beta decay.

Fig. 5. The Feynman diagram that leads to neutrinoless double beta decay with two Majoron emission.

Finally, it is worth pointing out, in the context of supersymmetric models, one can spontaneously break lepton number symmetry by giving a non-zero v.e.v.[76] to the superpartner of the neutrino, which due to supersymmetry carries lepton number 1. This gives rise to the doublet Majoron. The doublet Majoron contribution to the $Z$-width is half the contribution of a single neutrino species. The constraints on the sneutrino v.e.v. coming from astrophysics are very similar to those of the triplet v.e.v. and imply that $\langle\tilde{\nu}\rangle \leq$ keV. An important test of the doublet Majoron model will be the new process of double Majoron emission shown in fig. 5. It appears with a strength which can be roughly estimated to be about $G_F^2 \cdot g^2/\tilde{M}_Z$ and can lead to a large contribution to neutrinoless double beta decay for $M_{\tilde{Z}} \leq 100$ GeV. It is, therefore, important to calculate the energy spectrum of the electron in this process.

## 9. Conclusion:

To summarize, we have given an extensive review of the basic ideas that go into understanding the smallness of neutrino masses in general gauge and superstring inspired models. We see that various kinds of mass spectra can emerge in different models. Testing between these various possibilities may be possible if the MSW solution to the solar neutrino puzzle is backed up by future experiments and other considerations. For instance, this would rule out a low mass $m_{W_R}$ in an $SO(10)$ type model as well as the mechanisms (i) and (ii) in superstring inspired models.

An important question that we did not address in this article is the theoretical prediction of neutrino mixings. The existence of arbitrary parameters in the theories discussed precludes any reliable conclusion about this at the moment. We eagerly await new experimental developments regarding this as well as the whole question of neutrino masses.

## References

[ 1 ] J.C. Pati and A. Salam, Phys. Rev. D**10**, 275 (1974); R.N. Mohapatra and J.C. Pati, Phys. Rev. D**11**, 566 (1975); G. Senjanović and R.N. Mohapatra, Phys. Rev. D**12**, 1502 (1975). For a review, see R.N. Mohapatra in "Quarks, Leptons and Beyond," ed. by H. Fritzsch et al, Plenum (N.Y.), p. 219.

[ 2 ] R.N. Mohapatra and G. Senjanović, Phys. Rev. Lett. **44**, 912 (1980); Phys. Rev. D**23**, 165 (1981).

[ 3 ] For a review, see "Superstring Theory" Vols. I and II by M. Green, J. Schwarz and E. Witten, Cambridge University Press (1986).

[ 4 ] For a review of standard model and grandunification see "Unification and Supersymmetry: The Frontiers of Quark-Lepton Physics," by R.N. Mohapatra, Springer-Verlag (N.Y.) (1986).

[ 5 ] R. Cowsik and J. Mclelland, Phys. Rev. Lett. **29**, 699 (1972).

[ 6 ] S.S. Gershtein and Ya B. Zeldovich, JETP Lett. **4**, 120 (1966).

[ 7 ] B.W. Lee and S. Weinberg, Phys. Rev. Lett. **39**, 165 (1977); P. Hut, Phys. Lett. **69B**, 85 (1977); K. Sato and M. Kobayashi, Prog. Theor. Phys. **58**, 1775 (1977); M.I. Vysotskii, A.D. Dolgov and Ya. B. Zeldovich, JETP Lett. **26**, 188 (1977).

[ 8 ] D. Dicus, E. Kolb and V. Teplitz, Phys. Rev. Lett. **39**, 169 (1977).

[ 8a ] S. Sarkar and A.M. Cooper, Phys. Lett. **148B**, 347 (1984).

[ 9 ] Y. Chikashige, R.N. Mohapatra and R.D. Peccei, Phys. Lett. **98B**, 265 (1981).

[ 10 ] Y. Chikashige, R.N. Mohapatra and R.D. Peccei, Phys. Rev. Lett. **45**, 1926 (1980); G. Gelmini and J.W.F. Valle, **142B**, 181 (1984); A. Kumar and R.N. Mohapatra, Phys. Lett. **150B**, 191 (1985).

[ 11 ] M. Roncadelli and G. Senjanović, Phys. Lett. **107B**, 59 (1983); P. Pal, Nucl. Phys. B**227**, 237 (1983).

[ 12 ] R.N. Mohapatra and Palash Pal, Phys. Lett. **179B**, 105 (1986).

[ 13 ] J.K. Rowley, B.T. Cleveland and R. Davis, Jr. in "Solar Neutrino and Neutrino Astronomy," ed. M.L. Cherry (A.I.P. Conf. Proc. No. 126, New York, 1985).

[ 14 ] J.N. Bahcall, W.F. Huebner, S.H. Lubow, P.D. Parker and R.G. Ulrich, Rev. Mod. Phys. **54**, 767 (1982); J.N. Bahcall, B.T. Cleveland, R. Davis, Jr. and J.K. Rowley, Appl. J. **292**, 279 (1985), R. Bahcall and R.K. Ulrich, preprint (in preparation).

[ 15 ] B. Pontecorvo, Zh. Eksp. Teor. Fiz. **53**, 1717 (1967) Soviet. Phys. JETP **26**, 984 (1968); V. Gribov and B. Pontecorvo, Phys. lett. **28B**, 493 (1969).

[ 16 ] R.N. Mohapatra and J.W.F. Valle, Phys. Lett. 177B, 47 (1986).

[ 17 ] S.P. Mikheyev and A.Yu. Smirnov, Nuovo Cim. **9C**, 17 (1986); Yad. Fiz. **42**, 1441 (1985) [Soc. J. Nucl. Phys. **42**, 913 (1985)]; L. Wolfenstein, Phys. Rev. **D17**, 2369 (1978).

[ 18 ] For a recent review, see M. Fukugita, Kyoto Preprint, RIFP-718 (1987).

[ 19 ] J. Bahcall, N. Cabibbo and A. Yahil, Phys. Rev. Lett. **28**, 316 (1972).

[ 20 ] J. Bahcall, S. Petcov, S. Toshev and J.W.F. Valle, Phys. Lett. **181B**, 369 (1986).

[ 21 ] A. Cisneros, Applied Space Science **10**, 87 (1971).

[ 22 ] L.B. Okun, M.B. Voloshin and M.I. Vysotskii, Zh. Eksp. Teor. Fiz. **91**, 754 (1986) [Sov. Phys. JETP **64**, 446 (1986)].

[ 23 ] P. Sutherland, J.N. Ng, E. Flowers, M. Ruderman and C. Inman, Phys. Rev. **D13**, 2700 (1976).

[ 24 ] J.P. Morgan, Phys. Let. **102B**, 247 (1981).

[ 25 ] K. Hirata, et al., Phys. Rev. Lett. **58**, 1490 (1987).

[ 26 ] R.M. Bionta, et al., Phys. Rev. Lett. **58**, 1494 (1987).

[ 27 ] We give a very incomplete list of references on this subject: J.N. Bahcall and S.L. Glashow, Nature, **326**, 476 (1987); W.D. Arnett and J.L. Rosner, Phys. Rev. Lett. **58**, 1906 (1987); E.W. Kolb, A.J. Stebbing and M.S. Turner, Phys. Rev. **D35**, 3598 (1987).

[ 28 ] R. Barbieri and R.N. Mohapatra, Univ. of MD Preprint (1988); J. Cooperstein and J. Lattimer, Brookhaven Preprint (1988); S. Nussinov, G. Alexander, I. Goldman and A. Aharanov, Tel Aviv Preprint (1987); A. Dar, Tel Aviv Preprint (1987).

[ 29 ] A. Dar and S. Dado, Phys. Rev. Lett. **59**, 2368 (1987).

[ 30 ] For a recent review of all radiative neutrino decays, see, M. Roos, Proceedings of "international Conference on Neutrino Physics," ed. by H.V. Klapdor, Springer-Verlag (1988).

[ 31 ] Particle Data Table, Phys. Lett. **B** , April (1987).

[ 32 ] S. Petcov, Phys. Lett. **B1101**, 245 (1982); J.W.F. Valle, Phys. Rev. D**27**, 1672 (1983); M. Doi, T. Kotani, M. Kenmoku, and E. Takasugi, Phys. Rev. D**30**, 626 (1984). For a review, see B. Kayser, Comments in Nuc. and Part. Physics **14**, 69 (1985).

[ 33 ] M. Gell-Mann, P. Ramond and R. Slansky, in "Supergravity," ed. by D.Z. Freedman and P. Vanniuenhuizen, (North Holland 1980); T. Yanagida, KEK Lectures, (1979).

[ 34 ] For a recent review, see D. Caldwell, Proceedings of "XXIII$^{rd}$ International Conference on High Energy Physics," ed. S. Loken, World Scientific Publishing Co. (1986), p. 951; H. Ejiri, Proceedings of INS Symposium on "Neutrino Mass and Related Topics," ed. by T. Oshima (1988). A recent report of the discovery of the Majoron in neutrinoless double $\beta$-decay by F. Avignone et al., Proceedings of the APS-DPF meeting, Salt Lake City, Utah (1987) has been disputed by other expts on $(\beta\beta)_{o\nu}$-decay, see for instance E. Fiorini, Proceedings of "VIII$^{th}$ Workshop on Grandunification," ed. by K.C. Wali (World Scientific), 1988.

[ 35 ] R.N. Mohapatra and G. Senjanović, Phys. Rev. D**23**, 165 (1981).

[ 36 ] L. Durkin and P. Langacker, Phys. Lett. B**166**, 436 (1986); R. Robinett and J. Rosner, Phys. Rev. D**25**, 3036 (1982); V. Barger, E. Ma and K. Whisnant, Phys. Rev. D**26**, 2378 (1982).

[ 37 ] M.A.B. Bég, R.V. Budny, R.N. Mohapatra and A. Sirlin, Phys. Rev. Lett. **38**, 1252 (1977). For further discussion, see J. Maalampi, K. Mursula and M. Roos, Nuc. Phys. B**207**, 233 (1982).

[ 38 ] J. Carr **et al**, Phys. Rev. Lett. **51**, 627 (1983); D.P. Stoker, **et al.**, Phys. Rev. Lett. **54**, 1887 (1985).

[ 39 ] P. Herczeg, Los Alamos preprint LA-UR-85-2761 (1985).

[ 40 ] R.N. Mohapatra, New Frontiers in High Energy Physics, ed. A. Perlmutter and L. Scott (Plenum, New York, 1978), p. 337.

[ 41 ] J.F. Donoghue and B. Holstein, Phys. Lett. B**113**, 382 (1982).

[ 42 ] G. Beall, M. Bander and A. Soni, Phys. Rev. Lett. **48**, 848 (1982); R.N. Mohapatra, G. Senjanović and M.D. Tran, Phys. Rev. **D28**, 546 (1983); G. Ecker, W. Grimus and H. Neufeld, Phys. Lett. **127B**, 365 (1983); F.J. Gilman and M.H. Reno, Phys. Rev. **D29**, 937 (1984). Inclusion of QCD corrections improve this bound; see G. Ecker and W. Grimus, Z. Phys. C**30**, 293 (1986).

[ 43 ] I.I. Bigi and J.M. Frere, Phys. Lett. **110B**, 255 (1982).

[ 44 ] H. Abramowicz, **et al.**, Z. Phys. C**12**, 225 (1982).

[ 45 ] L. Wolfenstein, Phys. Rev. D**29**, 2130 (1984).

[ 46 ] E. Masso, Phys. Rev. Lett. **52**, 1956 (1984).

[ 47 ] D.O. Caldwell, **et al.**, Phys. Rev. D**33**, 2737 (1986); H. Ejiri, **et al.**, Nucl. Phys. A**448**, 27 (1986); E. Firoini, **et al.**, Phys. Lett. B**146**, 450 (1984); F.T. Avignone III, **et al.**, Phys. Rev. Lett. **54**, 2309 (1985).

[ 48 ] R.N. Mohapatra, Phys. Rev. D**34**, 909 (1986). For limits on $m_{\nu_R}$ from other considerations, see M. Gronau, C. Leung and J. Rosner, Phys. Rev. D**29**, 2539 (1984).

[ 49 ] For reviews, see, W. Haxton and G. Stephenson, Jr., Progress in Nuclear and Particle Physics, **12**, 409 (1984); J. Vergados, Phys. Rep. **133**, 1 (1986); M. Doi, T. Kotani and E. Takasugi, Prog. Theor. Phys. Supp. **83**, 1 (1985).

[ 50 ] R.N. Mohapatra, Proceedings of "The International Symposium on Weak and Electromagnetic Interactions in Nuclei," ed. H.V. Klapdor, Springer-Verlag (1986); S. Bertolini and J. Liu, CMU-HEP 87-09 (1987).

[ 51 ] D. Chang and R.N. Mohapatra, Phys. Rev. D**32**, 1248 (1985).

[ 52 ] D. Chang, R.N. Mohapatra and M.K. Parida, Phys. Rev. Lett. **52**, 1072 (1984).

[ 53 ] U. Chattopadhyay and P.B. Pal, Phys. Rev. D**34**, 3444 (1986).

[ 54 ] A. Kumar and R.N. Mohapatra, Phys. Lett. **150B**, 191 (1985); R.N. Mohapatra and P. Pal, Univ. of Massachusetts Preprint (1987).

[ 55 ] J.E. Kim, Phys. Rev. D**14**, 3000 (1976); M.A.B. Bég, W. Marciano and M. Ruderman, Phys. Rev. D**17**, 1395 (1978); J. Liu, Phys. Rev. **D35**, 3447 (1987); M.J. Duncan et al., Phys. Lett. **191B**, 304 (1987); G. Branco and J. Liu, Carnegie-Mellon Preprint (1988).

[ 56 ] L.B. Okun, M.B. Voloshin and M. Vysotski, Zh. Eksp. Teor. Fiz. **91**, 754 (1986); [Societ Phys. JETP **64**, 446 (1986)].

[ 57 ] D. Chang and R.N. Mohapatra, Phys. Rev. Lett. **58**, 1600 (1987); R.N. Mohapatra, Phys. Lett. **B** (to appear) (1988).

[ 58 ] R.N. Mohapatra and G. Senjanović, Zeit. für Phys. C**17**, 53 (1983); F. Stecker and Q. Shafi, Phys. Rev. Lett. **50**, 928 (1983); P. Langacker, R.D. Peccei and T. Yanagida, Mod. Phys. Lett. **A1**, 541 (1986).

[ 59 ] For reviews of strong CP-problem, see J.E. Kim, Phys. Rep. **150**, 1 (1987); H.Y. Cheng, Indiana Preprint IUHET-125 (1986).

[ 60 ] R.N. Mohapatra, Proceedings of "Eighth Workshop on Grandunification," ed. by K.C. Wali, (World Scientific Publishing, Singapore, 1987).

[ 61 ] R.N. Mohapatra, Phys. Rev. Lett. **56**, 561 (1986).

[ 62 ] J.W.F. Valle, private communication (1987).

[ 63 ] E. Witten, Nucl. Phys. B**268**, 79 (1986).

[ 64 ] R.N. Mohapatra and J.W.F. Valle, Phys. Rev. D**34**, 1642 (1986).

[ 65 ] S. Nandi and U. Sarkar, Phys. Rev. Lett. **56**, 564 (1986).

[ 66 ] A. Masiero, D.V. Nanopoulos and A.I. Sanda, Phys. Rev. Lett. **57**, 663 (1986); G.C. Branco and C.Q Geng, Phys. Rev. Lett. **58**, 969 (1987); E. Ma, Phys. Rev. Lett. **58**, 1047 (1987).

[ 67 ] M. Doi, T. Kotani, T. Kurimoto, H. Nishiura and E. Takasugi, Phys. Rev. **D37**, (1988).

[ 68 ] For a recent review of the properties of the Majoron, see, R.N. Mohapatra, Proceedings of the XXIII International Conference on High Energy Physics, ed. by S. Loken (World Scientific) p.

[ 69 ] Y. Chikashige, R.N. Mohapatra and R.D. Peccei, ref. 10. For further discussion of $\nu_{\mu,\tau} \rightarrow \nu_e + \chi$ decay, see J. Schecter and J.W.F. Valle, Phys. Rev. **D25**, 774 (1982).

[ 70 ] S.L. Glashow, Phys. Lett. **187B**, 367 (1987).

[ 71 ] G. Gelmini and M. Roncadelli, Phys. Lett. **99B**, 411 (1981).

[ 72 ] H. Georgi, S.L. Glashow and S. Nussinov, Nuc. Phys. **B193**, 297 (1981).

[ 73 ] M. Fukugita, S. Watamura and M. Yoshimura, Phys. Rev. Lett. **48**, 1522; D. Dearborn, D. Schramm and G. Steigman, Phys. Rev. Lett. **56**, 26 (1985).

[ 74 ] V. Barger, W.Y. Keung and S. Pakvasa, Phys. Rev. **D25**, 907 (1982).

[ 75 ] T. Goldman, E. Kolb and G. Stevenson, Phys. Rev. **D26**, 2503 (1982).

[ 76 ] C.S. Aulakh and R.N. Mohapatra, Phys. Lett. **119B**, 136 (1982). For further analysis of this model, see R. Santamaria and J.W.F. Valle, Phys. Rev. Lett. **59**, (1987). For a non-supersymmetric model with doublet Majoron, see, S. Bertolini and R. Santamaria, Carnegie-Mellon Preprint (1988).

# Double Beta Decay Experiments and Searches for Dark Matter Candidates and Solar Axions

*F.T. Avignone, III*[1] *and R.L. Brodzinski*[2]

[1] Department of Physics and Astronomy,
University of South Carolina, Columbia, SC 29208, USA
[2]Pacific Northwest Laboratory, Richland, WA 99352, USA

A brief discussion of the theoretical status of double beta decay is given. A number of experiments searching for double beta decay of $^{76}$Ge, $^{82}$Se, $^{100}$Mo, $^{136}$Xe and $^{150}$Nd are discussed. A brief review is given of published efforts to search for cosmic dark matter candidate particles and solar axions using the low background detector of a $^{76}$Ge double-beta-decay experiment.

## I. Double-Beta Decay

### 1.1 Introduction

Double-beta decay ($\beta\beta$-decay) is a second order weak process which converts two neutrons in a nucleus into two protons. Figure 1 shows the diagrams of three commonly considered decay modes. Figure 2 depicts the decay scheme and spectral shapes of the sum energy of the two electrons in these decays. The existence of $\beta\beta$-decay was first suggested by WIGNER as reported by GOEPPERT-MAYER in 1935 [1], but was not directly observed in a laboratory experiment until 1987. On August 30, 1987, MICHAEL MOE announced the results of his experiment at the annual American Chemical Society meeting in New Orleans. The $2\nu$ $\beta\beta$-decay half-life of $^{82}$Se was determined to be $(1.1^{+0.8}_{-0.3}) \times 10^{20}$ years using the University of California (Irvine) Time Projection Chamber (TPC). Years earlier, indirect experiments yielded half-lives for $^{128,130}$Te and $^{82}$Se using geochemical techniques to determine isotopic abundances and ore ages of rocks [2,3]. The current geochemical $^{82}$Se half-life, $(1.30\pm0.05) \times 10^{20}$ years, is in agreement with the recent value of MOE et al. [4]. Such half-lives make this process the slowest observed decay ($\lambda \sim 10^{-28}$sec$^{-1}$).

The renaissance of interest in $\beta\beta$-decay over the past decade is driven by the constraints certain decay modes can place on gauge field theories beyond the standard model. This chapter stands on the shoulders of many excellent recent reviews [5-9] and is intended to be an update.

Double-beta-decay with the emission of a neutrino from one neutron and its absorption by another, was first suggested by FURRY in 1939 [10]. This mode has, however, not yet been observed, although it has been the subject of several recent intense searches. The most interesting connection between $0\nu$ $\beta\beta$-decay and fundamental particle theory is that it can be engendered by Majorana neutrino mass, or explicit right-handed neutrino couplings to hadrons, or both. It was shown recently that these latter mechanisms will produce a vanishing amplitude unless at least one neutrino eigenstate has a non-zero Majorana mass. In the standard electroweak model, neutrinos are massless, while interactions which induce neutrino masses naturally arise in the minimal GUT of SO(10) for example [11]. Models which possess left-right symmetry, as well as lepton-hadron symmetry, in general have massive neutrinos; SO(10) is but one example. There are many realistic scenarios in which Majorana neutrino masses on the order of tenths to tens of eV arise in this class of models. This entire domain should be accessible to $0\nu$ $\beta\beta$-decay experiments in the near future. The sensitivities of these experiments as probes of Grand Unification, or as probes of any other exotic physics beyond the standard model, ultimately depend on the level to which background can be reduced and on a clear understanding of the nuclear structure involved. A chronology of counter experiments from 1948 to 1983 is given by HAXTON and STEPHENSON [See Table 3 of ref. 6]. It is interesting to recall how $0\nu$ $\beta\beta$-decay sensitivities have progressed over four decades from $\sim 3 \times 10^{15}$ y in 1948 to $\sim 5 \times 10^{23}$ y in 1987 and

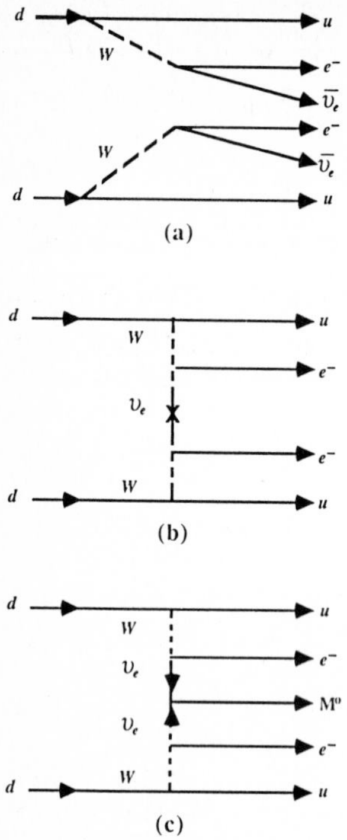


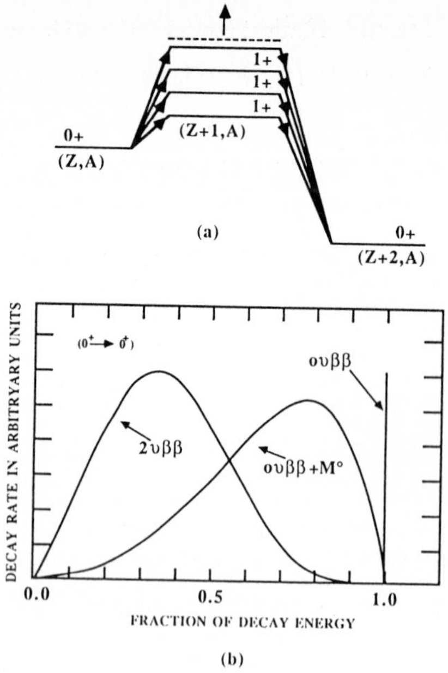


Fig. 1. Diagrams commonly associated with $\beta\beta$-decay modes: (a) ordinary lepton conserving $2\nu$ $\beta\beta$-decay, (b) $o\nu$ $\beta\beta$-decay with no other particles, and (c) $o\nu$ $\beta\beta$-decay with the emission of a goldstone boson (majoron).

Fig. 2. (a) Schematic decay diagram depicting $\beta\beta$-decay through 1+ states of the intermediate nucleus. (b) Theoretical shapes of the spectra of the total electron energy from the three decay modes shown in the diagrams of Fig. 1.

have changed by more than a factor of 100 since 1974. In fact, sensitivities of $2\nu$ $\beta\beta$-decay experiments have also improved by a factor of about 100 since 1967. There is no reason to believe that significant improvements are not still possible.

The time projection chamber (TPC) measurements of the electron spectrum from the $\beta\beta$-decay of $^{82}$Se [4], can in principle distinguish between $2\nu$ and $0\nu$ decay; however, the energy resolution achievable with a source large enough to obtain statistically significant data will be relatively poor (7-10%). The $^{76}$Ge experiments have the distinct advantage of excellent energy resolution ($\Delta E/E \sim 1.8 \times 10^{-3}$) at 2041 keV, the $\beta\beta$-decay energy. In addition, the source is the 7.78% abundant $^{76}$Ge in the detector itself and increases in sample volume will not necessarily affect the energy resolution. Radioactive impurities in the detector and cryostat are more easily identifiable and are less prone to be confused with $\beta\beta$-decay.

Double beta decay is clearly on the forefront of both nuclear physics and elementary particle physics. It has made its comeback mainly on the grounds of constraints it can place on the Majorana mass of the neutrino. The nucleus in that case is just a complicated laboratory. The requirement that the nuclear physics issues be well understood has brought many high-level theoretical efforts into being, and new nuclear structure issues have appeared.

## 1.2 General Theoretical Considerations

The semi-leptonic interaction for the $\beta$-decay of a d-quark is written as follows:

$$H = \frac{-G_F \cos \theta_C}{\sqrt{2}} \{ j_L^\mu [J^+_{L\mu} + \eta_{LR} J^+_{R\mu}] + j_R^\mu [\eta_{RL} J^+_{L\mu} + \eta_{RR} J^+_{R\mu}] \}, \quad (1)$$

where $j_L^\mu$ ($j_R^\mu$) are the components of left (right)-handed leptonic currents, $J_{L\mu}$ ($J_{R\mu}$) are the components of the left (right)-handed hadronic currents, and $\eta_{LR}$, $\eta_{RL}$ and $\eta_{RR}$ are their relative weightings. The Cabibbo angle is $\theta_c$, and the Fermi coupling coefficient, $G_F$, is $1.023 \times 10^{-5}/M_P^2$, where $M_P$ is the proton mass. In the above expression,

$$j_L^\mu \equiv \bar{e} \gamma^\mu (1-\gamma_5) \nu, \quad \text{and} \quad J^+_{L\mu} \equiv \bar{u} \gamma_\mu (1-\gamma_5) d, \quad (2)$$

where e and $\nu$ are the lepton fields and u and d are the quark fields. The corresponding right-handed currents are obtained by use of the projection operator $(1+\gamma_5)$. The left-and right-handed neutrino fields can be expanded in the complete basis of Majorana mass eigenstates $\{v_i(x)\}$ as follows:

$$\nu_R = \sum_{i=1}^{2n} U_i^R v_i, \quad \text{and} \quad \nu_L = \sum_{i=1}^{2n} U_i^L v_i, \quad (3)$$

where n is the number of $\nu$-generations and the transformation matrices U are those which diagonalize the neutrino mass matrix.

The parameters x, y and z, are introduced with the conventions of [6] and with a priori assumption of CP conservation. Accordingly,

$$x \equiv m_e^{-1} \left| \sum_{k=1}^{2n} \lambda^{CP} U_k^L U_k^L (m_\nu)_k \right|, \quad (4)$$

where $\lambda^{CP}$ is the CP eigenvalue of the mass eigenstate with subscript k. The physical quantity to which $0\nu$ $\beta\beta$-decay is sensitive is $\langle m_\nu \rangle_L \equiv x\, m_e$. A quantity $\langle m_\nu \rangle_R$ can be defined analogous to (4). The two parameters associated with right-handed neutrino couplings are:

$$y \equiv \eta_{RR} \left| \sum_{k=1}^{2n} U_k^L U_k^R \right|, \quad \text{and} \quad z \equiv \eta_{RL} \left| \sum_{k=1}^{2n} U_K^L U_K^R \right|. \quad (5)$$

The $\beta\beta$-decay rate, for the $0\nu$ mode, can be expressed in the following general form:

$$\omega(yr^{-1}) = \alpha_1(yr^{-1}) \{x^2 + \alpha_2 y^2 + \alpha_3 z^2 + \alpha_4 xy + \alpha_5 xz + \alpha_6 yz\}. \quad (6)$$

The quantities $\{\alpha_i\}$ depend on the nuclear and atomic wave functions and relevant matrix elements. The various values of these parameters, from recent calculations discussed below, are given in Table I.

TABLE I. Numerical values for the parameters $\alpha_i$ in (6) for the $\beta\beta$-decay of $^{76}$Ge.

| | ref. 6 | ref. 8 | ref. 12 | ref. 13 |
|---|---|---|---|---|
| $\alpha_1$ | $1.08\times10^{-13}$ | $1.12\times10^{-13}$ | $2.18\times10^{-13}$ | $1.21\times10^{-13}$ |
| $\alpha_2$ | 1.44 | 0.92 | 1.04 | 1.43 |
| $\alpha_3$ | 0.65 | 1.14 | 7016 | 30500 |
| $\alpha_4$ | -0.45 | -0.33 | -0.36 | -0.25 |
| $\alpha_5$ | -0.38 | -6.68 | 82.0 | 86.0 |
| $\alpha_6$ | -1.34 | -0.68 | -0.82 | -0.47 |

$\alpha_1$ has dimensions of $y^{-1}$; $\alpha_2$ through $\alpha_6$ are dimensionless

## 1.3 Right-Handed Currents and Majorana Neutrino Mass

Until now it has been customary to state that $o\nu$ $\beta\beta$-decay can be engendered by either non-zero Majorana neutrino mass or the existence of explicit right-handed neutrino couplings to quarks through right-handed $W^{\pm}$ bosons. Recently, KAYSER, PETCOV and ROSEN [14] showed that in the context of any gauge theory, a non-zero amplitude of $o\nu$ $\beta\beta$-decay unambiguously implies that at least one neutrino eigenstate must have a non-vanishing Majorana mass. This is a very important generalization of an earlier proof by DOI, KOTANI and TAKASUGI and the black box proof of SCHECHTER and VALLE [8].

The most rigorous argument hinges on the demonstration that the diagonalization of the most general possible mass matrix results in $<\bar{\nu}_{R'}|\nu_L>=0$; accordingly,

$$\sum_i U^R_{ej}\, U^L_{ej} = 0. \tag{7}$$

Using the general form for the neutrino propagator, one can show that the R-L amplitude for $o\nu$ $\beta\beta$-decay, driven by $\nu$ exchange, is proportional to

$$\sum_j I(m_j)\, U^R_{ej}\, U^L_{ej}, \tag{8}$$

where $m_j$ is the mass of the jth neutrino eigenstate and $I(m_j)$ is the usual integral over momentum exchange. In the case that all neutrino mass eigenstates are degenerate, including the case in which they are zero, $I(m_j)$ will factor out, and the amplitude corresponding to R-L coupling vanishes, while that corresponding to other couplings need not. In any case, if $\beta\beta$-decay with the emission of two electrons only is observed, it will constitute unambiguous evidence that at least one neutrino eigenstate has a non-zero Majorana mass.

## 1.4 Neutrinoless Double-Beta Decay with the Emission of Majorons

Neutrinoless ($0\nu$) $\beta\beta$-decay might also be engendered by a process in which the exchange Majorana neutrinos annihilate resulting in the emission of a Goldstone boson. CHIKASHIGE, MOHAPATRA and PECCEI [15] considered lepton number as a spontaneously broken global symmetry in the context of simple gauge models. Their model contained a new Higgs singlet to break lepton number symmetry, and heavy as well as light neutrino masses were generated. They introduced the name majoron for the resulting massless Goldstone boson. GELMINI and RONCADELLI [16] and, independently but subsequently, GEORGI, GLASHOW and NUSSINOV [17] proposed a model which contains a complex Higgs triplet in addition to the usual doublet. In this model, a non-zero vacuum expectation value of the triplet field spontaneously breaks B-L global symmetry, giving mass to light Majorana neutrinos and resulting in a real massless Goldstone boson, the majoron.

Neutrinoless $\beta\beta$-decay with the emission of a majoron was first discussed by GEORGI, GLASHOW, and NUSSINOV [17], and subsequently by VERGADOS [18], HAXTON and STEPHENSON [6], and DOI, KOTANI and TAKASUGI [8]. Using the results given in ref. 17, the ratio of the rate of $0\nu$ $\beta\beta$-decay with the emission of a majoron, $\Gamma(B)$, to that driven by Majorana $\nu$-mass, $\Gamma(m_\nu)$, can be expressed as follows:

$$\Gamma(B)/\Gamma(m_\nu) = 1.21 \times 10^{-3}\, g_{ee}^2 \left[\frac{m_e}{m_\nu}\right]^2 \epsilon^2\, R(\epsilon), \tag{9}$$

where

$$R(\epsilon) = \left\{\frac{\epsilon^4+14\epsilon^3+84\epsilon^2+210\epsilon+210}{\epsilon^4+10\epsilon^3+40\epsilon^2+60\epsilon+30}\right\}. \tag{10}$$

In (9), $g_{ee}$ is the coupling constant of majorons to electron-neutrinos, and $\epsilon$ is the $\beta\beta$-decay energy in units of electron rest mass energy. The numerical coefficient in (9) differs from those derived from the results given in ref. 18 and ref. 8 by factors of 8 and 4, respectively. The formulae from different articles should be used with great care in the analysis of data because the root of these discrepancies is not clearly understood. However, DOI and his coworkers [19] have addressed these differences and discussed this problem very thoroughly. The progress in searching for this interesting decay mode will be briefly discussed later.

### 1.5 Earlier Nuclear Structure Calculations

An overview of the relevant nuclear structure calculations was given earlier [20]. The recent developments concerning possible suppression of both $2\nu$ and $0\nu$ $\beta\beta$-decay will be dealt with briefly, because it is discussed at some length by K. MUTO and H. V. KLAPDOR elsewhere in this volume.

The most extensive microscopic nuclear structure calculations were the weak-coupling shell model calculations of HAXTON, STEPHENSON and STROTTMAN [21], who treated $^{76}$Ge, $^{82}$Se, $^{128}$Te and $^{130}$Te. In the cases of $^{76}$Ge and $^{82}$Se, the valence space included the $1g_{9/2}$, $1f_{5/2}$, $2p_{3/2}$ and $2p_{1/2}$ levels. The assumed closed core was $^{56}$Ni so that the $^{76}$Ge ground state consisted of 4p-,6n-holes, while the $^{76}$Se ground state consisted of 6p-,8n-holes. The wave functions were constructed with all possible combinations of proton- and neutron-holes. The Kuo matrix elements for the $^{56}$Ni core, were adjusted to fit 28 observed energy levels in the region with an RMS deviation of 270 keV. An improved update of these calculations appears in ref. 6. The results exhibited strong coherence in the density matrix, resulting in large Gamow-Teller matrix elements.

The Osaka group included the p-wave effect [8], as well as a weak magnetism correction to the nuclear current. They found these effects to have approximately equal contributions. It should be pointed out that the Osaka group used the nuclear matrix elements of HAXTON, STEPHENSON AND STROTTMAN so that the differences in their results are attributable to other effects.

Subsequently, TOMODA, FAESSLER, SCHMID and GRÜMMER [12], discussed neutrinoless $\beta\beta$-decay by describing the initial and final nuclear states by angular-momentum- and particle-number-projected Hartree-Fock-Bogoliubov wave functions. In this approach the energy was minimized after projection. For $^{76}$Ge the model space included the $1g_{9/2}$, $1f_{5/2}$, $2p_{3/2}$ and $2p_{1/2}$ levels. The p-wave contribution was included as well as relativistic corrections to the nuclear current including weak magnetism. They also accounted for the fact that the associated two-body operator acquires a finite range due to the finite dimensions of the nucleon. Properly accounting for this effect in the short range NN-correlations results in only a moderate reduction in the matrix element.

ZAMICK and AUERBACH [22] considered the $\beta\beta$-decays of $^{48}$Ca and $^{76}$Ge in the framework of the Nilsson model with pairing. Their calculations explain the slow decay rate of $^{48}$Ca in terms of the K-selection rule ($\Delta K = 0, 1$). Their result is $M_{GT} = 0.18$, in excellent agreement with the shell model calculations of HAXTON and STEPHENSON [6]

($M_{GT}$ = 0.19). These calculations were repeated using the full Nilsson wave functions with similar results [6] and are in quantitative agreement with the shell model calculations in an equivalent model-space.

Nuclear structure calculations of the $\beta\beta$-decays of $^{128,130}Te$, $^{82}Se$ and $^{76}Ge$ were published by GROTZ and KLAPDOR [23,24]. In this work, right-handed couplings were neglected and the nuclear matrix elements calculated in the framework of particle number-projected BCS wave functions with a two body interaction including pairing, spin-isospin, and quadrupole-quadrupole terms. The decay rate is expressed as:

$$w_{0\nu} = (\langle m_\nu\rangle^2/m_e^2)\; G^{0\nu}\; |1-\chi_F|^2 |M^{0\nu} R_0|^2, \tag{11}$$

where $G^{0\nu}$ is the phase space factor given in ref. 8 as 4.2 x $10^{-15}$ $y^{-1}$ for $^{76}Ge$. The Fermi decay branching ratio, $\chi_F$, was calculated as -0.24 and $R_0|M^{0\nu}|$ = 10.4 for this decay, where $M^{0\nu}$ is the Gamow-Teller matrix element. The final result is $\langle m_\nu\rangle^2$ = 2.5 x $10^{23}$ $y/T^{0\nu}_{1/2}$ ($^{76}Ge$) in electron volts squared. KLAPDOR and GROTZ [23] found that strong cancellations, from Gamow-Teller and quadrupole-quadrupole correlations, reduced the $2\nu$ decay matrix elements of the Te isotopes by more than a factor of 10. These cancellations were not found to effect $0\nu$ decay significantly; hence, they explained the discrepancy between the shell model predictions and the geochronological half-lives while strongly supporting the existence of large matrix elements in the case of $0^+ \to 0^+$, $0\nu$ $\beta\beta$-decay. Subsequent work by VOGEL and FISHER [25] includes the effects of pairing, static quadrupole deformation, spin-isospin polarization and the $\Delta_{33}$ isobar admixtures in a Random Phase Approximation analysis. Another extensive RPA treatment, with many useful tables, was given by GROTZ and KLAPDOR [23,24], who give theoretical half lives for $2\nu$ and $0\nu$ $\beta\beta$-decay candidates for all nuclei with $A \geq 70$.

## 1.6 Recent Quasiparticle Random Phase Approximation Calculations

In a recent paper, VOGEL and ZIRNBAUER [26] demonstrated that the Gamow-Teller matrix element, in the quasiparticle random phase approximation (QRPA), is sensitive to particle-particle interactions in the spin-isospin polarization force. They claim that this was neglected in previous calculations, and that values of the relevant coupling constant $g_{pp}$, consistent with $\beta$+ ft values in $^{94}Ru$, $^{96}Pd$, $^{148}Dy$, $^{150}Er$, and $^{152}Yb$, can lead to severe suppression of the Gamow-Teller matrix element for $2\nu$ $\beta\beta$-decay in a number of nuclei. Their two body matrix elements were calculated using a $\delta$-function interaction of the form $V = \sigma_1 \cdot \sigma_2\; \tau_1 \cdot \tau_2\; \delta(\vec{r}_1 - \vec{r}_2)$. In the interaction Hamiltonian, these potentials are weighted by empirical particle-hole and particle-particle coupling constants and the appropriate products of occupation and unoccupation probabilities calculated within the model.

The Gamow-Teller matrix element has the usual second order form:

$$M_{GT} = \sum \frac{\langle f|\vec{\sigma}\tau^+|k\rangle \langle k|\vec{\sigma}\tau^+|i\rangle}{E_k - (E_i - E_f)/2} \;. \tag{12}$$

The matrix elements $\langle 1_k^+|\vec{\sigma}\tau^+|0_i^+\rangle$ and $\langle 0_f^+|\vec{\sigma}\tau^+|1_k^+\rangle$ take the following form in this model:

$$\langle 1_k^+|\vec{\sigma}\tau^+|0_i^+\rangle = \sum_{pn} \langle p|\vec{\sigma}|n\rangle\, (U_p V_n X^k_{pn} + V_p U_n Y^k_{pn}), \tag{13}$$

and

$$\langle 0_f^+|\vec{\sigma}\tau^+|1_k^+\rangle = \sum_{pn} \langle p|\vec{\sigma}|n\rangle\, (V_p U_n X^k_{pn} + U_p V_n Y^k_{pn}). \tag{14}$$

In these equations V (U) are the occupation (unoccupation) probabilities and X and Y are QRPA amplitudes. In the framework of this model, it is clearly demonstrated [26] that in the case of $^{130}Te$, $M_{GT}$ increases from -0.1 (in units of $m_e c^2$) for $g_{pp}$=0 and

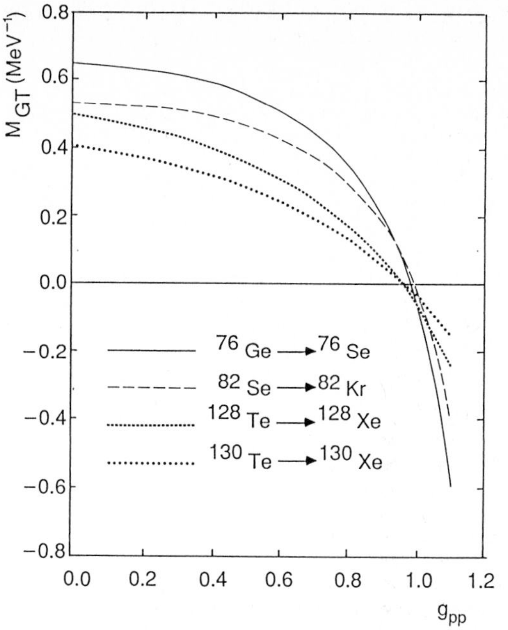


Fig. 3. Gamow-Teller matrix elements of $^{76}$Ge, $^{82}$Se, and $^{128,130}$Te calculated with QRPA for $2\nu$ $\beta\beta$-decay.

passes through 0 in the vicinity of $g_{pp}/g_{pair} \simeq 1$. These parameters are weighting factors for the pairing ($g_{pair}$), particle-hole ($g_{ph}$), and particle-particle ($g_{pp}$) terms. If this model and the chosen parameters are realistic, then the large discrepancy between theory and experiment for the total $\beta\beta$-decay rate, in the case of $^{130}$Te for example, might be explained.

A similar calculation was reported by CIVITARESE, FAESSLER and TOMODA [27]: however, the more realistic renormalized Bonn one-boson exchange potential was used. It was renormalized by solving the Bethe-Goldstone equation and parameterizing with $g_{pair}$, $g_{ph}$, and $g_{pp}$. Again there were values for $g_{pp}/g_{pair} \cong 1$ for which $2\nu$ $\beta\beta$-decay was suppressed in the decays of $^{76}$Ge, $^{82}$Se, $^{128}$Te and $^{130}$Te. Graphs of $M_{GT}^{2\nu}$ versus $g_{pp}$ for these isotopes are shown in Fig. 3. Muto and Klapdor (see K. Muto and H. V. Klapdor in this volume) have completed QRPA calculations of the $2\nu$ $\beta\beta$-decay of $^{76}$Ge, $^{82}$Se, and $^{128,130}$Te. They also use a realistic interaction (the Paris potential) and fix $g_{pp}=0.85 \pm 0.08$ using a large number of experimental $\beta+$ strengths. They also predict a suppression of $2\nu$ $\beta\beta$-decay [28], consistent with ref. [27].

## 1.7 QRPA Analyses Extended to Neutrinoless Double-Beta Decay

Very recently there have been two articles addressing $0\nu$ $\beta\beta$-decay, arriving at somewhat different conclusions [13,29]. In both cases, the calculations were performed in the Quasiparticle Random Phase Approximation; however, different nucleon-nucleon interactions were used. TOMODA and FAESSLER [13] used realistic effective interactions with a G-matrix derived from the Bonn potential. ENGEL, VOGEL, and ZIRNBAUER [29], on the other hand, accounted for short range correlations by the use of a parametrized function $p=1-e^{-\gamma r^2}(1-\gamma r^2)$ multiplying the spin and isospin operators in the Gamow-Teller and Fermi $\beta\beta$-decay matrix elements $M_{GT}^{\alpha}$ and $M_F^{\alpha}$, respectively. They used a $\delta$-function interaction for the nucleon-nucleon two body interaction of the nuclear model itself.

The results based on the approach of ref. 29 were significantly affected by the inclusion of short range correlations, while the inclusion of these effects by use of G-matrix of ref. 13 did not impact the calculated decay rates as much.

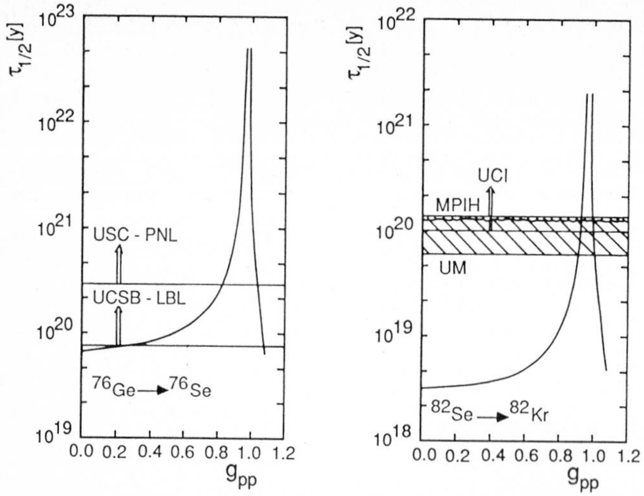


Fig. 4. Calculated $2\nu$ $\beta\beta$-decay half-lives of $^{76}Ge$ and $^{82}Se$ with QRPA [27] as a function of $g_{pp}$.

The calculations of ref. 29 were done with the approximation $y = z \equiv 0$. For comparison if one assumes $\langle m_\nu \rangle = 1$ eV and $x = y \equiv 0$ for both cases, then $T^{0\nu}_{1/2}(^{76}Ge) = 2.9 \times 10^{24}$ years [13], while $T^{0\nu}_{1/2}(^{76}Ge) = 2.7 \times 10^{25}$ years [29]. This factor of 9 difference partly results from the different residual interactions used and partly from the different values of the particle-particle interaction parameter; $g_{pp}/g_{pair} = 1.0$ [13] and $g_{pp}/g_{pair} = 1.47$ [28]. If one uses the graphs of $T^{2\nu}_{1/2}$ versus $g_{pp}$, taken from [27] and shown in Fig. 4, with the experimental half-life of $^{82}Se$ from the recent report by the UCI group [4], one obtains $g_{pp}/g_{pair} = 1.00 \pm 0.04$. For the calculations given in ref. 13, the value 1.0 appears to be appropriate.

At this point, in the framework of QRPA, a mechanism has been identified which suppresses $2\nu$ $\beta\beta$-decay. This could, in principal, also suppress the $0\nu$ mode; however, recent calculations demonstrate that this suppression may not be significant [13,28,29].

In summary, recent developments suggest the possible suppression of the $2\nu$ $\beta\beta$-decay rates. The results of the direct measurement of the $2\nu$ $\beta\beta$-decay half-life of $^{82}Se$ by the UCI group shows excellent consistency between the choice of $g_{pp}/g_{pair}$ by TOMODA et al. [13] and that obtained by intersecting the experimental half-life and theoretical predictions [27,28]. A similar comparison would be valuable for the model of ENGEL, VOGEL, and ZIRNBAUER [29]. This demonstrates the importance of directly observing $2\nu$ $\beta\beta$-decay in a number of isotopes.

## II. $^{76}Ge$ Double-Beta Decay Experiments

### 2.1 Introduction

The ingenious introduction of Ge detectors into the field of $\beta\beta$-decay more than twenty years ago was an innovation of FIORINI and his colleagues at the Institute Nazionale di Fisica Nucleare di Milano [30]. The first limit on the half-life for $0\nu$ $\beta\beta$-decay of $^{76}Ge$ was $T_{1/2} > 2 \times 10^{20}$ y, followed by several more experiments [31,32] which improved that limit by an order of magnitude. This early series of experiments demonstrated the importance of the careful selection of construction materials for the fabrication of the cryostat. They discovered the low content of primordial radioactivities contained in some copper. These efforts were paralleled by the low background studies of WOGMAN and BRODZINSKI [33]. In 1979, AVIGNONE and GREENWOOD suggested the use of a NaI(Tl), Compton suppression live shield for the search for $0\nu$ $\beta\beta$-decay of $^{76}Ge$ [34] and numerically calculated ideal suppression factors for the background sources and levels observed by the Milano group [32]. In 1983, the

USC and PNL groups joined efforts and set a new limit on $0\nu$ $\beta\beta$-decay of $^{76}$Ge of $T_{1/2}>1.7 \times 10^{22}$ y using a Compton-suppressed, low background intrinsic Ge detector in a laboratory on the Earth's surface. Shortly thereafter, BELLOTTI et al. improved the Milano limit to $T_{1/2}>5 \times 10^{22}$ y with a singles experiment [35]. During the same year, BOEHM and his coworkers at Cal Tech produced a very low background Ge detector, housed in a copper cryostat, and operated in a $4\pi$ anticosmic ray veto detector above ground [36]. This was a small detector without Compton suppression, and the limit set was only $T_{1/2}>1.0 \times 10^{22}$ y. The background in the region of the $\beta\beta$-decay energy (~2041 keV) was clearly lower than previous attempts and showed great promise for their planned larger detector. Other efforts by SIMPSON et al. [37], EJIRI et al. [38], LECCIA et al. [39], and CALDWELL et al. [40] have all served to pioneer this important class of experiments. The UCSB-LBL detector [40] is the largest volume and lowest background Compton-suppressed Ge detector and has produced the best limit to date on $0\nu$ $\beta\beta$-decay of $^{76}$Ge ($T_{1/2}>5 \times 10^{23}$ y). It is interesting to note that the new limit on the $0^+\to0^+$, $0\nu$ $\beta\beta$-decay half-life of $^{76}$Ge, set by CALDWELL et al. is a factor of 100 longer than the limit ($5 \times 10^{21}$ y) set by FIORINI et al. in 1973. A very clear understanding of the backgrounds remaining [20, 41-44] promises to bring significant further improvement in the near future.

## 2.2 The Milano (Mont Blanc Tunnel) Experiment

The details of this experiment and its history were given in 1986 in a lengthy article by BELLOTTI et al. [42]. Here, the major experimental issues and the final results will be highlighted.

The apparatus consisted of two Ge detectors in cryostats of specially selected materials. Their Ge-1 was a detector with a volume of 143 cm$^3$, corresponding to a fiducial volume of 117 cm$^3$. The end cap was made of titanium, and the cold finger of oxygen-free high-conductivity (OFHC) copper with a sapphire insulator in the detector holder in place of the usual boron nitride insulator. This detector was operated for a total of 2.361 yr in several shielding configurations. In one, it was surrounded by a 2.5-cm-thick band of Te to search for the $0^+\to2^+$, $0\nu$ $\beta\beta$-decays of $^{128}$Te and $^{130}$Te. In another, the detector was surrounded by 2.5 cm of OFHC copper and 25 cm of lead. The energy resolutions at 2041 keV and at 1482 keV were 2.4 keV and 2.0 keV, respectively. The Ge-2 detector had an active volume of 148 cm$^3$ which corresponds to a fiducial volume of 138 cm$^3$ by Monte Carlo calculation. The end cap was manufactured from OFHC copper 0.5 mm thick. The detector was operated for 0.844 yr with 3.5 cm of triple-distilled mercury, 4 cm of OFHC copper, and 25 cm of low-activity lead shielding. This detector was operated for a second period of 0.917 years with 5.0 cm of copper substituted for the mercury. The energy resolutions for this detector were 2.5 keV at 2041 keV, and 2.1 keV at 1482 keV. The expanded spectrum in the energy region of the $\beta\beta$-decay Q-value (2041 keV) is shown in Fig. 5.

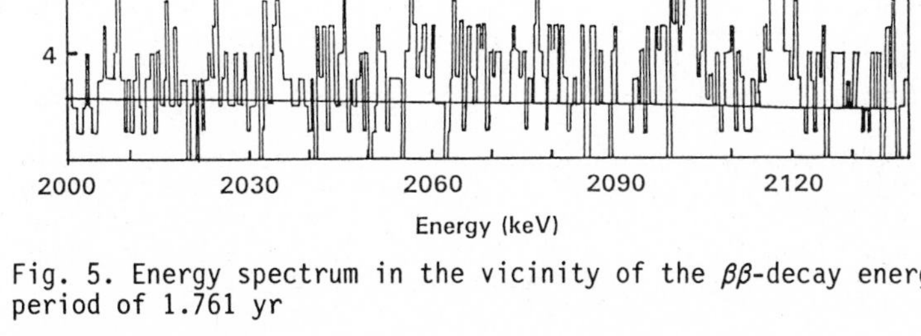

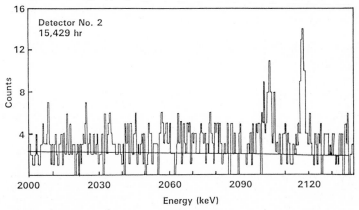

Fig. 5. Energy spectrum in the vicinity of the $\beta\beta$-decay energy taken with Ge-2 over a period of 1.761 yr

The spectra were subjected to maximum likelihood analyses with the following results: detector-1, $0\nu$ $\beta\beta$-decay mode, $T_{1/2}(0^+\to0^+) > 1.2 \times 10^{23}$ y (68%CL); $> 5.4 \times 10^{22}$ y (90%CL).

For detector-2, their analysis yielded $T_{1/2}^{0\nu}$ $(0^+\to0^+) > 2.3 \times 10^{23}$ y (68%CL) and $> 1.0 \times 10^{23}$ y (90%CL).

In this case the authors added the individual likelihood functions and obtained an overall limit:

$$T_{1/2}^{0\nu}\ (0^+\to0^+) > 3.3 \times 10^{23}\ \text{y (68\%CL)}.$$

This experiment has produced a large body of very useful data, however, it has been discontinued in favor of a $^{136}Xe$ $\beta\beta$-decay experiment.

## 2.3 The University of California, Santa Barbara-Lawrence Berkeley Laboratory (UCSB-LBL) Experiment

The UCSB-LBL $^{76}Ge$ $\beta\beta$-decay experiment incorporates the largest volume of germanium detectors at this time. (See ref. 40 for earlier results.) The most recent data derive from eight Ge detectors operated for different periods of time. The crystals average about 160 $cm^3$ of fiducial volume, or about 0.9 kg of natural Ge, operating inside of a 15-cm-thick NaI(Tl) shield. (See Fig. 6.)

The apparatus is located 200 meters underground in the powerhouse of the Oroville, California, Dam. This location is of more than passing significance. It experiences approximately $7 \times 10^6$ muons per $m^2$ per year, whereas the Kamioka site has $\sim 2 \times 10^4$, and the Homestake mine and the Frejus Tunnel $\sim 1 \times 10^3$, all in the same units. All Ge detectors which have been on the surface for a year or more will suffer significant cosmogenic production of $^{56,57,58}Co$, $^{65}Zn$, and $^{68}Ge$ [20]. It is also anticipated that $^{60}Co$ will be present to some extent. These are due mainly to energetic neutron reactions on $^{70}Ge$ in the detectors. The $^{56}Co$ and $^{58}Co$ isotopes are relatively short-lived but $^{57}Co$, $^{65}Zn$ and $^{68}Ge$ have half-lives of 272, 244 and 271 days, respectively. These isotopes appear to be decaying in the UCSB-LBL experiment, which means that only moderate depths may be necessary to achieve significant cosmogenic background reduction.

The main interfering component thus far appears to be the internal $\beta+$ spectrum of $^{68}Ga$, the equilibrium daughter of $^{68}Ge$. The $\beta+$ end point energy is 1.90 MeV, which sums with the 0.511 annihilation gamma-rays to produce a continuum up to 2.92 MeV. This component also appears to be decaying based on the decrease with time of the background in the 2041 keV region.

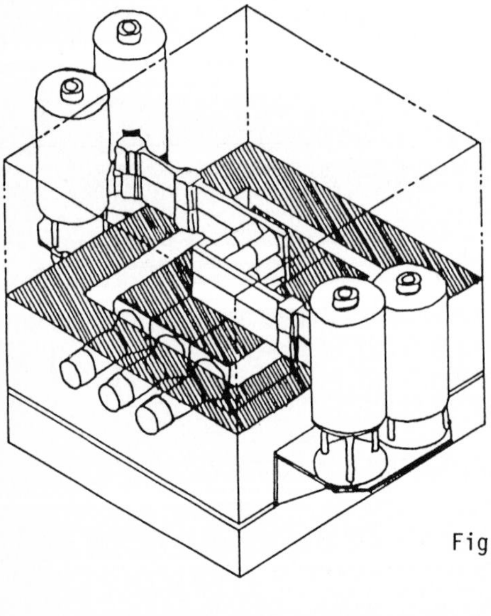

Fig. 6. The UCSB-LBL multi-crystal Ge detector and NaI(Tl) shield

Some of the background comes from primordial radioactivities as clearly demonstrated by $\gamma$-ray lines at 185.72, 911.07, 1001.03, 1460.75, and 2614.47 keV and the other expected lines associated with the decays of $^{235}U$, $^{228}Ac$, $^{234m}Pa$, $^{40}K$, and $^{208}Tl$. The last published background rate at 2041 keV for this experiment was 1.4 counts/keV/kg/yr, where the mass is that of the natural germanium in the fiducial volume of the detector. Even after the $^{68}Ge$ decays for several more half-lives, the primordial radioactivity will dominate and any improvement on the half-life will increase only as the square root of the counting time.

The data from this detector were subjected to a maximum likelihood analysis yielding, $T_{1/2}^{0\nu} > 5 \times 10^{23}$ y [40]. This limit is about twice as long as the recently published "world limit" [44], which included some UCSB-LBL data.

This detector has also been used to search for heavy neutrinos, $0\nu$ $\beta\beta$-decay with the emission of majorons, cold dark matter candidates, and other exotica.

## 2.4 Other promising Compton-Suppressed Ge detectors

There are three other Compton-suppressed Ge detectors which are significantly smaller than the UCSB-LBL detector, but which have made significant reductions in their backgrounds and promise interesting future results. One, which has an enriched $^{76}Ge$ crystal, will be discussed last.

### 2.4.1 The Frejus Tunnel Experiment

The Frejus Tunnel $^{76}Ge$ $\beta\beta$-decay experiment is a collaboration between French and Spanish Groups [45,46]. The system consists of four intrinsic Ge detectors with a total active volume of 417 $cm^3$ in a single cylindrical vacuum chamber 160 mm in diameter by 120 mm. The assembly is placed in a $4\pi$ annulus of hexagonal NaI(Tl) detectors, as shown in Fig. 7.

The NaI(Tl) scintillators are 13.5 cm across and 20.4 cm long. Fourteen crystals, manufactured to low background specifications, are wrapped in Teflon®. They have quartz optical windows and are contained in low-activity stainless steel shells 0.5 mm thick. Five more are packaged in aluminum and form the floor of the annulus. The detector's location in a laboratory in the Frejus tunnel experiences about $1 \times 10^3$ muons per $m^2$ per year, about 4 orders of magnitude fewer than in the Oroville Dam laboratory.

The most interesting result to date is the limit $T_{1/2}^{0\nu}$ $(0^+ \rightarrow 2^+) \gtrsim 2.1 \times 10^{22}$ y from an experiment requiring coincidences between the Ge detector and the NaI(Tl) annulus. A small bump was observed a few keV above the expected energy of 1482 keV for the decay to the first excited state of $^{76}Se$ at 559 keV. More details on this phenomenon will be given elsewhere in this volume.

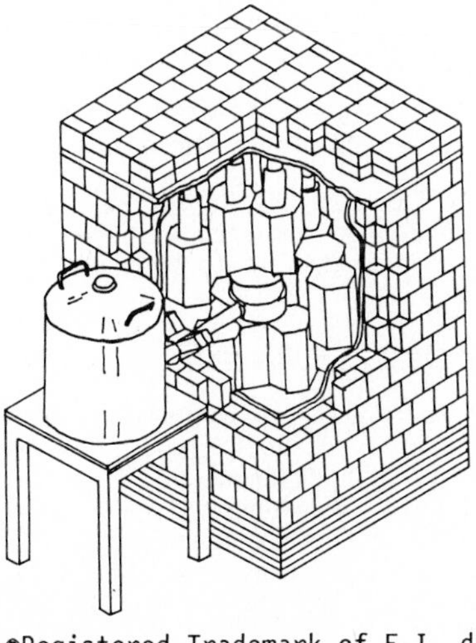

Fig. 7. The Frejus Tunnel, Compton-suppressed Ge spectrometer

®Registered Trademark of E.I. duPont de Nemours and Co., Inc., Wilmington, Delaware

A special low-background 200-$cm^3$ detector is being built for this experiment. A collaboration between PNL-USC and the University of Zaragoza, to better understand background in $^{76}Ge$ $\beta\beta$-decay experiments, will include searches for cold dark matter candidates, solar axions, and $(0^+ \rightarrow 2^+)$, $0\nu$ $\beta\beta$-decay.

2.4.2 The Osaka, Kamiokande Underground Laboratory Experiment

The engineering details of this experiment are given in an article by KAMIKUBOTA et al. [47]. The most recent published results appeared in a 1987 article by EJIRI et al. [38]. The detector, called ELEGANTS, is Compton-suppressed as shown in Fig. 8.

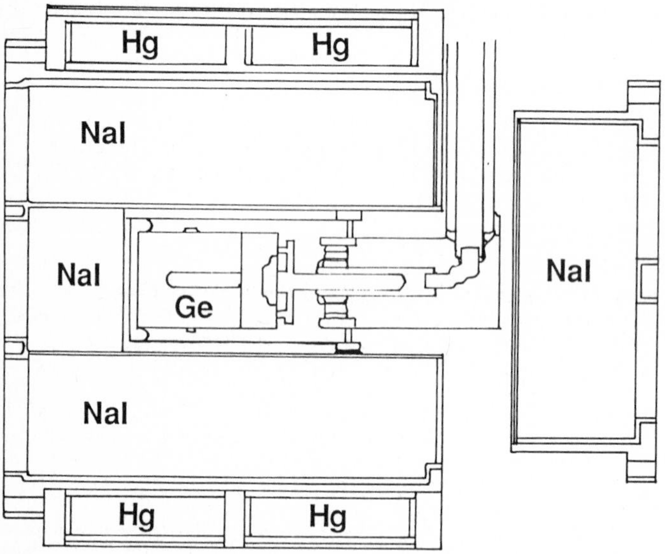


Fig. 8. The Osaka Compton-suppressed $^{76}Ge$ $\beta\beta$-decay ELEGANTS spectrometer

ELEGANTS comprises a 171-$cm^3$ intrinsic Ge detector placed in a 5-segment, NaI(Tl) annulus 25.4 cm in diameter and 30.48 cm in length. There is a NaI(Tl) crystal at each end of the annulus to form a $4\pi$ geometry. The Ge detector is surrounded by 2.7 cm of pure mercury, which is in turn surrounded by 15 cm of OFHC copper and 15 cm of lead. Radon is purged using the boiloff nitrogen gas which is piped to the central cavity. All materials were carefully selected for low radioactivity [47]. Even with these precautions, the $\gamma$-ray lines from $^{40}K$, $^{208}Tl$, $^{212}Pb$, $^{214}Pb$, $^{214}Bi$, $^{228}Ac$, $^{234m}Pa$, and $^{226}Ra$ are clearly visible in the spectrum. Turning on the annulus to reject background reduces the counting rate in all energy regions by more than a factor of 10.

The suppressed background was improved to ~$4.3 \times 10^{-4}$ counts/keV/hr in the 2041 keV region which converts to ~4 counts/keV/yr/kg and which can be compared with the rate 1.4 counts/keV/yr/kg of the UCSB-LBL experiment. This apparatus is to be used in a $^{100}Mo$ search for $\beta\beta$-decay described in ref. 47.

2.4.3 The ITEP-Erevan, Avansk-Mine, $^{76}Ge$ Enriched Experiment in the USSR

The advantages of $^{76}Ge$ enrichment by an order of magnitude over the natural abundance of 7.78% are obvious. Consider that the limit which can be set with the current PNL-USC 120 $cm^3$ fiducial volume detector, which presently has a background of $\lesssim 1$ count/keV/yr/kg, is $T^{0\nu}_{1/2}$ ($^{76}Ge$) $\gtrsim 2 \times 10^{23}$ y with Nt $\simeq 4 \times 10^{23}$ y, where N is the number of $^{76}Ge$ atoms and t is the counting time in years (~1 yr). If this detector was enriched to 85% $^{76}Ge$, the limit that could be set with the same background would be $T^{0\nu}_{1/2} \gtrsim 2.2 \times 10^{24}$ y. A number of such experiments throughout the world could set a limit $<m_\nu>\lesssim 0.1$ eV in the most conservative nuclear structure scenario, and eventually maybe even $\lesssim 0.03$ eV. The ultimate limit would, of course, depend on the numbers and volumes of such detectors and their background levels.

A very brief report on a 90-$cm^3$ Ge detector enriched to 85% $^{76}Ge$ and contained in a NaI(Tl) Compton suppression shield was presented by KIRPICHNIKOV at the July 1987 Underground Physics workshop at the Baksan Laboratory in the USSR. The configuration of the detector and shielding and a gross spectrum after only 436 hr of operation are shown in Fig. 9.

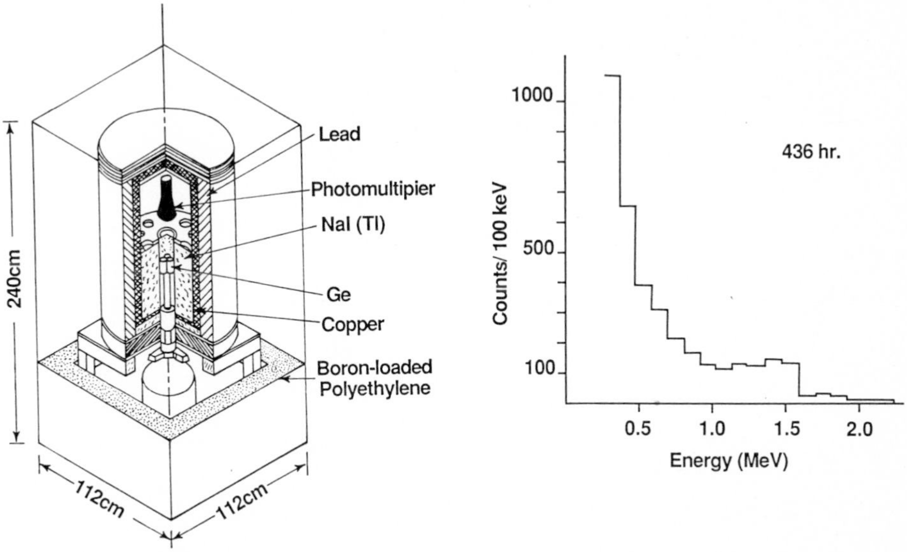


Fig. 9. Shielding configuration and gross spectrum of the 90 $cm^3$, 85% enriched $^{76}Ge$ ITEP-Erevan experiment

After 652 hrs of counting, there was one count in the interval 2035-2045 keV or less than 2 counts to a level of confidence of 68%. The quantity Nt is 2.53 x $10^{23}$ y, hence the implied limit is $T^{0\nu}_{1/2}$ ($^{76}Ge$) $\gtrsim 0.9 \times 10^{23}$ y. The background is ~2.8 counts/keV/yr/kg, which can be compared to that of the 1987 level of the UCSB-LBL experiment (1.4 counts/keV/yr/kg) and the new PNL-USC level (1.0 counts/keV/yr/kg).

This spectrometer will not be powerful for studying $2\nu$ $\beta\beta$-decay and $o\nu$ $\beta\beta$-decay with majoron emission because the background below ~1500 keV is at least a factor of 22 higher than contemporary experiments. The combination, however, of enrichment and ultralow background singles counting technology would indeed represent a more powerful tool than now exists, and should be encouraged.

## 2.5 The PNL-USC $^{76}Ge$ $\beta\beta$-Decay Experiment

Many of the details of the development of this experiment already appear in print [20, 41,44]. For purposes of comparison to experiments already discussed, the current mean background level is < 1 count/keV/yr/kg in the 2041 keV region which is somewhat lower than the Compton-suppressed spectrometers and about the same as the lowest background Milano experiment. The background in the present PNL-USC experiment is dominated by $^{68}Ga$ in this energy region. The shielding configuration and gross spectra under various conditions are shown in Fig. 10.

The most recent modifications to the experimental configuration resulted in substantial reductions in both the low- and high-energy regions of the spectrum. The drop at energies below ~500 keV was due to replacement of the indium contact ring with one made of 448-year-old lead. The dramatic decrease above 3000 keV was due to the removal of the solder in proximity to the detector. There was no improvement in the

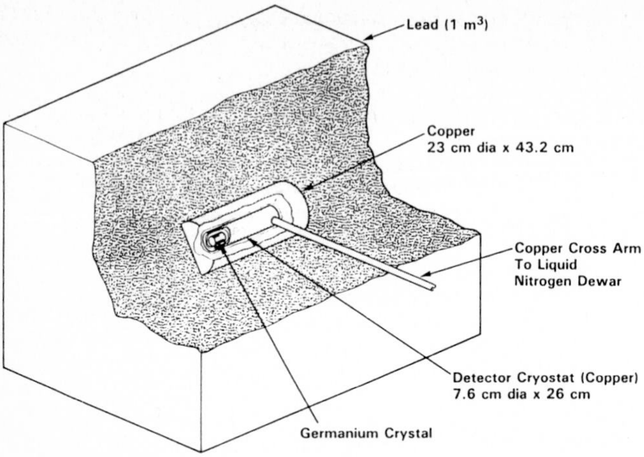


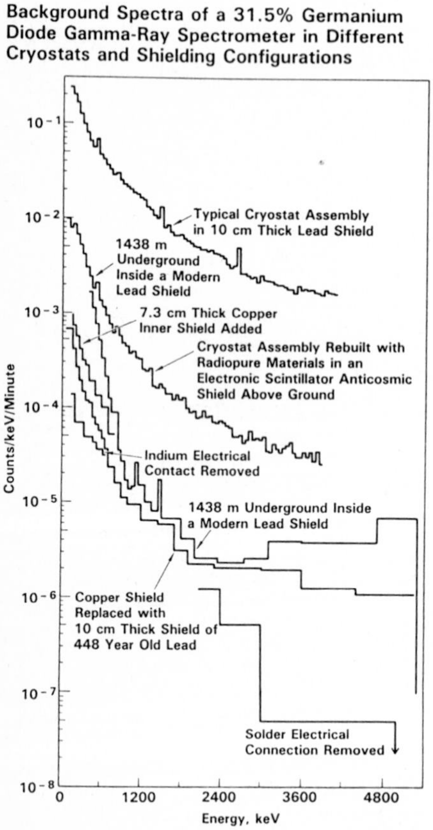


Fig. 10. PNL-USC detector configuration and spectra under various conditions

average background in the energy interval between about 800 keV and 2000 keV. This region is now dominated by $^{54}$Mn, $^{58}$Co, $^{65}$Zn, $^{68}$Ge and other cosmic-ray-produced radionuclides, because the detector was on the surface for almost 8 months for renovation. The earlier experiments suffered significantly lower background from cosmogenic isotopes in the crystal itself because of its long history of containment in neutron moderators, followed by several years in the Homestake mine. A current spectrum is shown in Fig. 11. The spectrum is remarkably free of $\gamma$-ray lines from primordial radioactivity. The striking feature of Fig. 11 is that there are no obvious lines above the 1461 keV $\gamma$-ray from the decay of $^{40}$K.

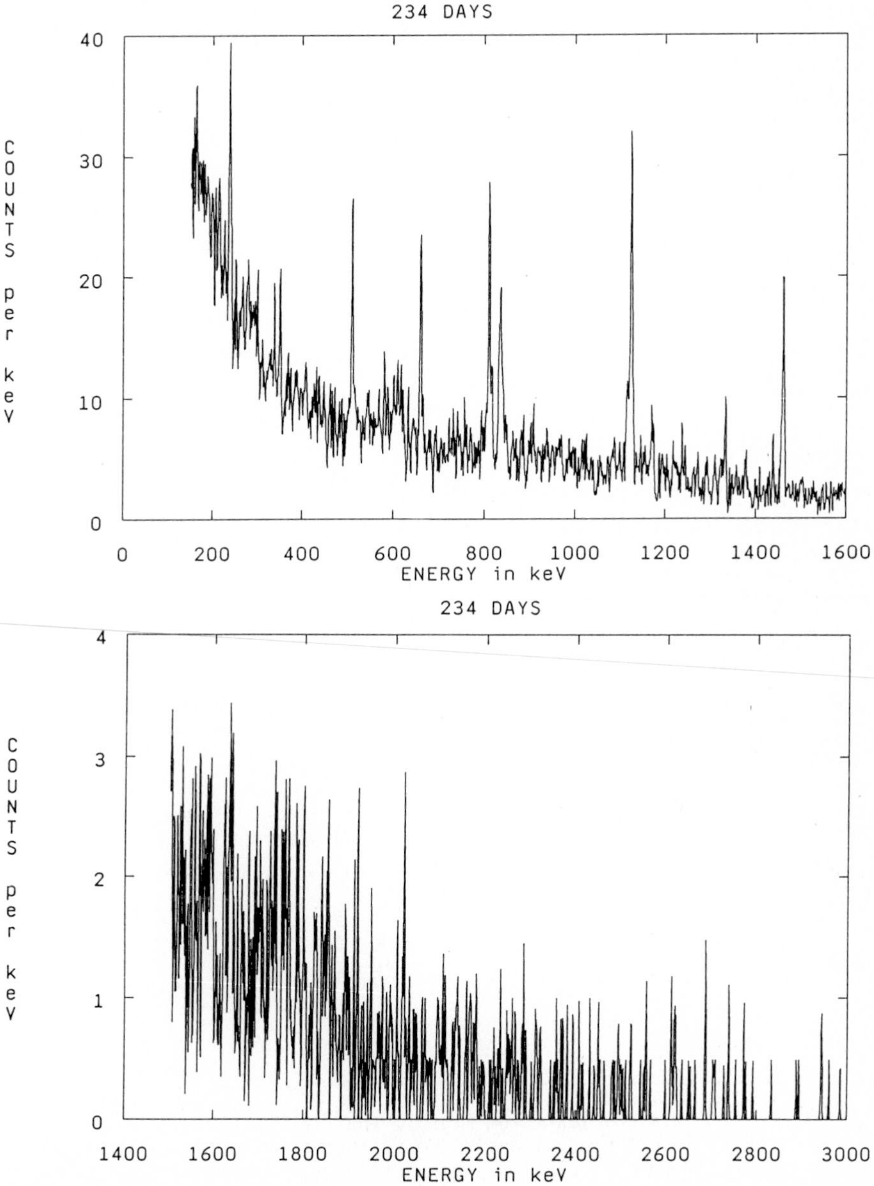


Fig. 11. High-resolution background spectrum from the PNL-USC 125 cm$^3$ fiducial volume detector accumulated for 234 days

In reviewing the subject of material selection, an abbreviated list of typical construction materials and their concentrations of the isotopes $^{208}$Tl and $^{214}$Bi, from the thorium and uranium chains, and for $^{40}$K is presented in Table II below.

TABLE II. Primordial radionuclide concentrations in various materials used in fabrication of radiation detector systems

| Materials | Radionuclide concentration in dis/min per kg $^{208}$Tl | $^{214}$Bi | $^{40}$K |
|---|---|---|---|
| Aluminum | 7-200 | <4-2000 | <20-1000 |
| Copper (grade 101) | <0.03 | <0.05 | <0.5 |
| Epoxy | 50-4000 | 80-53,000 | <1000-72,000 |
| Indium | <1 | <3 | <20 |
| Lead | <0.02 | <0.04 | <0.1 |
| Molecular sieve | 400-500 | 1000-3000 | 8000-9000 |
| Mylar, aluminized | 100 | 200 | <2000 |
| Printed circuit board | 2000 | 4000 | 4000 |
| Solder | <0.3 | <0.8 | <10 |
| Steel, Stainless | <2 | <6 | <200 |
| Steel, pre-WW II | <0.5 | <0.9 | <10 |
| Teflon | <0.3 | <1-7 | <20 |
| Wire, Teflon coated | <4 | <1 | <20 |

The importance of avoiding certain materials - for example, some types of aluminum, epoxies, molecular sieve and printed circuit board - is obvious. The impact of careful material selection is reflected in Table III below. The primordial radioactivity levels underwent reductions of factors between 1.6 x $10^3$ and 5.7 x $10^4$ by

TABLE III. Comparison of primordial radioactivity levels in the background of the Ge spectrometer before and after rebuilding with radiopurity selected materials

| Primordial Radionuclide | Gamma ray energy (keV) | Count rate before rebuilding (counts/h) | Count rate after rebuilding (counts/h) | Improvement factor |
|---|---|---|---|---|
| $^{235}$U | 185.72 | 73.0 | <0.0013 | >57,000 |
| $^{228}$Ac($^{232}$Th) | 911.07 | 9.0 | <0.00087 | >10,000 |
| $^{234m}$Pa($^{238}$U) | 1001.03 | 3.4 | <0.00080 | >4,300 |
| $^{40}$K | 1460.75 | 22.0 | 0.014 | 1,600 |
| $^{208}$Tl($^{228}$Th) | 2614.47 | 1.0 | 0.00051 | 2,000 |

material selection. It is also clear from Fig. 11 that the background counting rate of the PNL-USC detector is not predominantly due to these radioactivities. The dominant remaining background is due to cosmogenic radioactivities in the Ge crystal itself: primarily $^{68}$Ge, $^{65}$Zn, $^{58}$Co and $^{56}$Co with half-lives of 271, 244, 71, and 78 days, respectively. The $^{68}$Ge leads to a $\beta$+ and annihilation continuum out to 2920 keV. The $^{65}Zn$ decays by electron capture and produces a peak at 1124 keV, due to the sum of $\gamma$-ray and x-ray energies, and a continuum due to partial escape of the photons. The cobalt isotopes play little role after a few months.

To mitigate this background, enough Ge ore was recently mined to produce 20 kg of metal after purification and zone refinement. This process was completed in a few days, and the Ge ingots were rushed underground into the Homestake mine to await final purification and fabrication into large, ~200 $cm^3$, detectors. This process should be completed in a short enough time to produce only a few percent of the saturation level of $^{68}Ge$. The $^{65}Zn$ and Co isotopes should be removed up to the final zone refinement. In addition, the copper cryostat parts have been electroformed to reduce the $^{60}Co$ induced by energetic cosmic ray neutrons via the reaction $^{63}Cu(n,\alpha)^{60}Co$.

It is hoped that two ~200-$cm^3$ Ge crystals, low in cosmogenic background, and installed in the electroformed copper cryostat, will be operational by the end of 1987. It is anticipated that the low background level above 3000 keV, shown in Fig. 10, will extend below 2000 keV, allowing a sensitivity of at least 6 to 10 times that of the present prototype for $0\nu$ $\beta\beta$-decay. This would allow a half-life limit of $>10^{24}$yr to be established in one year of counting, and will produce extremely valuable data to combine with other experiments of similar sensitivities. Since it will be a singles experiment, it will allow a far more sensitive search for the continua from $2\nu$ $\beta\beta$-decay and from $0\nu$ $\beta\beta$-decay with majoron emission. These subjects will be discussed in more detail later.

## 2.6 Other Non-Compton-Suppressed $^{76}Ge$ $\beta\beta$-Decay Experiments

There are two other important single counting experiments which have the potential to make major contributions in the near future. One is the assembly of ~208 $cm^3$ detectors of SIMPSON et al. [37], which was last reported in the literature in 1984. A large body of data (Nt~6.6 x $10^{23}$ y) was made available in 1986 to assist in producing the "world limit" [44]. These background data were comparable to the other four experiments used in that analysis. New, high-quality results are expected in the not too distant future.

The other experiment is that of BOEHM and coworkers. This detector was originally above ground in a lead shield placed inside a $4\pi$ cosmic ray shield [36] and was then moved to the St. Gotthard tunnel laboratory. There, it had a 21-cm-thick copper shield inside of a lead shield. The background count rate at ~2000 keV was slightly lower than 2 counts/keV/yr/kg. The background below this energy increased rapidly due, in part, to the cosmogenic $^{60}Co$ activity in the copper. The $^{68}Ga$ background is also very evident in this detector. A large-volume (~1120 $cm^3$) version of this detector is already operating and will produce a valuable body of data on $0\nu$ $\beta\beta$-decay of $^{76}Ge$ in the near future. If the thick copper shield is used in the new version, the detector will, like its prototype, be limited in the quality of data obtainable on $2\nu$ $\beta\beta$-decay and $0\nu$ $\beta\beta$-decay with the emission of a majoron. This subject will be discussed in the next section.

## 2.7 Searches for Neutrinoless $\beta\beta$-Decay with the Emission of a Majoron

The theoretical issues involved in this subject were discussed in Section 1.4. In 1986 CALDWELL et al. [43] placed the limit $T_{1/2}^{0\nu,m} > 6 \times 10^{20}$ y on this mode of decay in $^{76}Ge$. References to earlier limits in other isotopes can be found in ref. 43. In January of 1987, the PNL-USC group announced the existence of an anomalous bump in the continuum of their $^{76}Ge$ $\beta\beta$-decay experiment, which was suggestive of $0\nu$ $\beta\beta$-decay with the emission of a massless or near massless third particle [48]. The corresponding half-life was $(6\pm1) \times 10^{20}$ y with certain assumptions in the analysis. Since then, two articles have appeared in the literature which place limits of about twice that on the process. The first was from the St. Gotthard Laboratory experiment of BOEHM and coworkers [49] and the second was by CALDWELL et al. [40]. This issue will be impossible to resolve without further, more sensitive experiments.

## III. Selenium-82 Double-Beta Decay: The First Laboratory Observation of $\beta\beta$-decay

In the fall of 1987, M. K. MOE and his co-workers announced the first direct observation of $\beta\beta$-decay in a laboratory experiment; they measured $T_{1/2}^{2\nu}$ ($^{82}$Se) = $(1.1^{+0.8}_{-0.3}) \times 10^{20}$ y using a TPC [4]. The shell model prediction for this half-life is $0.64 \times 10^{20}$ y [50]. This factor of 1.7 difference between theory and experiment does not seem to indicate severe suppression of this decay.

The experimental technique is described in the Proceedings of the conference "The Time Projection Chamber," held in Vancouver, British Columbia in 1983 [51]. The UCI TPC consists of an octagonal chamber with an 84-cm major diameter and a length of about 20 cm. The source was 14 g of 97% isotopically enriched $^{82}$Se metal contained between ultra-thin aluminized polyester sheets. The source thickness corresponds to about 7 mg/cm$^2$. A schematic diagram is shown in Fig. 12. Large Helmholtz coils provide an axial magnetic induction field of 700 gauss, uniform to 1% over the active volume. The spacial resolution afforded by the wire array at the end of the chamber is 5mm for both the x and y directions.

The chamber contains a gas mixture of 93% He and 7% $C_3H_8$, at an absolute pressure of an atmosphere. Two electrons from an event at the source plane create helical ionization tracks in the chamber that then drift toward the position-sensitive wire arrays. The longest transit time of the ionization electrons is about 20 microseconds. The 10-cm drift distance is separated into 20 equal time intervals, resulting in a time resolution for axial motion of $\sim 10^{-6}$ s which corresponds to an axial spatial resolution of 5 mm. The energies of electrons emerging from the source plane at angles of $\geq 45°$ relative to the axial fields can be determined with a precision of a few percent.

The materials used in the construction of the chamber were carefully selected for low radioactivity and for compatibility with the gas. The apparatus is shielded by a lead house, 15 cm thick on the ends and 10 cm thick on the sides, and is surrounded by a $4\pi$, multi-wire proportional counter, cosmic-ray veto. Cosmic-rays are electronically vetoed up to 30 microseconds prior to the prompt TPC pulse.

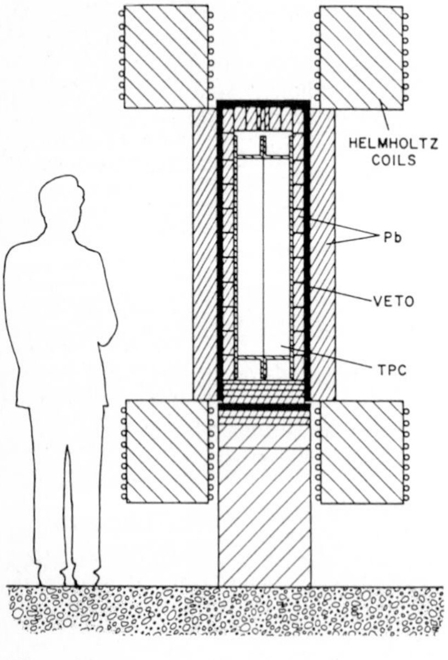


Fig. 12a.
University of California, Irvine TPC, showing a cross section of the chambers, wire detectors, and shielding.

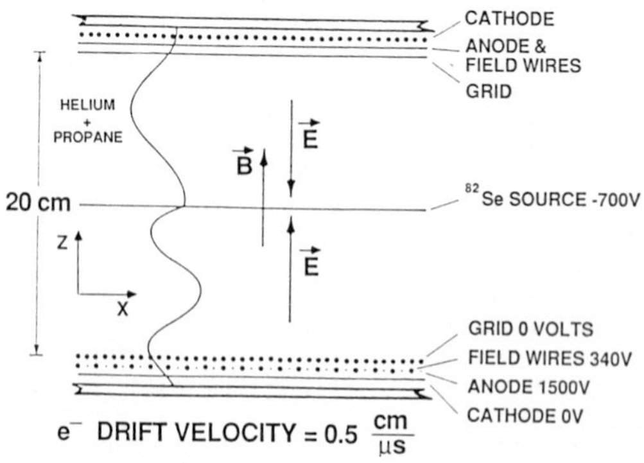


Fig. 12b.
Side view of the UCI TPC showing the source plane, $\vec{E}$ and $\vec{B}$ fields, and wire chambers.

The chamber was operated for 7960 hours live time. The dead time for the entire system was ~10%. Data from the 600 channels are recorded as "hits" with x and y wires on both ends of the chamber identified, as well as the relative times of the pulses. In this way, the ionization tracks can be reconstructed and fit to helices of a given pitch and radius.

The backgrounds were studied in detail and found to be dominated by two sources. One was the Moeller scattering of single electrons in the source plane. The second was the beta - internal conversion ($\beta$-IC) cascades in the decay of $^{208}$Tl. Similar cascades from $^{214}$Bi in the uranium series and $^{212}$Bi from the thorium series can easily be identified in the TPC by the associated $\beta$-$\alpha$ cascades. The only significant background in the event energy region of 1.3 to 2 MeV was found to be Moeller scattering. The energy spectral shapes of both the $\beta$-IC and Moeller scattering events were measured in the TPC. At the end of the 20-microsecond drift time, the system searches for the 164-microsecond alpha particle decay of $^{214}$Po following $^{214}$Bi beta decay.

The events, not vetoed by cosmic-ray pulses or by $\alpha$ pulses, were subjected to four software cuts: their sum kinetic energies must be greater than 800 keV; their single kinetic energies must be greater than 150 keV; events must consist of ionization tracks emerging from the same point on the source plane, with no further activity for one millisecond; and the electrons must be emitted on the opposite sides of the source plane.

There are many events below 1.1 MeV in excess of the expected $2\nu$ $\beta\beta$-decay rate. These events are associated with $\beta$-IC cascades from $^{214}$Pb which was shown to be deposited on the source plane from decay of $^{222}$Rn in the TPC gas. It was also found that a portion of the radon comes from the Be-Cu used for the grid, cathode, and field wires. Electron energy calibrations were obtained using internal conversion lines in the decay of $^{208}$Tl following injection of $^{220}$Rn into the chamber.

A total of 46 events between 1.3 and 2.0 MeV were collected in 7960 h, which satisfied the conditions of the software cuts discussed above. These events were assumed to be from three sources: $2\nu$ $\beta\beta$-decay, Moeller scattering of electrons, and $\beta$-IC cascades from $^{208}$Tl. The spectral shapes of Moeller scattering and $\beta$-IC cascades were measured in the TPC, and that of $2\nu$ $\beta\beta$-decay was predicted. A three-parameter maximum likelihood analysis was made, where the fractional weightings of each of the three spectral components were the parameters. The results of the analysis is that the most likely value of the $2\nu$ $\beta\beta$-decay mixture is 0.78, that for Moeller scattering is 0.22, and that for the $^{208}$Tl $\beta$-IC background is 0.00. Therefore, 35.9 of the 46 events are from $2\nu$ $\beta\beta$-decay, while 10.1 are due to Moeller scattering.

The probability for $2\nu$ $\beta\beta$-decay events to be detected in the TPC, and then to survive the data cuts discussed above, is 6.2%. The half-life is determined from the simple relation,

$$T_{1/2} = \frac{\ln(2)\epsilon N t}{n},$$

where N is the total number of $^{82}$Se atoms in the sample, t is the counting time, $\epsilon$ is the total efficiency (6.2%), and n is the total number of observed events (35.9) attributed to $2\nu$ $\beta\beta$-decay. The best value is $1.1 \times 10^{20}$ y, with a 68% confidence level range of $(0.8\text{-}1.9) \times 10^{20}$ y. This result agrees with the geochemical (geochronological) values given by Kirsten [3], $(1.30\pm0.05) \times 10^{20}$ y, and by Manuel [52], $(1.0\pm0.4) \times 10^{20}$ y.

## IV. Molybdenum-100 Double-Beta Decay Experiments

The search for various modes of $\beta\beta$-decay in $^{100}$Mo has a number of advantages. The decay energy is 3033 keV, and the larger phase space factor significantly enhances the decay rate over that of $^{76}$Ge, for example, if the nuclear matrix elements are identical. ELLIOT, HAHN, and MOE searched for the decay modes in $^{100}$Mo using the UCI TPC [53]. The Osaka group [54] and the Lawrence Berkeley Laboratory-Mt. Holyoke College-University of New Mexico group [55] have both been developing experiments consisting of layers of lithium-drifted silicon, Si(Li), detectors and thin molybdenum foils.

In Table 1 of ref. 55, the ratios of theoretical half-lives for the $\beta\beta$-decay modes of $^{76}$Ge to those of $^{100}$Mo are presented. The ratios of the decay rates, $\omega(^{100}\mathrm{Mo})/\omega(^{76}\mathrm{Ge})$, based only on the kinematic factors, are 10, 21, 15 and 72 for $(0^+\rightarrow0^+)$, $0\nu$ $\beta\beta$-decay driven by $\langle m_\nu\rangle$, $(0^+\rightarrow2^+)$ $0\nu$ $\beta\beta$-decay, $0\nu$ $\beta\beta$-decay with the emission of a majoron, and $2\nu$ $\beta\beta$-decay, respectively.

If the nuclear matrix elements of GROTZ and KLAPDOR are used in the case of $2\nu$ $\beta\beta$-decay, this ratio becomes 120 [23]. If, on the other hand, the nuclear matrix elements of VOGEL and ZIRNBAUER [26] are used, the ratio is between 170 and 640. Hence, there is no question that $^{100}$Mo experiments could yield the best measurements of $2\nu$ $\beta\beta$-decay if backgrounds can be reduced sufficiently.

## 4.1 The UCI $^{100}$Mo $\beta\beta$-Decay Experiment

The general features of the UCI TPC were discussed in Section III. There were actually two generations of the TPC, TPC1, and TPC2. TPC2 is essentially the same as TPC1 except that it was built with lower-background materials.

The internal conversion lines of $^{207}$Pb, following the electron capture decay of $^{207}$Bi, were used to measure the energy resolution and calibrate TPC1. These characteristics were assumed to be the same for TPC2. The resolution was ~175 keV at 1 MeV, so that the sum-energy resolution at 2 MeV is about $0.175/\sqrt{2}$ = 12.4%.

The source was a 30-cm by 29.2-mg/cm$^2$ foil that was 99.9%-pure, natural-abundance molybdenum. It contained $1.58 \times 10^{22}$ atoms of $^{100}$Mo and $2.44 \times 10^{22}$ atoms of $^{92}$Mo.

As described earlier, an important property of the TPC for the rejection of background is its ability to detect $\alpha$ particles from $^{214}$Po and thereby allow the elimination of interference from the $\beta$ - internal conversion electron pairs from the parent $^{214}$Bi. It is important to accurately know the probability that the $\alpha$ particle will escape from the source. This was calculated to be 25.3%. The Compton electron production rate in the $^{82}$Se source had been accurately determined [4], and the $\gamma$-ray background in the chamber had been reduced by rebuilding the chamber. The Mo foil was found to have a much higher rate of single-electron emissions than the $^{82}$Se source, although the $\gamma$ background was lower. The highest energy $\beta$ branch of $^{214}$Bi has an endpoint energy of 3.28 MeV, the highest in the uranium and thorium chains. It was assumed that the majority of electrons between 2.2 and 3.4 MeV leaving the molybdenum source were from $^{214}$Bi. There were 285 single-electron events; only 60 had accompanying $\alpha$ particles. Assuming all were from the $\beta$ decay of $^{214}$Bi, the $\alpha$-escape probability is 21.0±2.4%. This agrees with the theoretical prediction assuming that the $^{214}$Bi is distributed homogeneously throughout the foil rather than on the surface. It was further assumed that the U and Th chains in the foil were in equilibrium, which may not be valid, and it was concluded that the contaminations are 60 ppb U and 30 ppb Th.

The chamber was operated for 252.7 live hours. The sum spectrum was corrected by subtracting the $^{214}$Bi background using the calculated $\alpha$-escape efficiency. The results for the various decay modes are:

$$T_{1/2}(^{100}\mathrm{Mo};2\nu;0^+\rightarrow0^+)\gtrsim6.8 \times 10^{17} \text{ y (68\% CL)},$$

$$T_{1/2}(^{100}\mathrm{Mo};\ 0\nu M^\circ;0^+\rightarrow0^+)\gtrsim7.5 \times 10^{18} \text{ y},$$

$$T_{1/2}(^{100}\mathrm{Mo};0\nu;0^+\rightarrow0^+)\gtrsim1.3 \times 10^{19} \text{ y},$$

$$T_{1/2}(^{92}\mathrm{Mo};2\nu;\beta^+\text{-EC})\gtrsim2.3 \times 10^{17} \text{ y, and}$$

$$T_{1/2}(^{92}\mathrm{Mo};0\nu;\beta^+\text{-EC})\gtrsim2.7 \times 10^{18} \text{ y}.$$

## 4.2 The Osaka $^{100}$Mo $\beta\beta$-decay Experiment

The ELEGANTS spectrometer, discussed earlier, was converted into a live-shielded stack of Si(Li) detectors and 10 molybdenum foils, each 44 mm$^2$ in area by 50 mg/cm$^2$ in thickness and isotopically enriched to 94.5% $^{100}$Mo (See Fig. 13). The data were based

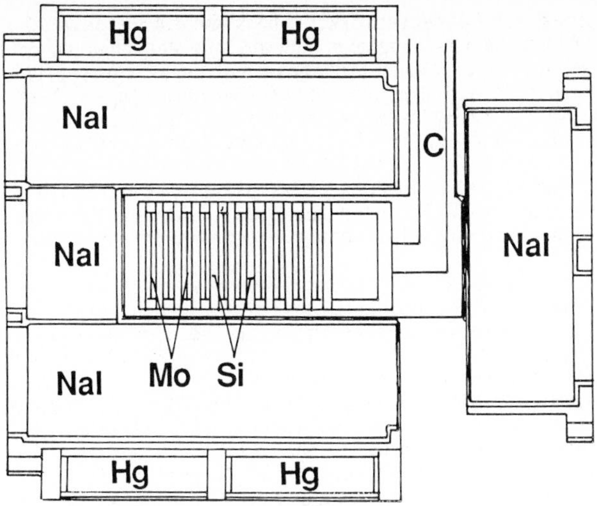


Fig. 13. The new Osaka ELEGANTS spectrometer for the search for $^{100}$Mo ββ-decay.

on only 4 foils which had relatively low levels of uranium and thorium contamination and a total of 1.88 x $10^{22}$ $^{100}$Mo nuclei. The Si(Li) detectors are 1500 mm$^2$ by 4 mm deep and are liquid nitrogen cooled. Their resolutions are between 6 and 10 keV in the energy range of interest. The total effective energy resolution is approximately 80 keV FWHM, due mainly to the thickness of the molybdenum foils. The events selected as candidates for $\beta\beta$-decay require that two adjacent detectors record signals simultaneously.

The detector was operated for 1866 h with an electronic discriminator set at 200 keV. The $\beta\beta$-decay response functions and efficiencies were calculated by Monte Carlo and included the energy losses and multiple scattering in the foils and detectors, but neglected re-entrance of backscattered electrons into the detectors.

The limiting $2\nu$ $\beta\beta$-decay spectrum is expressed as $N_t^{2\nu}\cdot\epsilon^{2\nu}(E)$, where $\epsilon(E)$ is the response function for the sum electron energy E and $N_t$ is the total number of events. The upper limit on $N_t^{2\nu}$ was constrained so that $N_t^{2\nu}\cdot\epsilon^{2\nu}(E)$ <u>does not exceed</u> the observed number of events in any energy bin. For $0\nu$ $\beta\beta$-decay with the emission of majorons, peak and background regions were set at 2.1-2.6 MeV and 2.7-3.2 MeV, respectively. The dominant backgrounds in these regions were assumed to be from $\beta$-$\alpha$ coincidences in the decay chain $^{212}Bi\rightarrow{}^{212}Po\rightarrow{}^{208}Pb$ and were also assumed to be energy independent. $N_t^{2\nu}$ were 829 and 873 at the 68% and 90% CL, respectively. Corresponding $N_t^{0\nu,M}$ are 45 and 66. The calculated efficiencies are $\epsilon^{2\nu}$ = 0.08 and $\epsilon^{0\nu,M}$ = 0.12. The resulting limits are:

$$T_{1/2}(^{100}Mo;\ 0\nu;\ 0^+\rightarrow 0^+) \gtrsim 1.9\ (1.0) \times 10^{20}\ y,$$

$$T_{1/2}(^{100}Mo;\ 0\nu;\ 0^+\rightarrow 2^+) \gtrsim 2.9\ (1.7) \times 10^{19}\ y,$$

$$T_{1/2}(^{100}Mo;\ 0\nu,M;\ 0^+\rightarrow 0^+) \gtrsim 0.7\ (0.5) \times 10^{19}\ y, \text{ and}$$

$$T_{1/2}(^{100}Mo;\ 2\nu;\ 0^+\rightarrow 0^+) \gtrsim 2.6\ (2.4) \times 10^{17}\ y,$$

to levels of confidence of 68% (90%).

Using the nuclear structure calculations of KLAPDOR and GROTZ [24], the authors deduce the limits: $\langle m_\nu \rangle \lesssim 13$ and 18 eV at confidence levels of 68% and 90%, respectively, and at corresponding confidences $|g_{ee}| \lesssim 1.0 \times 10^{-3}$ and $1.2 \times 10^{-3}$ for the coupling of majorons to electron neutrinos.

The Osaka group is developing another concept for the detection of the $\beta\beta$-decay modes of $^{100}$Mo. This detector, called ELEGANTS V, consists of a combination of drift chambers, plastic scintillators, and NaI(Tl) detectors. The source will be a 70-cm by 70-cm, by 60-mg/cm$^2$-thick foil located at the midplane and will be enriched to contain $1.7 \times 10^{24}$ $^{100}$Mo atoms. Drift chambers will be located above and below the source plane to provide particle identification and trajectory information. The active volume of each chamber is 100 cm by 100 cm by 8 cm, with 6 layers for X and Y sensitivity. Each layer will consist of 80 cells that will provide 10- to 20-mm position accuracy.

The $\beta$ particle energies will be measured with 20 sets of plastic scintillators, 100 cm by 10 cm by 1.5 cm thick. The counters will cover about 85% of the total solid angle subtended by the source. The energy resolution for electrons is estimated to be ~250 keV at 3000 keV.

Twenty sets of 10-cm by 10-cm by 100-cm NaI(Tl) scintillators will be used as a veto for cosmic rays and $\gamma$ rays, as an anticoincidence shield for ground state to ground state $\beta\beta$-decay modes, and as a coincidence detector for the accompanying $\gamma$ ray in the decay to excited states.

OFHC copper bricks and low-activity lead will be used to shield against $\gamma$ rays. The system will be airtight to eliminate background due to radon gas. The projected background rate near the endpoint energy is 0.05 counts/keV/y. This detector, if it functions anywhere near its projections, could provide an extremely accurate measurement of $2\nu$ $\beta\beta$-decay. The broad resolution function will preclude it from being a highly effective $0\nu$ $\beta\beta$-decay experiment if there is any background at all.

### 4.3 The LBL-Mt. Holyoke-UNM $^{100}$Mo $\beta\beta$-decay Experiment

Another interesting detector for the observation of $\beta\beta$-decay of $^{100}$Mo has been under development by the Lawrence Berkeley Laboratory - Mount Holyoke College - University of New Mexico collaboration [55]. The detector consists of 40 Si(Li) detectors, each of which is 7.6 cm in diameter and 1.5 mm thick. Thin $^{100}$Mo foils are placed in the 1-mm gaps between detectors (See Fig. 14). This geometry provides for coincidence measurements and allows the subtraction of background not associated with the foils by operating with and without the source. The detectors are in a titanium cryostat and cooled to approximately 120 °K with liquid nitrogen. The apparatus is shielded by 25.4 cm of lead inside 5.08 to 10.16 cm of boron-loaded polyethylene, which in turn is inside 61 cm of wax. All materials near the detector were assayed for radioactive impurities using a Ge spectrometer.

The source was enriched to 97% $^{100}$Mo at the Oak Ridge National Laboratory and was mixed with formvar, chloroform and cyclohexanone to form a slurry, which was dried in

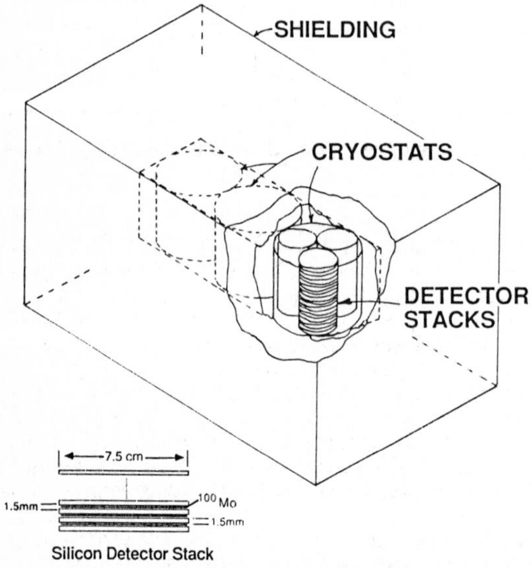


Fig. 14. A sketch of the LBL - Mt. Holyoke - UNM $^{100}$Mo $\beta\beta$-decay experiment.

a thin layer to form flat sheets, supported by nylon structures resembling tennis rackets.

The experiment is located in the Consolidated Silver (Consil) mine in Osburn, Ohio, at a depth of 4000 feet, which is equivalent to 3350 meters of water. The detector was operated in a variety of shielding configurations, with and without Mo foils. Moving the detector underground reduced the background in the 2- to 3-MeV region by factors of 10 to 15. The authors attribute a peak in the spectra at 5.3 MeV to $\alpha$ particles from the decay of $^{210}$Po on the surface of solder as discussed by Brodzinski et al. [56]. Although there are events all the way out to 10 MeV, the background above 5.3 MeV was greatly reduced when the detector was moved underground.

The detector has been operating for several months without molybdenum. Another $7.8 \times 10^{23}$ atoms of $^{100}$Mo have been obtained, and after the detector has been operated for 6 months, with equal numbers of Mo foils and formvar foils, this experiment should be able to set limits of

$$T_{1/2}(^{100}\text{Mo};\ 2\nu;\ 0^+\to 0^+) \gtrsim 5 \times 10^{19}\ \text{y}$$

and

$$T_{1/2}(^{100}\text{Mo};\ 0\nu,\text{M};\ 0^+\to 0^+) \gtrsim 4 \times 10^{20}\ \text{y}$$

to a 68% level of confidence. These values are equivalent to $3.5 \times 10^{21}$ y and $6 \times 10^{21}$ y, respectively, for the $2\nu$ and $0\nu$,M decays of $^{76}$Ge, assuming the nuclear matrix elements are the same.

## 4.4 Summary and Conclusions Concerning $^{100}$Mo $\beta\beta$-decay Experiments

The results from the three $^{100}$Mo $\beta\beta$-decay experiments are summarized in Table 14.

Table 14. Lower limits on the $\beta\beta$-decay of $^{100}$Mo (y)

| Mode† | [53] | [54] | [55] |
|---|---|---|---|
| $0\nu$ | $1.3 \times 10^{19}$ | $1.9 \times 10^{18}$ | ---- |
| $2\nu$ | $6.8 \times 10^{17}$ | $2.5 \times 10^{17}$ | $4 \times 10^{18}$ |
| $0\nu$,M | $7.5 \times 10^{18}$ | $7 \times 10^{18}$ | $2.1 \times 10^{20}$ |

†All decay modes here are $0^+\to 0^+$ decays. Values correspond to 68% confidence levels.

In addition, ELLIOT et al. [53] set limits of $2.3 \times 10^{17}$ y and $2.7 \times 10^{18}$ y on the $2\nu$ and $0\nu$, positron - electron-capture decay modes of $^{92}$Mo. Earlier NORMAN [57] had set corresponding limits of $3 \times 10^{17}$ y for both of these modes.

The future Osaka experiment, called ELEGANTS V, shows promise of producing extremely sensitive results with very low background, if it performs as predicted. The projected background of 0.05 counts/keV/y in the vicinity of the total decay energy for $1.4 \times 10^{25}$ atoms of $^{100}$Mo converts to 0.0022 counts/keV/y/mole, which can be compared to 0.94 counts/keV/y/mole ($^{76}$Ge) at the end-point energy already achieved in Ge detectors. The anticipated background for Ge detectors free from cosmogenic background is ~0.02 counts/keV/y/mole ($^{76}$Ge) in a natural Ge detector and ~0.002 for a detector enriched to 75% $^{76}$Ge. The Mo-Si(Li) stack detectors have energy resolutions 20 to 60 times worse than $^{76}$Ge experiments, which will seriously affect the sensitivity for measuring the $0^+\to 0^+$, $0\nu$ $\beta\beta$-decay mode. They should, however, be able to provide excellent data for $2\nu$ $\beta\beta$-decay, which is of great importance.

## V. Xenon-136 and $^{150}$Nd Double-Beta Decay Experiments

Another promising experiment involves the $\beta\beta$-decay of $^{136}$Xe. The natural isotopic abundance is 8.86%, and the $\beta\beta$-decay energy is 2478±5 keV [58]. The first $^{136}$Xe $\beta\beta$-decay search is credited to De'MUNARI and MAMBRIANI [59,60]. The four experiments discussed here involve Multi-element Proportional Counters (MEPC), Time Projection Chambers (TPC), and High Pressure Ionization Chambers.

### 5.1 The University of Milano Multi-element Proportional Counter

In the Milano MEPC the wires run parallel to the axis of the cylindrical pressure tank [58] and form a honeycomb structure of 61 hexagonal cells, 80 cm long and 2.5 cm on a side. The cathode of each cell is made up of 24 Cu-Be wires 100 microns in diameter. Each anode is a gold-plated tungsten wire 20 microns in diameter. The tank holds 100 liters of Xe gas at a pressure of 10 bar. The fiducial volume is 46 liters, which corresponds to $1.13 \times 10^{24}$ nuclei of $^{136}$Xe when natural Xe is used.

A spectrum obtained in 148.92 live hours of counting was recently reported [58]. The two peaks from the $^{60}$Co $\gamma$-rays were observed, probably from the steel end flanges of the pressure vessel. The energy resolution of the 1332-keV $\gamma$-ray peak was 5%. After the software cuts, which require a coincidence signal from a contiguous cluster of 3 to 8 elements, an upper limit of 7 counts, to a confidence level of 90%, was observed for the $0^+ \to 0^+$, $0\nu$ $\beta\beta$-decay peak which corresponds to $T_{1/2}(^{136}\text{Xe};\ 0\nu;\ 0^+ \to 0^+) \gtrsim 1.7 \times 10^{21}$ y. Using a theoretical estimate for the matrix element, the authors conclude $\langle m_\nu \rangle \lesssim 13.3$ eV. They also place a limit of $T_{1/2}(^{136}\text{Xe};\ 0\nu,\text{M};\ 0^+ \to 0^+) \geq 1.6 \times 10^{19}$ y, which corresponds to $|g_{ee}| \lesssim 9.8 \times 10^{-4}$ for the coupling of majorons to electron neutrinos. The matrix element used to deduce these limits on $\langle m_\nu \rangle$ and $|g_{ee}|$ is an estimate, and no comparison can be made to experiments involving other isotopes.

The Milano group projects half-life sensitivities of $10^{23}$ y, $10^{21}$ y and $10^{21}$ y for the $0\nu$, $2\nu$, and $0\nu$,M $\beta\beta$-decay modes of $^{136}$Xe in one year of counting. Xenon gas, isotopically enriched to 60% in $^{136}$Xe, was recently prepared, which will increase the sensitivity by a factor of at least 8, and could well result in the reduction of background due to $^{85}$Kr.

### 5.2 The California Institute of Technology Time-Projection Chamber

A xenon TPC has been built to search for the neutrinoless $\beta\beta$-decay of $^{136}$Xe with a design half-life sensitivity of $8 \times 10^{22}$ y in one year of counting [61]. Using an estimate of the nuclear matrix element, this corresponds to $\langle m_\nu \rangle \lesssim 1.9$ eV.

The detector has an XY crossed wire chamber containing the anode and grid wires at one end and the cathode at the other very similar to the UCI TPC. Field shaping rings line the sides of the cylindrical chamber. The major differences between this chamber and the UCI TPC are that the source gas permeates the chamber and there is no axial magnetic field; hence helical tracks cannot be reconstructed, which makes positrons and electrons indistinguishable. The main discrimination is against single-electron events because $\beta\beta$-decay will produce two charge fields at the ends. The energy is measured by collecting the total charge during proportional multiplication from a set of connected anode wires. The trajectory is reconstructed much the same way as in the UCI TPC.

Gamma-ray backgrounds measured inside shielding in the St. Gotthard Tunnel have been used to estimate that if the TPC does not add a significant $\gamma$-ray flux of its own, the total background at 2.5 MeV will be ~833 counts/MeV/y. The authors estimate an efficiency of 20% and an energy resolution of 5%. From these parameters they conclude that they can set a limit of $T_{1/2}(^{136}\text{Xe};\ 0\nu;\ 0^+ \to 0^+) \gtrsim 8 \times 10^{22}$ y in one year of counting.

The TPC was originally operated as a detector using Ar-$CH_4$ gas at various pressures; it was later filled with Xe-$CH_4$.

## 5.3 The Institute of Nuclear Research Moscow-Cromenius University High- Pressure Proportional Counter Experiment

A 61-cell, high-pressure, high-resolution gas proportional counter is under development to measure $^{136}$Xe $\beta\beta$-decay. Recently, a report on the test characteristics of a single cell has been published [62]. Each of the cells is formed by 25 tungsten cathode wires 27 cm long and arranged in a hexagonal array about a 3.6-cm inner diameter cylinder with a gold-plated tungsten wire in the center. The design pressure is 20 atmospheres. A gas purification system, based on a Ni/$SiO_2$ reactor, was found to induce detectable levels of $^{222}$Rn. In the final experiment, it is anticipated that gas reactors of pure metals (Ca,Tl), or some physical technique, will be used for purification. This appears to be an unsolved technical problem at present.

The test cell was operated at pressures from 1 to 30 atmospheres to determine the dependence of the gas amplification factor and energy resolution, for argon and xenon and mixtures, on the applied voltage. It was tested with anodes made of 10-$\mu$m-diameter gold-plated tungsten wire and of 12- and 20-$\mu$m-diameter Ni-Cr alloy wires. The background rates were also determined.

The counter was calibrated with $\gamma$ rays from the decay of $^{109}$Cd. The data taken with pure Xe gas resulted in energy resolutions of 12.2%, 8.7% and 7.1% for the 29.8 keV Xe $K_\alpha$ x-ray line, the 58.2 keV first escape peak, and the 88 keV full-energy $\gamma$-ray line, respectively. The energy dependence of the resolution function is proportional to $1/\sqrt{E}$, which results in a resolution of 2% at 1 MeV and 1.3% at 2.5 MeV.

## 5.4 The INR, Moscow High-Pressure Ionization Chamber $^{136}$Xe Experiment

A simple, high-pressure, single-cell, high-resolution detector was operated for 113 live hours at a pressure of 3 MPa of natural xenon [63] and resulted in a $\beta\beta$-decay limit of $T_{1/2}(^{136}Xe;\ 0\nu;\ 0^+ \to 0^+) \gtrsim 5.5 \times 10^{19}$ y. The chamber has a cylindrical geometry, and the anode, wound with 100-$\mu$m beryllium bronze wire in 10-mm steps, is located at the midplane. Grids on each side of the anode are 8 cm from the midplane. The grid wires have 2-mm gaps. The inner diameter of the rings, on which the anode and grids are wound, is 15.8 cm. The cathodes are solid disks at the ends of the gas chamber.

Almost all of the chamber parts are made of titanium with insulators made of quartz. The working pressure is 3 MPa, and the mass of Xe in the fiducial volume is 627 g. A higher electron drift velocity was obtained by adding 0.8% $H_2$ by volume to the Xe. The FWHM energy resolutions for $^{137}$Cs, $^{22}$Na, and $^{88}$Y uncollimated $\gamma$ rays were 5.2% (662 keV), 3.5% (1275 keV), and 2.7% (1836 keV) for a pure gas mixture of 99.2% Xe + 0.8% $H_2$. A gas purification system with a Ni/$SiO_2$ absorbent was found to introduce significant quantities of $^{222}$Rn.

The detector is located in the Baksan Neutrino Laboratory at a depth equivalent to 850 meters of water for the absorption of cosmic-ray muons, and is shielded by 10 cm of OFHC copper and 15 cm of lead. The background underground is more than four orders of magnitude lower at 2.5 MeV than on the Earth's surface.

## 5.5 Baksan Multidimensional Scintillation $^{150}$Nd Experiment

A complete description of the Baksan Neutrino Observatory was given recently by POMANSKY [64]. One of the programs in that laboratory involves unique $\beta\beta$-decay searches using multidimensional counting with low background scintillation detectors and isotopically enriched sources. A recent report by KLIMENKO, POMANSKY and SMOLNIKOV [65] describes a search for $\beta\beta$-decay using a separated source of $^{150}$Nd.

The detector consists of four layers of plastic scintillators, 50 cm by 25 cm by 5 cm thick. Each is connected to four phototubes via 5-cm-long light pipes. The source is 50 mg/cm$^2$ thick isotopically enriched $Nd_2O_3$ powder, equivalent to 50.5 g of $^{150}$Nd and is placed between the inner layers of the detectors with no extra material between source and detector. The energy resolution of each detector is approximately 200 keV at 1 MeV.

The detector cavity consists of 0.5 m of low radioactivity concrete outside 0.5 m of dunite rock. The detectors are surrounded by 5 cm of high-purity tungsten inside 10 cm of plexiglas, which is surrounded by 20 cm of OFHC copper. Limits on the half-lives of four $\beta\beta$-decay modes of $^{150}$Nd were deduced by measuring energy-correlated,

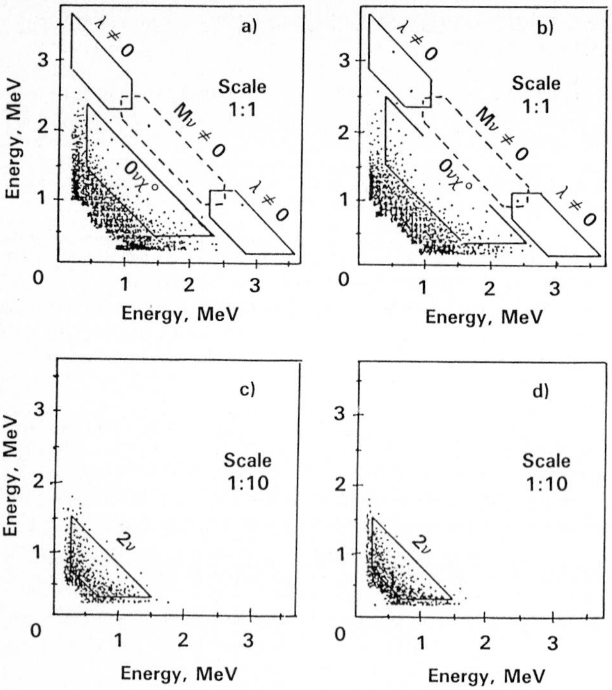


Fig. 15. Scatter plots of the data from 120 h of counting with the Baksan $^{150}$Nd $\beta\beta$-decay experiment. Figures (a) and (c) are source-out backgrounds for the source-in data shown in (b) and (d).

two-detector coincidence spectra. Source-out spectra were subtracted from source-in spectra. The data are presented as scatter plots on the energy vs. energy plane as shown in Fig. 15.

The detector array was operated for 2000 h of live time with the source and for an equal time with the source removed. Fig. 15 shows data after only 120 h of operation. The limits are to a 90% level of confidence.

$$T_{1/2}(^{150}\mathrm{Nd};\ 2\nu;\ 0^+ \to 0^+) \gtrsim 1.8 \times 10^{19}\ \mathrm{y},$$

$$T_{1/2}(^{150}\mathrm{Nd};\ 0\nu,\mathrm{M};\ 0^+ \to 0^+) \gtrsim 1.0 \times 10^{20}\ \mathrm{y},$$

$$T_{1/2}(^{150}\mathrm{Nd};\ 0\nu;\ 0^+ \to 0^+) \gtrsim 1.7 \times 10^{21}\ \mathrm{y},$$

$$T_{1/2}(^{150}\mathrm{Nd};\ 0\nu;\ 0^+ \to 2^+) \gtrsim 1.1 \times 10^{21}\ \mathrm{y},$$

Extraction of limits on $\langle m_\nu \rangle$, $\eta_{RL}$, $\eta_{RR}$, and $g_{ee}$ depend strongly on assumptions made about the nuclear matrix elements since there are no detailed calculations available.

The plastic scintillators are currently placed inside large NaI(Tl) detectors to increase the sensitivity of the apparatus to the $0^+ \to 2^+$ decay mode. In a private discussion, A. A. SMOLNIKOV stated that a very preliminary small positive result had been obtained for this mode, and that extensive tests and improvements were being conducted to determine what background, if any, might account for such an observation.

## VI. Applications of Ge Detectors in Astrophysics

Recently two efforts have been published in which the PNL/USC detector has been used to place limits on the masses of weakly interacting cold dark matter candidates and on the coupling constant of Dine-Fishler-Srednicki (DFS) axions to electrons [66,67]. In this section these results will be reviewed and the reader will be referred to the original articles for a more complete discussion and references to the original literature. Subsequent to the PNL/USC work, the UCSB-LBL group has placed similar limits on dark matter candidates with their experiments and is also examining new approaches.

## 6.1 Limits on Cold Dark Matter Candidates

The history of this problem will not be repeated here, but, it must be remembered that, in the conventional wisdom of astrophysics there is a significant portion of the matter in the universe which is non-luminous and might be weakly interacting massive particles as suggested by Whitten and which he calls WIMPs. The Ge detector experiment involves coherent scattering of the WIMPs off of Ge nuclei via $Z^\circ$ exchange. The recoiling Ge nuclei generate a signal by creating electron-hole (e-h) pairs. This article will concentrate on the experimental issues and on the results. For more details, see ref. 66.

The measurement of the nuclear recoil due to the scattering of WIMPs requires a detector with a low energy threshold and excellent background rejection. Because of its low band gap (0.69 eV at 77°K) and high efficiency for converting electronic energy loss to e-h pairs (2.96 eV per e-h at 77°K), germanium is probably the best suited material for a semiconducting WIMP detector. In late 1985, the threshold on the PNL/USC detector was reduced to an equivalent incident electron energy of 4 keV, which permitted detection of Ge nuclei-WIMP scattering with recoil energies greater than 15 keV. Ten weeks of low-energy data are shown in Fig. 16. Interpretation of the data requires knowledge of the relative efficiency factor (REF), which is the ratio of e-h pairs produced by an electron, to the number produced by a recoiling Ge nucleus.

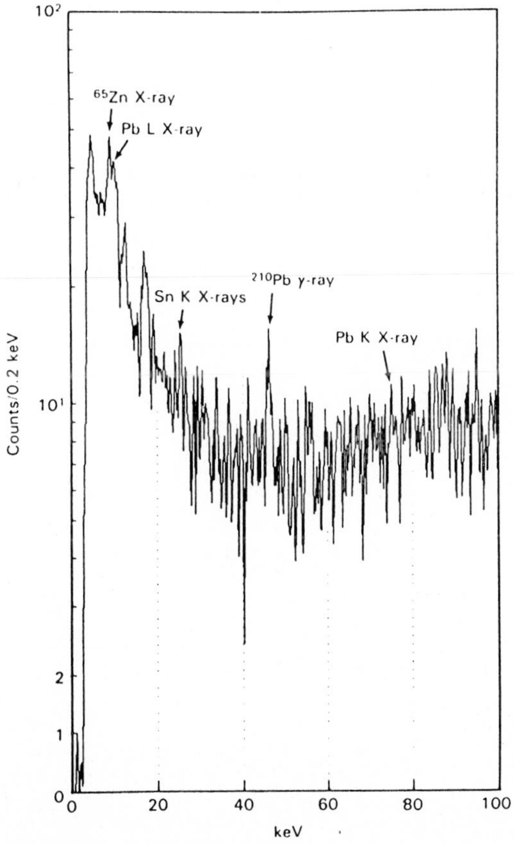


Fig. 16. Ten weeks of data from the low energy portion of the Ge detector spectrum

The particles that comprise the halo are assumed to have a velocity distribution function, $f(\bar{v})$, with an RMS of 250 km/s and a maximum of 550 km/s while the halo, like the galactic spheroid, slowly rotates with a velocity of 80 km/s [68]. The maximum halo velocity may be higher [69] in which case the quoted limits extend to lower masses. The predicted detection rate of recoils having energy T, $R_p(T)$, was calculated according to,

$$R_p(T) = n_x \Delta T \int \frac{d\sigma}{dT}(v,T) f(\bar{v}) v d^3v, \qquad (16),$$

where $\Delta T$ is the range of recoil energies detected in a given channel and $n_x$ is the local density of dark matter. The integral was evaluated over all velocity phase space. The cross section as a function of recoil energy, $d\sigma/dT$, for an incident WIMP depends upon its mass, $m_x$, and velocity, v, according to

$$\frac{d\sigma}{dT} \simeq \frac{G_F^2 m_N c^2}{8\pi v^2} [Z(1-4 \sin^2 \theta_W) - N]^2 \left[1 + (1 - T/E)^2 - \frac{m_N T + m_x^2}{E^2}\right] \qquad (17)$$

$$\bullet \exp(-m_N T R^2/3).$$

In (17), $m_N$ is the mass of the nucleus, Z and N are the number of protons and neutrons respectively, T is the recoil energy, $G_F$ is the weak coupling constant, $\theta_W$ is the Weinberg angle and $E = m_x(c^2 + v^2/2)$. When the de Broglie wavelength, corresponding to the momentum transferred in the recoil, is smaller than the nucleus, the assumption of a coherent interaction with a point-like mass is no longer valid, and the finite size of the nucleus must be included in the calculation. The exponential factor in (17) is a nuclear form factor derived with the assumption of a gaussian density distribution of nucleons. For small energy transfers, E<15 keV, the WIMP interacts with a point-like nucleus, and there is no loss of coherence.

The halo model used in the calculation is conservative. If the maximum halo velocity is chosen closer to the local escape velocity of ~750 km/s [70], the lower limit of the excluded mass range decreases. Figure 17 shows the limits obtained for standard as well as a "perverse" halo model in which all of the dark matter particles were assumed to lie on purely circular orbits. This model, which yields a lower limit of 35 GeV for the excluded mass range, is in contradiction with all existing galaxy formation theories, which predict isotropic or radial velocity distributions.

Since the predicted count rate also depends upon WIMP density, limits on the density of interacting dark matter particles in the halo can be obtained. Figure 17 also shows these limits for particles with spin independent (s.i.) Z° exchange interactions for the standard isotropic halo model and for the model that assumes that

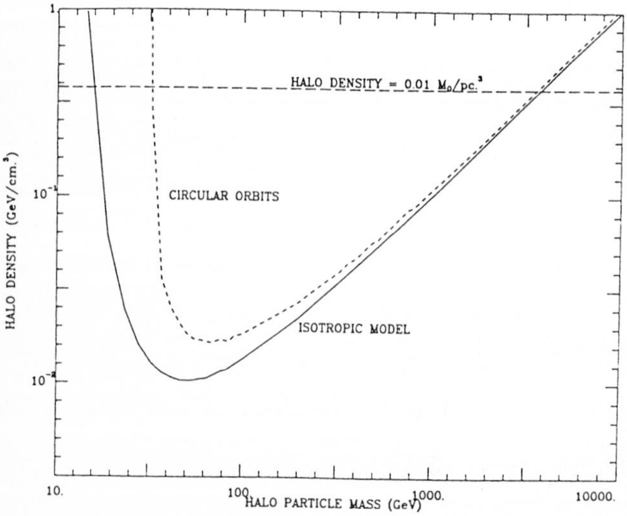


Fig. 17. Maximum halo density consistent with the observed count rate

all of the particles are on purely circular orbits. Both models include the sun's motion relative to the galactic halo. Figure 17 can be used for other s.i. vectorial interactions by multiplying the vertical axis by the ratio $(\sigma^{s.i.}_{weak}/\sigma_{s.i.})$. For example, for neutral technibaryons this ratio is approximately 0.1.

Under the assumption that the ratio of total mass of dark matter to the total mass of baryonic matter in our galaxy, $F_{galaxy} = M_{\nu}/M_{baryon}$, does not differ from the cosmological ratio, the bounds on the local density of massive stable Dirac neutrinos can be translated into bounds on their cosmological density. Figure 18 shows the upper limits of $F_{galaxy}$ under two sets of assumptions: first, all the halo mass is baryonic, and second, $M_{baryon}$ includes only the observed baryons (stars, gas, dust, etc.). The cosmological ratio of stable Dirac neutrinos to baryons, $F_{cosmological}$, was calculated using an analytic solution to the BOLTZMANN equation [71] to find $\rho^{\nu}_{cosm.}$ and the largest value for $\rho^{baryon}_{cosm.}$ consistent with bounds from big-bang nucleosynthesis [72]. The existence of these particles is, therefore, excluded for masses larger than 20 GeV except in a narrow mass range near the $Z°$ resonance at $m_{\nu} = m_{Z°}/2$.

The detector background has a smooth continuum as well as the narrow-line components. The low energy peaks are primarily due to the presence of $^{210}Pb$ in a solder connection. The solder was removed, and the radioactive shield was upgraded by the use of 448-year-old lead in place of the copper which has measurable amounts of cosmogenic radioactivity. Thus, the background in some areas of the spectrum has been reduced more than a factor of ten. The energy threshold was set at 4 keV because of noise at lower energies. The shapes of the low energy x-ray lines suggest that $\Delta E(FWHM) \approx 500$ eV in this energy region. In the next round of experiments it may be possible to extend the sensitivity to masses of $\gtrsim 8$ GeV.

The most important results are shown in Fig. 19, where the range of mass and cross-section of particles excluded as main components of the halo are presented. The ratio $g/g_w$ is defined as $(\sigma/\sigma_{weak})^{\frac{1}{2}}$ where $\sigma_{weak}$ is the cross section for standard heavy

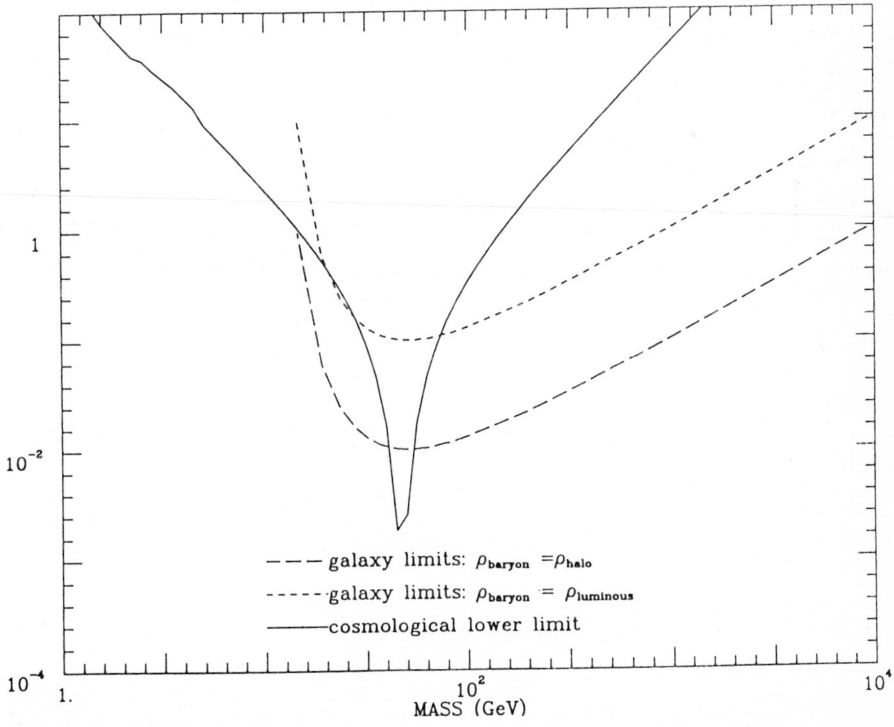


Fig. 18. Maximum ratio of total Dirac neutrino mass to total baryonic mass in our galaxy

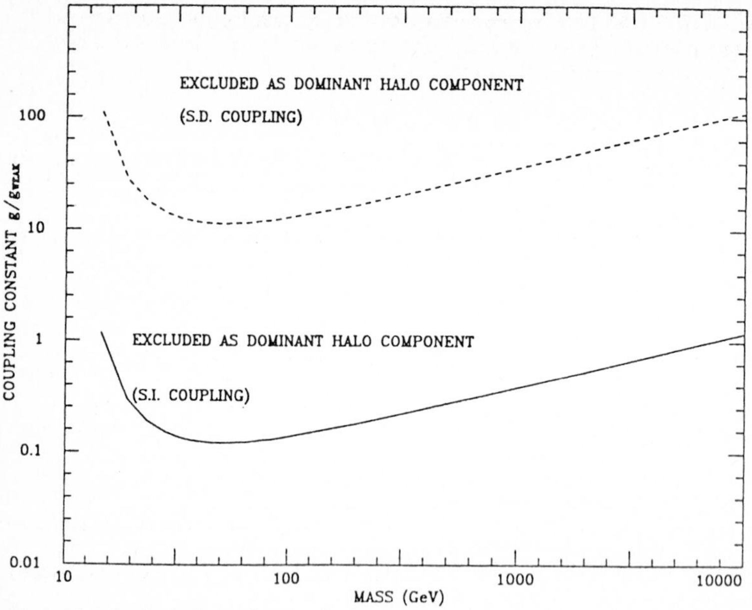


Fig. 19. Excluded regions of the coupling constant-mass plane lie above the curves

Dirac (s.i.) or Majorana spin dependent (s.d.) neutrino interactions. The validity of the experimental bound extends to cross sections on the order of $10^{-28}$ $cm^2$. The halo cannot be composed of particles that interact with nuclei through s.i. interactions whose coupling constant (normalized to the coupling of massive Dirac neutrinos to baryons) lies above the solid line. Nor can the halo be composed of particles that interact with nuclei through s.d. interactions whose coupling constant (normalized to the coupling of massive Majorana neutrinos to baryons) lies above the dashed line.

## 6.2 Laboratory Limits on Solar Axions

It is well known that the strong CP problem can be solved by the introduction of a light pseudoscalar axion. Its mass and couplings are inversely proportional to the vacuum expectation value (VEV) which breaks the Peccei-Quinn U(1) symmetry [73]. The CP problem is not sensitive to the magnitude of the particular VEV, but the detectability of the axion is sensitive, through its mass and coupling constants. With few exceptions, the standard axion has been ruled out by a large number of experiments [74]. It has been recently pointed out [75] that a short lived ($\tau \sim 10^{-13}$s) axion has not been ruled out and may explain the anomalous positron and electron peaks in heavy-ion collisions.

Atomic enhancements can be exploited when axions interact with bound electrons [76] in a process analogous to the photoelectric effect, which is enhanced by factors of $\sim 10^3$ when the photon energy is near the electron binding energy. This process is called the axioelectric effect. Such effects are expected to be large for solar axions, since their energy should be comparable to atomic energies; the average solar temperature is near 1 keV. In the dipole approximation with axion energies $\omega \ll m_e$ (in natural units h = c = 1), we have

$$\sigma_{\text{axioelectric}} = \frac{\alpha(\text{axion})}{\alpha(\text{em})} \left(\frac{\omega}{2m_e}\right)^2 \sigma_{\text{photoelectric}}, \qquad (18)$$

and

$$\alpha(\text{axion}) = (2x_e' m_e/F)^2 \frac{1}{4\pi}, \qquad (19)$$

where $\alpha(em) \cong (137)^{-1}$ and $x_e'$ is a constant of order unity and which Srednicki argues is greater than one in the DFS model. F is defined by the axion-electron interaction Lagrangian

$$L = 2x_e' \frac{m_e}{F} a\bar{e}i\gamma_5 e, \tag{20}$$

where a is the axion field. Equation (18) includes all Coulomb effects for the nonrelativistic electron. The axion mass can be related to F by

$$m_{axion} \cong 7.2 \text{ eV } \left[\frac{10^7 \text{ GeV}}{F}\right]. \tag{21}$$

The most reliable theoretical lower bound may be placed on F by requiring that the solar bremsstrahlung axion luminosity not exceed the photon luminosity. Such a large axion luminosity would imply that the sun is significantly younger than ~$4.5 \times 10^9$ years, the age of the oldest known meteorites. This gives

$$F/2x_e' \gtrsim 1.08 \times 10^7 \text{ GeV}. \tag{22}$$

Motivated by this bound, the axionization cross sections per kg for C, Si, Ge and Pb were calculated with (18) for $F/2x_e' = 10^7$ GeV and are plotted in Fig. 20. It is clear that the detector should have the lowest possible background and an energy threshold of the order of 1 keV; this is possible with semiconducting detectors. Because of their low threshold energy, Ge detectors can make use of the huge enhancement in the axioelectric cross section. It has been shown [60] that the axioelectric event rate for solar DFS axions could exceed by 4 to 5 orders of magnitude the published design capabilities of bolometric detectors. Because of its low band gap and high efficiency for converting electronic energy loss to e-h pairs, germanium detectors are probably the best suited particle detectors for DFS axions.

In Fig. 21, the number of events per kg per day for germanium are plotted against the incoming axion energy for $F/2x_e' = 0.5 \times 10^7$ GeV (solid line) and $F/2x_e' = 10^7$ GeV (dashed line). The major contribution to the eventrate comes from a narrow band between 1 keV and 10 keV. This is because both the solar axion flux and the axioelectric cross section peak in this region. In the following, the expected rates are compared with the count rates observed in the ultralow background germanium spectrometer described earlier. Three months of data were accumulated, of which six weeks were low in noise. Also plotted in Fig. 21 are some of the experimental points (crosses) for $\omega \cong 4$ keV. The statistical error on these data is estimated at ± 25%. From this, the experimental bound

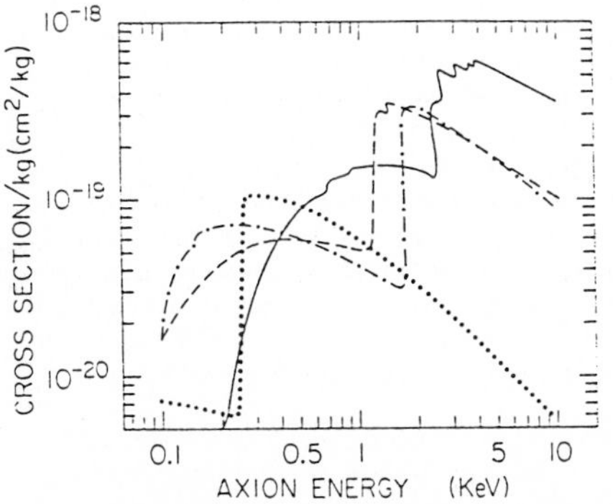


Fig. 20. Axioelectric cross sections for C (dots), Si (dot-dash), Ge (dash) and Pb (solid)

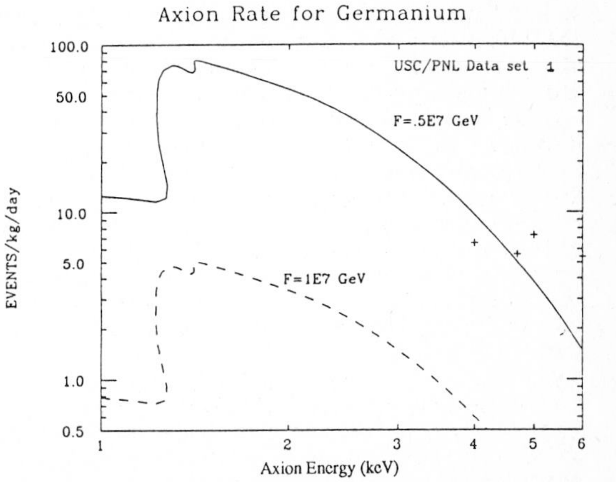


Fig. 21. Solar axion events per kg per day for Ge, $F/2x_e = 0.5 \times 10^7$ GeV (solid), $1.0 \times 10^7$ GeV (dashed). The crosses are data points

$$F/2x_e' \gtrsim 0.5 \times 10^7 \text{ GeV} \quad (23)$$

may be deduced. For $2x_e' = 1$ the laboratory bound on the DFS axion mass is

$$m_a \lesssim 15 \text{ eV}. \quad (24)$$

Bounds similar to (24) are obtained for any light pseudoscalars or light scalars that couple directly to electrons. Familons [77] and singlet majorons [15-17], associated with right-handed neutrinos, have couplings similar to (20) where $F/2x_e'$ is replaced by a model dependent coupling constant and F is the large global horizontal symmetry breaking scale. Triplet majorons, associated with left-handed neutrinos, appear if lepton number is a global symmetry spontaneously broken (up to this point the same applies to singlet majorons) at a scale $v_T$ (the vacuum expectation value of a triplet Higgs field) small with respect to the electroweak scale. From the coupling of majorons, M, to a pseudoscalar electron current we have

$$L = 2\sqrt{2}\, G_F v_T m_e M \bar{e} i \gamma_5 e. \quad (25)$$

By comparison with (20), the bound analogous to (23) becomes

$$v_T \lesssim 6.9 \text{ MeV}. \quad (26)$$

There is a problem of self-consistency here, however. A laboratory bound for axions $F/2x_e' \succ 0.5 \times 10^7$ GeV has been given. Suppose $F/2x_e'$ were indeed $0.5 \times 10^7$ GeV. Then the solar axion luminosity, $\ell_a$, would be approximately four times the solar photon luminosity, $\ell_\gamma$. To calculate the axion flux, a model of the sun was used which was dominated by QED, weak and nuclear processes; axions were assumed unimportant for solar dynamics. This might not be the case if $\ell_a \cong 4\,\ell$, and therefore the laboratory bound might be self-inconsistent. However, a bound stronger by a factor of 2 to 3 would be self-consistent. Thus, future improvements in our laboratory bounds are crucial.

The DFS bound $F/2x_e' \gtrsim 0.5 \times 10^7$ GeV is a laboratory bound relying on a realistic model of the sun, the closest and best understood star. Limits on majorons have also been displayed ($v_T \lesssim 6.9$ MeV for triplet majorons) and for most familons which couple directly to electrons. (For models in which familons couple dominantly to quarks, this bound does not apply.) This laboratory bound does not rely on a detailed understanding of the dynamics and evolutions of red giants, white dwarfs, neutron stars or other stars as do the more sophisticated theoretical bounds.

Using the axioelectric effect, semiconducting Ge detectors could eventually set limits $F/2x_e' > 10^8$ GeV ($v_T < 0.3$ MeV) or, more exciting, see solar axions or other light bosons. The discovery of these particles would allow us to study physics at energies beyond the reach of accelerators and provide us with a new laboratory tool to study the interior of stars.

The results reported in Sections 6.1 and 6.2 are from work in collaboration with S. P. Ahlen, A. K. Drukier, G. Gelmini, D. N. Spergel, S. Dimopoulos, G. D. Starkman, and B. W. Lynn. The details and more complete reference to the literature on the subject are given in ref. [66] and [67].

A new detector of 200 $cm^3$, incorporating the low background features of the PNL-USC detector and low noise characteristics, has been manufactured for the University of Zaragoza in a cooperative venture. Copper cryostat parts near the crystal have been electroformed to avoid background from $^{60}Co$, and the germanium has been protected from energetic neutrons as much as possible, to avoid cosmogenic radioactivity from $^{68}Ge$, $^{65}Zn$, and Co isotopes in the crystal. Two more such detectors will be operable in the Homestake Mine in early 1988. New dark matter and solar axion search data should be available in only a few months from this writing.

Both the PNL/USC and the UCSB-LBL collaborations are attempting to design and operate smaller detectors with lower energy thresholds to place stronger mass constraints on dark matter candidates.

## VII. Summary and Conclusions Concerning $\beta\beta$-Decay

There have been many recent exciting developments in $\beta\beta$-decay, both theoretical and experimental. The question of $0\nu$ $\beta\beta$-decay suppression by particle-particle interactions appears to be generally resolved, and experiments may still place sensitive limits on $\langle m_\nu \rangle$. The first direct observation of $2\nu$ $\beta\beta$-decay, by ELLIOTT, HAHN and MOE [4] was reported in September 1987 with the result $T^{2\nu}_{1/2}(^{82}Se) = (1.1^{+0.8}_{-0.3}) \times 10^{20}$ years which is less than a factor of two longer than shell model predictions [6].

Several Ge detector experiments have produced interesting new limits on the $0\nu$ $\beta\beta$-decay half life of $^{76}Ge$. These were, in order of their publication, $T_{1/2} > 3.3 \times 10^{23}$ yr [42], $T_{1/2} > 3.1 \times 10^{23}$ yr [44], and $T_{1/2} > 5 \times 10^{23}$ yr [40]. In addition, preliminary results from the first enriched $^{76}Ge$ detector were announced in 1987. A combination of vast improvements in low background techniques and enrichment could result in extremely tight limits on the majorona mass of the neutrino (~0.03 eV).

A major step has been made in the theoretical understanding of the retardation of $2\nu$ $\beta\beta$-decay rates, and present experiments can now be analyzed consistently (See MUTO and KLAPDOR, this volume).

The controversy which arose over the announcement of an anomalous bump in the spectrum of the PNL-USC experiment [40,48,49], has led to a deeper understanding of the continuum background of several of the Ge $\beta\beta$-decay experiments, which has resulted in increased sensitivity in searches for the $2\nu$ $\beta\beta$-decay of $^{76}Ge$, as well as $0\nu$ $\beta\beta$-decay with the emission of a majoron.

Accelerated progress in $\beta\beta$-decay experiments has led to improvement of the sensitivity of low-background Ge detectors for a host of other applications.

The most dramatic improvement would be use of ~85% $^{76}Ge$ enriched material in an ultralow-background experiment which, though worthwhile, is extremely costly. In searches for $2\nu$ $\beta\beta$-decay of $^{76}Ge$, the technique of deep mining Ge, protecting it from energetic cosmic ray neutrons during the preparation of the detector, and then returning it underground, is expected to produce a new generation of sensitivity.

This subject is experiencing an exciting stage of rapid development, and the levels of sensitivity which would allow direct observation of $2\nu$ $\beta\beta$-decay of $^{76}Ge$ should be achieved in the near future.

First generation low-background Ge detector searches for cold dark matter candidates for the non-luminous mass of the galactic halo have been performed. Low-energy threshold Ge detectors could decrease the lower detectable mass-limit of WIMPs to 5-10 GeV [66]. Larger arrays of lower, background Ge detectors could improve the laboratory bounds on solar axions to $F/2X_e > 10^8$ GeV.

## References


1. M. Goeppert-Mayer: Phys. Rev. 48, 512 (1935)
2. T. Kirsten, H. Richter, E. Jessberger: Phys. Rev. Lett. 50, 475 (1983)
3. T. Kirsten: Proc. Int. Symposium on Nucl. Beta Decay and Neutrinos, Osaka 1986, eds. T. Kotani, H. Ejiri, and E. Takasugi, (World Scientific, Singapore 1986), p.81
4. S.R. Elliott, A.A. Hahn, M.K. Moe: Phys. Rev. Lett. 59, 2020 (1987)
5. H. Primakoff, S.P. Rosen: Ann. Rev. Nucl. Part. Sci. 31, 145 (1981)
6. W.C. Haxton, G.J. Stephenson, Jr.: Prog. Part. Nucl. Phys. 12, 409 (1984)
7. M.G. Shchepkin: Sov. Phys. Usp. 27, 555 (1984)
8. M. Doi, T. Kotani, E. Takasugi: Prog. Theor. Phys. Suppl. 83, 1 (1985)
9. J.D. Vergados: Phys. Repts. 133, 1 (1986)
10. W.H. Furry: Phys. Rev. 56, 1184 (1939)
11. H.M. Georgi, D.V. Nanopoulos: Nucl. Phys. B155, 52 (1979)
12. T. Tomoda, A. Faessler, K.W. Schmid, F. Grümmer: Nucl. Phys. A452, 591 (1986)
13. T. Tomoda, A. Faessler: Phys. Lett. B199, 475 (1987)
14. For a brief report on the theorem of Kayser, Petkov and Rosen see B. Kayser: Proc. XXIII Int. Conf. on High Energy Physics, ed. S.C. Loken, (World Scientific, Singapore) p.945 (1987); see also J. Schechter, J.W.F. Valle: Phys. Rev. D25, 2951 (1982) and Ref. 8
15. Y. Chicashige, R.N. Mohapatra, R.D. Pecci: Phys. Lett. 98B, 265 (1981)
16. G.B. Gelmini, M. Roncadelli: Phys. Rev. Lett. 99B, 411 (1981)

17. H.M. Georgi, S.L. Glashow, S. Nussinov: Nucl. Phys. B193, 297 (1981)
18. J.D. Vergados: Phys. Lett. 109B, 96 (1982)
19. M. Doi, T. Kotani, E. Takasugi: Osaka preprint OS-GE-97-07 (1987)
20. F.T. Avignone, R.L. Brodzinski, J.C. Evans, Jr., W.K. Hensley, H.S. Miley, J.H. Reeves: Phys. Rev. C34, 666 (1986)
21. W.C. Haxton, G.J. Stephenson, Jr., D. Strottman: Phys. Rev. D25, 2360 (1981)
22. L. Zamick, N. Auerbach: Phys. Rev. C26, 2185 (1982)
23. K. Grotz, H.V. Klapdor: Phys. Lett. 153B, 1 (1985); Nucl. Phys. A460, 359 (1986); H.V. Klapdor, K. Grotz: Phys. Lett. 142B, 323 (1984)
24. K. Grotz, H. V. Klapdor: Phys. Lett. 157B, 242 (1985)
25. P. Vogel, P. Fisher: Phys. Rev. C32, 1362 (1985); K. Grotz and H.V. Klapdor: Nucl. Phys. A460, 395 (1986)
26. P. Vogel, M.R. Zirnbauer: Phys. Rev. Lett. 57, 3148 (1986)
27. O. Civitarese, A. Faessler, T. Tomoda: Phys. Lett. B194, 11 (1987)
28. K. Muto, H. V. Klapdor: Phys. Lett. 201B, 420 (1988)
29. J. Engel, P. Vogel, M.R. Zirnbauer: Phys. Rev. C37, 731 (1988)
30. E. Fiorini, A. Pullia, G. Bertolini, F. Cappellani, G. Restelli: Phys. Lett. 25B, 602 (1967)
31. E. Fiorini, A. Pullia, G. Bertolini, F. Cappellani, G. Restelli: Lett. Nuovo Cimento 3, 149 (1970)
32. E. Fiorini, A. Pullia, G. Bertolini, F. Cappellani, G. Restilli: Lett. Nuovo Cimento 13A, 747 (1973)
33. N.A. Wogman and R.L. Brodzinski: Nucl. Instrm. Methods 109, 277 (1973); IEEE Trans. Nucl. Sci. NS-20, 73 (1973); R.L. Brodzinski: Radiochem. Radioanal. Letters 13, 375 (1973)
34. F.T. Avignone, III, Z.D. Greenwood: Nucl. Instr. and Methd. 160, 493 (1979)
35. E. Bellotti, O. Cremonesi, E. Fiorini, C. Liguori, A. Pullia, P.P. Sverzellati, L. Zanotti: Phys. Lett. B121, 72 (1983)
36. A. Forster, H. Kwon, J.K. Markey, F. Boehm, H.E. Henrikson: Phys. Lett. 138B, 301 (1984)
37. J.J. Simpson, P. Jagam, J.L. Campbell, H.L. Malm, B.C. Robertson: Phys. Lett. 53, 141 (1984)
38. H. Ejiri, N. Kamikubota, Y. Nagai, T. Nakamura, K. Okada, T. Shibata, T. Shima, N. Takashasi, W. Watanabe: J. Phys. G13, 839 (1987)
39. F. Leccia, Ph. Hubert, D. Dassie, P. Mennrath, M.M. Villard, A. Morales, J. Morales, R. Nunez-Lagos: Lett. Nuovo Cimento 78A, 50 (1983)
40. D.O. Caldwell, R.M. Eisberg, D.M. Grumm, D.L. Hale, M.S. Witherell, F.S. Goulding, D.A. Landis, N.W. Madden, D.F. Malone, R.H. Pehl, A.R. Smith: Phys. Rev. Lett. 59, 419 (1987)
41. R.L. Brodzinski, D.P. Brown, J.C. Evans, Jr., W.K. Hensley, J.H. Reeves, N.A. Wogman, F.T. Avignone, III, H.S. Miley: Nucl. Instr. and Methd. A239, 207 (1985)
42. E. Bellotti, O. Cremonesi, E. Fiorini, C. Liguori, A. Pullia, P.P. Sverzellati, L. Zanotti: Lett. Nuovo Cimento 95A, 1 (1986)
43. D.O. Caldwell, R.M. Eisberg, D.M. Grumm, D.L. Hale, M.S. Witherell, F.S. Goulding, D.A. Landis, N.N. Madden, D.F. Malone, R.H. Pehl, A.R. Smith: Phys. Rev. D33, 2737 (1986)
44. F.T. Avignone, III, R.L. Brodzinski, H.S. Miley, J.H. Reeves: Phys. Rev. D35, 1713 (1987)
45. Ph. Hubert, F. Leccia, D. Dassie, P. Mennrath, M.M. Villard, A. Morales, J. Morales, R. Nuñez-Lagos, J.A. Villar: Lett. Nuovo Cimento 85A, 19 (1985)
46. A. Morales, J. Morales, R. Nuñez-Lagos, J. Puimedón, J.A. Villar, D. Dassie, Ph. Hubert, F. Leccia, P. Mennrath, M.M. Villard, J. Chevallier, B. Hass: Proc. XXIII Int. Conf. on High Energy Physics, ed. S.C. Loken, World Scientific, Singapore (1987) p.948
47. N. Kamikubota, H. Ejiri, T. Shibata, Y, Nagai, K. Okada, T. Watanabe, T. Irie, Y. Itoh, T. Nakamura, N. Takahasi: Nucl. Instr. and Methd. A245, 379 (1986)
48. F.T. Avignone, III, R.L. Brodzinski, H.S. Miley, J.H. Reeves: Proc. APS, Div. Part. Fields (Salt Lake City, UT, January, 1987), AIP Conf. Proc. (AIP, New York); Proc. Telemark IV: Neutrino Masses and Neutrino Astrophysics (Ashland, WI, March, 1987) World Scientific Publishers, Singapore

49. P. Fisher, F. Boehm, E. Bovet, J.-P. Egger, K. Gabathuler, H. Henrikson, J.-L. Vuilleumier: Phys. Lett. B192, 460 (1987)
50. W.C. Haxton: Proc. Int. Symp. on Nucl. Beta Decay and Neutrinos, ed. T. Kotani, H. Ejiri and E. Takasugi, World Scientific, Singapore, (1987) p. 225
51. M.K. Moe, A.A. Hahn, H.E. Brown: AIP Conference Proceedings 108, 37 (1984)
52. O.K. Manuel: Proc. Int. Symp. on Nucl. Beta Decay and Neutrinos, ed. T. Kotani, H. Ejiri and E. Takasugi, World Scientific (1986) p. 71
53. S.R. Elliot, A.A. Hahn, M.K. Moe: Phys. Rev. C36, 2129 (1987)
54. H. Ejiri, N. Kamikubota, Y. Nagai, T. Okada, T. Shibata, T. Shima, N. Takashashi, J. Tanaka, T. Taniguchi, T. Watanabe: Proc. TELEMARK IV, ed. V. Barger et al., World Scientific (1987) p. 281
55. M. Alston-Garnjost, B.L. Dougherty, R.W. Kenney, J.M. Krivicich, R.D. Tripp, H.W. Nicholson, B.D. Dieterle, C.P. Leavitt: Proc. TELEMARK IV, ed. V. Barger et al., World Scientific (1987) p. 298 (An update was given at the 1988 Moriond Conference by B.D. Dieterle)
56. R.L. Brodzinski, J.H. Reeves, F.T. Avignone, H.S. Miley: Nucl. Instr. and Meth. A254, 472 (1987)
57. E.B. Norman: Phys. Rev. C31, 1937 (1985)
58. E. Bellotti, D. Camin, O. Cremonesi, E. Fiorini, C. Liguori, S. Ragazzi, L. Rossi, P.P. Sverzallate, L. Zanotti: Proc. TELEMARK IV, ed. V. Barger et al. World Scientific (1987) p. 307
59. G.M. De'Munari, G. Mambriani: Suppl. Nuovo Cim. 19, 314 (1961)
60. G.M. De'Munari, G. Mambriani: Nuovo Cim. 33A, 299 (1976)
61. M.Z. Iqbal, H.E. Henrikson, L.W. Mitchell, B.M.G. O'Callaghan, J. Thomas, H. T-K Wong: Proc. TELEMARK IV, ed. V. Barger et al., World Scientific (1987) p. 317
62. V.V. Kuzimov, V.N. Novikov, B.V. Pritichenko, A.A. Pomansky, P. Povinec, R. Janik: Nucl. Instr. and Meth. B17, 452 (1986)
63. A.S. Barabash, A.A. Golubev, O.V. Kazachenko, V.V. Kuzminov, V.M. Lobashev, V.M. Novikov, B.M. Ovchimnikov, A.A. Pomansky, B.E. Stern: Nucl. Instr. and Meth. B17, 450 (1986)
64. A.A. Pomansky: Nucl. Instr. and Meth. B17, 406 (1986)
65. A.A. Klimenko, A.A. Pomansky, A.A. Smolnikov: Nucl. Instr. and Meth. B17, 445 (1986)
66. F.T. Avignone, III, R.L. Brodzinski, S. Dimopoulos, G.D. Starkman, A.K. Drukier, D.N. Spergel, G. Gelmini, B.W. Lynn: Phys. Rev. D35, 2752 (1987)
67. S.P. Ahlen, F.T. Avignone, III, R.L. Brodzinski, A.K. Drukier, G. Gelmini, D.N. Spergel: Phys. Lett. B195, 603 (1987)
68. J. Bahcall, S. Casertano: Kinematics and Density of the Galactic Spheroid, IAS preprint 11/85
69. D.Z. Freedman: Phys. Rev. D9, 1389 (1974); D. Tubbs, D.N. Schramm: Astrophys. J. 201, 467 (1975)
70. J. Caldwell, J.P. Ostriker: Astrophys. J. 251, 61 (1981)
71. J. Bernstein, L.S. Brown, G. Feinberg: Phys. Rev. D32, 3261 (1985)
72. J. Yang, M.S. Turner, G. Steigman, D.N. Schramm, K.A. Olive: Astrophys. J. 281, 492 (1984)
73. R.D. Peccei, H.R. Quinn: Phys. Rev. Lett. 3B, 1440 (1977); Phys. Rev. D16, 1791 (1977); S. Weinberg: Phys. Rev. Lett. 40, 223 (1978); F. Wilczek, Phys. Rev. Lett. 40, 279 (1978)
74. A. Zehnder: SIN Report No. PR-83-03 (1983)
75. N.C. Mukhopadhyay, A. Zehnder: Phys. Rev. Lett. 56, 206 (1986)
76. S. Dimopoulos, G.D. Starkman, B.W. Lynn: Mod. Phys. Lett. 1, 491 (1986)
77. D.B. Reiss: Phys. Lett. B115, 217 (1982); F. Wilczek: Phys. Rev. Lett. 49, 1549 (1982); G.B. Gelmini, S. Nussinov, T. Yanagida: Nucl. Phys. B219, 31 (1983)

# Double Beta Decay, Neutrino Mass and Nuclear Structure

*K. Muto*† *and H.V. Klapdor*

Max-Planck-Institut für Kernphysik, Saupferchweg 1,
D-6900 Heidelberg, Fed. Rep. of Germany

## 1. Introduction

Nuclear double beta ($\beta\beta$) decay is one of the rarest processes in nature, with a half-life of the order of $\sim 10^{20}$ years or longer [Kir 67,68,69, Pri 81, Hax 84, Doi 85, Kir 86, Ver 86, Gro 86, Ell 87, Mut 88a]. It can become observable only when a nucleus cannot undergo ordinary single $\beta$ decay because of energy conservation, or of very strong suppression of energetically allowed transitions by a large spin difference between the parent and daughter states (for example, $^{48}$Ca). All $\beta\beta$ emitters and their daughter nuclei have even numbers of protons and neutrons, since the pairing force, which is a prominent feature of the nucleon-nucleon interaction acting between like nucleons, makes even-even nuclei more bound relative to the neighbouring odd-odd nuclei. As an example the mass spectrum for A = 76 isobars is shown in Fig. 1.1. For $^{76}$Ge, $\beta^-$ decay to $^{76}$As and $\beta^+$/EC to $^{76}$Ga are both forbidden, and consequently $\beta\beta$ decay to $^{76}$Se is the only possible decay mode. In heavy-mass regions, even if a nucleus is stable against $\beta^\pm$ decay, the $\alpha$-particle emission channel is usually open and $\beta\beta$ decay is hardly observable. (This will affect nuclei beyond $^{204}$Hg in Table 5.3.)

A $\beta\beta$ decay is expected to take place mainly via the two decay modes of the two-nucleon mechanism (for other mechanisms, see sect. 3.4): the two-neutrino ($2\nu$) mode in which two neutrinos are emitted together with two electrons, and the neutrinoless ($0\nu$) mode in which no neutrino leaves the nucleus,

$$2\nu \text{ mode}: \quad (A,Z) \to (A,Z+2) + 2e^- + 2\bar{\nu} \tag{1.1a}$$

$$0\nu \text{ mode}: \quad (A,Z) \to (A,Z+2) + 2e^- . \tag{1.1b}$$

The $2\nu$ mode (Fig. 1.2a) resembles two successive single $\beta^-$ transitions with the exception that the intermediate states are virtual (see Fig. 3.1). It is described by second-order perturbation of the weak interaction within the standard electroweak model [Gla 61, Wei 67, Sal 68] irrespective of the nature of the neutrino, Dirac ($\nu \neq \nu^c$) or Majorana ($\nu = \nu^c$), massless or massive ($\nu^c$ denotes the charge-conjugate state (antiparticle)). On the other hand, the $0\nu$ mode (Fig. 1.2b) is a lepton-number non-conserving process, and is forbidden in the Standard Model. It occurs when the neutrino emitted at one vertex can be absorbed by the nucleon at the other vertex. Therefore two conditions must be satisfied: (i) the neutrino is a Majorana particle, and (ii) the neutrino has a finite mass and/or there exists a right-handed component in the charged weak interaction. It should be noted that observation of $0\nu$ $\beta\beta$ decay *always* implies that the electron neutrino is a massive Majorana particle [Sch 82, Nie 84, Tak 84]. Variants of $0\nu$ $\beta\beta$ decay other than eq.(1.1b), among them decays implying the emission of a majoron, are discussed in sect. 3.4.

Historically, $\beta\beta$ decay was regarded as a sensitive test of the Dirac/Majorana character of the neutrino. The $2\nu$ mode was first proposed by Goeppert-Mayer in 1935 [Goe 35], who estimated, on the basis of Fermi's $\beta$ decay theory [Fer 34], half-lives of the order of $10^{20}$ years. In 1937, Majorana developed a new theory of the neutrino [Maj 37], a two-component theory in contrast to Dirac's

---

† On leave of absence from Department of Physics, Tokyo Institute of Technology, Japan.

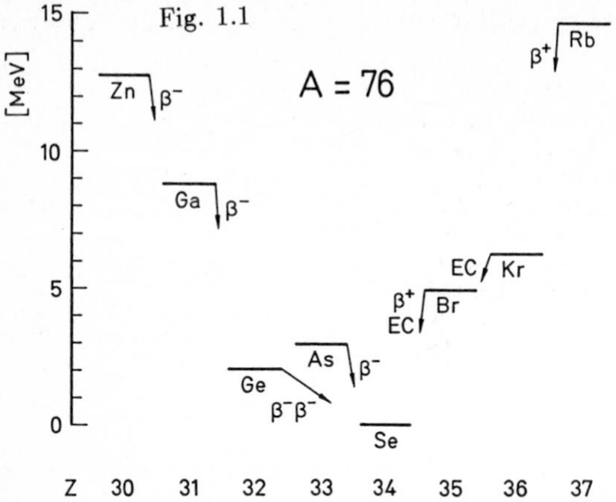


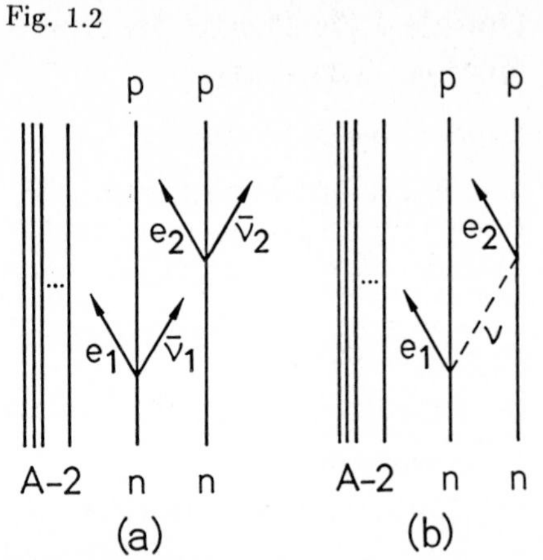


Fig. 1.1 Mass spectrum of isobars with mass number A = 76. Decay modes of the ground states are given. The $\beta^-\beta^-$ decay is the only possible mode for $^{76}$Ge.

Fig. 1.2 The two modes of $\beta\beta$ decay possible in the two-nucleon mechanism; (a) the two-neutrino ($2\nu$) mode, and (b) the neutrinoless ($0\nu$) mode.

four-component theory. Furry suggested in 1939 the possibility of the $0\nu$ mode [Fur 39] in which a Majorana neutrino is exchanged between two nucleons. The half-lives, calculated for various possible weak interactions, were shorter than the $2\nu$ decay half-lives by large factors, $10^5$ or more. Experimental efforts in the late 1940's and early 1950's (the first experiment devoted to search for the $0\nu$ mode was performed just 40 years ago [Fir 49]) to look for the favoured $0\nu$ decay resulted in negative evidence. The lower limits on half-lives exceeded by far the theoretical values for the $0\nu$ mode. This result implied that the electron neutrino is a Dirac particle, and suggested the introduction of lepton number, which distinguishes the neutrino from the antineutrino as a conserved quantity. In 1955, Davis tried a neutrino capture experiment [Dav 55], using antineutrinos $\bar{\nu}$ from fission products in a reactor,

$$\text{(Reactor) } n \to p + e^- + \bar{\nu}$$
$$\nu + {}^{37}\text{Cl} \to {}^{37}\text{Ar} + e^- . \qquad (1.2)$$

If the antineutrino $\bar{\nu}$ and the neutrino $\nu$ are, in fact, identical, the second reaction should be possible and $^{37}$Ar should have been observed. However, this reaction did not show up, apparently confirming the Dirac character of the neutrino.

Soon after the Davis experiment, it was found that parity is maximally violated in the weak interaction, and its V-A form was established [Lee 56, Wu 57]. This implies that, as a massless (anti)neutrino has a definite helicity, the reaction of eq.(1.2) is forbidden even if the neutrino is a Majorana particle, i.e., if $\nu = \nu^c$. The neutrino emitted in the first reaction has positive helicity (right-handed) and the neutrino to be absorbed in the second reaction should have negative helicity (left-handed). This is also the case for the $0\nu$ $\beta\beta$ decay. Thus, the interest in $\beta\beta$ decay waned after the discovery of parity violation and the establishment of lepton-number conservation. At the same time the question about the Dirac or Majorana character of the neutrino was reopened.

The development of modern gauge theories has revitalised the study of $\beta\beta$ decay. The standard electroweak theory unifies the electromagnetic and weak interactions in the framework of a $SU(2)_L \times U(1)$ gauge theory, and is quite successful in describing the known experimental data. This model contains no right-handed neutrinos, and the left-handed neutrino in the theory must be massless. The same is true for the minimal SU(5) model. However, massive neutrinos naturally appear in grand unified theories based on the SO(10) gauge group and in some superstring models (see [Lan 81,86,88, Moh 86,88]). They also imply the existence of both left-handed and right-handed Majorana neutrinos and of right-handed components of the weak interaction, but provide little information on precise values of the neutrino masses. The predicted values for light neutrinos lie anywhere between $10^{-11}$ eV and $\sim$ eV (see Langacker, Mohapatra in this volume).

The $0\nu\ \beta\beta$ decay is the most sensitive test of the existence of Majorana neutrinos, and sets, at present, the most stringent limits on the (electron) neutrino mass. Thus $0\nu\ \beta\beta$ decay experiments belong to the most important present-day underground physics experiments. Observation of $0\nu\ \beta\beta$ decay would be equivalent to uncovering new physics beyond the standard theory. The mass scales of grand unified theories probed by a finite neutrino mass are far beyond direct reach by present accelerators. On the other hand, reliable deductions of neutrino masses from measured $0\nu\ \beta\beta$ decay rates depend decisively on reliable calculations of nuclear matrix elements.

## 2. Majorana neutrinos and grand unified theories

Let us consider left-handed and right-handed fermion fields, $\psi_L$ and $\psi_R$. They transform under the proper Lorentz transformation as (0,1/2) and (1/2,0), respectively, where $(j_1,j_2)$ is the irreducible representation of the Lorentz group SO(1,3) [Wei 64]. For the sake of transparency, we define[1] the charge-conjugate fields as $(\psi_L)^c = C\bar{\psi}_L^T$ and $(\psi_R)^c = C\bar{\psi}_R^T$, with the charge-conjugation matrix $C$. Here, $\psi^T$ denotes the transpose of $\psi$, and $\bar{\psi} = \psi^\dagger\gamma^0$. In the Weyl representation of $\gamma$ matrices,

$$\gamma^0 = \begin{bmatrix} 0 & 1 \\ 1 & 0 \end{bmatrix}, \qquad \gamma^k = \begin{bmatrix} 0 & -\sigma_k \\ \sigma_k & 0 \end{bmatrix}, \qquad \gamma_5 = \begin{bmatrix} 1 & 0 \\ 0 & -1 \end{bmatrix},$$
$$C = i\gamma^2\gamma^0 = \begin{bmatrix} -i\sigma_2 & 0 \\ 0 & i\sigma_2 \end{bmatrix}. \tag{2.1}$$

The charge-conjugate field $(\psi_L)^c$ ($(\psi_R)^c$) behaves like $\psi_R$ ($\psi_L$), in the sense that it is a right-handed (left-handed) field and transforms as (1/2,0) ((0,1/2)) under a Lorentz transformation. Therefore $\psi_L$ can mix with $(\psi_R)^c$, and $\psi_R$ with $(\psi_L)^c$. In terms of these four fields, the most general mass Lagrangian which is Hermitian and Lorentz invariant is

$$L_m = -\tfrac{1}{2}\overline{(\psi_L)^c} m_L^M \psi_L - \tfrac{1}{2}\bar{\psi}_R m_R^M (\psi_R)^c - \bar{\psi}_R m^D \psi_L + h.c. \tag{2.2}$$

($h.c.$ denotes the hermitian conjugate of the three preceding terms) or in matrix form

$$L_m = -\tfrac{1}{2}\left[\, \overline{(\psi_L)^c} \quad \bar{\psi}_R \,\right] M \begin{bmatrix} \psi_L \\ (\psi_R)^c \end{bmatrix} + h.c. \tag{2.3}$$

$$M = \begin{bmatrix} m_L^M & m^D \\ m^D & m_R^M \end{bmatrix}. \tag{2.4}$$

A Dirac mass $m^D$ results from the coupling of the independent left-handed and right-handed fields, whereas Majorana masses $m_L^M$ and $m_R^M$ come from the coupling of fields with their charge-conjugate fields. The mass coupling scheme is illustrated in Fig. 2.1.

Charged fermions are forbidden to couple via the Majorana mass because of charge conservation, and have only Dirac masses. For example (Fig. 2.1b), the electron with a negative charge does not couple with the positively charged positron. Coupling is allowed only between the left-handed and right-handed components of the electron (positron) by the Dirac mass term,

$$L_m = -\bar{e}_R m^D e_L + h.c. = -m^D \bar{e}e. \tag{2.5}$$

The Dirac mass term is invariant with respect to the global gauge transformation $e_{L,R} \to e_{L,R}\exp(i\phi)$. This invariance implies conservation of the lepton number: the electron can be distinguished from its charge-conjugate field, namely the electron is a Dirac particle.

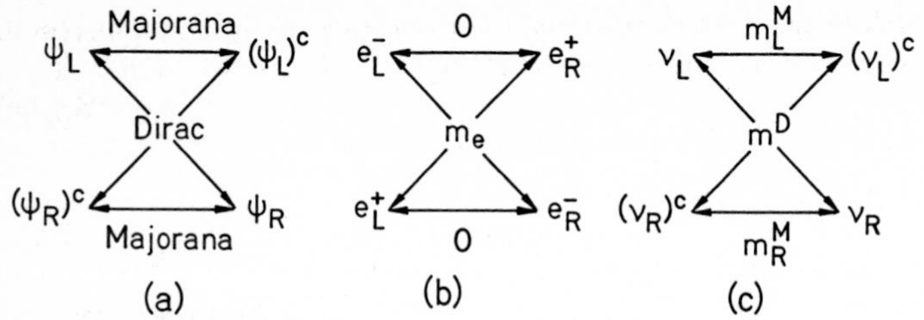


Fig. 2.1 Coupling schemes of fermion fields through Majorana and Dirac masses. (a) The general coupling scheme for the left- and right-handed fields and their charge conjugate fields. (b) The coupling scheme for electrons as an example of charged fermions. (c) The coupling scheme for neutrinos. Only neutrinos can have the general coupling via both Majorana and Dirac masses.

The neutrino is the only fermion that can have the general mass matrix consisting of Dirac as well as of Majorana mass terms (Fig. 2.1c), and can have the Majorana self-conjugate character, $\psi = \psi^c$. Mass eigenstates are obtained by diagonalizing the mass matrix, eq.(2.4). For simplicity, we assume here that CP is conserved. Then the mass matrix is real (symmetric), and is diagonalized by the orthogonal matrix $O$ as

$$\begin{bmatrix} \nu_L \\ (\nu_R)^c \end{bmatrix} = O \begin{bmatrix} N'_{1L} \\ N'_{2L} \end{bmatrix} \tag{2.6a}$$

$$O^T M O = \begin{bmatrix} s_1 m_1 & 0 \\ 0 & s_2 m_2 \end{bmatrix} \tag{2.6b}$$

where $m_j > 0$ and $s_j$ is a sign factor which can be $+1$ or $-1$. The sign factors can be absorbed in the fields and the mixing matrix [Sch 80, Doi 83b] by introducing an additional phase factor which is $i$ for $s_j = -1$ (CP odd) and 1 for $s_j = +1$ (CP even). When $s_1 = -1$ and $s_2 = +1$,

$$\begin{bmatrix} N_{1L} \\ N_{2L} \end{bmatrix} = \begin{bmatrix} i & 0 \\ 0 & 1 \end{bmatrix} \begin{bmatrix} N'_{1L} \\ N'_{2L} \end{bmatrix} \tag{2.7}$$

$$U_\nu = O \begin{bmatrix} -i & 0 \\ 0 & 1 \end{bmatrix} = \begin{bmatrix} U \\ V^* \end{bmatrix} \tag{2.8}$$

$$\begin{bmatrix} \nu_L \\ (\nu_R)^c \end{bmatrix} = U_\nu \begin{bmatrix} N_{1L} \\ N_{2L} \end{bmatrix}. \tag{2.9}$$

The mass Lagrangian is then written as

$$L_m = -\frac{1}{2} \sum_{j=1}^{2} m_j \bar{N}_j N_j \tag{2.10}$$

with

$$N_j = N_{jL} + (N_{jL})^c. \tag{2.11}$$

Obviously, the mass eigenstate $N_j$ has the mass $m_j$ and satisfies the Majorana self-conjugate condition, $N_j = (N_j)^c$. This mass term is not invariant under any global gauge transformation of the neutrino fields, and can change the lepton number by two units.

One can have the following special cases in eq.(2.4):

a) The "see-saw" mechanism [Yan 79, Gel 79], see also [Moh 80, Doi 85] and Langacker in this volume.

Assuming $m_L^M = 0$, $m_R^M \gg m^D$, with $m^D$ of the order of the charged lepton or quark masses, i.e $m_L^M \ll m^D \ll m_R^M$, diagonalization of the mass matrix leads to the following mass eigenvalues

$$m_1 \sim (m^D)^2/m_R^M \qquad \text{(light neutrino)} \tag{2.12a}$$

$$m_2 \sim m_R^M \qquad \text{(heavy neutrino).} \tag{2.12b}$$

In general, the right-handed Majorana mass $m_R^M$ is assumed to be produced when the left-right symmetry is broken, while the Dirac mass $m^D$ is induced when the $SU(2)_L\times$ U(1) symmetry is broken. The "see-saw" mechanism proves to be useful in reducing the large Dirac mass of the neutrino automatically present in the SO(10) model (see below) by an interplay between the large Dirac mass terms and an even larger Majorana mass term. It explains "automatically" the small mass of the observed left-handed neutrino in comparison with the charged fermion masses. The condition $m_R^M \geq 10^3$ GeV would lead to $m_1 \leq 1$ eV. The neutrino in this case is a Majorana particle.

b) Pseudo-Dirac neutrino [Wol 81, Pet 82, Doi 83b, Leu 83, Val 83a,83b, Wyl 83].
For a pseudo-Dirac neutrino, $m_L^M = m_R^M \ll m^D$. Diagonalization of the mass matrix yields two Majorana neutrinos with slightly different masses, $m_{1,2} = m^D \pm m^M$ and opposite CP signs. In this case the interaction neutrino is predominantly a Dirac particle with a very slight admixture of Majorana character. Even if the masses are not small, the pseudo-Dirac neutrinos could suppress $0\nu\ \beta\beta$ decay rates by a possible cancellation between the contributions from the almost degenerate masses. Therefore, this suppression mechanism could become attractive, if the effective neutrino mass deduced from $0\nu\ \beta\beta$ decay differs considerably from that measured in tritium decay. In these two processes, we see effective neutrino masses corresponding to different superpositions of the mass eigenstates (see v. Feilitzsch in this volume).

c) Dirac neutrino
In the limit of $m_L^M = m_R^M = 0$, a Dirac neutrino appears. It can be constructed with a pair of Majorana neutrinos which have the degenerate masses, $m_1 = m_2 = m^D$,

$$\psi_L = (N_{1L} + iN_{2L})/\sqrt{2} \tag{2.13a}$$

$$\psi_R = ((N_{1L})^c + i(N_{2L})^c)/\sqrt{2} \tag{2.13b}$$

where

$$N_k = \begin{bmatrix} -i\sigma_2\eta_k^* \\ \eta_k \end{bmatrix} \quad \text{or} \quad \begin{bmatrix} \xi_k \\ i\sigma_2\xi_k^* \end{bmatrix}. \tag{2.14}$$

The two-component fields $\eta_k$ and $\xi_k$ are left-handed and right-handed fields, respectively.

The one flavour mass Lagrangian, eq.(2.4), can easily be generalized to a system consisting of n left-handed neutrinos $\nu_{lL}$ and n right-handed neutrinos $\nu_{lR}$ ($l$ = e, $\mu$, ...), corresponding to n flavours. In this case $\nu_{lL}$ and $\nu_{lR}$ are n-component column vectors. Correspondingly, the mass matrix is 2n $\times$ 2n ( and symmetric),

$$M = \begin{bmatrix} m_L^M & (m^D)^T \\ m^D & m_R^M \end{bmatrix} \tag{2.15}$$

(each submatrix is n $\times$ n). With the following transformation and mass eigenstates,

$$\begin{bmatrix} \nu_L \\ (\nu_R)^c \end{bmatrix} = U_\nu N_L \tag{2.16}$$

$$N = N_L + (N_L)^c \tag{2.17}$$

the mass term is expressed as

$$L_m = -\tfrac{1}{2}(\bar{N_L})^c D_\nu N_L + h.c. = -\tfrac{1}{2}\bar{N} D_\nu N \tag{2.18}$$

where $D_\nu$ is a diagonal matrix with real, positive mass eigenvalues.

The existing species of neutrinos and their fundamental properties depend on the choice of the gauge model (see Langacker, Mohapatra in this volume, and also [Gro 88]). The standard electroweak theory of Glashow, Weinberg and Salam [Gla 61, Wei 67, Sal 68] is based on $SU(2)_L \times U(1)$ gauge invariance. The multiplets of leptons belonging to the first generation are shown in Fig. 2.2a. The left-handed components of the leptons form a doublet of the group $SU(2)_L$, while the right-handed component of the charged lepton forms a singlet. The left-handed and right-handed quarks are assumed to form multiplets separately from the leptons, and as a result the lepton and baryon numbers are conserved separately. There is no right-handed neutrino in the standard theory. This means that the left-handed neutrino must be massless. One can see from Fig. 2.1 that the Dirac mass coupling does not exist in the absence of right-handed neutrinos, and the Majorana mass should vanish because it violates lepton number conservation.

In spite of the great success of the standard theory, it seems incomplete and inadequate in many respects. The theory has in its minimal version 21 free parameters which have to be fixed to specify the theory. The global symmetry implying separate lepton- and baryon-number conservation cannot be considered as fundamental in the context of gauge theories.

The grand unified theories (GUT's) represent an attempt to unify the electroweak and strong interactions of leptons and quarks in the framework of gauge theories. The simplest GUT is the minimal SU(5) theory of Georgi and Glashow [Geo 74]. The fermions of each generation are assigned to the $(\bar{5}+10)$ representation of the gauge group, Fig. 2.2b. In this theory, superheavy gauge bosons, $X$ and $Y$, are predicted, in addition to the vector bosons $W$ and $Z$ of the standard theory. In the SU(5) model, baryon number $B$ and lepton number $L$ are not conserved separately, but the combination $B-L$ is. However, as in the standard theory, the right-handed neutrino is absent, and the neutrino is massless. In the minimal SU(5) model, a right-handed neutrino again does not exist, leading to a vanishing Dirac mass; on the other hand, SU(5) invariant Majorana couplings are not possible with the Higgs content of the minimal model. An interesting consequence of the $B$ non-conservation is the possible decay of the proton. Recent theoretical estimates based on the SU(5) models [Lan 81,86] lead to a half-life of $\tau_p \simeq 10^{28} - 10^{32}$ years, the decay mode $p \to e^+ + \pi^0$ being predicted as the strongest decay branch with a branching ratio of 40-60 %. The experimental value $\tau(p \to e^+ + \pi^0) > 4 \times 10^{32}$ years [Mey 86] rules out the minimal SU(5) model.

Massive neutrinos appear quite naturally in the SO(10) GUT's [Geo 75, Fri 75]. This is the minimal left-right symmetric grand unified model that gauges the $B-L$ symmetry (see [Moh 86]). All fermions of one generation form a 16-dimensional spinor representation of SO(10), which consists of the $\bar{5}$- and 10-dimensional SU(5) representations and one additional neutral fermion, which has to be interpreted as the right-handed neutrino $\nu_R$. Fig. 2.2c shows the multiplet of the first generation. The Dirac mass term shows up here, since both left-handed and right-handed neutrinos exist. In addition, since $B-L$ is not necessarily conserved, the Majorana mass terms do not vanish. Thus, in the SO(10) theories, neutrinos are expected to have the general mass Lagrangian of eq. (2.3). Assuming the right-handed neutrino $\nu_R$ represented by the SU(5) singlet to

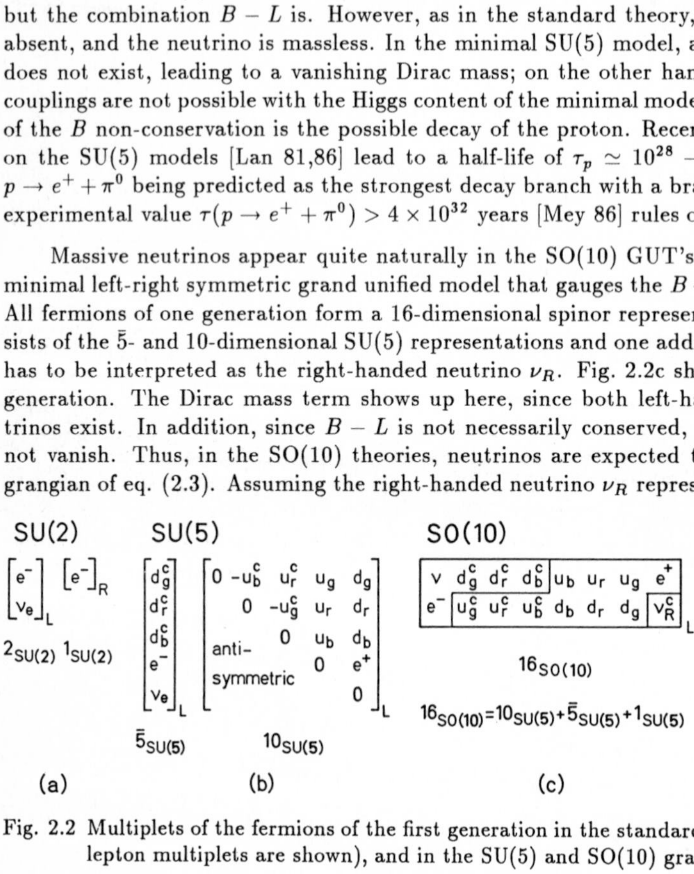


Fig. 2.2 Multiplets of the fermions of the first generation in the standard electroweak theory (only the lepton multiplets are shown), and in the SU(5) and SO(10) grand unified theories.

be very heavy — which is consistent with its non-observation — the see-saw mechanism leads in the way described above to the phenomenologically desired small mass of the observed left-handed neutrino. This mechanism predicts, neglecting for simplicity the possible mixing between different generations, that the mass of the left-handed (light) neutrino is proportional to the square of the fermion (quark or lepton) mass from the same generation, $m_1 \simeq (m^D)^2/m_R^M$ (see eq.(2.12a)),

$$m_{\nu_e} : m_{\nu_\mu} : m_{\nu_\tau} = m_u^2 : m_c^2 : m_t^2 \quad \text{or} \quad = m_e^2 : m_\mu^2 : m_\tau^2. \tag{2.19}$$

In some models, typical values of the electron neutrino mass are around $10^{-1}$ eV and 1 eV (Langacker, Mohapatra in this volume, and [Moh 81a, Cha 85]); in general, however, depending on the type of the see-saw mechanism applied (GUT see-saw, intermediate see-saw, $SU(2)_L\times SU(2)_R\times U(1)$ see-saw, ... ), i.e. on the mass of the right-handed heavy neutrino, the predicted masses range between $10^{-11}$ eV and about 1 eV.

It should be noted that spontaneous breaking of the global $B-L$ symmetry would — in addition to allowing for neutrinoless $\beta\beta$ decay — require the existence of a new particle, the majoron [Gel 81, Chi 81, Geo 81], which should be emitted in one of the $\beta\beta$ decay modes (see sect. 3.4).

Finally, left-right symmetric models involve, as well as left-handed gauge bosons, right-handed gauge bosons which mediate right-handed (V+A) weak interactions.

(footnote 1)
This is not the usual charge conjugation, because under the so defined operation the handedness of the field changes. The usual definition of charge conjugation for the handed fields would be: $(\psi_{L/R})^c = C\,(\bar{\psi}_{R/L})^T$.

## 3. Decay rates of $2\nu$ and $0\nu$ $\beta\beta$ decay

### 3.1. Effective charged weak interaction

We adopt the following effective weak interaction consisting of the vector (V) and axial-vector (A) currents,

$$H_W = \frac{G}{\sqrt{2}}[j_L \cdot J_L^\dagger + \kappa\, j_L \cdot J_R^\dagger + \eta\, j_R \cdot J_L^\dagger + \lambda\, j_R \cdot J_R^\dagger] + h.c. \tag{3.1}$$

Here, $G$ is the Fermi coupling constant, and $\kappa$, $\eta$ and $\lambda$ are mixing parameters of the right-handed currents. The leptonic currents are defined as

$$j_L^\mu = \bar{e}\gamma^\mu(1-\gamma_5)\nu_{eL}, \qquad j_R^\mu = \bar{e}\gamma^\mu(1+\gamma_5)\nu'_{eR} \tag{3.2}$$

$$\nu_{eL} = \textstyle\sum_j U_{ej} N_{jL}, \qquad \nu'_{eR} = \sum_j V_{ej} N_{jR} \tag{3.3}$$

where $N_j$ is a Majorana neutrino with the mass eigenvalue $m_j$. This expression includes the case of Dirac neutrinos, since a Dirac neutrino can be composed of a pair of Majorana neutrinos with degenerate masses (see sect. 2). The hadronic currents are expressed in terms of quarks in a form similar to eq.(3.2), and can be reduced to currents for nucleons. The hadronic currents are modified by higher-order perturbations from the strong and electroweak interactions; as a consequence the A part is renormalized (the effective coupling constant is denoted by $g_A$), and in addition weak magnetism ($g_W$) and pseudo-scalar ($g_P$) terms are induced. In the non-relativistic impulse approximation, the hadronic currents in the form of operators acting on the nucleons can be written as

$$J_{L/R}^{0\dagger}(\mathbf{x}) = \sum_n t_{-n}(g_V \mp g_A C_n)\delta(\mathbf{x}-\mathbf{r}_n) \tag{3.4a}$$

$$J_{L/R}^{k\dagger}(\mathbf{x}) = \sum_n t_{-n}(\pm g_A \sigma_n^k - g_V D_n^k)\delta(\mathbf{x}-\mathbf{r}_n). \tag{3.4b}$$

The isospin operator $t_-$ converts a neutron into a proton, $\sigma$ and $\mathbf{r}$ are spin and position operators, respectively, and the sum $n$ runs over all nucleons in a nucleus. The nucleon recoil terms, which are first-order terms of the non-relativistic expansion, have the following forms,

$$C_n = [(\mathbf{P}_n + \mathbf{P}'_n)\cdot\sigma_n - \frac{g_P}{g_V}(E_n - E'_n)(\mathbf{P}_n - \mathbf{P}'_n)\cdot\sigma_n]/(2M) \tag{3.5a}$$

$$\mathbf{D}_n = [(\mathbf{P}_n + \mathbf{P}'_n) - (1 - 2M\frac{g_W}{g_V})i\sigma_n \times (\mathbf{P}_n - \mathbf{P}'_n)]/(2M) \tag{3.5b}$$

where $(E_n, \mathbf{P}_n)$ and $(E'_n, \mathbf{P}'_n)$ are the initial and final nucleon four-momenta, and $M$ the nucleon mass. Values of the coupling constants are $G = 1.16637 \times 10^{-5}$ GeV$^{-2}$, $g_V = 0.9737$, $g_A/g_V = -1.254$, $\mu_\beta = 1 - 2Mg_W/g_V = 4.706$ [PDG 86] and $g_P/g_A = 2M/m_\pi^2$ with $m_\pi$ being the pion mass.

3.2. Decay rate formulae for the $2\nu$ mode

The $2\nu$ mode $\beta\beta$ decay can be described in second order perturbation of the weak interaction as successive Gamow-Teller transitions via virtual intermediate $1^+$ states (see Fig. 3.1). Since the energy released by the decay is of the order of 1 MeV (the largest $Q$-value $T_0$ for $\beta\beta$ decay is that of $^{48}$Ca with $T_0 = 4.27$ MeV) and is shared by the four leptons in the continuum, the electrons and neutrinos can be assumed to be emitted in S-wave states. Consequently higher-order terms can be neglected: the higher partial waves of the leptons, and finite wave-length effects. Furthermore, in what follows, we neglect the nucleon recoil terms and contributions from the right-handed currents.

Let us first consider $0_i^+ \to 0_f^+$ transitions, which have been extensively studied both experimentally and theoretically. In second order perturbation theory, the inverse half-life is given by [Gro 83a, Kla 84, Doi 85] (for a derivation see [Gro 88])

$$[T_{1/2}^{2\nu}(0^+ \to 0^+)]^{-1} = \frac{a^{2\nu}}{\ln 2}\int_{m_e}^{T_0+m_e} F_0(Z,e_1)k_1e_1\,de_1 \int_{m_e}^{T_0+2m_e-e_1} F_0(Z,e_2)k_2e_2\,de_2$$
$$\times \int_0^{T_0+2m_e-e_1-e_2} \nu_1^2\nu_2^2\,d\nu_1 \sum_{a,a'} A_{aa'} \tag{3.6}$$
$$(T_0 = e_1 + e_2 - 2m_e + \nu_1 + \nu_2)$$

$$a^{2\nu} = \frac{G^4 g_A^4}{32\pi^7 m_e} \tag{3.7}$$

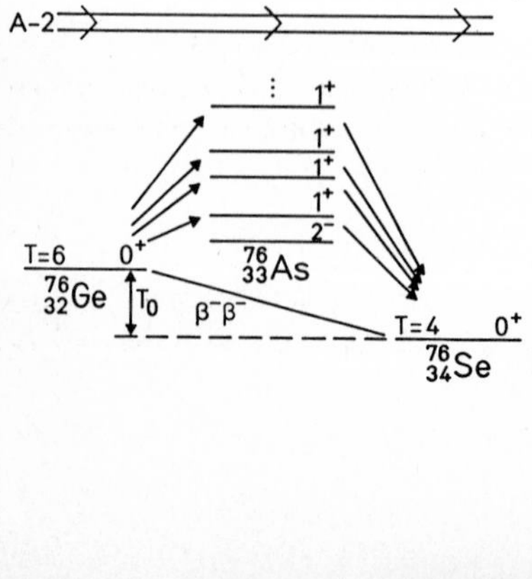

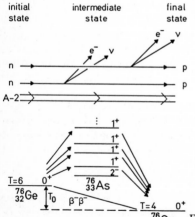

Fig. 3.1 Diagram and illustration of the $2\nu$ $\beta\beta$ decay of $^{76}$Ge.

where $k = \sqrt{e^2 - m_e^2}$ is the electron momentum, and $F_0(Z, e)$ is the Fermi function which takes into account the distortion of the electron wave functions due to the charge $Z$ of the daughter nucleus. (The energies of the leptons are denoted here simply by $e_i$, $\nu_i$.) The nuclear Gamow-Teller transition matrix elements and the energy denominators in the perturbation expression appear in $A_{aa'}$,

$$A_{aa'} = \langle 0_f^+ \| t_-\sigma \| 1_a^+ \rangle \langle 1_a^+ \| t_-\sigma \| 0_i^+ \rangle \langle 0_f^+ \| t_-\sigma \| 1_{a'}^+ \rangle \langle 1_{a'}^+ \| t_-\sigma \| 0_i^+ \rangle \times \frac{1}{3}(K_a K_{a'} + L_a L_{a'} + \frac{1}{2} K_a L_{a'} + \frac{1}{2} L_a K_{a'}) \tag{3.8}$$

where

$$K_a = \frac{1}{E_a + e_1 + \nu_1 - E_i} + \frac{1}{E_a + e_2 + \nu_2 - E_i} \tag{3.9a}$$

$$L_a = \frac{1}{E_a + e_1 + \nu_2 - E_i} + \frac{1}{E_a + e_2 + \nu_1 - E_i}. \tag{3.9b}$$

The quantity $A_{aa'}$ depends on the energies of both the intermediate nuclear states and the leptons. (For the reduced matrix elements we adopt the Edmonds convention [Edm 57].)

When defining the nuclear matrix element $M_{GT}^{2\nu}$ as

$$M_{GT}^{2\nu} = \sum_a \frac{\langle 0_f^+ \| t_-\sigma \| 1_a^+ \rangle \langle 1_a^+ \| t_-\sigma \| 0_i^+ \rangle}{E_a + T_0/2 + m_e - E_i} \tag{3.10}$$

replacing the lepton energy $e + \nu$ in eq.(3.9) by $T_0/2 + m_e$, the nuclear part of eq.(3.6) is separated from kinematical factors, and considerations of nuclear structure effects can focus on the nuclear matrix element $M_{GT}^{2\nu}$. The half-life is then

$$T_{1/2}^{2\nu} = [G^{2\nu} |M_{GT}^{2\nu}|^2]^{-1} \tag{3.11}$$

where $G^{2\nu}$ is the phase space integral. The replacement of the lepton energy by $T_0/2 + m_e$ is in general a good approximation, since the integrand of eq.(3.6) becomes large when an electron and neutrino pair carries a half of the maximum available kinetic energy $T_0$. However, the approximation of eq.(3.11) fails when an almost complete cancellation between terms in eq.(3.10) occurs. In this case, $M_{GT}^{2\nu}$ is merely a measure of the nuclear transition strength, and the integration over the lepton energies must be carried out explicitly for each intermediate nuclear state.

The closure approximation, which was sometimes employed in earlier calculations, introduces an additional approximation. That is to replace the intermediate state energies $E_a$ by some average value $\langle E_a \rangle$, and then complete the sum by closure, $\sum_a |1_a^+\rangle\langle 1_a^+| = 1$. The nuclear matrix element, eq.(3.10), is then reduced to

$$M_{GT}^{2\nu} = -\frac{[M_{GT}^{2\nu}]_c}{\langle E_a \rangle + T_0/2 + m_e - E_i} \tag{3.12}$$

with the closure matrix element

$$[M_{GT}^{2\nu}]_c = \sum_{m,n} \langle 0_f^+ \| t_{-m} t_{-n} \sigma_m \cdot \sigma_n \| 0_i^+ \rangle. \tag{3.13}$$

In this approximation only wave functions of the initial and final $0^+$ states are needed, and a tedious calculation of the many intermediate states is avoided. The validity of the closure approximation depends simply on the choice of $\langle E_a \rangle$. It is clear that it fails badly, if the matrix elements on the right side of eq.(3.10) have different signs and there is a significant cancellation in the sum. This is usually more or less the case (see e.g. Figs. 5.1 and 5.4).

The vector part (V) of the weak interaction gives no (or an extremely small) contribution to the $2\nu$ mode. The vector part of the hadronic current, eq.(3.4a), contains only the isospin operator $t_-$ of the Fermi transition, and the sum over the nucleons results in the isospin lowering operator for the nucleus, $\sum_n t_{-n} = T_-$. The operator $T_-$ cannot change the isospin of nuclear states, and isospin is normally a good quantum number. The initial and final states of $\beta\beta$ decay have, however, isospins different by two units (see e.g. Fig. 3.1). A finite matrix element for the Fermi transition results mainly from isospin violating terms which are brought into the nuclear structure model by simplification (i.e., approximation) of the calculation.

In even-even nuclei, a $2^+$ state appears at low excitation energy, usually as the first excited state, because of strong proton-neutron quadrupole correlations. When the $2^+$ state in the daughter nucleus lies energetically lower than the parent state, $0_i^+ \to 2^+$ $\beta\beta$ decay may occur. The decay rate can be obtained from that of the $0_i^+ \to 0_f^+$ transition, eq.(3.6), by small changes; the maximum available kinetic energy $T_0$ should be interpreted as that of the $0_i^+ \to 2^+$ decay, and $A_{aa'}$ is replaced by

$$A_{aa'} = \frac{1}{3}\langle 2^+\|t_-\sigma\|1_a^+\rangle\langle 1_a^+\|t_-\sigma\|0_i^+\rangle\langle 2^+\|t_-\sigma\|1_{a'}^+\rangle\langle 1_{a'}^+\|t_-\sigma\|0_i^+\rangle \times(K_a - L_a)(K_{a'} - L_{a'}). \tag{3.14}$$

The energy denominators $K_a$ and $L_a$ are defined in eq.(3.9). In contrast to the $0_i^+ \to 0_f^+$ decay, differences of $K_a$ and $L_a$ appear. The leading terms cancel each other,

$$K_a - L_a \simeq \frac{2(\nu_2 - \nu_1)(e_2 - e_1)}{(E_a + T_0/2 + m_e - E_i)^3}. \tag{3.15}$$

The cubic factor in the denominator and the asymmetry in the electron and neutrino energies indicate strong suppression of the $0_i^+ \to 2^+$ decay, and moreover the smaller $T_0$ implies additional kinematical hindrance of the decay. As before the nuclear matrix element

$$M_{GT}^{2\nu}(0_i^+ \to 2^+) = \sum_a \frac{\langle 2^+\|t_-\sigma\|1_a^+\rangle\langle 1_a^+\|t_-\sigma\|0_i^+\rangle}{(E_a + T_0/2 + m_e - E_i)^3}. \tag{3.16}$$

is introduced and the half-life factorized as

$$T_{1/2}^{2\nu}(0_i^+ \to 2^+) = [G_2^{2\nu}|M_{GT}^{2\nu}(0_i^+ \to 2^+)|^2]^{-1} \tag{3.17}$$

where $G_2^{2\nu}$ is the phase space integral. (The lepton energy factor $2(\nu_2 - \nu_1)(e_2 - e_1)$ in eq.(3.15) should be included in the integral.)

In the closure approximation, the nuclear matrix element is reduced to

$$M_{GT}^{2\nu}(0_i^+ \to 2^+) \simeq \frac{[M_{GT}^{2\nu}(0_i^+ \to 2^+)]_c}{(\langle E_a\rangle + T_0/2 + m_e - E_i)^3} \tag{3.18}$$

with the closure matrix element

$$[M_{GT}^{2\nu}(0_i^+ \to 2^+)]_c = \sum_{m,n}\langle 2^+\|t_{-m}t_{-n}[\sigma_m \times \sigma_n]^{(2)}\|0_i^+\rangle \tag{3.19}$$

where the square bracket denotes the coupling of the two spin operators (spherical tensors of rank 1) to a spherical tensor operator of rank 2.

It is known that the Gamow-Teller strength is quenched by the excitation of subnuclear degrees of freedom in $\beta$ decay (see. e.g., [Gro 83b,86]). To take into account virtual excitations of the $\Delta$ resonance, one can include the $\Delta N^{-1}$ space in a simple quark model treatment of the nucleons [Boh 81, Gro 83b,86]. Examples of $\beta\beta$ transition amplitudes involving $\Delta N^{-1}$ configurations instead of nucleonic particle-hole excitations are shown in Fig. 3.2. The effect on the $\beta\beta$ transition rates is of the order of 30 % and is discussed in detail in [Gro 86], see also Table 5.4.

### 3.3. Decay rate formulae for the $0\nu$ mode

Fig. 3.3 shows a schematic diagram of the $0\nu$ mode in the two-nucleon mechanism. (Here again there is also the possibility of $\Delta N^{-1}$ excitations, see Fig. 3.2.) An antineutrino is created at

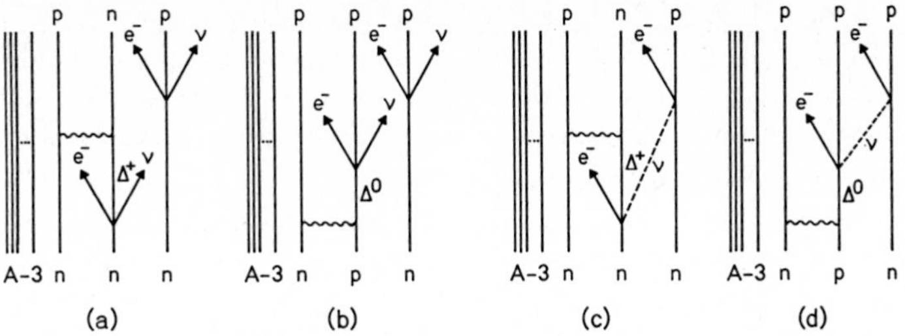


Fig. 3.2 Diagrams involving $\Delta N^{-1}$ excitations; (a) in the intermediate state, (b) in the initial state for the $2\nu$ mode, (c) and (d) the corresponding diagrams for the $0\nu$ mode. The wavy line represents the strong interaction.

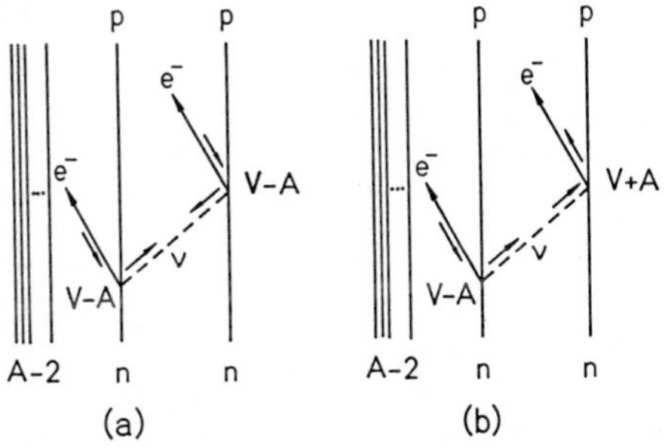


Fig. 3.3 Diagrams of the neutrinoless $\beta\beta$ decay in the two-nucleon mechanism. At each vertex are indicated the handedness of the weak interaction, V-A (left-handed) or V+A (right-handed), and the main helicity components of the leptons by short arrows.

one vertex, $n \to p + e^- + \bar{\nu}$, and a neutrino is to be absorbed at the other vertex, $\nu + n \to p + e^-$. If the (anti)neutrino has zero mass, helicity is a good quantum number. The antineutrino has positive helicity (right-handed), while the neutrino has negative helicity (left-handed), and consequently the $0\nu$ $\beta\beta$ decay is forbidden if only a left-handed (V-A) weak interaction is acting. For massive neutrinos (Fig. 3.3a), the opposite helicity component can be mixed, and the small right-handed (left-handed) component of the neutrino (antineutrino) matches the main right-handed (left-handed) component of the antineutrino (neutrino), resulting in the occurrence of $0\nu$ $\beta\beta$ decay. The helicity mismatch between the main components for the left-handed interaction is also avoided by a possible right-handed (V+A) interaction mediated by right-handed weak gauge bosons (Fig. 3.3b).

Since the formalism for deriving the $0\nu$ decay rate is rather complex we refer for completeness to [Doi 81,83a,85, Gro 88] and concentrate here on some essential points.

The leptonic part of the $0\nu$ transition operator is

$$\begin{aligned} T_{lep} &\sim \bar{e}(x)\gamma_\rho \frac{1 \pm \gamma_5}{2} N_j(x)\bar{e}(y)\gamma_\sigma \frac{1 \pm \gamma_5}{2} N_k(y) \\ &= -\bar{e}(x)\gamma_\rho \frac{1 \pm \gamma_5}{2} N_j(x) N_k(y)^T \frac{1 \pm \gamma_5}{2} \gamma_\sigma^T \bar{e}(y)^T \\ &= -i\delta_{jk} \int \frac{d^4q}{(2\pi)^4} \frac{e^{-iq(x-y)}}{q^2 - m_j^2 + i\epsilon} \bar{e}(x)\gamma_\rho \frac{1 \pm \gamma_5}{2} (q^\mu \gamma_\mu + m_j) \frac{1 \pm \gamma_5}{2} \gamma_\sigma e^c(y) \end{aligned} \tag{3.20}$$

where a contraction between the neutrino operators $N_j$ and $N_k^T$ has been made, which is allowed for a Majorana neutrino, and in the last step the neutrino propagator $S_F(x - y)$ substituted,

$$S_F(x-y) = \int \frac{d^4q}{(2\pi)^4} \frac{e^{-iq(x-y)}}{q^2 - m_j^2 + i\epsilon}(q^\mu \gamma_\mu + m_j) \tag{3.21}$$

Here, $q^2 = \omega^2 - \mathbf{q}^2$ with $\mathbf{q}$ being the three-momentum of the neutrino and $\omega = \sqrt{\mathbf{q}^2 + m_j^2}$ the neutrino energy. By performing the integration over the neutrino energy, the factor $(q^2 - m_j^2)^{-1}$ in eq.(3.20) is replaced by its residue $\pi/\omega$. Therefore the leptonic part of the transition operator includes the integration over the virtual neutrino momentum,

$$T_{lep} \sim \frac{1}{(2\pi)^3} \int \frac{d\mathbf{q}}{\omega} e^{-\mathbf{q}\cdot(\mathbf{x}-\mathbf{y})} \bar{e}(\mathbf{x}) \gamma_\rho \frac{1 \pm \gamma_5}{2} (\omega\gamma^0 - \mathbf{q}\cdot\gamma + m_\nu) \frac{1 \pm \gamma_5}{2} \gamma_\sigma e^c(\mathbf{y}). \tag{3.22}$$

Using the relations,

$$(1-\gamma_5)(\omega\gamma^0 - \mathbf{q}\cdot\gamma + m_\nu)(1-\gamma_5) = 2m_\nu(1-\gamma_5) \tag{3.23a}$$

$$(1-\gamma_5)(\omega\gamma^0 - \mathbf{q}\cdot\gamma + m_\nu)(1+\gamma_5) = 2(\omega\gamma^0 - \mathbf{q}\cdot\gamma)(1+\gamma_5) \tag{3.23b}$$

we can see that only the neutrino mass term contributes when the left-handed current acts at both vertices (or the right-handed current at both vertices but its contribution can be neglected), eq.(3.23a). On the other hand, the interference between left-handed (at one vertex) and right-handed (at the other vertex) currents allows a transition originating from the energy and momentum of the virtual neutrino, eq.(3.23b).

The integration over the neutrino momentum in the leptonic sector leads to a "neutrino potential" (eq.(3.39)) which acts on the nuclear wave function. (In the $2\nu$ mode the leptonic and hadronic parts are almost separate, except for the energy denominator of the perturbation theory.) The potential, which represents the exchange of the neutrino between the two nucleons, is a function of the momentum transferred by the neutrino, or, in coordinate space, of distance $|\mathbf{r}_1 - \mathbf{r}_2|$ between the two nucleons. Such a potential induces transitions leading to states of various multipolarity $(J^\pi)$ in the intermediate odd-odd nucleus. This can be seen by a multipole decomposition of the potential into products of two operators which, respectively, act on the two nucleons. For instance, in the case of the Gamow-Teller matrix element eq.(3.37a),

$$H(|\mathbf{r}_1 - \mathbf{r}_2|)\sigma_1 \cdot \sigma_2 = \sum_{J,L} (-1)^{L+1-J}(2L+1) \int_0^\infty dq q^2 v(q)$$
$$\times (j_L(qr_1)[\sigma_1 \times Y_L(\hat{\mathbf{r}}_1)]^{(J)}) \cdot (j_L(qr_2)[\sigma_2 \times Y_L(\hat{\mathbf{r}}_2)]^{(J)}) \tag{3.24}$$

where $j_L(qr)$ is the spherical Bessel function of rank $L$, $Y_L(\hat{\mathbf{r}})$ the spherical harmonic of rank $L$, and $v(q)$ the Fourier transform of the potential $H(|\mathbf{r}_1 - \mathbf{r}_2|)$. When we consider a matrix element of eq.(3.24) between the initial $0_i^+$ and final $0_f^+$ states of $\beta\beta$ decay, and insert intermediate states $\sum_{J,M} |JM\rangle\langle JM| = 1$ between the operator with index 1 and that with index 2, the operator $j_L(qr)[\sigma \times Y_L(\hat{\mathbf{r}})]^{(J)}$ leads to intermediate states with the angular momentum $J$ and parity $\pi = (-1)^L$.

In the presence of a neutrino potential, wave functions at short internucleon distances must be treated carefully, in particular for the zero-range potential of the nucleon recoil matrix element (see, eqs. (3.45) and (3.46)). In the calculation of the nuclear matrix elements, nuclear wave functions consisting of (antisymmetrized) products of single-particle wave functions have usually been used, in which the short-range correlations are not taken into account. These correlations arise from the strong, short-range, repulsive components of the nucleon-nucleon interaction, which prevent two nucleons from coming close to each other. If the short-range correlations are taken into account, a matrix element of the zero-range potential vanishes completely. However, if we take into account the fact that the nucleon has a finite extention in space, the matrix element is not affected drastically by the short-range correlations. One can introduce the short-range correlations by multiplying the uncorrelated two-nucleon wave function with a function $f(|\mathbf{r}_1 - \mathbf{r}_2|)$

$$|\phi(\mathbf{r}_1)\phi(\mathbf{r}_2)\rangle \longrightarrow f(|\mathbf{r}_1-\mathbf{r}_2|)|\phi(\mathbf{r}_1)\phi(\mathbf{r}_2)\rangle, \qquad (3.25)$$

for which the functional form

$$f(r) = 1 - e^{-ar^2}(1-br^2) \qquad (3.26)$$

with $a = 1.1$ fm$^{-2}$ and $b = 0.68$ fm$^{-2}$ [Mil 76] has usually been used. The finite nucleon-size effects are taken into account by replacing the vector and axial-vector coupling constant by dipole form factors in momentum space [Ver 81,83],

$$g_V \longrightarrow g_V(\frac{\Lambda^2}{\Lambda^2+q^2})^2, \qquad g_A \longrightarrow g_A(\frac{\Lambda^2}{\Lambda^2+q^2})^2 \qquad (3.27)$$

with typically $\Lambda = 850$ MeV. This amounts to smearing out the delta function in eq.(3.4) with the replacement

$$\delta(\mathbf{x}-\mathbf{r}_n) \longrightarrow \frac{\Lambda^3}{8\pi}\exp(-\Lambda|\mathbf{x}-\mathbf{r}_n|). \qquad (3.28)$$

In $0\nu$ $\beta\beta$ decay one can apply the closure approximation which in this case is good, in contrast to $2\nu$ $\beta\beta$ decay. The reason is that the average energy of intermediate states appears in the neutrino potential, eq.(3.39), in the combination $\omega + \bar{E}$, where $\bar{E} = \langle E_a\rangle + T_0/2 + m_e - E_i$; this is typically $\sim 10$ MeV, while the energy of the virtual neutrino is typically $\omega \sim 1/\bar{r} \sim 100$ MeV (with $\bar{r}$ being the mean distance between two nucleons), i.e. much larger than $\bar{E}$.

3.3.1. $0^+ \rightarrow 0^+$ transitions

The inverse half-life of the $0\nu$ mode $0_i^+ \rightarrow 0_f^+$ $\beta\beta$ decay is given by [Doi 85]

$$[T_{1/2}^{0\nu}(0_i^+ \rightarrow 0_f^+)]^{-1} = C_{mm}(\frac{\langle m_\nu\rangle}{m_e})^2 + C_{\eta\eta}\langle\eta\rangle^2 + C_{\lambda\lambda}\langle\lambda\rangle^2$$
$$+C_{m\eta}\langle\eta\rangle\frac{\langle m_\nu\rangle}{m_e} + C_{m\lambda}\langle\lambda\rangle\frac{\langle m_\nu\rangle}{m_e} + C_{\eta\lambda}\langle\eta\rangle\langle\lambda\rangle. \qquad (3.29)$$

Effective values of the neutrino mass and right-handed parameters are defined by

$$\langle m_\nu\rangle = \textstyle\sum_j m_j U_{ej}^2 \qquad (3.30a)$$
$$\langle\eta\rangle = \eta\textstyle\sum_j U_{ej}V_{ej} \qquad (3.30b)$$
$$\langle\lambda\rangle = \lambda\textstyle\sum_j U_{ej}V_{ej}. \qquad (3.30c)$$

The sum extends over light neutrino mass eigenstates, $m_j < 10$ MeV. Terms including $\kappa$ (see eq.(3.1)) are neglected, since $\kappa$ always appears in the $0\nu$ $\beta\beta$ decay amplitude in the combination $1 \pm \kappa$ and we expect $\kappa \ll 1$. The coefficients $C_{xy}$ include the nuclear matrix elements and phase space integrals (index $0\nu$ on the matrix elements is suppressed),

$$C_{mm} = (M_{GT} - M_F)^2 G_1 \qquad (3.31a)$$
$$C_{\eta\eta} = M_{2+}^2 G_2 + \frac{1}{9}M_{1-}^2 G_4 - \frac{2}{9}M_{1-}M_{2+}G_3 + M_P^2 G_8 - M_P M_R G_7 + M_R^2 G_9 \qquad (3.31b)$$
$$C_{\lambda\lambda} = M_{2-}^2 G_2 + \frac{1}{9}M_{1+}^2 G_4 - \frac{2}{9}M_{1+}M_{2-}G_3 \qquad (3.31c)$$
$$C_{m\eta} = (M_{GT} - M_F)(M_{2+}G_3 - M_{1-}G_4 - M_P G_5 + M_R G_6) \qquad (3.31d)$$
$$C_{m\lambda} = -(M_{GT} - M_F)(M_{2-}G_3 - M_{1+}G_4) \qquad (3.31e)$$
$$C_{\eta\lambda} = -2M_{2+}M_{2-}G_2 + \frac{2}{9}(M_{1+}M_{2+} + M_{1-}M_{2-})G_3 - \frac{2}{9}M_{1-}M_{1+}G_4 \qquad (3.31f)$$

with the combinations

$$M_{1\pm} = M'_{GT} \pm 3M'_F - 6M_T, \qquad (3.32a)$$
$$M_{2\pm} = M_{GT\omega} \pm M_{F\omega} - \frac{1}{9}M_{1\pm}. \qquad (3.32b)$$

The phase space integrals are

$$G_i = \frac{a^{0\nu}}{(m_e R)^2 \ln 2} \int_{m_e}^{T_0+m_e} F_0(Z,e_1)k_1e_1F_0(Z,e_2)k_2e_2g_i\, de_1 \qquad (3.33)$$

$$(e_1 + e_2 - 2m_e = T_0)$$

$$a^{0\nu} = \frac{G^4 g_A^4 m_e^4}{32\pi^5} \qquad (3.34)$$

with

$$g_1 = 1 \qquad (3.35a)$$

$$g_2 = \frac{1}{2}(\frac{e_1 - e_2}{m_e})^2 \frac{e_1e_2 - m_e^2}{e_1e_2} \qquad (3.35b)$$

$$g_3 = \frac{(e_1 - e_2)^2}{e_1e_2} \qquad (3.35c)$$

$$g_4 = \frac{2}{9}\frac{e_1e_2 - m_e^2}{e_1e_2} \qquad (3.35d)$$

$$g_5 = \frac{4}{3}[x_P \frac{m_e(T_0 + 2m_e)}{2e_1e_2m_eR} - \frac{e_1e_2 + m_e^2}{e_1e_2}] \qquad (3.35e)$$

$$g_6 = \frac{4}{m_eR}\frac{T_0m_e + 2m_e^2}{e_1e_2} \qquad (3.35f)$$

$$g_7 = \frac{16}{3}\frac{1}{m_eR}[x_P\frac{e_1e_2 + m_e^2}{2e_1e_2m_eR} - \frac{m_e(T_0 + 2m_e)}{e_1e_2}] \qquad (3.35g)$$

$$g_8 = \frac{2}{9}\frac{1}{(m_eR)^2}([x_P^2 + (2m_eR)^2]\frac{e_1e_2 + m_e^2}{e_1e_2} - 4m_eRx_P\frac{m_e(T_0 + 2m_e)}{e_1e_2}) \qquad (3.35h)$$

$$g_9 = (\frac{4}{m_eR})^2\frac{e_1e_2 + m_e^2}{2e_1e_2}. \qquad (3.35i)$$

The quantity which characterizes the P-wave emission of the electron is denoted by $x_P$,

$$x_P = 3\alpha Z + R(T_0 + 2m_e) \qquad (3.36)$$

where $\alpha$ is the fine structure constant and $R$ the nuclear radius.

The nine types of nuclear matrix elements of the $0_i^+ \to 0_f^+$ $0\nu$ $\beta\beta$ decay are summarized in Table 3.1. The first two elements originate from the neutrino mass term with left-handed interaction at both vertices, two electrons being emitted in S-wave states. The double Fermi transition ($M_F$) is possible in $0\nu$ decay, in contrast to $2\nu$ decay, because natural parity states with $\pi = (-1)^J$ are allowed for the nuclear intermediate states in the presence of the neutrino potential. The next two matrix elements in Table 3.1 are from the $\omega$-term of the left-right interference, (V-A) interaction at one vertex and (V+A) at the other vertex. The scalar property of the neutrino energy results in a similar structure for the transition operator to that of the previous elements, with a different neutrino potential. On the other hand, the neutrino three-momentum is a vector and has negative parity. This odd parity can be compensated, for the $0^+ \to 0^+$ transition with no parity change, either by P-wave emission of one of the two electrons ($M'_{GT}$, $M'_F$, $M_T$ and $M_P$), or by the nuclear part, i.e., the nucleon recoil term $M_R$.

Explicit forms of the closure matrix elements are

$$M_{GT}^{0\nu} = \langle H(r)\sigma_m \cdot \sigma_n \rangle \qquad (3.37a)$$

$$M_F^{0\nu} = \langle H(r)\rangle(\frac{g_V}{g_A})^2 \qquad (3.37b)$$

$$M_{GT\omega}^{0\nu} = \langle H_\omega(r)\sigma_m \cdot \sigma_n \rangle \qquad (3.37c)$$

Table 3.1. Nature of the transition operators for the neutrino, electron and nuclear parts of $0_i^+ \rightarrow 0_f^+$ $0\nu$ $\beta\beta$ decay. In the second column, L•L denotes the left-handed interaction at both vertices, and L•R left-handed interaction at one vertex and right-handed interaction at the other. The parity of each operator is presented in parentheses.

| Matrix Element | Vertex | Neutrino ($\pi$) | Electron ($\pi$) | Nucleus ($\pi$) |
|---|---|---|---|---|
| $M_{GT}$ | L•L | $m_\nu$ (+) | SS (+) | $\sigma_1 \cdot \sigma_2$ (+) |
| $M_F$ | | | | $1_1 \cdot 1_2$ (+) |
| $M_{GT\omega}$ | L•R | $\omega$ (+) | SS (+) | $\sigma_1 \cdot \sigma_2$ (+) |
| $M_{F\omega}$ | | | | $1_1 \cdot 1_2$ (+) |
| $M'_{GT}$ | | q (−) | SP (−) | $\sigma_1 \cdot \sigma_2$ (+) |
| $M'_F$ | | | | $1_1 \cdot 1_2$ (+) |
| $M_T$ | | | | $S_{12}$ (+) |
| $M_P$ | | | | $i(\sigma_1 - \sigma_2) \cdot (\hat{r} \times \hat{r}_+)$ (+) |
| $M_R$ | | | SS (+) | $\hat{r} \cdot (\sigma_1 \times D_2 + D_1 \times \sigma_2)$ (−) |

$S_{12} = (\sigma_1 \cdot \hat{r})(\sigma_2 \cdot \hat{r}) - \frac{1}{3}(\sigma_1 \cdot \sigma_2)$

$$M_{F\omega}^{0\nu} = \langle H_\omega(r) \rangle (\frac{g_V}{g_A})^2 \tag{3.37d}$$

$$M_{GT}^{'0\nu} = \langle H'(r) \sigma_m \cdot \sigma_n \rangle \tag{3.37e}$$

$$M_F^{'0\nu} = \langle H'(r) \rangle (\frac{g_V}{g_A})^2 \tag{3.37f}$$

$$M_T^{0\nu} = \langle H'(r)[(\sigma_m \cdot \hat{\mathbf{r}})(\sigma_n \cdot \hat{\mathbf{r}}) - \frac{1}{3}\sigma_m \cdot \sigma_n] \rangle \tag{3.37g}$$

$$M_P^{0\nu} = \langle H'(r) \frac{r_+}{2r} i(\sigma_m - \sigma_n) \cdot (\hat{\mathbf{r}} \times \hat{\mathbf{r}}_+) \rangle (\frac{g_V}{g_A}) \tag{3.37h}$$

$$M_R^{0\nu} = \langle H'(r) \frac{R}{2r} \hat{\mathbf{r}} \cdot (\sigma_m \times \mathbf{D}_n + \mathbf{D}_m \times \sigma_n) \rangle (\frac{g_V}{g_A}) \tag{3.37i}$$

where

$$\mathbf{r} = \mathbf{r}_m - \mathbf{r}_n, \qquad r = |\mathbf{r}|, \qquad \hat{\mathbf{r}} = \mathbf{r}/r,$$

$$\mathbf{r}_+ = \mathbf{r}_m + \mathbf{r}_n, \qquad r_+ = |\mathbf{r}_+|, \qquad \hat{\mathbf{r}}_+ = \mathbf{r}_+/r_+.$$

The simplified form

$$\langle O_{mn} \rangle = \sum_{m,n} \langle 0_f^+ \| t_{-m} t_{-n} O_{mn} \| 0_i^+ \rangle \tag{3.38}$$

is used in the expressions of eq.(3.37). The scalar recoil term $C_n$ of eq.(3.5a) does not contribute under the assumption that finite wave-length effects can be neglected.

The neutrino potentials, which are functions of $r$, are defined by

$$H(r) = \frac{R}{2\pi^2}\int \frac{d\mathbf{q}}{\omega}\frac{1}{\omega+\bar{E}}e^{i\mathbf{q}\cdot\mathbf{r}} \tag{3.39a}$$

$$H_\omega(r) = H(r) - \bar{E}\frac{R}{2\pi^2}\int \frac{d\mathbf{q}}{\omega}\frac{1}{(\omega+\bar{E})^2}e^{i\mathbf{q}\cdot\mathbf{r}} \tag{3.39b}$$

$$H'(r) = -r\frac{d}{dr}H(r). \tag{3.39c}$$

The factor $\omega^{-1}$ comes from the neutrino propagator (see eq.(3.22)), and $(\omega+\bar{E})^{-1}$ is the energy denominator of the perturbation theory. The radial form of the potential $H(r)$ is shown in Fig. 3.4 for some values of the neutrino mass. It behaves as $\sim 1/r$ for light neutrinos with $m_\nu < 10$ MeV, or as a Yukawa potential $e^{-m_\nu r}/r$ for heavy neutrinos. The dependence of the potential on the average energy of intermediate states is shown in Fig. 3.5. (The nuclear radius $R$ has been introduced in the definition of the neutrino potentials, eq.(3.39), in order to make the potentials dimensionless. This is compensated by a factor $R^{-2}$ in the phase space integrals, eq.(3.33).)

When there are no right-handed components of the weak interaction, the $0\nu$ $\beta\beta$ decay originates only from the finite neutrino mass. The $0\nu$ decay rate is then proportional to the square of the effective neutrino mass,

$$[T_{1/2}^{0\nu}(0_i^+ \to 0_f^+)]^{-1} = (M_{GT}^{0\nu} - M_F^{0\nu})^2(\frac{\langle m_\nu\rangle}{m_e})^2 G_1. \tag{3.40}$$

The phase space factor $G_1$ is defined by eqs. (3.33), (3.34) and (3.35a). The two nuclear matrix elements relevant to the neutrino mass, eqs. (3.37a) and (3.37b), are more explicitly expressed as

$$M_{GT}^{0\nu} = \sum_{m,n}\langle 0_f^+\|t_{-m}t_{-n}H(|\mathbf{r}_m - \mathbf{r}_n|)\sigma_m\cdot\sigma_n\|0_i^+\rangle \tag{3.41a}$$

$$M_F^{0\nu} = \sum_{m,n}\langle 0_f^+\|t_{-m}t_{-n}H(|\mathbf{r}_m - \mathbf{r}_n|)\|0_i^+\rangle(\frac{g_V}{g_A})^2 \tag{3.41b}$$

with the neutrino potential defined by eq.(3.39a). Knowing the values of the phase space factor and the nuclear matrix elements, one can deduce from an observed $0\nu$ decay half-life the effective value of the neutrino mass

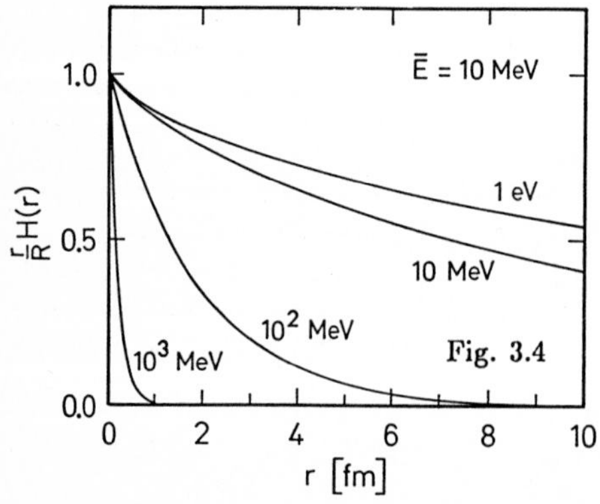


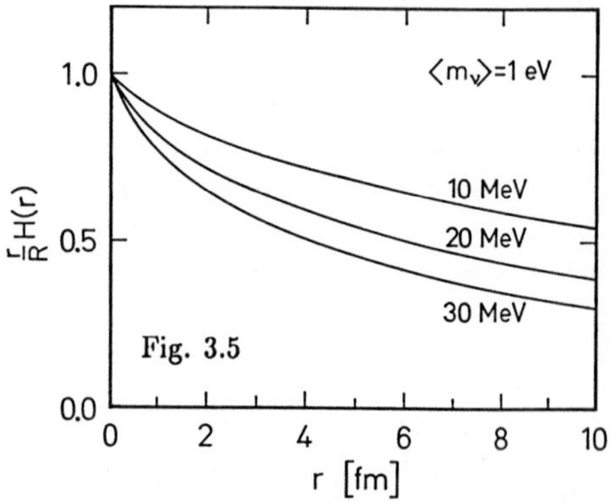


Fig. 3.4 Radial shape of the neutrino potential $H(r)$ for neutrino masses 1 eV, 10 MeV, $10^2$ MeV and $10^3$ MeV. An average energy of the intermediate states $\bar{E} = 10$ MeV is assumed. The neutrino potential is almost completely identical for neutrino masses $0 < m_\nu < 1$ MeV.

Fig. 3.5 Radial shape of the neutrino potential $H(r)$ for average energies of the intermediate states $\bar{E}$ = 10, 20 and 30 MeV. The neutrino mass is assumed to be 1 eV.

$$\langle m_\nu \rangle = \sum_j m_j U_{ej}^2 \tag{3.42}$$

where the sum runs over light neutrinos with mass eigenvalues $m_j < 10$ MeV. This factorization of the effective neutrino mass is possible for light neutrinos. Although the radial form of the neutrino potential $H(r)$ depends on the mass of the neutrino which is exchanged between two nucleons, the dependence is negligibly small for light neutrinos with masses $0 < m_j < \bar{E}$, where the average energy of intermediate nuclear states $\bar{E}$ is typically $\sim 10$ MeV, see Fig. 3.4.

If a heavy neutrino contributes to the $0\nu$ decay, one must take into account the dependence of the neutrino potential on the neutrino mass. Then the nuclear matrix elements have an $m_j$ dependence, and the expression for the half-life should be modified to

$$[T_{1/2}^{0\nu}(0_i^+ \to 0_f^+)]^{-1} = [\sum_j m_j U_{ej}^2 (M_{GT}^{0\nu}(m_j) - M_F^{0\nu}(m_j))]^2 \frac{G_1}{m_e^2}. \tag{3.43}$$

For two Majorana neutrinos with masses $m_1$ and $m_2$, one being light and the other heavy, $m_1 \ll \bar{E} < m_2$, the effective value of the neutrino mass which one deduces from an experimental half-life is expressed as [Hal 83, Ver 83]

$$\langle m_\nu \rangle = |m_1 \cos^2\theta - m_2 \sin^2\theta \frac{M_{GT}^{0\nu}(m_2) - M_F^{0\nu}(m_2)}{M_{GT}^{0\nu}(0) - M_F^{0\nu}(0)}| \tag{3.44}$$

where $\theta$ is the angle of mixing between the two neutrinos. The ratio of the nuclear matrix elements in the second term of eq.(3.44) depends sensitively on the short-range correlations (see the discussion above, eq.(3.25)), since the contributions to $M_{GT}^{0\nu}(m_2)$ and $M_F^{0\nu}(m_2)$ come from the short-distance region $|\mathbf{r}_1 - \mathbf{r}_2| < 1/m_2$. This ratio also depends on the mass number, and consequently the values of $\langle m_\nu \rangle$ deduced should be different for various $\beta\beta$ emitters (see, e.g., the application in [Gro 86]).

Among the nuclear matrix elements which originate from the right-handed currents, i.e., the V+A interaction at one vertex and the V-A interaction at the other vertex, the matrix element $M_R$ would give the largest contribution to the $0\nu$ decay rate. This matrix element comes from the three-momentum part of the neutrino propagator (see eq.(3.22) and Table 3.1). The negative parity of the momentum is compensated by the negative parity operator on the nuclear side, the nucleon recoil term, eq.(3.5b), which is the small relativistic amplitude of the nucleon wave function. (The negative parity can also be compensated on the electron side [Hax 84], but the corresponding matrix element is smaller than $M_R$ by two orders of magnitude [Tom 86].) The transition operator corresponds to a forbidden single $\beta$ decay. Such a forbidden transition, where the neutrino is a real particle, is highly hindered compared to an allowed Gamow-Teller $\beta$ decay due to the small factor $q/Mc \leq 1/100$, where $Mc = 940$ MeV/c. However, the neutrino is virtual in the $0\nu$ $\beta\beta$ decay, and the neutrino momentum is only limited by the requirement that the nucleons are confined in the nucleus, i.e., $q \leq 2k_F \sim 560$ MeV/c, with $k_F$ being the Fermi momentum of a nucleon in the nucleus. The ratio $q/Mc$ is then of the order of unity and is not small any more. The recoil matrix element is enhanced also by the large coupling constant $\mu_\beta = 4.7$, and together with the large phase factor $G_9$ it dominates the coefficient $C_{\eta\eta}$. Therefore, the matrix element $M_R^{0\nu}$ is the most important one for a determination of the mixing parameter $\eta$ of the right-handed leptonic current.

The nucleon recoil term $\mathbf{D}_n$ of eq.(3.5b) involves $\mathbf{P}'_n - \mathbf{P}_n$, the difference between the nucleon momenta before and after the weak interaction vertex. The factor can be replaced by the momentum $\mathbf{q}$ which the virtual neutrino carries, by neglecting the small momentum of the emitted electron, $\mathbf{q} = \mathbf{P}_n - \mathbf{P}'_n$. The replacement gives rise to an additional $\mathbf{q}$ which is to be included in the momentum integral of the neutrino potential. With the resulting potential, the recoil matrix element $M_R^{0\nu}$ of eq.(3.37i) is rewritten as [Tom 86],

$$M_R^{0\nu} = \langle H_{RC}(r) \sigma_m \cdot \sigma_n \rangle + (\text{non} - \text{central} \quad \text{terms}) \tag{3.45}$$

$$H_{RC}(r) = \frac{2R^2}{3M}\mu_\beta[4\pi\delta(\mathbf{r}) - \frac{2}{\pi}\frac{\bar{E}}{r^2} + \bar{E}^2 H(r)]. \tag{3.46}$$

The zero-range potential singular at the origin yields a leading contribution to the recoil matrix element $M_R^{0\nu}$, and the matrix element has to be evaluated by taking account of both short-range correlations and nucleon finite size effects, see eqs. (3.25) and (3.27).

3.3.2. $0^+ \to 2^+$ transitions

As for the $2\nu$ decay, a $0^+ \to 2^+$ transition could occur for the $0\nu$ decay. In this case, the neutrino mass term does not contribute, and the main contribution of the right-handed (V+A) current arises when one electron is emitted in S-wave and the other in P-wave (j=3/2) because of angular momentum conservation.

The inverse half-life is

$$[T_{1/2}^{0\nu}(0_i^+ \to 2^+)]^{-1} = C_{2\eta\eta}\langle\eta\rangle^2 + C_{2\lambda\lambda}\langle\lambda\rangle^2 + C_{2\eta\lambda}\langle\eta\rangle\langle\lambda\rangle \tag{3.47}$$

The coefficients $C_{2xy}$ contain the nuclear matrix elements and phase space integrals,

$$C_{2\eta\eta} = M_\eta^2 G_+ + M_\eta'^2 G_- \tag{3.48a}$$

$$C_{2\lambda\lambda} = M_\lambda^2 G_+ \tag{3.48b}$$

$$C_{2\eta\lambda} = -2M_\eta M_\lambda G_+. \tag{3.48c}$$

The nuclear matrix elements are given by

$$M_\eta = -\frac{2}{3}M_1 - \frac{\sqrt{6}}{3}M_2 - \frac{\sqrt{6}}{9}M_3 + \frac{\sqrt{14}}{3}M_4 \tag{3.49a}$$

$$M_\eta' = \sqrt{2}M_6 \tag{3.49b}$$

$$M_\lambda = -\frac{2}{3}M_1 + \frac{\sqrt{6}}{3}M_2 + \frac{\sqrt{6}}{9}M_3 + \frac{\sqrt{14}}{3}M_4 + M_5 \tag{3.49c}$$

$$M_1 = \langle H'(r)[\sigma_m \times \sigma_n]^{(2)}\rangle \tag{3.50a}$$

$$M_2 = \langle H'(r)C_2(\hat{\mathbf{r}})\rangle \tag{3.50b}$$

$$M_3 = \langle H'(r)C_2(\hat{\mathbf{r}})(\sigma_m \cdot \sigma_n)\rangle \tag{3.50c}$$

$$M_4 = \langle H'(r)[C_2(\hat{\mathbf{r}})[\sigma_m \times \sigma_n]^{(2)}]^{(2)}\rangle\rangle \tag{3.50d}$$

$$M_5 = \langle H'(r)[C_2(\hat{\mathbf{r}}) \times (\sigma_m + \sigma_n)]^{(2)}\rangle \tag{3.50e}$$

$$M_6 = \langle H'(r)\frac{r_+}{r}[\hat{\mathbf{r}}_+ \times [\hat{\mathbf{r}}\times(\sigma_m - \sigma_n)]^{(1)}]^{(2)}\rangle \tag{3.50f}$$

where $C_2(\hat{\mathbf{r}}) = \sqrt{4\pi/5}Y_2(\hat{\mathbf{r}})$ is the modified spherical harmonics of rank 2. The phase space integrals $G_\pm$ are obtained by replacing $g_i$ in eq.(3.33) by $g_\pm$,

$$g_\pm = \frac{e_1 e_2 \pm m_e^2}{6 e_1 e_2}[(\frac{p_1}{m_e})^2 \frac{F_1(Z,e_1)}{F_0(Z,e_1)} + (\frac{p_2}{m_e})^2 \frac{F_1(Z,e_2)}{F_0(Z,e_2)}]. \tag{3.51}$$

The first (second) term of eq.(3.51) corresponds to P-wave emission of the first (second) electron, the other electron being emitted in S-wave.

3.4. Non-nucleonic mechanisms

We have considered in the previous sections $2\nu$ and $0\nu$ $\beta\beta$ transitions only in terms of two-nucleon processes in the sense that two different nucleons decay — including the possibility of virtual $\Delta$ excitations. In this subsection we briefly discuss other possible mechanisms.

### 3.4.1. The $N^*$ mechanism

The $N^*$ mechanism is characterized by the fact that the same particle undergoes two successive decay steps, in contrast to the two-nucleon process in which two different nucleons decay (see Figs. 3.1, 3.2 and 3.3). Since the charge has to be changed by two units, this process involves necessarily the $\Delta$ resonance, but it should not be confused with $\Delta$ excitations in the two-nucleon process. The $N^*$ mechanism was first introduced by Primakoff and Rosen for the $0\nu$ mode [Pri 69] and later for the $2\nu$ mode by Smith, Picciotto and Bryman [Smi 73]. Diagrams are shown in Fig. 3.6. The $\beta\beta$ decay takes place through the $\Delta$ isobar (intrinsic spin-parity $J^\pi = 3/2^+$, isospin 3/2) which is admixed by the strong interaction into the initial and final nuclear states. The $N^*$ mechanism could be expected to be important for the $0\nu$ mode, since the neutrino is emitted and absorbed by the same hadron and therefore the range of neutrino propagation is much shorter than for the two-nucleon mechanism. In the $N^*$ mechanism, however, $0^+ \to 0^+$ transitions are forbidden [Doi 83a, Hax 81]. The hadronic current acts only on the intrinsic parts of the $\Delta$ and the nucleon in the static approximation, and the $\Delta J = 0$ operator of the $0^+ \to 0^+$ transition cannot connect the nucleon with $J^\pi = 1/2^+$ and the $\Delta$ with $J^\pi = 3/2^+$. (This transition occurs if higher partial waves ($l \geq 2$) of the lepton are taken into account, but is highly suppressed.) The selection rules of the two-nucleon and $N^*$ mechanisms are summarized in Table 3.2.

Recently, Tomoda calculated the $0^+ \to 2^+$ $0\nu$ $\beta\beta$ decay of $^{76}$Ge, and showed the importance of the $N^*$ mechanism particularly for the nuclear matrix element relevant to the determination of the right-handed current mixing parameter $\lambda$ (the term of $j_R \cdot J_R$, see eq.(3.1)) [Tom 88].

Haxton and Stephenson pointed out another diagram of the $N^*$ mechanism, Fig. 3.7, which allows the $0^+ \to 0^+$ transition through both the $m_\nu$ and V+A parts [Hax 84]. However, this process is of second order in the coupling between $\Delta$ and nucleon, and the decay rate is suppressed by the

Table 3.2. Selection rules for $\beta\beta$ decay in the two-nucleon mechanism and the N* mechanism (in the static quark model). Circles, crosses and triangles denote dominant, forbidden and suppressed transitions, respectively.

| | $2\nu$ mode | | $0\nu$ mode | | | |
|---|---|---|---|---|---|---|
| | | | $m_\nu$ | | $\omega$, q | |
| | 2n | N* | 2n | N* | 2n | N* |
| $0^+ \to 0^+$ transition | ○ | × | ○ | × | ○ | × |
| $0^+ \to 2^+$ transition | △ | △ | × | × | ○ | ○ |

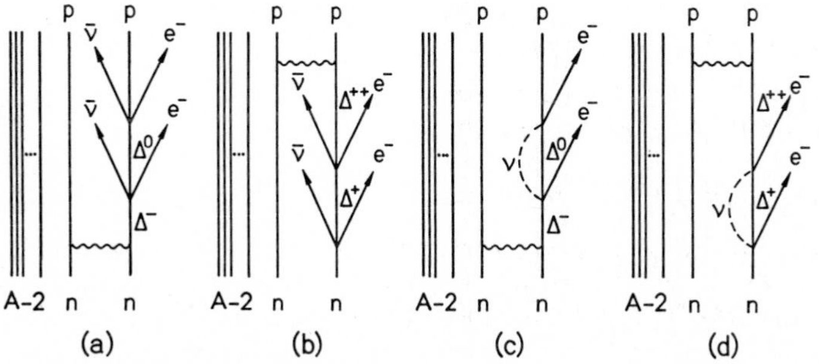


Fig. 3.6 Diagrams of the $\beta\beta$ decay amplitudes in the $N^*$ mechanism for the $2\nu$ mode, (a) and (b), and for the $0\nu$ mode, (c) and (d). The wavy line represents the strong interaction.

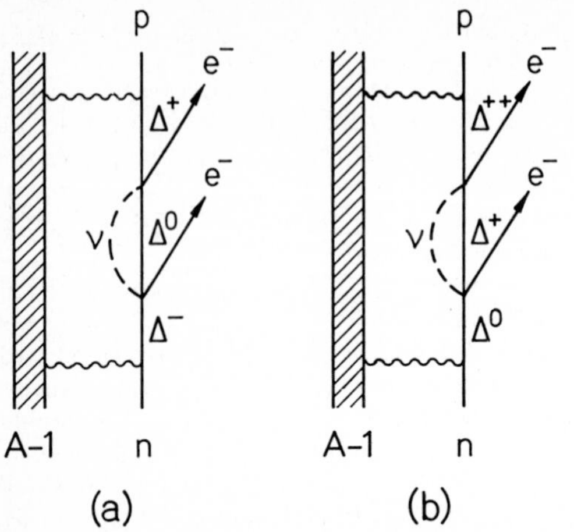


Fig. 3.7 The $0\nu$ mode $\beta\beta$ decay in the higher-order $N^*$ mechanism. The two wavy lines, which represent the strong interaction, couple to either one nucleon or two nucleons of the A-1 nucleons.

factor $P(\Delta)^2 \sim 10^{-4}$, where $P(\Delta)$ is the probability of finding a $\Delta$ in a nucleus. The $0^+ \to 0^+$ transition is also possible through various higher nucleon resonances with $J^\pi = 1/2$ [Hax 81].

### 3.4.2. The pion exchange mechanism

Since the nuclear interaction can be described in terms of pions and heavier mesons, these mesons could explicitly participate in the $\beta\beta$ decay. Diagrams of the pion exchange mechanism are shown in Fig. 3.8. Vergados considered the diagram 3.8c ($0\nu$ mode) and showed that, if the

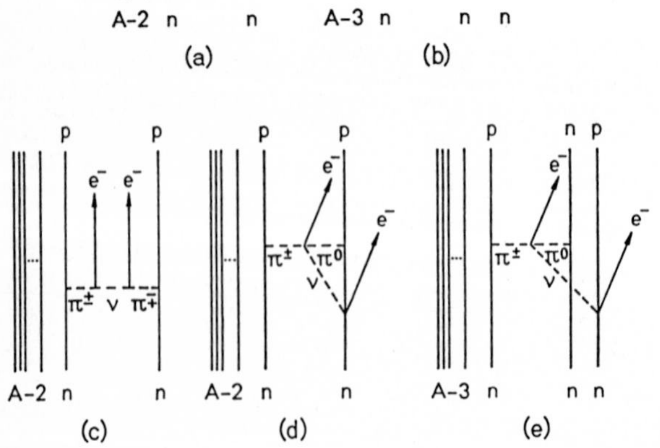

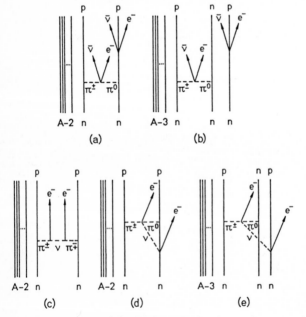

Fig. 3.8 Diagrams of the $\beta\beta$ decay in the pion exchange mechanism; (a) and (b) for the $2\nu$ mode; (c), (d) and (e) for the $0\nu$ mode.

mixing between light and heavy neutrinos is not negligible, the double charge exchange of pions in flight between two nucleons makes an important contribution to the $0\nu$ $\beta\beta$ decay mediated by a heavy Majorana neutrino [Ver 82]. There are no estimates for the other meson exchange processes.

### 3.4.3. $0\nu$ decay through Higgs bosons

Mohapatra and Vergados proposed a $0\nu$ mode via Higgs bosons without virtual neutrinos [Moh 81b,86], as shown in Fig. 3.9. This process does not involve a Majorana neutrino, but proceeds by decay of a doubly charged Higgs boson to electrons. Contributions of this mechanism are, however, negligible because of the weak coupling of the physical Higgs particle to quarks [Sch 82] and furthermore because of the suppressed nuclear matrix element relevant for this process [Hax 82b].

### 3.4.4. $0\nu$ mode with majoron and familon emission

Georgi, Glashow and Nussinov pointed out the $0\nu$ mode associated with majoron emission [Geo 81, Gel 82]. The majoron is a light pseudo-scalar Goldstone boson arising from the spontaneous breaking of the global $B - L$ symmetry [Chi 80,81, Gel 81]. The diagram is shown in Fig. 3.10. For the effective interaction Hamiltonian

$$H_W = \frac{G}{\sqrt{2}} j_L \cdot J_L^\dagger + \frac{i}{2} \sum_{j,k} g_{jk} \bar{N}_j \gamma_5 N_k \phi_M \tag{3.52}$$

the inverse half-life for a $0_i^+ \to 0_f^+$ transition is given by [Doi 85]

$$[T_{1/2}^{0\nu,B}(0_i^+ \to 0_f^+)]^{-1} = |\langle g_B \rangle|^2 F^{0\nu,B} |M_{GT}^{0\nu} - M_F^{0\nu}|^2. \tag{3.53}$$

Here, $\langle g_B \rangle$ is the effective value of the coupling between the majoron and neutrinos,

$$\langle g_B \rangle = \sum_{j,k} g_{jk} U_{ej} U_{ek} \tag{3.54}$$

and the two nuclear matrix elements in eq.(3.53) are the same ones that appear in the $0\nu$ decay of the two-nucleon mechanism, eqs. (3.37a) and (3.37b). The phase space integral is given by

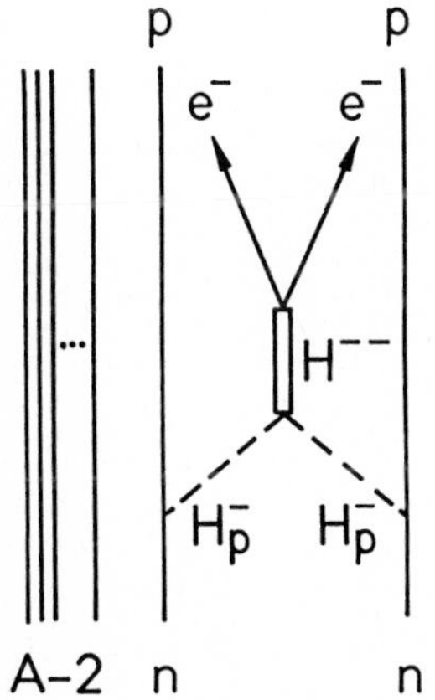


Fig. 3.9

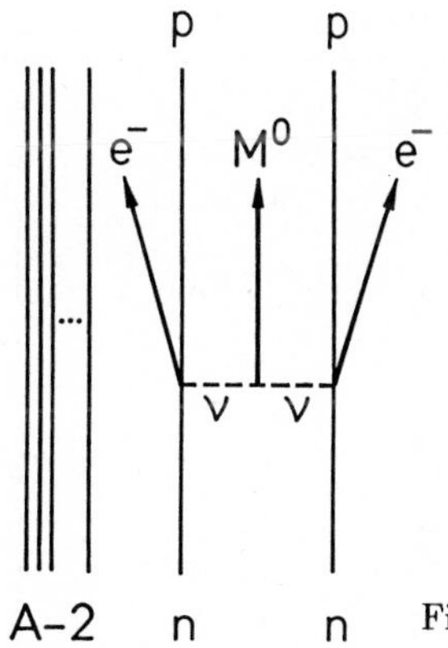


Fig. 3.10

Fig. 3.9 $\beta\beta$ decay through Higgs bosons. $H_p^-$ represents the physical Higgs boson. The neutrino does not participate in this mechanism.

Fig. 3.10 $0\nu$ $\beta\beta$ decay with majoron emission. The virtual Majorana neutrino in flight emits the majoron.

$$F^{0\nu,B} = \frac{G^4 g_A^4}{128\pi^7 R^2} \int_{m_e}^{T_0+m_e} F(Z,e_1)k_1e_1\, de_1 \int_{m_e}^{T_0+2m_e-e_1} F(Z,e_2)k_2e_2k_M\, de_2 \tag{3.55}$$

where $k_M = T_0 - (e_1 + e_2 - 2m_e)$ is the momentum of the emitted majoron. A detailed examination of this process for $0^+ \to 0^+$ as well as for $0^+ \to 2^+$ transitions is given by [Doi 87]. The latter process is found to be suppressed by of the order of $10^{-7}$ compared to $0^+ \to 0^+$ transitions.

This process may become an important background in the electron spectrum in $\beta\beta$ decay counter experiments. The electron sum-energy spectra are illustrated in Fig. 3.11. Compared to the $2\nu$ decay which leads to five-body final states (two electrons, two neutrinos and the residual nucleus), the sum-energy spectrum of the four-body majoron emission is shifted to higher energies. The sum-energy of the $0\nu$ electrons should be identical to the $\beta\beta$ decay $Q$ value, $e_1 + e_2 - 2m_e = T_0$.

A similar situation and similar results are expected for the familon, the Nambu-Goldstone boson associated with spontaneous breaking of the horizontal (family) symmetry [Rei 82, Wil 82, Gel 83].

## 4. Nuclear structure models used in calculations of $\beta\beta$ decay

We have seen in the preceding sections that for the calculation of $2\nu$ and $0\nu$ $\beta\beta$ decay rates and particularly for the deduction of the neutrino mass or of information on right-handed currents from measured $0\nu$ decay rates, reliable calculations of the contributing nuclear matrix elements are the essential prerequisite.

Several approaches with different nuclear structure models have been used in the calculation of $\beta\beta$ decay nuclear matrix elements. In spite of many efforts, until recently there was a long-standing problem namely that calculations systematically overestimated most $\beta\beta$ decay rates of the $2\nu$ mode. It was argued that the nuclear theory could have a poor predicticable power also for the $0\nu$ mode, and that therefore the upper limits on the neutrino mass and the mixing parameters of right-handed currents, eqs. (3.29) and (3.47), deduced from experimental half-lives of the $0\nu$ mode might have considerable uncertainties. Non-nuclear explanations of the problem were investigated [Gro 84], but ultimately a nuclear structure solution was found. The suppression mechanism of the $2\nu$ decay rates has been clarified in recent studies [Kla 84, Gro 85,86, Vog 86, Civ 87, Mut 88a,88c] by recognizing the importance of ground state correlations for the calculated decay rates and by using a nuclear model based on the quasiparticle random phase approximation (QRPA). In the

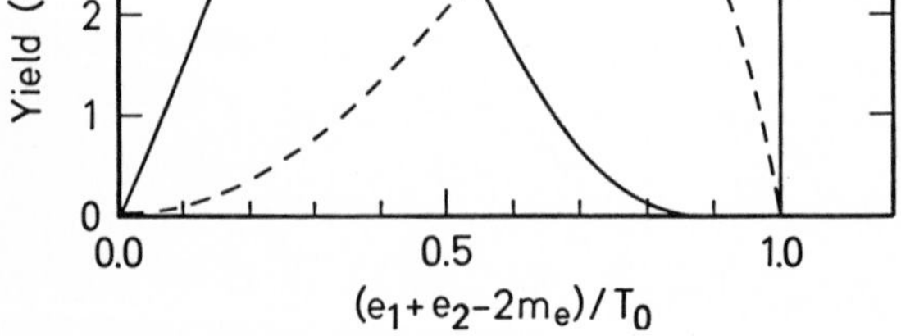

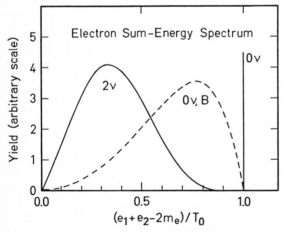

Fig. 3.11 Sum-energy spectra of the two electrons emitted in the $\beta\beta$ decay for the $2\nu$ and $0\nu$ modes and for the $0\nu$,B mode ($0\nu$ mode associated with majoron emission).

next section we discuss the general features of the different types of correlations which play a decisive role in $\beta\beta$ decay. In sections 4.2 - 4.4 we present general characteristic features of nuclear models used in the calculation of $\beta\beta$ decay. In sect. 5 we then give the results of calculations of $\beta\beta$ decay rates based on these models.

### 4.1. Ground state correlations

Nuclear states are not well described by a single Slater determinant in a microscopic model, even for the $0^+$ ground state of an even-even nucleus. Fig. 4.1 shows the configuration for the ground states of parent and daughter nuclei for the $\beta\beta$ decay of $^{76}$Ge in the simplest shell model. The $2\nu$ $\beta\beta$ decay is forbidden for these ground state wave functions, since Gamow-Teller transitions are allowed only between orbits with the same orbital angular momentum belonging to the same oscillator major shell. The situation is the same for most nuclei which undergo $\beta\beta$ decay, except for some light nuclei such as $^{48}$Ca. In the latter case wave functions of the simplest shell model configuration yield on the other hand a much too large nuclear matrix element of $\beta\beta$ decay. In order to obtain a better description of $\beta\beta$ transitions, configurations including particle-hole excitations have to be taken into account in the nuclear wave functions. Such admixtures in the ground states are in general called ground state correlations. There exist different types of correlations.

The shell model automatically takes into account any type of correlation, if the model space is large enough and a realistic residual interaction is used. But such a shell model calculation is practically impossible in most cases, and the model space must be truncated drastically, see sect. 4.2. It is, therefore, reasonable to include explicitly correlations which result from prominent features of the nucleon-nucleon interaction and are important for the transition under consideration.

The pairing force is the most prominent part of the residual nuclear interaction acting between like nucleons (between two protons or between two neutrons). This force makes two nucleons in the same orbit form a pair coupled to $J^\pi = 0^+$. These correlations are called pairing correlations. Because of the attractive character of the pairing force (the matrix element is negative), all even-even nuclei have the ground state spin-parity $J^\pi = 0^+$ and are energetically lower than the neighbouring nuclei with odd numbers of protons and neutrons.

The pairing correlations are taken into account in the BCS formalism, see sect. 4.3. Fig. 4.2 compares nucleon distributions among orbits for the cases without and with pairing correlations. When there are no correlations (Fig. 4.2a), low-lying orbits are completely filled and high-lying

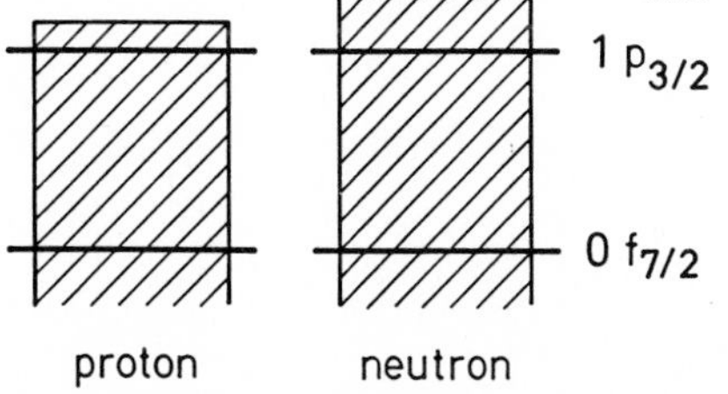

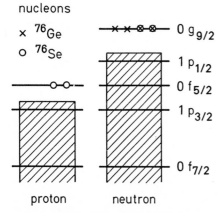

Fig. 4.1 Configurations in the simplest shell model for the ground states of $^{76}$Ge and $^{76}$Se (the initial and final states of $^{76}\text{Ge} \to ^{76}\text{Se}$ $\beta\beta$ decay, respectively). The hatched area indicates fully occupied orbits.

orbits are empty (one orbit nearest to the Fermi energy can be partly filled). On the other hand, the pairing interaction smears out the distribution of nucleons (Fig. 4.2b). The occupation probability changes from 1 to 0 in orbits near the Fermi energy. All nucleons are paired to $J^\pi = 0^+$. This state is the vacuum for quasiparticles, which are defined by the Bogoliubov transformation, eq.(4.4). The pairing correlations give rise to a finite decay rate for $2\nu$ $\beta\beta$ decay which is forbidden for the simplest shell model configurations (Fig. 4.1). They are therefore the origin of the most important collective effect to enhance $\beta\beta$ decay rates. The calculated half-lives, however, are much shorter than the experimental values by up to several orders of magnitude, when including *only* the pairing correlations (see Table 5.4 and Fig. 5.4).

The RPA (random phase approximation) formalism takes into account the correlations in the ground state which are directly connected with the nuclear excitation considered. There are several versions of RPA, depending on the mode of excitation under consideration and/or on the nucleus [Rin 80]. Here we mention the proton-neutron quasiparticle RPA (QRPA), see sect. 4.4. This RPA model describes charge-changing transitions which are associated with a change of the atomic number by one unit, such as Gamow-Teller transitions, and is expressed in terms of quasiparticle degrees of freedom. The ground state is the vacuum for the QRPA phonon, eq.(4.11). The ground state wave function is illustrated in Fig. 4.3. The main component is the BCS ground state with no quasiparticles (Fig. 4.2b), and the leading admixtures are four-quasiparticle states. When the RPA equation (eq.(4.12)) is solved for a $J^\pi$ mode, the four-quasiparticle state consists of two pairs of proton- and neutron-quasiparticles which both have spin-parity $J^\pi$ as indicated in Fig. 4.3,

$$|RPA\rangle \sim |-\rangle + \sum_{pnp'n'} \alpha_{pn,p'n'} |pn(J^\pi), p'n'(J^\pi); 0^+\rangle + \cdots \tag{4.1}$$

where $|-\rangle$ denotes the BCS ground state. Since, by the charge-changing transition, a creation of a $pn$ pair from the main component and a destruction of one of the pairs in the four-quasiparticle components can lead to the same $pn$ pair state in the odd-odd nucleus, the transition strength could be strongly affected by the inclusion of the ground state correlations through interference between the two transition amplitudes, see Fig. 4.6. For the calculation of $2\nu$ $\beta\beta$ decay, which proceeds by successive Gamow-Teller transitions through $1^+$ intermediate states, the RPA equation is solved for $J^\pi = 1^+$. In this case the wave function admixtures given in eq.(4.1), the spin-isospin correlations, play an important role for the suppression of the decay rate. The particle-particle interaction (see

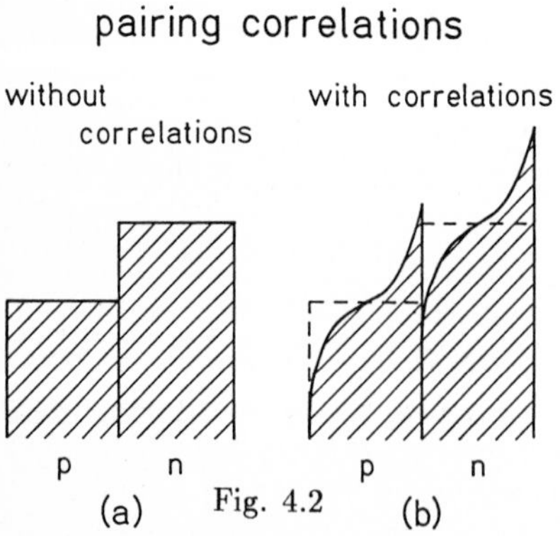


Fig. 4.2

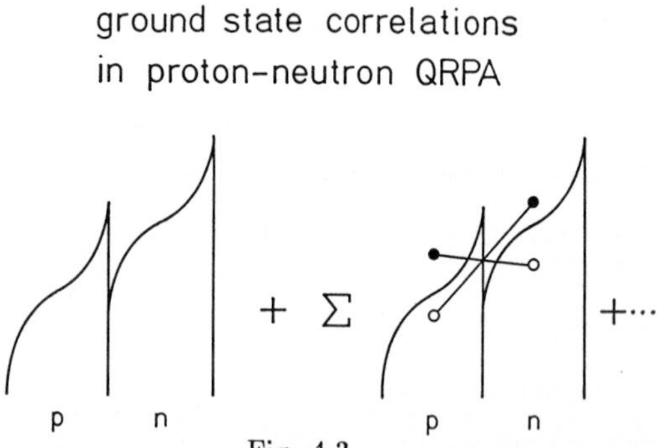


Fig. 4.3

Fig. 4.2 Distributions of nucleons among single-particle orbits in a nucleus; (a) without pairing correlations (the simplest shell model), (b) with pairing correlations.

Fig. 4.3 Ground-state wave function in the proton-neutron quasiparticle RPA. The line connecting circles, which denote quasiparticles, indicates angular momentum coupling of a proton-neutron pair. Both pairs have the same spin-parity $J^\pi$.

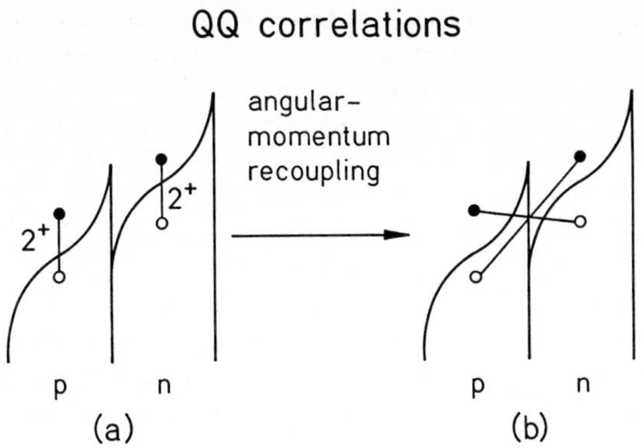


Fig. 4.4 Quadrupole-quadrupole (QQ) correlations in the ground-state wave function. (a) The dominant four-quasiparticle components of the QQ correlations consist of two protons coupled to $2^+$ and two neutrons coupled to $2^+$. (b) The quadrupole coupling can be expressed, by angular momentum recoupling, as a linear combination of two proton-neutron pairs which both have the same $J^\pi$.

eq.(4.14) and the following discussion) enhances the spin-isospin correlations, namely enhances the amplitudes $\alpha_{pn,p'n'}$ in eq.(4.1).

The quadrupole-quadrupole (QQ) force, which is the strongest component of the proton-neutron interaction, yields the QQ correlations in nuclear wave functions. Because of the QQ interaction, even-even nuclei have a low-lying $2^+$ state, usually as the first excited state. The RPA ground state for the $2^+$ mode excitation in the same nucleus (without a change of the atomic number, in contrast to charge-changing transitions) has the structure

$$|RPA\rangle \sim |-\rangle + \sum_{pp'nn'} \alpha'_{pp',nn'} |pp'(2^+), nn'(2^+); 0^+\rangle + \cdots \tag{4.2}$$

(cf. eq.(4.1) for the charge-changing mode). Klapdor and Grotz showed that the QQ correlations in the ground states suppress $\beta\beta$ decay rates of the $2\nu$ mode [Kla 84, Gro 86]. This can be understood by the angular momentum recoupling, see also Fig. 4.4,

$$\begin{aligned} &|pp'(2^+), nn'(2^+); 0^+\rangle \\ &= \sum_J (-1)^{p+n'-2-J} \sqrt{5(2J+1)}\, W(pp'nn'; 2J) |pn(J^\pi), p'n'(J^\pi); 0^+\rangle. \end{aligned} \tag{4.3}$$

The QQ correlations thus include components which affect Gamow-Teller transitions ($J^\pi = 1^+$).

## 4.2. Shell model

All microscopic models are based on the shell model. Nucleons in a nucleus are assumed to move in a single-particle (s.p.) potential, which is generated by the nuclear and Coulomb forces acting between them, and interact with each other via the residual interaction. A nuclear wave function of the many-particle system is expressed as a linear combination of antisymmetrized, orthonormal basis states which are products of the s.p. wave functions. Amplitudes of the nuclear wave function are computed by diagonalizing the effective Hamiltonian consisting of the s.p. term and the residual interaction. A set of basis states is determined by specifying an inert core and valence s.p. orbits. The s.p. orbits in the core are fully occupied by nucleons, and its nucleons do not participate in nuclear transitions. The remaining nucleons are distributed among the valence orbits in accordance with the Pauli exclusion principle.

Application of conventional shell model techniques is limited to light $\beta\beta$ emitters. As already mentioned, for heavy nuclei the model space and configurations have to be seriously truncated and the collective effects important in $\beta\beta$ decay can only partly be included. It is difficult or even impossible to treat explicitly the intermediate $1^+$ spectrum, which could be compared with results of (p,n) experiments, within this approach. This has only been done for $^{48}$Ca which, besides $^{46}$Ca, is the lightest $\beta\beta$ emitter.

For the states of parent ($^{48}$Ca), daughter ($^{48}$Ti) and intermediate ($^{48}$Sc) nuclei, $^{40}$Ca is expected to be a good inert-core nucleus, and the additional eight nucleons can be restricted to an $f_{7/2}$ - $p_{3/2}$ - $p_{1/2}$ - $f_{5/2}$ space [Hax 84]. The smallest model space $(f_{7/2})^8$ was assumed in earlier calculations. Zamick and Auerbach used the pure configuration for qualitative discussion on the hindrance of $2\nu$ decay by the K selection rule in the Nilsson-pairing scheme [Zam 82]. It was shown later, however, that a few nucleons have to be excited from $f_{7/2}$ to the higher-lying orbits [Sko 83, Tsu 84, Wu 85, Mut 86].

In its application to the $\beta\beta$ decay of heavier nuclei, the shell model approach has the difficulty that the Hamiltonian matrix becomes much too large to be handled. For instance, if four protons and fourteen neutrons are distributed among the valence orbits $p_{3/2}$, $p_{1/2}$, $f_{5/2}$ and $g_{9/2}$ (this situation corresponds to $^{76}$Ge), the number of linearly independent basis states with $J^\pi = 0^+$ is 210,772; and for a larger model space consisting of full $3\hbar\omega$ and $4\hbar\omega$ oscillator major shells, the number exceeds $10^{21}$. Thus, one is forced to reduce the number of basis states very drastically but preserving as much as possible the components important for the phenomena under consideration.

Haxton, Stephenson and Strottman [Hax 82a,84] chose the weak-coupling model in the calculation of nuclear matrix elements for the $\beta\beta$ decay of $^{76}$Ge, $^{82}$Se and $^{128,130}$Te. They first solved eigenvalue equations for proton and neutron systems separately, and took from each system about fifty states with the smallest energies. Then the ground-state wave functions of both parent and daughter nuclei of the $\beta\beta$ decay were calculated by diagonalizing the proton-neutron residual interaction in the proton-neutron coupled states. The number of basis states with $J^\pi = 0^+$ is reduced to the order of 100. These authors had the problem of treating properly and explicitly the intermediate $1^+$ spectrum. The closure approximation with its inherent problems in the case of $2\nu$ decay (see sect. 3.2) was used in these calculations. In addition, they did not include the spin-orbit partners, for example $f_{7/2}$ for $f_{5/2}$, which play an important role for Gamow-Teller transitions.

### 4.3. Quasiparticle models

The quasiparticle picture is attractive, since it defines a new vacuum for quasiparticles and nuclear wave functions are described in terms of a small number of quasiparticle degrees of freedom (for the formalism see, e.g., [Sol 76, Rin 80]). Quasiparticles are introduced by the Bogoliubov transformation

$$a_k^\dagger = u_k c_k^\dagger - v_k c_{\bar{k}}$$
$$a_{\bar{k}}^\dagger = u_k c_{\bar{k}}^\dagger + v_k c_k \tag{4.4}$$

where $a_k^\dagger$ and $c_k^\dagger$ are quasiparticle and particle (nucleon) creation operators, respectively, and $\bar{k}$ is a conjugate (usually time-reversed) state of $k$. The quasiparticle operators obey the fermion anti-commutation relations as the particle operators do. The occupation amplitudes $u_k$ and $v_k$ satisfy $u_k^2 + v_k^2 = 1$, $v_k^2$ being the occupation probability of a nucleon in the orbits $k$ and $\bar{k}$. These coefficients of the Bogoliubov transformation can be calculated in the BCS approximation by the constrained variational principle

$$\delta\langle BCS|H - \lambda\hat{N}|BCS\rangle = 0 \tag{4.5a}$$
$$\langle BCS|\hat{N}|BCS\rangle = N. \tag{4.5b}$$

The Fermi energy $\lambda$ is introduced as a Lagrange multiplier, $N$ is the nucleon number and $\hat{N}$ the number operator. The variational principle yields a set of equations

$$\tilde{\epsilon}_k = \bar{\epsilon}_k + \sum_{k'>0} \frac{1}{2} v_{k'}^2 (\langle kk'|V|kk'\rangle + \langle \bar{k}k'|V|\bar{k}k'\rangle) - \lambda \tag{4.6}$$

$$\Delta_k = -\sum_{k'>0} u_{k'} v_{k'} \langle k\bar{k}|V|k'\bar{k}'\rangle \tag{4.7}$$

$$v_k^2 = \frac{1}{2}\left(1 - \frac{\tilde{\epsilon}_k}{\sqrt{\tilde{\epsilon}_k^2 + \Delta_k^2}}\right) \tag{4.8}$$

where $\bar{\epsilon}_k$ and $\epsilon_k = \sqrt{\tilde{\epsilon}_k^2 + \Delta_k^2}$ are single-particle and quasiparticle energies, respectively. These equations are solved self-consistently with the particle-number condition, eq.(4.5b),

$$2\sum_{k>0} v_k^2 = N. \tag{4.9}$$

The BCS ground state defined by $|BCS\rangle = \prod_k a_k a_{\bar{k}} |\rangle$ is obviously the vacuum for the quasiparticles, $a_k|BCS\rangle = 0$ (here, $|\rangle$ is the vacuum for nucleons). In terms of nucleons, it is a superposition of states with different nucleon numbers,

$$|BCS\rangle \sim |\rangle + \sum_{k\geq 0} \frac{v_k}{u_k} c_k^\dagger c_{\bar{k}}^\dagger |\rangle + \frac{1}{2} \sum_{kk'\geq 0} \frac{v_k v_{k'}}{u_k u_{k'}} c_k^\dagger c_{\bar{k}}^\dagger c_{k'}^\dagger c_{\bar{k}'}^\dagger |\rangle + \cdots \tag{4.10}$$

and each state consists of pairs of nucleons coupled to $J^\pi = 0^+$ (see also Fig. 4.2). The quasiparticle model thus takes into account the pairing correlations which result from the prominent, strong pairing forces acting between like nucleons. The ground state of an even-even nucleus is the quasiparticle vacuum, and excited states are described to the lowest order by two-quasiparticle states.

The program VAMPIR (or MONSTER) of the Tübingen group [Sch 84a,84b] is a generalized version of the quasiparticle model. The authors developed these computer codes on the basis of the Hartree-Fock-Bogoliubov formalism; i.e., the s.p. states are calculated self-consistently and quasiparticles are defined by a Bogoliubov transformation more general than eq.(4.4). The nucleon number is restored by the particle-number projection, and the rotational invariance which is violated by the use of a deformed basis is restored by angular-momentum projection. The wave functions generated by this method have been used in the calculation of $\beta\beta$ decay rates [Tom 85,86,88].

Klapdor and Grotz [Kla 84, Gro 85a,85b,86] calculated half-lives of both $2\nu$ and $0\nu$ (the mass term) $\beta\beta$ decay for all possible $\beta\beta$ emitters with A $\geq$ 70 (see Table 5.3). They adopted two types of models. One is a QRPA calculation with a schematic Gamow-Teller force, which is the part of the proton-neutron interaction most important for Gamow-Teller transitions (the particle-particle component is neglected, see eq.(4.14)). The other is an extended nucleon-number-projected BCS calculation with a residual Gamow-Teller force, in a model space consisting of zero- and four-quasiparticle states, as well as an additional quadrupole-quadrupole (QQ) force, which is another important part of the proton-neutron interaction. These calculations included further virtual excitations of the $\Delta$ resonance, the Gamow-Teller force being extended to a generalized spin-isospin force in a simple quark model. Therefore, wave functions of the $0^+$ ground states are linear combinations of eq.(4.2) and the second term of eq.(4.1) with $J^\pi = 1^+$ including $\Delta$-quasiparticle components. These authors showed that the QQ ground state correlations reduce the $2\nu$ decay rates by up to an order of magnitude compared with those of the QRPA calculation, and lead to a remarkable improvement in the agreement of the calculated $2\nu$ $\beta\beta$ decay rates with experimental results (see Fig. 5.2). These calculations led to two important results. First they demonstrated the large sensitivity of the calculated $\beta\beta$ decay rates to various types of ground state correlations. Second, more generally, these calculations showed the way to the solution of the above mentioned long-standing problem in the calculation of $2\nu\ \beta\beta$ rates: The QQ force, which had a strongly suppressing effect on the $2\nu\ \beta\beta$ matrix elements, enhances the component in the ground state wave functions with $J^\pi = 1^+$ (second term in eq.(4.1), see eq.(4.3)). The QQ force, on the other hand, is contained as a constituent in the particle-particle interaction of the $J^\pi = 1^+$ mode, which enhances further the amplitude $\alpha_{pn,p'n'}$ of the spin-isospin correlations (see sects. 4.4 and 5.1.2).

It was, therefore, natural to investigate the particle-hole and particle-particle forces (eq.(4.14)) separately. This is possible and has been done in the most recent calculations [Vog 86, Civ 87, Tom 87, Mut 88a,88c] in which a particle-particle component was included in the QRPA calculations in addition to the particle-hole component (which was the only one considered in previous RPA calculations). This led to a final clarification of the suppression mechanism of $2\nu\ \beta\beta$ decay rates. The formalism of the QRPA including the particle-particle interaction, therefore, is described in some detail in the next section.

## 4.4. Quasiparticle random phase approximation (QRPA) including the particle-particle interaction

The proton-neutron QRPA was developed for the description of charge-changing excitations, such as Gamow-Teller transitions, from the $0^+$ ground state of an even-even nucleus $(A, Z)$ to states in the odd-odd nucleus $(A, Z \pm 1)$ [Hal 67, Mut 88b]. In the derivation of QRPA formulae one makes two main assumptions as in the usual RPA: (i) one assumes boson commutation relations for creation and annihilation operators of a proton-neutron pair, (ii) one neglects $a^\dagger a$ terms in the matrix elements of the QRPA ground state, $\langle RPA|a^\dagger_k a_{k'}|RPA\rangle = 0$.

The QRPA phonons with angular momentum $J$ and its z-component $M$ are defined by

$$A^\dagger_\omega(JM) = \sum_{pn}(X^{pn,J}_\omega [a^\dagger_p a^\dagger_n]^{(J)}_M - Y^{pn,J}_\omega(-1)^{1+J+M}([a^\dagger_p a^\dagger_n]^{(J)}_{-M})^\dagger) \tag{4.11}$$

where indices $p$ and $n$ distinguish between proton and neutron states, and the square bracket denotes angular-momentum coupling. The energy eigenvalues $\omega$ and forward- and backward-going amplitudes X and Y (see Fig. 4.6) are obtained by solving the QRPA equation

$$\begin{bmatrix} A & B \\ -B & -A \end{bmatrix}\begin{bmatrix} X \\ Y \end{bmatrix} = \omega \begin{bmatrix} X \\ Y \end{bmatrix} \tag{4.12}$$

with the normalization

$$\sum_{pn}((X^{pn,J}_\omega)^2 - (Y^{pn,J}_\omega)^2) = 1. \tag{4.13}$$

Elements of the submatrices are explicitly

$$\begin{aligned} A_{pn,p'n'} &= \delta(pn,p'n')(\epsilon_p + \epsilon_n) \\ &\quad +(u_p v_n u_{p'} v_{n'} + v_p u_n v_{p'} u_{n'})\langle pn^{-1}|V|p'n'^{-1}\rangle_J \\ &\quad +(u_p u_n u_{p'} u_{n'} + v_p v_n v_{p'} v_{n'})\langle pn|V|p'n'\rangle_J \end{aligned} \tag{4.14a}$$

$$\begin{aligned} B_{pn,p'n'} &= -(u_p v_n v_{p'} u_{n'} + v_p u_n u_{p'} v_{n'})\langle pn^{-1}|V|p'n'^{-1}\rangle_J \\ &\quad +(u_p u_n v_{p'} v_{n'} + v_p v_n u_{p'} u_{n'})\langle pn|V|p'n'\rangle_J. \end{aligned} \tag{4.14b}$$

The quasiparticle energies $\epsilon_j$ and the occupation amplitudes $u_j$ and $v_j$ are obtained in the BCS calculation. The matrix elements of particle-particle $\langle pn|V|p'n'\rangle$ and particle-hole $\langle pn^{-1}|V|p'n'^{-1}\rangle$ interactions are related to each other by the Pandya transformation,

$$\begin{aligned} &\langle pn^{-1}|V|p'n'^{-1}\rangle_J \\ &\quad = -(-1)^{p+n+p'+n'}\textstyle\sum_{J'}(2J'+1)W(pnp'n';JJ')\langle pn'|V|p'n\rangle_{J'}. \end{aligned} \tag{4.15}$$

The particle-particle and particle-hole interactions are reduced in the limit of the nucleon picture, $u_j v_j \to 0$, to interactions acting between two-particle (or two-hole) states and between particle-hole states, respectively. This can be seen from eq.(4.14) (see also Fig. 4.5). In the nucleon picture of a doubly closed shell nucleus, the orbits occupied by nucleons have $v_j = 1$ ($u_j = 0$) and the orbits with no nucleons have $u_j = 1$ ($v_j = 0$). The former can be called hole orbits and the latter particle orbits, because when a particle-hole pair is excited, the particle occupies a particle orbit and the hole is created in a hole orbit. The particle-particle interaction is always associated with the product $u_p u_n$ (or $v_p v_n$), which means that both proton and neutron occupy particle (or hole) orbits. On the other hand, the particle-hole interaction is associated with $u_p v_n$ or $v_p u_n$, namely one of the proton and neutron is in a particle orbit and the other in a hole orbit.

The QRPA ground state is the vacuum with respect to the phonons, $A_\omega(JM)|RPA\rangle = 0$, and is expressed as a superposition of states with quasiparticle number 0, 4, 8, ...

$$|RPA\rangle = n_0 \exp[\textstyle\sum_{pn,p'n'} Z_{pn,p'n'} [a_p^\dagger a_n^\dagger]^{(J)} \cdot [a_{p'}^\dagger a_{n'}^\dagger]^{(J)}] \, |BCS\rangle \tag{4.16}$$

where the normalization factor $n_0$ and the matrix of the ground state correlations $Z$ are given by

$$n_0 = |\det X|^{-1/2} \tag{4.17}$$

$$Z = YX^{-1}. \tag{4.18}$$

The main component of the ground state has no quasiparticles (BCS ground state), and the next leading terms are four-quasiparticle states, which, in the nucleon picture, are two particle - two hole states. The probability of the (RPA-type) ground state correlations is given by $1 - \langle RPA|BCS\rangle^2 = 1 - n_0^2$.

Reduced matrix elements of charge-changing transition operators $t_\pm f_M^{(J)}$, from the $0^+$ ground state (the RPA vacuum) of an even-even nucleus to states with $J^\pi$ (one-phonon state) in the odd-odd nucleus, are given by

$$\langle \omega J \| t_- f^{(J)} \| 0^+ \rangle = \textstyle\sum_{pn} \langle p \| f^{(J)} \| n \rangle (X_\omega^{pn,J} u_p v_n - Y_\omega^{pn,J} v_p u_n) \tag{4.19a}$$

$$\langle \omega J \| t_+ f^{(J)} \| 0^+ \rangle = \textstyle\sum_{pn} \langle p \| f^{(J)} \| n \rangle (X_\omega^{pn,J} v_p u_n - Y_\omega^{pn,J} u_p v_n). \tag{4.19b}$$

For example, in case of the $0_i^+ \to 0_f^+$ $2\nu$ $\beta\beta$ decay (see Fig. 4.6), the former corresponds to the Gamow-Teller transition from the parent $0_i^+$ state to an intermediate $1^+$ state, and the latter to the inverse of the Gamow-Teller transition from the intermediate state to the daughter $0_f^+$ state. The $X$ term of eq.(4.19a) ((4.19b)) is associated with $u_p v_n$ ($v_p u_n$), implying excitation of a proton particle - neutron hole ( neutron particle - proton hole ) pair. On the other hand, the amplitude $Y$ is defined in eq.(4.11) with two quasiparticle annihilation operators. Therefore, the $Y$ terms of eq.(4.19) originate from destruction of one of the particle-hole pairs in the RPA ground state, and depend on the degree of ground state correlations.

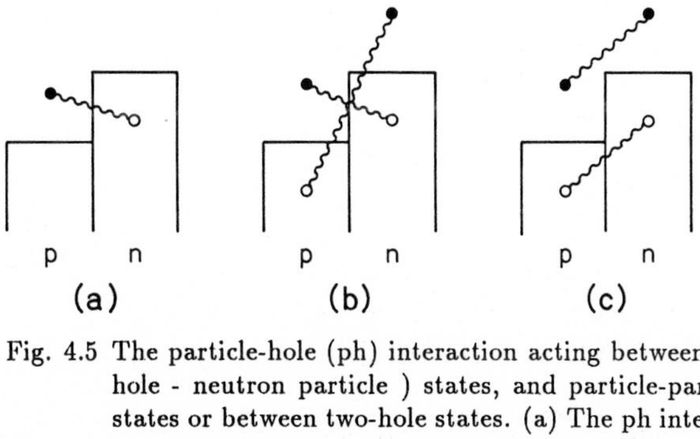


Fig. 4.5 The particle-hole (ph) interaction acting between proton particle - neutron hole ( or proton hole - neutron particle ) states, and particle-particle (pp) interaction between two-particle states or between two-hole states. (a) The ph interaction for a particle-hole state. (b) The ph interaction for a two particle - two hole state, and (c) the pp interaction for the same state.

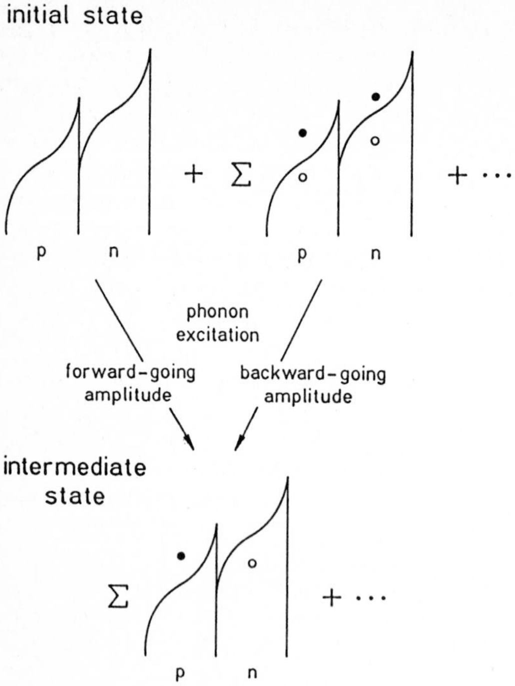


Fig. 4.6 A phonon excitation from the ground state (QRPA vacuum) to a one-phonon state in the proton-neutron QRPA. The dominant contribution of the forward-going amplitude term (from the zero-quasiparticle state to a proton-neutron pair state) interferes with the backward-going amplitude term (from four-quasiparticle states to the pair state). In the case of $\beta\beta$ decay, this phonon excitation corresponds to the transition from the initial (final) $0^+$ state to an intermediate state.

In the QRPA formalism, the nuclear matrix element of a $0_i^+ \to 0_f^+$ $2\nu$ $\beta\beta$ decay is expressed as (see eq.(3.10))

$$M_{GT}^{2\nu} = \sum_{a,b} \frac{\langle 0_f^+ \| t_-\sigma \| 1_b^+ \rangle \langle 1_b^+ | 1_a^+ \rangle \langle 1_a^+ \| t_-\sigma \| 0_i^+ \rangle}{E_a + T_0/2 + m_e - E_i}. \tag{4.20}$$

The amplitudes of Gamow-Teller transitions are obtained from eq.(4.19) by substituting $f^{(J)} = \sigma$ $(J^\pi = 1^+)$,

$$\langle 1_a^+ \| t_-\sigma \| 0_i^+ \rangle = \sum_{p,n} \langle p \| \sigma \| n \rangle (X_a^{pn} u_p v_n - Y_a^{pn} v_p u_n) \tag{4.21a}$$

$$\langle 0_f^+ \| t_-\sigma \| 1_b^+ \rangle = \sum_{p,n} \langle p \| \sigma \| n \rangle (\bar{X}_b^{pn} \bar{v}_p \bar{u}_n - \bar{Y}_b^{pn} \bar{u}_p \bar{v}_n). \tag{4.21b}$$

The overlap of intermediate $1^+$ states is given by

$$\langle 1_b^+ | 1_a^+ \rangle = \sum_{pn} (\bar{X}_b^{pn} X_a^{pn} - \bar{Y}_b^{pn} Y_a^{pn}). \tag{4.22}$$

The quantities without (with) bar are defined with respect to the initial $0_i^+$ (final $0_f^+$) state.

The $0\nu$ $\beta\beta$ matrix elements of a $0_i^+ \to 0_f^+$ decay (see eqs. (3.37), (3.38) and (3.41)) are calculated with the following formulae,

$$\langle O_{mn}\rangle = \sum_{pnp'n'J\pi J'} \langle 0_f^+\|[c_{p'}^\dagger \times \tilde{c}_{n'}]^{(J)\dagger}\|J_b^\pi\rangle\langle J_b^\pi|J_a^\pi\rangle\langle J_a^\pi\|[c_p^\dagger \times \tilde{c}_n]^{(J)}\|0_i^+\rangle$$

$$\times(-1)^{p+n+J+J'}(2J'+1)W(pnp'n';JJ')\langle pp'|t_{-1}t_{-2}O_{12}|nn'\rangle_{J'} \quad (4.23)$$

$$\langle J_a^\pi\|[c_p^\dagger \times \tilde{c}_n]^{(J)}\|0_i^+\rangle = \sqrt{2J+1}(X_a^{pn}u_pv_n - Y_a^{pn}v_pu_n) \quad (4.24a)$$

$$\langle 0_f^+\|[c_{p'}^\dagger \times \tilde{c}_{n'}]^{(J)\dagger}\|J_b^\pi\rangle = \sqrt{2J+1}(\bar{X}_b^{pn}\bar{v}_p\bar{u}_n - \bar{Y}_b^{pn}\bar{u}_p\bar{v}_n) \quad (4.24b)$$

where $\tilde{c}_{jm} = (-1)^{j+m}c_{j-m}$ and the overlap is defined by eq.(4.22).

We make now an estimation of the order of the effects of the particle-particle interaction on the ground state correlations and charge-changing transitions. This is important because as mentioned the success of the QRPA calculations of $2\nu$ $\beta\beta$ decay was achieved basically by including the particle-particle interaction. (It is noted that the particle-particle interaction is not an extra one being added to the QRPA model. It appears naturally by the Bogoliubov transformation, in addition to the particle-hole counterpart. But it was usually neglected in earlier calculations of $\beta\beta$ decay.) First of all, for nuclei with a large neutron excess, we expect that $u_p$ and $v_n$ are normally significantly larger than $v_p$ and $u_n$,

$$u_p \sim v_n \sim l, \qquad v_p \sim u_n \sim s, \qquad s^2 \ll l^2. \quad (4.25)$$

Then, the elements of QRPA equation (4.14) are expressed as

$$A_{pn,p'n'} \sim \delta(pn,p'n')(\epsilon_p+\epsilon_n)$$
$$+(l^4+s^4)\langle pn^{-1}|V|p'n'^{-1}\rangle + 2l^2s^2\langle pn|V|p'n'\rangle \quad (4.26a)$$
$$B_{pn,p'n'} \sim -2l^2s^2\langle pn^{-1}|V|p'n'^{-1}\rangle + 2l^2s^2\langle pn|V|p'n'\rangle. \quad (4.26b)$$

The submatrix $A$ is a matrix between main forward-going amplitudes $X$ (or between backward-going amplitudes $Y$), and is therefore responsible for the mixing of various particle-hole states in the odd-odd nucleus reached by the charge-changing transition. For example, for the $J^\pi = 1^+$ mode the mixing leads to a concentration of $\beta^-$ Gamow-Teller strength to the giant resonance (see Figs. 5.4 and 5.7). The elements of $A$ are dominated by the particle-hole interaction and quasiparticle energies. The contribution of the particle-particle interaction is much smaller, by the factor $s^2/l^2$, implying that this interaction has little effect on the mixing of particle-hole states in the intermediate nucleus. On the other hand, the submatrix $B$ is important for the ground state correlations since it connects $X$ and $Y$. An increase of the backward-going amplitudes $Y$ means enhancement of the ground state correlations (see eqs. (3.18) and (3.16)). The elements of $B$ consist of both interaction terms of the same order, $\sim l^2s^2$. This indicates that the particle-particle interaction, being added coherently to the particle-hole interaction, can significantly enhance the ground state correlations. The degree of enhancement depends on the nature of particle-hole and particle-particle interactions in the $J^\pi$ channel.

The estimation of the effect on the transition amplitudes, eq.(4.19), clarifies a substantial difference between transitions induced by $t_-f^{(J)}$ and $t_+f^{(J)}$. Let us take $\beta^\pm$ Gamow-Teller transitions to $1^+$ states as an example. (Gamow-Teller strength distributions calculated in the QRPA model are compared in Fig. 5.7 for different strengths of the particle-particle interaction.) With the large factor $l$ and the small factor $s$, the transition amplitudes are simplified such that

$$\beta^- : \langle\omega,1^+\|t_-\sigma\|0^+\rangle \sim Xl^2 - Ys^2 \quad (4.27a)$$
$$\beta^+ : \langle\omega,1^+\|t_+\sigma\|0^+\rangle \sim Xs^2 - Yl^2. \quad (4.27b)$$

For $\beta^-$ transitions ($n \to p$), the amplitude is dominated by the $X$ term, which results from excitation of a proton particle - neutron hole pair. The $Y$ term gives a very small contribution suppressed by the factor $s^2$ and furthermore by the ratio of $Y$ over $X$. Normally the forward-going amplitudes $X$ are much larger than the backward-going amplitudes $Y$, eq.(4.13). Therefore $\beta^-$ transitions are insensitive to the ground state correlations, i.e., insensitive to the particle-particle interaction. The factors multiplied with $X$ and $Y$ are exchanged in $\beta^+$ transitions ($p \to n$).

When the particle-particle interaction is switched off, the $X$ term dominates the transition, even if one considers the enhancement factor $l^2$ of the $Y$ term since the ground state correlations are negligibly small. But, the particle-particle interaction enhances the ground state corrlations appreciably, and makes the $Y$ term more important. The two terms are of the same order for $J^\pi = 1^+$, and $\beta^+$ transitions are strongly suppressed. Generally, in the $1^+$ channel, the particle-hole interaction acting between particle-hole pairs is repulsive, while the particle-particle interaction is attractive. This characteristic feature results in a destructive interference between the $X$ and $Y$ terms. Thus, the particle-particle interaction plays a fundamental role in the suppression of $\beta^+$ transition strength. This conclusion can be understood also by the following simple argument. In a nucleus with a large neutron excess, most s.p. $\beta^+$ transitions ($p \to n$) are blocked by the Pauli exclusion principle in the absence of ground state correlations (the neutron orbits which can be reached by the Gamow-Teller transitions are fully occupied). When ground state correlations exist, vacancies in the neutron hole orbits can be found, as well as occupations in proton particle orbits, with the consequent occurrence of $\beta^+$ amplitudes which by destructive interference then reduce the total $\beta^+$ strength.

## 5. Results of calculations of $\beta\beta$ decay rates

### 5.1. Two-neutrino mode

#### 5.1.1. Calculations till 1986

The main points of the development till 1986 were

(1) the observation [Gro 83,86, Kla 84, Tsu 84] that the closure approximation which was used in most earlier work such as that of [Hax 82a,84] cannot give reliable predictions of $2\nu$ $\beta\beta$ decay rates. Figs. 5.1 and 5.4 are examples showing the importance of the signs of the transition matrix elements (see discussion in sect. 3). The closure approximation is, however, expected to be reliable in the calculation of $0\nu$ $\beta\beta$ decay.

(2) the observation that the ground state correlations play the decisive role in the prediction of $\beta\beta$ decay rates of the $2\nu$ mode, while the neutrino mass term of the $0\nu$ mode is much less

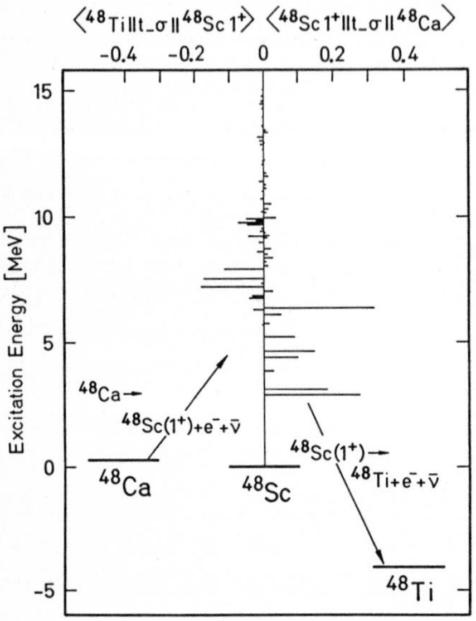


Fig. 5.1 Products of matrix elements of successive Gamow-Teller transitions through $1^+$ intermediate states for the $2\nu$ $\beta\beta$ decay, $^{48}Ca \to ^{48}Ti$, calculated in the shell model (from [Tsu 84]).

affected by the inclusion of the correlations [Kla 84, Gro 85,86]. The reason why details of the nuclear structure affect $M^{2\nu}_{GT}$ and $M^{0\nu}_{GT}$ ($M^{0\nu}_{F}$) so differently is discussed in sect. 5.2.2.

The unreliability of the closure approximation can be seen in the $2\nu$ $\beta\beta$ decay of $^{48}$Ca. The shell model approach is expected to work well for this doubly magic nucleus. Table 5.1 shows results of shell model calculations by [Hax 82a, Zam 82, Sko 83, Tsu 84]. The four calculations yielded similar values of the closure matrix element (eq.(3.13)). The difference in the half-life between [Tsu 84] and [Hax 82a] is traced to the use of the closure approximation by the latter authors. These authors estimated the average energy of intermediate states from the strength distribution of $^{48}$Ca $\rightarrow$ $^{48}$Sc Gamow-Teller excitations. On the other hand, for the matrix elements shown in Fig. 5.1, Tsuboi, Muto and Horie [Tsu 84] performed the phase space integral, eq.(3.6), explicitly for each intermediate state without assuming the closure approximation. The signs of the matrix elements for high-lying intermediate states are generally opposite to those for low-lying states. The sum in $M^{2\nu}_{GT}$, eq.(3.10), contains the energy denominator, while the closure matrix element, eq.(3.13) is a sum with equal weight ( =1). The two sums involve different interference of those matrix elements. Klapdor and Grotz [Kla 84] used the same procedure as Tsuboi et al. for heavier nuclei and obtained for the average energies, also about half the values of Haxton et al. [Hax 84].

Fig. 5.2 and Table 5.2 show the most intensively investigated experimental and theoretical (till end of 1986) half-lives of $2\nu$ $\beta\beta$ decay. The theoretical values are taken from systematic calculations, with the shell model [Hax 82a], the number-projected BCS [Kla 84, Gro 86], and the QRPA with a particle-hole Gamow-Teller force [Gro 85,86, Vog 85]. It is seen that for the heavier nuclei the discrepancy between experiment and calculation is rather large, up to two orders of magnitude for the calculations of [Hax 82a] and [Vog 85]. The calculations of [Kla 84, Gro 86] lead here to a remarkable improvement, but obviously cannot yet reproduce the full degree of observed suppression of the $2\nu$ $\beta\beta$ decay rates.

A comment has to be made here concerning the experimental half-lives given in Table 5.2. The $2\nu$ character is clear for the half-lives obtained by counter experiments for $^{76}$Ge, $^{82}$Se, $^{100}$Mo, $^{136}$Xe and $^{150}$Nd. The recent counter measurement by Elliott, Hahn and Moe [Ell 87] yielded a half-life for $^{82}$Se consistent with geochemical measurements [Kir 86, Man 86]. The values for $^{128,130}$Te are, on the other hand, only from geochemical determination [Kir 86, Man 86], which cannot differentiate between different decay modes. However, according to the phase space argument discussed below, the $\beta\beta$ decay of $^{130}$Te is clearly dominated by the $2\nu$ mode.

Table 5.1. Half-lives, closure matrix elements and average energies of intermediate states of the $2\nu$ $\beta\beta$ decay of $^{48}$Ca, calculated in the shell model. The experimental lower limit of the half-life is given in the last line.

| References | $T^{2\nu}_{1/2}$ ($10^{19}$ y) | $(M^{2\nu}_{GT})_c$ | $\bar{E}$ (MeV) |
|---|---|---|---|
| Calculation | | | |
| Hax 82 | 2.70 | 0.444 | 7.72 |
| Zam 82 | | 0.360 | |
| Sko 83 | 4.1 | 0.250 | 4.6 |
| Tsu 84 | 0.61 | 0.464 | 3.80 |
| Experiment | | | |
| Bar 70 | >3.6 | | |

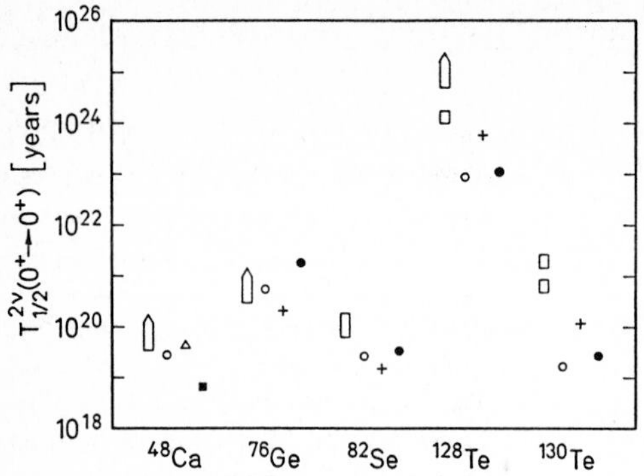


Fig. 5.2 Comparison of experimental and calculated half-lives of $2\nu$ $\beta\beta$ decay. Experimental values are denoted by an open square, or an open arrow when only the lower limit is known. The theoretical values are from [Hax 84] (open circle), [Sko 83] (triangle), [Tsu 84] (filled square), [Gro 86] PBCS (cross) and [Vog 85] (filled circle).

Table 5.2. Comparison of $0^+_i \to 0^+_f$ $2\nu$ $\beta\beta$ half-lives calculated with different nuclear structure models (see text). Recent experimental values are also presented.

| References | $^{76}$Ge | $^{82}$Se | $^{128}$Te | $^{130}$Te | $^{100}$Mo | $^{136}$Xe | $^{150}$Nd |
|---|---|---|---|---|---|---|---|
| | $10^{21}$y | $10^{20}$y | $10^{24}$y | $10^{21}$y | $10^{18}$y | $10^{19}$y | $10^{19}$y |
| Calculation | | | | | | | |
| Hax 82 | 0.415 | 0.262 | 0.088 | 0.017 | -- | -- | -- |
| Vog 85 | 1.8 | 0.33 | 0.11 | 0.023 | 2.1 | 2.3 | 0.22 |
| Gro 86 | | | | | | | |
| QRPA (without pp force) | 0.11 | 0.048 | 0.12 | 0.019 | 1.8 | 6.0 | 0.05 |
| PBCS (GT) | 0.13 | 0.051 | 0.062 | 0.015 | | | |
| (GT + QQ) | 0.22 | 0.15 | 0.57 | 0.12 | | 3.3 | |
| Experiment | | | | | | | |
| | >0.3[a] | 1.30±0.05[b] | >5[b] | 1.5-2.75[b] | >6.2[e] | >2.1[f] | >2.4[g] |
| | | 1.0 ± 0.4[c] | 1.4±0.4[c] | 0.7±0.2[c] | | | |
| | | $1.1^{+0.8}_{-0.3}$[d] | | | | | |

[a]Avi 86, [b]Kir 86, [c]Man 86, [d]Ell 87, [e]Smo 88, [f]Bar 87, [g]Kli 86.

The phase space argument for the branching ratio of $^{130}$Te decay between $2\nu$ and $0\nu$ modes is as follows: The main assumption is that the nuclear matrix elements are assumed to be almost identical for $^{128}$Te and $^{130}$Te, which seems reasonable, since these nuclei have the same number of protons and the numbers of neutrons differ only by two units. Under this assumption, the half-life

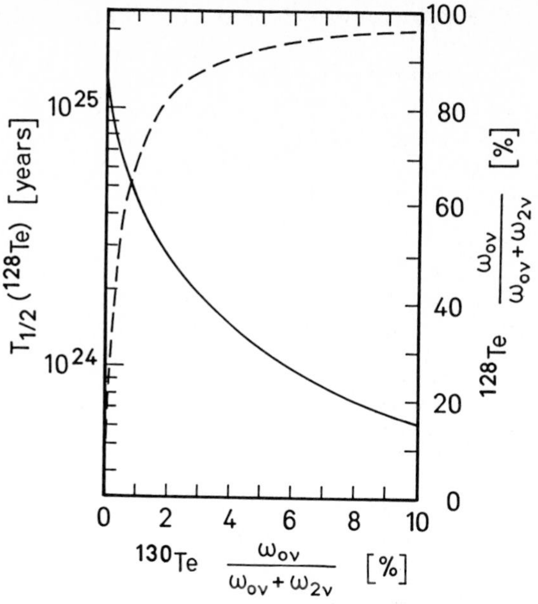


Fig. 5.3 The half-life (solid line) and the branching ratio (dashed line) of the $^{128}Te \rightarrow {}^{128}Xe$ $\beta\beta$ decay as function of the branching ratio of the $^{130}$Te decay under the assumption that the corresponding $0\nu$ and $2\nu$ matrix elements are equal to each other for these isotopes. The experimental half-life $T_{1/2}(^{130}Te) = 2.55 \times 10^{21}$ y from the geochemical measurement [Kir 86] is used.

and branching ratio of $^{128}$Te and $^{130}$Te are related by the phase space integrals. Fig. 5.3 shows the relation between the branching ratio of $^{130}$Te and the half-life and branching ratio of $^{128}$Te, for $T_{1/2}(^{130}Te) = 2.55 \times 10^{21}$ years [Kir 86]. The observed half-lives of $^{128}$Te, $T_{1/2}(^{128}Te) > 5 \times 10^{24}$ years [Kir 86] and $T_{1/2}(^{128}Te) = (1.4 \pm 0.4) \times 10^{24}$ years [Man 86], set the limit on the branching ratio of $^{130}$Te. The $\beta\beta$ decay of $^{130}$Te is therefore dominated by the $2\nu$ mode, the branching ratio of the $0\nu$ mode being of the order of 1% or smaller. However, the branching ratio of $^{128}$Te is not restricted by this phase space argument. (Generally, the $0\nu$ mode is more favoured for nuclei with a smaller $Q_{\beta\beta}$ value ($T_0$) because of the different dependence on $T_0$ for the $0\nu$ and $2\nu$ modes, respectively.)

Klapdor and Grotz [Kla 84, Gro 85a,85b,86] calculated half-lives of $2\nu$ and $0\nu$ $\beta\beta$ decay for all possible $\beta\beta$-decaying nuclei with $A \geq 70$, using two types of models. Details of the models are explained in sect. 4.3. The calculated half-lives are listed in Table 5.3. Although being systematically (in particular the QRPA results) somewhat too short, they may still provide a useful guide line for experimental planning. These authors also investigated the sensitivity of the calculated $\beta\beta$ decay rates to various types of ground state correlations and increasing complexity of the wave functions. Table 5.4 (from [Gro 86]) shows the sensitivity of the matrix elements to various types of correlations in the case of $2\nu$ $\beta\beta$ decay (see also Table 5.2 for the half-lives). Table 5.6 demonstrates the insensitivity of the $0\nu$ $\beta\beta$ decay rates to the ground state correlations. Using the projected BCS approach, these authors first took into account the quadrupole-quadrupole correlations and $\Delta N^{-1}$ excitations in the initial and final ground states of $0_i^+ \rightarrow 0_f^+$ transitions in addition to a residual Gamow-Teller interaction. As an example, transition amplitudes for the $\beta\beta$ decay of $^{130}$Te are shown in Fig. 5.4. This investigation suggested a clue for solving the problem of the significant discrepancies between theoretical and experimental half-lives of the $2\nu$ mode. The largest values of the closure matrix elements are obtained for the BCS wave functions. Significant reductions of the

Table 5.3. Calculated half-lives for $2\nu$ and $0\nu$ double beta decay for double beta emitters with $A \geq 70$ (from [Gro 86]). The QRPA calculations were performed neglecting the pp force. The $0\nu$ half-lives are neutrino mass dependent and the given figures correspond to $\langle m_\nu \rangle = 1$ eV. The $T_{1/2}^{2\nu *}$ and $T_{1/2}^{0\nu *}$ are calculated using the PBCS model including spin-isospin and quadrupole-quadrupole forces (see text). The $T_{1/2}^{0\nu}$ are relatively rough estimates, for their derivation see eqs. (44), (45) in [Gro 86]. $T_o$ denotes the $\beta\beta$ Q value.

| | $T_0$ | QRPA $T_{1/2}^{2\nu}$ | BCS (GT+QQ) $T_{1/2}^{2\nu *}$ | $T_{1/2}^{0\nu} \times \langle m_\nu \rangle^2$ |
|---|---|---|---|---|
| | (MeV) | (y) | (y) | (y × eV$^2$) |
| $^{70}$Zn | 1.00 | $1.9 \times 10^{22}$ | | $7.6 \times 10^{23}$ |
| $^{76}$Ge | 2.04 | $1.1 \times 10^{20}$ | $2.2 \times 10^{20}$ | $2.6 \times 10^{23}$* |
| $^{80}$Se | 0.136 | $9.0 \times 10^{28}$ | | $6.7 \times 10^{25}$ |
| $^{82}$Se | 3.01 | $4.5 \times 10^{18}$ | $1.5 \times 10^{19}$ | $9.5 \times 10^{22}$* |
| $^{86}$Kr | 1.25 | $3.5 \times 10^{22}$ | | $5.0 \times 10^{24}$ |
| $^{94}$Zr | 1.15 | $4.1 \times 10^{21}$ | | $6.2 \times 10^{23}$ |
| $^{96}$Zr | 3.35 | $5.2 \times 10^{17}$ | | $1.6 \times 10^{22}$ |
| $^{98}$Mo | 0.11 | $1.6 \times 10^{29}$ | | $7.3 \times 10^{25}$ |
| $^{100}$Mo | 3.03 | $1.8 \times 10^{18}$ | | $3.3 \times 10^{22}$ |
| $^{104}$Ru | 1.30 | $1.8 \times 10^{21}$ | | $5.0 \times 10^{23}$ |
| $^{110}$Pd | 2.01 | $5.0 \times 10^{19}$ | | $1.3 \times 10^{23}$ |
| $^{114}$Cd | 0.54 | $2.7 \times 10^{24}$ | | $1.7 \times 10^{25}$ |
| $^{116}$Cd | 2.81 | $8.3 \times 10^{18}$ | | $1.7 \times 10^{23}$ |
| $^{122}$Sn | 0.36 | $1.4 \times 10^{26}$ | | $3.6 \times 10^{25}$ |
| $^{124}$Sn | 2.28 | $9.3 \times 10^{19}$ | | $3.1 \times 10^{23}$ |
| $^{128}$Te | 0.87 | $1.2 \times 10^{23}$ | $5.7 \times 10^{23}$ | $9.8 \times 10^{23}$* |
| $^{130}$Te | 2.53 | $1.9 \times 10^{19}$ | $1.2 \times 10^{20}$ | $4.6 \times 10^{22}$* |
| $^{134}$Xe | 0.84 | $5.1 \times 10^{22}$ | $2.5 \times 10^{23}$ | $8.7 \times 10^{23}$* |
| $^{136}$Xe | 2.48 | $6.0 \times 10^{19}$ | $3.3 \times 10^{19}$ | $3.0 \times 10^{23}$* |
| $^{142}$Ce | 1.41 | $2.8 \times 10^{21}$ | $4.1 \times 10^{20}$ | $4.7 \times 10^{23}$* |
| $^{146}$Nd | 0.06 | $2.9 \times 10^{30}$ | | $5.6 \times 10^{25}$ |
| $^{148}$Nd | 1.93 | $2.5 \times 10^{19}$ | | $1.1 \times 10^{23}$ |
| $^{150}$Nd | 3.37 | $4.8 \times 10^{17}$ | | $2.4 \times 10^{22}$ |
| $^{154}$Sm | 1.25 | $9.5 \times 10^{20}$ | | $2.4 \times 10^{23}$ |
| $^{160}$Gd | 1.73 | $4.4 \times 10^{19}$ | | $6.4 \times 10^{22}$ |
| $^{170}$Er | 0.66 | $6.6 \times 10^{22}$ | | $9.1 \times 10^{23}$ |
| $^{176}$Yb | 1.08 | $1.1 \times 10^{21}$ | | $2.5 \times 10^{23}$ |
| $^{186}$W | 0.49 | $3.2 \times 10^{23}$ | | $1.2 \times 10^{24}$ |
| $^{192}$Os | 0.41 | $1.7 \times 10^{24}$ | | $1.6 \times 10^{24}$ |
| $^{198}$Pt | 1.04 | $1.2 \times 10^{22}$ | | $1.6 \times 10^{24}$ |
| $^{204}$Hg | 0.41 | $4.6 \times 10^{25}$ | | $2.6 \times 10^{25}$ |
| $^{232}$Th | 0.85 | $1.6 \times 10^{20}$ | | $3.8 \times 10^{22}$ |
| $^{238}$U | 1.15 | $2.2 \times 10^{19}$ | | $2.4 \times 10^{22}$ |

values are found when the spin-isospin force (without the particle-particle component) is switched on and later the QQ force. The spin-isospin force, which is diagonalized in a zero- plus four-quasiparticle space, induces admixtures of four-quasiparticle components in the ground state, i.e., ground state correlations of the spin-isospin mode $|pn(1^+), p'n'(1^+); 0^+\rangle$ (eq.(4.1)), which suppress the decay rates. The calculated half-lives are quite consistent with those of the QRPA calculation with the same force (see Table 5.2), in which the four-quasiparticle admixtures are taken into account in the RPA formalism. The inclusion of the QQ force further reduces the decay rates by large factors of up to an order of magnitude. The QQ force gives rise to the QQ correlations of eq.(4.2), and as a consequence, as one can see by the angular-momentum recoupling of eq.(4.3), enhances the spin-isospin ground state correlations. This implies a large sensitivity of the calculated $2\nu$ $\beta\beta$

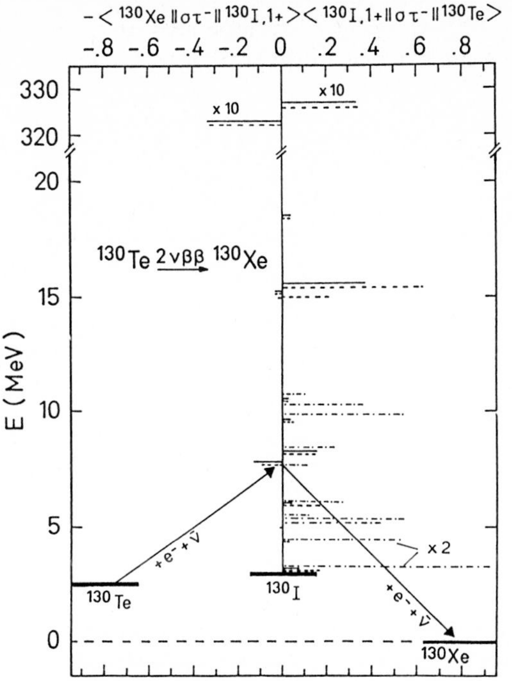


Fig. 5.4 Amplitudes for the $2\nu$ $\beta\beta$ transition $^{130}$Te $\rightarrow$ $^{130}$Xe calculated with pairing force (dashed-dotted lines) and pairing plus Gamow-Teller forces including $\Delta$-hole mixing (dashed lines). The solid lines show the final result including also quadrupole-quadrupole forces (from [Gro 86]).

Table 5.4. Closure matrix elements $(M_{GT}^{2\nu})_c$ calculated in the PBCS treatment in various approaches of increasing complexity from left to right (from [Gro 86]).

| | Only pairing | Pairing + $H_{\sigma\tau}$ (no $\Delta$h mixing) | Pairing + $H^{\Delta}_{\sigma\tau}$ (with $\Delta$h mixing) | Pairing + $H^{\Delta}_{\sigma\tau}$ + $H_{QQ}$, without higher order phonon contributions | Pairing + $H^{\Delta}_{\sigma\tau}$ + $H_{QQ}$ including higher order phonon contributions |
|---|---|---|---|---|---|
| $^{128}$Te | 6.34 | 2.00 | 1.77 | 0.61 | 0.52 |
| $^{130}$Te | 5.86 | 1.75 | 1.55 | 0.56 | 0.48 |
| $^{82}$Se | 5.52 | 2.83 | 2.33 | 1.84 | 1.38 |
| $^{76}$Ge | 7.05 | 3.76 | 3.26 | 2.64 | 1.93 |

The first column gives the large matrix elements in the simple pairing model. These are reduced step by step including the spin-isospin force first in the nucleonic and then also in the $\Delta$ nucleon hole space and further by including the $2^+$ phonon correlations. The matrix elements in the last column include also higher order g.s. correlations.

half-lives to the ground state correlations of the spin-isospin mode. The ground state correlations of the spin-isospin mode are naturally enhanced by including the particle-particle component of a residual interaction. This is discussed in the next section.

### 5.1.2. QRPA calculations including the particle-particle interaction

The suppression mechanism of $2\nu$ $\beta\beta$ matrix elements was finally clarified by recent calculations by Vogel and Zirnbauer [Vog 86], Civitarese, Faessler and Tomoda [Civ 87], and Muto and Klapdor [Mut 88a]. All of these calculations are based on the QRPA approach, but with different effective interactions; zero-range force [Vog 86], a realistic force derived from the Bonn one-boson-exchange potential [Civ 87], and another realistic force from the Paris potential [Mut 88a]. The conclusion of these calculations was that the particle-particle interaction enhances the ground state correlations beyond those considered earlier [Kla 84, Gro 86] and as a result, the $2\nu$ matrix elements are further strongly suppressed. For a reliable prediction of the half-life, the strength of the particle-particle interaction is crucial. Muto and Klapdor [Mut 88a], and later Muto, Bender and Klapdor [Mut 88c] taking a similar procedure, fixed the strength to the Gamow-Teller strengths observed in $\beta^+$ decay of a number of nuclei. These authors showed that the half-lives calculated with the fixed interaction strength are consistent with the experimental values, with a remarkable improvement over the earlier calculations. Therefore we will follow the calculation of [Mut 88a,88c].

Half-lives of $0_i^+ \rightarrow 0_f^+$ transitions were calculated for the $\beta\beta$ emitters $^{76}$Ge, $^{82}$Se, $^{100}$Mo, $^{128,130}$Te, $^{136}$Xe, $^{150}$Nd. The model space is assumed to include two oscillator major shells, for example full $3\hbar\omega$, $4\hbar\omega$ shells for $A = 76$ and $A = 82$ systems and full $4\hbar\omega$, $5\hbar\omega$ and $i_{13/2}$, $i_{11/2}$ shells for $A = 150$. Single-particle energies are computed by solving the eigenvalue equation for a nucleon in the Coulomb-corrected Woods-Saxon potential [Boh 69]. The depth of the Woods-Saxon central potential is modified by adding a term depending on the orbital angular momentum, $-0.05l(l+1)$ MeV, in order to better reproduce observed single (quasi)particle spectra. The occupation amplitudes, $u_j$ and $v_j$, which define the quasiparticle basis are calculated in the BCS approximation. Then the QRPA equation for $J^\pi = 1^+$ is solved for both $0_i^+$ and $0_f^+$ ground states. The G-matrix of the Paris potential [Vin 79,Lac 80] fitted to a sum of Yukawa terms [Ana 83] is adopted as nucleon-nucleon residual interaction. The same interaction is used consistently for the BCS and RPA calculations. The Paris potential and the Bonn potential [Hol 81] are constructed from a phase-shift analysis of nucleon-nucleon scattering.

The realistic interaction could be renormalized in a nucleus, and additionally in the restricted model space. The renormalization could differ in the BCS and RPA approximations. In order to adjust the interaction strengths, four renormalization factors which multiply the corresponding interaction matrix elements are introduced. These factors, however, are not expected to deviate much from unity. In the BCS calculation, the renormalization factor of the pairing interaction, $g_{pair}^p$ or $g_{pair}^n$, which multiplies the matrix elements of eq.(4.7), is fixed for proton and neutron systems separately so as to reproduce experimental even-odd mass differences [Wap 85]. The values of $g_{pair}^p$ and $g_{pair}^n$ thus fixed are close to unity, as expected, with deviations of at most 30%. In the RPA two factors are introduced, $g_{ph}$ multiplying the particle-hole interaction, which is mainly responsible for mixing of particle-hole pairs in the odd-odd intermediate nucleus, and $g_{pp}$ for the particle-particle interaction, which is important for the ground state correlations of the parent and daughter nuclei. By multiplying these renormalization factors, eq.(4.14) becomes

$$\begin{aligned} A_{pn,p'n'} &= \delta(pn,p'n')(\epsilon_p+\epsilon_n) \\ &+g_{ph}(u_p v_n u_{p'} v_{n'} + v_p u_n v_{p'} u_{n'})\langle pn^{-1}|V|p'n'^{-1}\rangle_J \\ &+g_{pp}(u_p u_n u_{p'} u_{n'} + v_p v_n v_{p'} v_{n'})\langle pn|V|p'n'\rangle_J \end{aligned} \tag{5.1a}$$

$$\begin{aligned} B_{pn,p'n'} &= -g_{ph}(u_p v_n v_{p'} u_{n'} + v_p u_n u_{p'} v_{n'})\langle pn^{-1}|V|p'n'^{-1}\rangle_J \\ &+g_{pp}(u_p u_n v_{p'} v_{n'} + v_p v_n u_{p'} u_{n'})\langle pn|V|p'n'\rangle_J . \end{aligned} \tag{5.1b}$$

The particle-hole interaction strength is first adjusted so that the RPA calculation reproduces the excitation energy of the Gamow-Teller giant resonance observed in charge-exchange $(p,n)$ reactions on the $\beta\beta$ emitters [Mad 87]. The fitted mass-number-dependent strength $g_{ph} = 1 + 0.002 \cdot A$ is consistently used in the calculation of $\beta\beta$ decay and also in $\beta^+$ decay for the adjustment of $g_{pp}$.

The nuclear matrix elements of $\beta\beta$ decay are calculated with the fixed interaction strengths but as a function of $g_{pp}$, which is not so far adjusted. Fig. 5.5 shows the behaviour of the matrix elements. As $g_{pp}$ is increased, the matrix element decreases and crosses through zero at around $g_{pp} = 1$. Since $g_{pp}$ is not expected to be far from unity, it is a nice feature of the calculation with the realistic interaction that the matrix element is strongly suppressed at $g_{pp} \sim 1$, which implies a long half-life consistent with experiment. However, it is seen that the matrix elements decrease so rapidly near the crossing point that the value of $g_{pp}$ is critical in the prediction of half-lives with satisfactory reliability. The rapid decrease of the matrix elements comes from the enhancement of the ground state correlations with increasing particle-particle interaction strength. Fig. 5.6 shows the percentage of ground state correlations in the ground state wave functions as function of $g_{pp}$. When the particle-particle interaction is switched off ($g_{pp} = 0$), the percentage is smaller than 1%. It increases exponentially as a function of $g_{pp}$ and reaches about 10% at $g_{pp} \sim 1$.

The particle-particle interaction strength $g_{pp}$ is fixed from $\beta^+$ decay observed in nuclei with $A = 60 - 170$. $\beta^+$ decay, corresponding to the inverse of the second transition in $\beta\beta$ decay, $1^+ \to 0^+_f$, is sensitive to $g_{pp}$, while, as discussed in sect. 4.4, $\beta^-$ transitions are much less sensitive to the ground state correlations, since most of the $\beta^-$ transitions come from the excess neutrons or from spin-flip transitions which are allowed even if there are no ground state correlations. As an example of this difference in the sensitivity to the particle-particle interaction in $\beta^-$ and $\beta^+$ decay, the transitions relevant for the $^{82}Se \to {}^{82}Kr$ $\beta\beta$ decay are shown in Fig. 5.7. The transitions

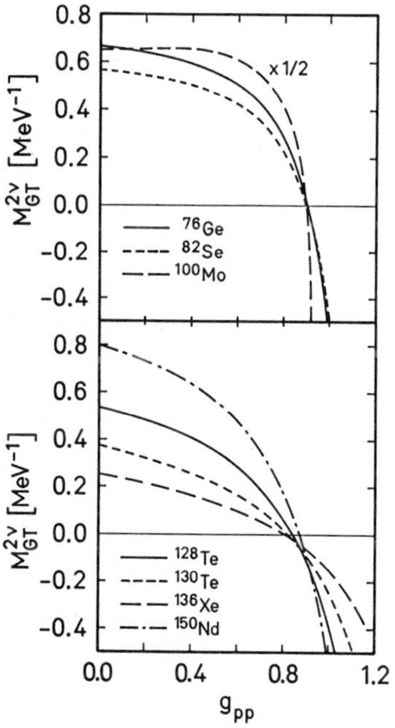


Fig. 5.5

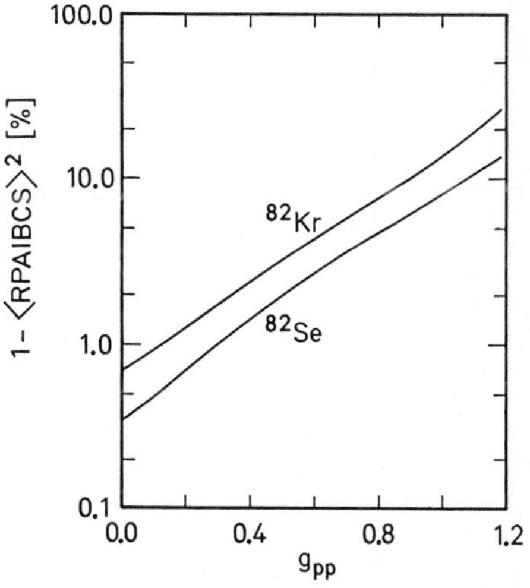


Fig. 5.6

Fig. 5.5 Nuclear matrix elements of the $0^+ \to 0^+$ $2\nu$ $\beta\beta$ decay calculated as a function of the particle-particle interaction strength of the proton-neutron quasiparticle RPA.

Fig. 5.6 Percentages of the spin-isospin correlations in the QRPA ground-state wave functions as function of $g_{pp}$ for $^{82}Se$ and $^{82}Kr$.

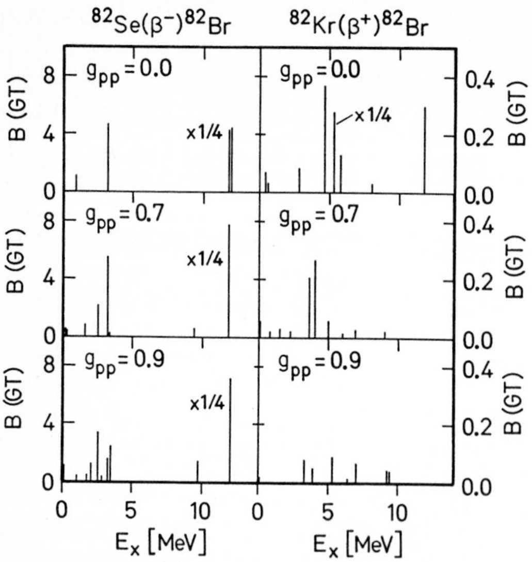


Fig. 5.7 Comparison of Gamow-Teller strength distributions for $\beta^-$ and $\beta^+$ excitations relevant for $\beta\beta$ decay of $^{82}$Se, for three values of the particle-particle interaction strength.

shown there are the ones occuring in the $\beta\beta$ nuclear matrix element. The Gamow-Teller strength distributions are plotted for three values of $g_{pp}$; $g_{pp} = 0$ (the particle-particle interaction is switched off), $g_{pp} = 0.7$, and $g_{pp} = 0.9$, where the $2\nu$ $\beta\beta$ nuclear matrix element vanishes almost completely. It is obvious that the $\beta^+$ strength is drastically quenched by the particle-particle interaction, while in the $\beta^-$ sector only a small fraction of strength is shifted from the giant resonance region to low-lying states.

For some thirty $\beta^+$-unstable even-even nuclei with $A$ = 60-170 (from Table of Isotopes [Led 78]), the Gamow-Teller strength in the $Q_\beta$ window is calculated in the QRPA as a function of $g_{pp}$, and then the value of $g_{pp}$ with which the calculation reproduces the observed $B(GT)$ values is determined. This procedure is shown in Fig. 5.8 for four examples. The values of $g_{pp}$ thus obtained are summarized in Fig. 5.9. Semi-magic nuclei such as $^{94}$Ru and $^{148}$Dy are excluded from this analysis of $g_{pp}$. In these nuclei, a s.p. transition $\pi g_{9/2} \to \nu g_{7/2}$ or $\pi h_{11/2} \to \nu h_{9/2}$ dominates the $\beta^+$ strength, whereas in non-magic nuclei various s.p. transitions contribute. From the viewpoint of the *random phase* assumption, $\beta^+$ decay of the latter group is expected to be well described by the QRPA model. In addition, the $\beta^+$ strengths of the semi-magic nuclei are rather sensitive to the occupation probability of the orbits relevant to the dominant s.p. transitions, i.e., sensitive to s.p. energies of these orbits and to the effective interaction. The fitted values of $g_{pp}$ shown in Fig. 5.9 show a general tendency that $g_{pp}$ decreases as the neutron number increases and show an additional increase after the magic numbers, $N$ = 28, 50 and 82. Therefore, the following form is assumed for describing the behaviour of the fitted $g_{pp}$ values, as a function of the neutron number of the parent nucleus,

$$g_{pp}(N) = a - bN + \frac{c}{N - d} \tag{5.2}$$

where $a$ and $b$ are common for all nuclei but $c$ and $d$ are different for each range of $N$, $30 \le N < 50$, $50 < N < 82$, $82 < N \le 98$. Values of the constants are determined using a least-squares fitting procedure. The fitted curves are plotted in Fig. 5.9; the standard deviation is $\delta g_{pp} = 0.039$.

For the calculated half-lives of $2\nu$ $\beta\beta$ decay, we show two numbers; the lower limit of the half-life in the $1\sigma$ range of $g_{pp}$, and the average in the same range. The average half-life is defined

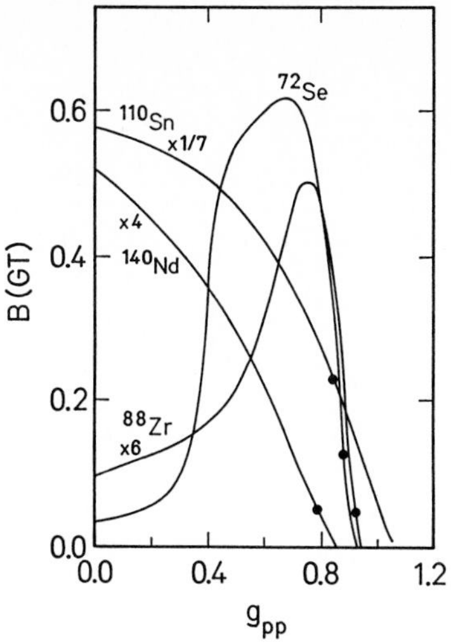


Fig. 5.8

Fig. 5.9

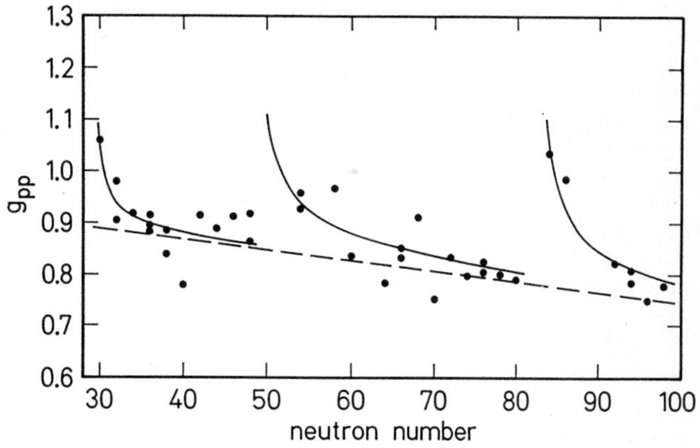


Fig. 5.8 The fitting procedure of the particle-particle interaction strength $g_{pp}$ to the experimental Gamow-Teller strength of single $\beta^+$ decay. The experimental value is presented by a dot on the theoretical curve calculated as a function of $g_{pp}$.

Fig. 5.9 The values of the particle-particle interaction strength, with which a QRPA calculation reproduces experimental Gamow-Teller $\beta^+$-decay strength. The solid line denotes the function of eq.(5.2) fitted to the $g_{pp}$ values presented by dots, the dashed line the linear term of the function.

by $T_{1/2}^{2\nu} = ln2/\bar{\omega}_{2\nu}$, where $\bar{\omega}_{2\nu}$ is obtained by simply averaging the decay rate over the $1\sigma$ range,

$$\bar{\omega}_{2\nu} = \frac{1}{2\delta g_{pp}} \int_{g_{pp}(N)-\delta g_{pp}}^{g_{pp}(N)+\delta g_{pp}} \omega_{2\nu}(g_{pp}) dg_{pp} \tag{5.3}$$

It should be noted that the nuclear matrix element of $2\nu$ decay goes through zero at a $g_{pp}$ lying within the range of one standard deviation from the optimum value corresponding to the neutron number of the daughter nucleus, implying a possible infinite half-life. The half-lives thus calculated are listed in Table 5.5, with the experimental values in the last row, and are also given in Fig. 5.10. The theoretical half-lives are consistent with the existing experimental data, and the large discrepancies of the earlier calculations are removed.

To conclude, a solution of the long-standing problem of understanding $2\nu$ $\beta\beta$ decay half-lives is found in the nuclear structure calculation. The success is based on the fine tuning of the strength of the particle-particle interaction. This interaction is automatically taken into account, for instance, in a shell model calculation. However, although being simplified compared to such a sophisticated model, the QRPA approach gives insight into and clarifies the role of the particle-particle interaction for the suppression of the $2\nu$ $\beta\beta$ decay rates precisely because of this simplification.

It is noted, however, that we have taken into account only the leading term in the calculation of $2\nu$ $\beta\beta$ decay rates. When the leading term is strongly suppressed, higher order terms and other mechanisms could become more important.

5.1.2.1. Limitations of the QRPA approach

The calculations described in section 5.1.2. revealed limitations of the QRPA approach [Mut 88c]. While the results obtained for the nuclei $^{76}$Ge, $^{82}$Se, $^{128,130}$Te and $^{136}$Xe are considered to be reliable, such limitations can be clearly seen in the behaviour of the nuclear matrix

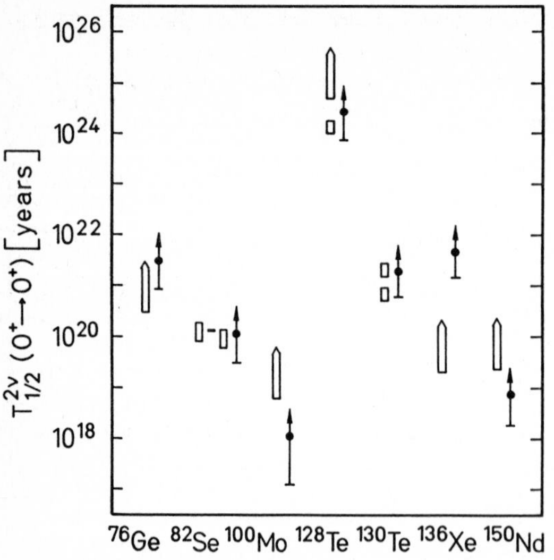


Fig. 5.10 Comparison of experimental and calculated half-lives of $0^+ \to 0^+$ $2\nu$ $\beta\beta$ decay. The experimental values are given by open squares or open arrows for lower limits. The theoretical half-lives, calculated in the proton-neutron QRPA model [Mut 88c], are presented by arrows. The arrow starts from the lower limit and the circle on it denotes the average (see text).

Table 5.5. Comparison of calculated [Mut 88c] and experimental half-lives of $0^+ \to 0^+$ $2\nu$ $\beta\beta$ decay. The half-lives evaluated at $g_{pp} = 0$ (no particle-particle interaction), and the lower limit and average in the $1\sigma$ range of $g_{pp}$ (see text) are presented.

| | $^{76}$Ge | $^{82}$Se | $^{100}$Mo | $^{128}$Te | $^{130}$Te | $^{136}$Xe | $^{150}$Nd |
|---|---|---|---|---|---|---|---|
| | $10^{21}$y | $10^{20}$y | $10^{18}$y | $10^{24}$y | $10^{21}$y | $10^{21}$y | $10^{19}$y |
| Calculation | | | | | | | |
| $g_{pp} = 0$ | 0.070 | 0.029 | 0.22 | 0.020 | 0.0069 | 0.015 | 0.0061 |
| lower limit | 0.822 | 0.306 | 0.125 | 0.723 | 0.594 | 1.43 | 0.189 |
| average | 2.99 | 1.09 | 1.13 | 2.63 | 1.84 | 4.64 | 0.737 |
| Experiment | >0.3[a] | 1.30±0.05[b] | >6.2[e] | >5[b] | 1.5-2.75[b] | >0.021[f] | >2.4[g] |
| | | 1.0 ±0.4[c] | | 1.4±0.4[c] | 0.7±0.2[c] | | |
| | | $1.1^{+0.8}_{-0.3}$[d] | | | | | |

[a]Avi 86, [b]Kir 86, [c]Man 86, [d]Ell 87, [e]Smo 88, [f]Bar 87, [g]Kli 86

element for the decay of $^{100}$Mo, for which the calculated half-lives are somewhat uncertain. The matrix element, as $g_{pp}$ is increased, first increases and then very rapidly decreases, passing through zero (see Fig. 5.5), and after the crossing the RPA equation tends to a collapse. The collapse occurs, when the pp interaction enhances the ground-state correlations too much and pushes down the energy of the lowest eigenstate such that the depressed eigenvalue finally becomes imaginary.

The ground-state correlations of, say, more than 50% obviously contradict the main assumption of the RPA formalism. Therefore, even if the RPA equation is stable, the probable values of $g_{pp}$ could enter an unreasonable region of the RPA solution. The lower limit of the half-life for $^{100}$Mo is evaluated at a point just before the collapse, and consequently both lower-limit and 'average' would likely be underestimated.

The calculated half-lives of $^{100}$Mo (and $^{150}$Nd) also suffer from another origin of uncertainties. This nucleus is located in a region where the $\beta\beta$ decay is dominated by few strong single-particle transitions in $g_{9/2} - g_{7/2}$ $(h_{11/2} - h_{9/2})$ orbits. It means that the half-lives are affected rather sensitively by a change of occupation amplitudes of the relevant orbits, i.e., by a change of single-particle energies and/or that of the pairing interactions. These uncertainties would probably also result in an overestimation of the decay rate for the transitions strongly suppressed by the destructive interference between terms with opposite signs.

Finally, we point out that, in general, the $2\nu$ $\beta\beta$ decay rate should not vanish, although the $2\nu$ decay nuclear matrix element of the QRPA calculation vanishes completely at a $g_{pp}$ value in its $1\sigma$ range. The matrix element, with $F_\mu = t_-\sigma_\mu$,

$$M_{GT}^{2\nu} = -\langle 0_f^+ | \sum_\mu (-1)^\mu F_\mu \frac{1}{H - (M_i + M_f)/2} F_{-\mu} | 0_i^+ \rangle$$

can be expressed in two alternative forms

$$\begin{aligned} M_{GT}^{2\nu} &= \langle 0_f^+ | \sum_\mu (-1)^\mu \left\{ F_\mu [F_{-\mu}, \frac{1}{H - (M_i + M_f)/2}] + \frac{1}{(M_i - M_f)/2} F_\mu F_{-\mu} \right\} | 0_i^+ \rangle \\ &= \langle 0_f^+ | \sum_\mu (-1)^\mu \left\{ -[F_{-\mu}, \frac{1}{H - (M_i + M_f)/2}] F_\mu - \frac{1}{(M_i - M_f)/2} F_\mu F_{-\mu} \right\} | 0_i^+ \rangle \end{aligned}$$

where relations $H|0_i^+\rangle = M_i|0_i^+\rangle$ and $H|0_f^+\rangle = M_f|0_f^+\rangle$ are used for the second terms, which have the same magnitude and opposite signs. Therefore, the nuclear matrix element is given by

$$M_{GT}^{2\nu} = \frac{1}{2} \langle 0_f^+ | \sum_\mu (-1)^\mu [F_\mu, [F_{-\mu}, \frac{1}{H - (M_i + M_f)/2}]] | 0_i^+ \rangle \tag{5.4}$$

This expression holds model independently. The Hamiltonian contains spin-isospin dependent components and they yield non-vanishing contributions through the double commutator to the nuclear matrix element. In the QRPA calculation, the crossing of $M_{GT}^{2\nu}$ through zero happens because of the artificial change of $g_{pp}$ as a free parameter, resulting in a violation of the Pandya transformation, eq. (4.15), which connects pp and ph interaction matrix elements.

### 5.2. Neutrinoless mode

#### 5.2.1. Calculations of nuclear matrix elements

The $0\nu$ $\beta\beta$ decay nuclear matrix elements have been calculated using different nuclear models. Estimations of $0\nu$ decay half-lives for all possible nuclei with $A \geq 70$ [Gro 86] are given in Table 5.3. Table 5.6 compares half-lives calculated with different nuclear models and assuming a neutrino mass $\langle m_\nu \rangle = 1$ eV (contributions of the right-handed currents are neglected). These half-lives are also shown in Fig. 5.11, in which $\langle m_\nu \rangle = 2.1$ eV is used. When this value is chosen the experimental lower limit of the half-life of $^{76}$Ge, $T_{1/2}^{0\nu}(^{76}\text{Ge}) > 4.7 \times 10^{23}$ years [Cal 86], is reproduced with the nuclear matrix elements of the QRPA calculation with both particle-hole and particle-particle interactions [Tom 87]. Comparing the half-lives shown in Table 5.6 and Fig. 5.11, the four shell model calculations for $^{48}$Ca yield half-lives quite consistent with each other. The shell model calculation for heavier nuclei by [Hax 84] gives half-lives similar to those of the QRPA calculation [Tom 87] for $^{76}$Ge and $^{82}$Se. The discrepancies for $^{128,130}$Te are about a factor of three. The half-lives calculated by Klapdor and Grotz [Kla 84, Gro 86] are systematically shorter than those of the QRPA and shell model calculations, probably because only Gamow-Teller and quadrupole-quadrupole correlations are taken into account.

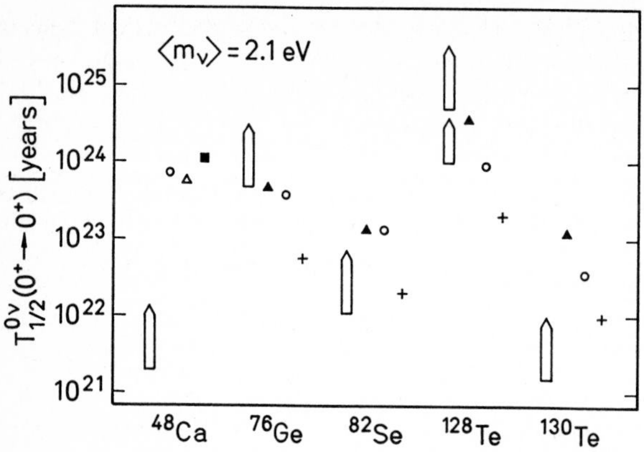


Fig. 5.11 Comparison of experimental and theoretical half-lives of $0^+ \to 0^+$ $0\nu$ $\beta\beta$ decay. The experimental lower limits are given by open arrows. For the theoretical values $\langle m_\nu \rangle = 2.1$ eV is assumed, and contributions of right-handed currents are neglected. Nuclear matrix elements are taken from [Hax 84] (open circle), [Sko 83] (open triangle), [Mut 86] (filled square), [Tom 87] (filled triangle), [Gro 86] PBCS (cross).

Table 5.6. Calculated half-lives of $0\nu$ $\beta\beta$ decay for an effective neutrino mass $\langle m_\nu \rangle = 1$ eV. Contributions of right-handed currents are neglected. Present experimental half-lives are given in the last line.

| References | $^{48}$Ca | $^{76}$Ge | $^{82}$Se | $^{128}$Te | $^{130}$Te | $^{134}$Xe | $^{136}$Xe | $^{142}$Ce |
|---|---|---|---|---|---|---|---|---|
| | $10^{24}$y | $10^{24}$y | $10^{23}$y | $10^{24}$y | $10^{23}$y | $10^{23}$y | $10^{23}$y | $10^{23}$y |
| Sko 83 | 2.6 | | | | | | | |
| Wu 85 | 3.2 | | | | | | | |
| Mut 86 | 5.4 | | | | | | | |
| Hax 84 | 3.2 | 1.8 | 5.8 | 4.0 | 1.6 | | | |
| Gro 86 | | | | | | | | |
| pairing | | 0.17 | 0.68 | 0.62 | 0.27 | | | |
| pairing + $H_{\sigma\tau}+H_{QQ}$ | | 0.26 | 0.95 | 0.98 | 0.46 | 8.7 | 3.0 | 4.7 |
| Tom 86 | | 0.91 | | | | | | |
| Tom 87 | | 2.2 | 6.0 | 9.6 | 5.2 | | | |
| Experiment | >0.002[a] | >0.47[b] | >0.11[c] | >5[d] | >0.015[d] | | >0.012[e] | |

[a]Bah 70, [b]Cal 86, [c]Ell 86, [d]Kir 86, [e]Bar 87,

Tomoda and Faessler [Tom 87] performed a QRPA calculation of the $0\nu$ decay, using a realistic interaction, the G-matrix of the Bonn one-boson exchange potential [Hol 81], and consequently taking into account all possible modes of ground state correlations. (Some details of the results are discussed in sect. 5.2.2.) However, they did not fix the strength of the particle-particle interaction. Comparing the behaviour of the $2\nu$ $\beta\beta$ decay matrix elements, which are calculated with the same interaction [Civ 87], with those of [Mut 88a,88c], in which the interaction strength is fixed to $\beta^+$

decay strength of many nuclei with $60 \le A \le 170$ (see sect. 5.1.2), we estimate the most probable value of $g_{pp}$ for the Bonn potential to be very close to unity. The coefficients $C_{xy}$ of the $0\nu$ decay rate (eqs. (3.29) and (3.31)) evaluated at $g_{pp}$ =1 are listed in Table 5.7.

Vogel and his collaborators have calculated $\beta\beta$ decay rates of the $2\nu$ and $0\nu$ (neutrino mass term) modes in the QRPA with schematic zero-range forces [Vog 86, Eng 88]. The results are qualitatively consistent with those of the QRPA calculations with realistic interactions by [Civ 87, Tom 87, Mut 88a,88c], but show a considerable suppression of $0\nu$ decay rates, compared to [Tom 87], of up to two orders of magnitude. Such long half-lives have two main origins. First, these authors obtained a stronger particle-particle interaction, by fixing its strength to $\beta^+$ decay strengths for a few semi-magic nuclei. We have seen in sect. 5.1.2 that larger $g_{pp}$ values are required for semi-magic nuclei than for open-shell nuclei (most $\beta\beta$ emitters belong to the latter class). The $0\nu$ decay matrix element $M_{GT}^{0\nu}$ of Fig. 5.12 calculated with the Bonn potential, for which $g_{pp}$ =1.0 seems to be optimum, would vanish at a slightly larger value $g_{pp} \sim 1.15$. Secondly, the use of zero-range forces has various consequences. It leads to a set of occupation amplitudes $u_j$ and $v_j$ (see sect. 4.3) which is significantly different from that of a realistic interaction, and as a result the calculated half-lives are long already at $g_{pp} = 0$. The adjustment of the zero-range force strengths, which was done for the $J^\pi = 1^+$ mode, does not necessarily mean correct force strengths for the other multipoles, which are also important for the calculation of the $0\nu$ $\beta\beta$ decay matrix element. With such adjustment the zero-range force could only reproduce the short-range character of a residual interaction, but would underestimate long-range parts. Furthermore, these authors adopted a quenched axial-vector coupling constant, $g_A = 1.0$, which additionally suppresses decay rates by $(1.254)^4 = 2.5$.

The theoretical $\beta\beta$ decay half-lives of the $2\nu$ [Mut 88a,88c] and $0\nu$ [Tom 87] modes are consistent with the phase space argument for the branching ratios of $^{128}$Te and $^{130}$Te (see sect. 5.1.1).

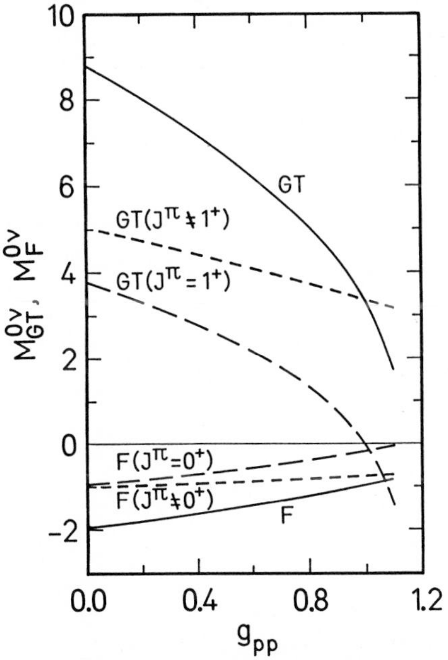


Fig. 5.12 The nuclear matrix elements (neutrino mass term) of the $0\nu$ $\beta\beta$ decay of $^{76}$Ge calculated in the proton-neutron QRPA as function of the particle-particle interaction strength (solid lines). Contributions through $1^+$ ($0^+$) intermediate states for $M_{GT}^{0\nu}$ ($M_F^{0\nu}$) and those through the other states are shown separately by long-dashed and short-dashed lines, respectively (from [Tom 87]).

Table 5.7. The coefficients [$y^{-1}$] of $0^+ \to 0^+$ $0\nu$ $\beta\beta$ decay rates calculated in the QRPA model including the particle-particle interaction with $g_{pp} = 1$ (from [Tom 87]).

| | $^{76}$Ge | $^{82}$Se | $^{128}$Te | $^{130}$Te |
|---|---|---|---|---|
| $C_{mm}$ | $1.21 \times 10^{-13}$ | $4.36 \times 10^{-13}$ | $2.73 \times 10^{-14}$ | $5.05 \times 10^{-13}$ |
| $C_{\eta\eta}$ | $3.69 \times 10^{-9}$ | $1.33 \times 10^{-8}$ | $1.18 \times 10^{-9}$ | $1.92 \times 10^{-8}$ |
| $C_{\lambda\lambda}$ | $1.74 \times 10^{-13}$ | $1.31 \times 10^{-12}$ | $7.27 \times 10^{-15}$ | $1.16 \times 10^{-12}$ |
| $C_{m\eta}$ | $2.08 \times 10^{-11}$ | $5.96 \times 10^{-11}$ | $7.58 \times 10^{-12}$ | $8.18 \times 10^{-11}$ |
| $C_{m\lambda}$ | $-6.04 \times 10^{-14}$ | $-2.68 \times 10^{-13}$ | $-6.48 \times 10^{-15}$ | $-2.88 \times 10^{-13}$ |
| $C_{\eta\lambda}$ | $-1.14 \times 10^{-13}$ | $-9.46 \times 10^{-13}$ | $-3.32 \times 10^{-15}$ | $-7.70 \times 10^{-13}$ |

Also the partial half-life $T^{0\nu}_{1/2}(^{130}\mathrm{Te}) \sim 1 \times 10^{23}$ years evaluated at $\langle m_\nu \rangle = 2.1$ eV indicates the branching ratio of the $0\nu$ mode to be at most a few percent of the total half-life from the geochemical measurement [Kir 86]. The ratio could be smaller since the effective value of the neutrino mass is taken from the experimental $0\nu$ decay half-life of $^{76}$Ge [Cal 86], which is only a lower limit.

### 5.2.2. Effects of ground state correlations

In sect. 5.1.2 it is shown that the QRPA calculation with carefully adjusted strengths of the residual interaction yields $2\nu$ $\beta\beta$ half-lives quite consistent with the existing experimental data, with a remarkable improvement over the earlier calculations. The $2\nu$ decay rates are strongly suppressed by the ground state correlations which are enhanced by the particle-particle interaction. This finding raises the question whether the matrix elements of the $0\nu$ mode are also drastically affected by the particle-particle interaction and more generally by ground state correlations.

The recent QRPA calculation by Tomoda and Faessler [Tom 87] of the $0\nu$ decay using a realistic interaction with both particle-hole and particle-particle components led to a negative answer to the question. Figs. 5.12 and 5.13 show the dependence of the matrix elements on the strength of the particle-particle interaction, for the $0\nu$ $\beta\beta$ decay of $^{76}$Ge. $M^{0\nu}_{GT}$ and $M^{0\nu}_{F}$ are relevant for the extraction of the neutrino mass $\langle m_\nu \rangle$, and $M^{0\nu}_{R}$ is most important for the determination of the mixing parameter $\langle \eta \rangle$ of the right-handed leptonic current. (These figures are reproduced from Fig. 1 of [Tom 87] by taking into account differences in the definitions of these matrix elements.)

The matrix elements $M^{0\nu}_{GT}$ and $M^{0\nu}_{F}$ (see Fig. 5.12) are not affected as much as $M^{2\nu}_{GT}$ by the inclusion of the particle-particle interaction. (A similar tendency was obtained for the quadrupole-quadrupole interaction, see sect. 5.2.1 and Table 5.6.) The reason why details of the nuclear structure affect $M^{0\nu}_{GT}$ and $M^{2\nu}_{GT}$ so differently is found in the different radial dependence of the involved transition operators:

$$M^{0\nu}_{GT} = \sum_{m,n} \langle 0^+_f \| t_{-m} t_{-n} \sigma_m \cdot \sigma_n H(|\mathbf{r}_m - \mathbf{r}_n|) \| 0^+_i \rangle \tag{5.5a}$$

$$M^{2\nu}_{GT} = \sum_{m,n} \langle 0^+_f \| t_{-m} t_{-n} \sigma_m \cdot \sigma_n \| 0^+_i \rangle \tag{5.5b}$$

Due to the neutrino potential $H(|\mathbf{r}_m - \mathbf{r}_n|)$ occuring in the $0\nu$ matrix element with its radial dependence, $\sim |\mathbf{r}_m - \mathbf{r}_n|^{-1}$, the decay of two neutrons being close together is favoured. Alternatively, the presence of the neutrino potential in the $0\nu$ decay makes transitions possible through intermediate

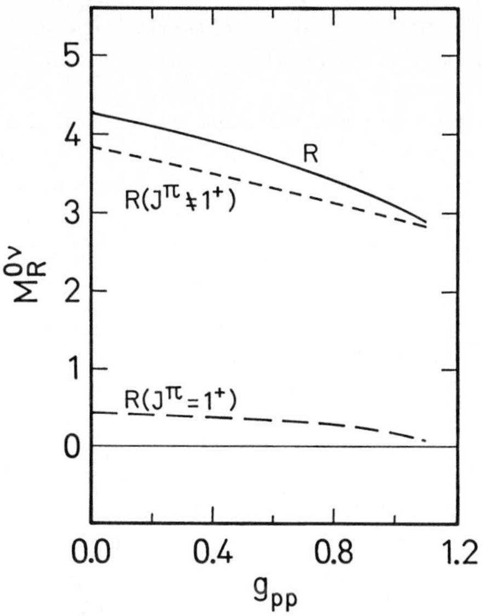


Fig. 5.13 The nucleon recoil matrix element for the $^{76}$Ge $0\nu$ decay calculated in the proton-neutron QRPA as function of the particle-particle interaction strength (solid line). This element is dominated by transitions through intermediate states with $J^\pi \neq 1^+$ (short-dashed line) while contributions through $1^+$ states are small (long-dashed line) (from [Tom 87]).

states with various multipoles ($J^\pi = 0^+$ is forbidden for $M_{GT}^{0\nu}$, and $\pi = (-1)^{J+1}$ are forbidden for $M_F^{0\nu}$), while in the case of $2\nu$ decay only $J^\pi = 1^+$ is allowed for the intermediate states. It is therefore interesting to decompose the $0\nu$ matrix element $M_{GT}^{0\nu}$ ($M_F^{0\nu}$) into contributions through $1^+$ ($0^+$) intermediate states and those through the other states. These contributions are separately plotted in Fig. 5.12 by dashed and dotted lines, respectively. The dashed line behaves quite similarly to the $2\nu$ matrix element, passing through zero at $g_{pp} \sim 1$. On the other hand, the dotted line is rather constant against a change of $g_{pp}$. Compared with the values at $g_{pp} = 0$, the matrix elements at $g_{pp} = 1$ are reduced by a factor of 2 - 3, i.e., the $0\nu$ decay rate (when right-handed current contributions are neglected) is suppressed by a factor of about 6, which is much less than in case of the $2\nu$ decay rates.

The insensitivity of $M_{GT}^{0\nu}$ to the strength of the particle-particle interaction, compared to the drastic suppression effects on the $2\nu$ decay rates, indicates negative evidence for the sometimes used procedure of scaling of $0\nu$ matrix elements. The scaling method, which was applied in order to set a conservative limit on the neutrino mass [Hax 84, Doi 85], had been used before the suppression mechanism of the $2\nu$ mode was found, and at a time when the calculations systematically overestimated the $2\nu$ $\beta\beta$ decay rates.

The behaviour of the matrix element $M_R^{0\nu}$ as a function of $g_{pp}$ is shown in Fig. 5.13. This matrix element is much less affected by the increase of the particle-particle interaction strength. In this case, the decomposition clarifies that the contribution through $1^+$ intermediate states is much smaller than that through the other multipoles. The main contribution to $M_R^{0\nu}$ comes from the zero-range part of the neutrino potential $H_{RC}(r)$, eq.(3.46). The zero-range potential favours transitions through intermediate states with high multipoles, compared to a finite range potential. — Generally, as the range of a neutrino potential becomes large, components of high momentum-transfer are suppressed, and low-multipole transitions are favoured. As the range tends to infinite, i.e. no radial dependence, which corresponds to the $2\nu$ transition and is not realized in the $0\nu$ decay, only the lowest possible multipole is allowed. — The observation that higher-multipole transitions dominate the $0\nu$ decay matrix elements and are almost not affected by the increase of $g_{pp}$, in contrast to the $2\nu$ decay matrix element, indicates that corresponding ground state correlations are not affected by the particle-particle interactions. It thus reflects basically some well-known feature of the fundamental nature of the nucleon-nucleon (proton-neutron) interaction, namely the particle-particle interaction in the $J^{\pi} = 1^{+}$ (the $0^{+}$ and highest multipolarity) channel is strongly attractive and that in other multipole channels is less attractive or slightly repulsive.

## 6. Present limits on the neutrino mass, right-handed current parameters and the majoron coupling constant from $\beta\beta$ decay

The $0\nu$ $\beta\beta$ decay rate of the $0^{+} \to 0^{+}$ transition is a quadratic function of the effective values of three quantities — the neutrino mass $\langle m_{\nu} \rangle$ and the mixing parameters of the right-handed currents $\langle \eta \rangle$ and $\langle \lambda \rangle$ (see eq.(3.29)). If the decay rate is measured by experiment, the admissible solutions for these quantities form the surface of a closed volume in a three dimensional space with one of the above quantities for one axis. Since only a lower limit of the half-life is measured, the solutions lie inside the volume. Fig. 6.1 shows cross sections of the volume for the cases when one quantity is put to zero, for the nuclear matrix elements of $^{76}$Ge decay [Tom 87] with the experimental half-life $T_{1/2}^{0\nu}(^{76}\text{Ge}) > 4.7 \times 10^{23}$ years [Cal 86]. As an example, Fig. 6.1a indicates the allowed region for $\langle m_{\nu} \rangle$ and $\langle \eta \rangle$ when $\langle \lambda \rangle = 0$. The upper limits of the effective values are summarized in Table 6.1.

Table 6.2 compares the upper limits of the neutrino mass for the nuclear matrix elements calculated with various nuclear structure models. Here, contributions of the right-handed currents are neglected (as a result the values of [Tom 87] in this table are slightly smaller than those in Table 6.1). For a given nucleus, a smaller neutrino mass indicates a larger nuclear matrix element. Among the present experimental results, the $0^{+} \to 0^{+}$ $0\nu$ decay of $^{76}$Ge and $^{128}$Te sets the smallest upper limits on the neutrino mass $\langle m_{\nu} \rangle \sim 2$ eV. It has to be kept in mind that the deduced neutrino masses are effective masses, i.e. linear combinations of mass eigenvalues of Majorana neutrinos (see eq.(3.30a)). This means that the true (electron) neutrino mass might in principle still be substantially larger than the limit for the effective mass deduced from $\beta\beta$ decay (see sect. 2). Possible interference effects between light and heavy neutrinos have been discussed in [Gro 86] (see also Langacker in this volume). In most Majorana neutrino models, however, one expects $\langle m_{\nu e} \rangle \approx m_{\nu e}$.

Haxton et al. calculated contributions of the right-handed currents [Hax 84]. Their nuclear matrix elements, for the experimental half-lives shown in Table 6.2, give effective values of $\langle \lambda \rangle$ (the

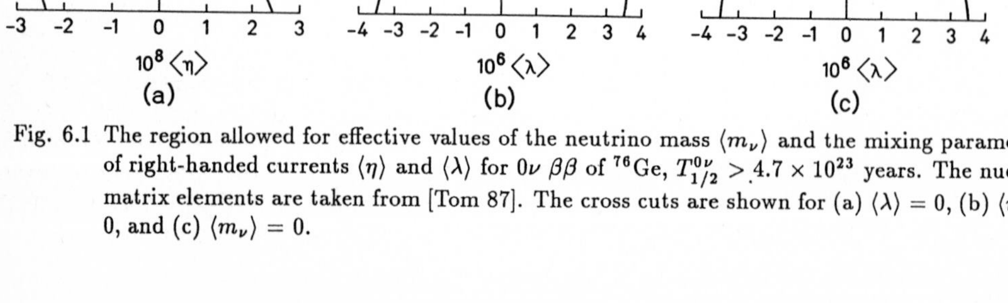

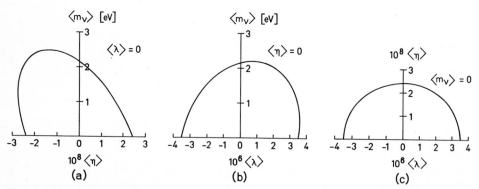

Fig. 6.1 The region allowed for effective values of the neutrino mass $\langle m_{\nu} \rangle$ and the mixing parameters of right-handed currents $\langle \eta \rangle$ and $\langle \lambda \rangle$ for $0\nu$ $\beta\beta$ of $^{76}$Ge, $T_{1/2}^{0\nu} > 4.7 \times 10^{23}$ years. The nuclear matrix elements are taken from [Tom 87]. The cross cuts are shown for (a) $\langle \lambda \rangle = 0$, (b) $\langle \eta \rangle = 0$, and (c) $\langle m_{\nu} \rangle = 0$.

Table 6.1. The upper limits of the effective values of the neutrino mass and right-handed current coupling constants deduced from the experimental lower limits of $0^+ \rightarrow 0^+$ $0\nu$ $\beta\beta$ decay half-lives given in the last line (from [Tom 87]).

| | $^{76}$Ge | $^{82}$Se | $^{128}$Te | $^{130}$Te |
|---|---|---|---|---|
| $\langle m_\nu \rangle$ [eV] | <2.5 | <8.2 | <1.9 | <21 |
| $\langle \eta \rangle$ | $<2.8 \times 10^{-8}$ | $<9.0 \times 10^{-8}$ | $<1.8 \times 10^{-8}$ | $<2.1 \times 10^{-7}$ |
| $\langle \lambda \rangle$ | $<3.6 \times 10^{-6}$ | $<8.5 \times 10^{-6}$ | $<5.5 \times 10^{-6}$ | $<2.4 \times 10^{-5}$ |
| $T^{0\nu}_{1/2}$[y] | $>4.7 \times 10^{23}$ [a] | $>1.1 \times 10^{22}$ [b] | $>5 \times 10^{24}$ [c] | $>1.5 \times 10^{21}$ [c] |

[a]Cal 86, [b]Ell 86, [c]Kir 86

Table 6.2. Upper limits of the effective neutrino mass [eV] deduced from experimental $0^+ \rightarrow 0^+$ $0\nu$ $\beta\beta$ decay half-life limits (last line) for the nuclear matrix elements of various nuclear structure calculations. Contributions of right-handed currents are neglected.

| Refs. | $^{48}$Ca | $^{76}$Ge | $^{82}$Se | $^{128}$Te | $^{130}$Te | $^{100}$Mo | $^{136}$Xe | $^{150}$Nd |
|---|---|---|---|---|---|---|---|---|
| Sko 83 | <36 | | | | | | | |
| Wu 85 | <40 | | | | | | | |
| Mut 86 | <52 | | | | | | | |
| Hax 84 | <40 | <1.8 | <7.3 | <0.90 | <10 | | | |
| Gro 86 | | <0.7 | <2.9 | <0.43 | <5.4 | <8.66* | <13.3 | <3.22* |
| Tom 86 | | <1.4 | | | | | | |
| Tom 87 | | <2.0 | <7.4 | <1.4 | <19 | | | |
| $T^{0\nu}_{1/2}$[y] | $>2\times10^{21}$ [a] | $>5.\times10^{23}$ [b] | $>1.1\times10^{22}$ [c] | $>5\times10^{24}$ [d] | $>1.5\times10^{21}$ [d] | $>4.4\times10^{20}$ [e] | $>1.7\times10^{21}$ [f] | $>2.3\times10^{21}$ [g] |

[a]Bah 70, [b]Cal 87, [c]Ell 86, [d]Kir 86, [e]Smo 88, [f]Ale 88, [g]Kli 86.
*rough estimate, see Table 5.3.

parameter for the coupling of right-handed leptonic current and right-handed hadronic current $j_R \cdot J_R^\dagger$, see eq.(3.1)) of the same order as the corresponding values deduced by [Tom 87]. However, the effective values of $\langle \eta \rangle$ (the parameter for $j_R \cdot J_L^\dagger$) of [Tom 87], which are of the order of $10^{-8}$, are smaller than those of [Hax 84] by two orders of magnitude, because the latter authors did not take into account the nucleon recoil matrix element $M_R^{0\nu}$ which is most important for the determination of $\langle \eta \rangle$.

As for $0\nu$ $\beta\beta$ decay, no evidence exists at present for neutrinoless $\beta\beta$ decay associated with majoron emission (see Figs. 3.10 and 3.11). In Table 6.3 the observed lower limits of the half-lives for this process are given together with the deduced upper limits on the coupling constant $\langle g_B \rangle$ defined in eq.(3.54). The phase space factors $G_B$ used are from [Doi 87], the $0\nu$ decay matrix elements $(M_{GT}^{0\nu} - M_F^{0\nu})^2$ are taken from [Mut 86, Tom 87, Gro 86]. The deduced upper limits of the coupling constant are of the order $10^{-5} - 10^{-4}$.

Table 6.3. Effective values of the coupling constant $<g_B>$ of the majoron to Majorana neutrinos, deduced from experimental limits on $0\nu$,B $\beta\beta$ decay half-lives. The nuclear matrix elements and the phase space factors $G_B$ are also presented.

| | $^{48}$Ca | $^{76}$Ge | $^{82}$Se | $^{128}$Te | $^{130}$Te | $^{100}$Mo | $^{150}$Nd |
|---|---|---|---|---|---|---|---|
| $T_{1/2}^{0\nu,B}$[y] | $>8.6\times10^{20}$ [a] | $>1.4\times10^{21}$ [b] | $>4.4\times10^{20}$ [c] | $>5\times10^{24}$ [d] | $>1.5\times10^{21}$ [d] | $>3.3\times10^{20}$ [e] | $>1.3\times10^{20}$ [f] |
| $(M_{GT}^{0\nu}-M_F^{0\nu})^2$ | 0.746 [g] | 19.1 [h] | 15.4 [h] | 14.9 [h] | 11.2 [h] | 173 [i] | 51.9 [i] |
| $G_B[y^{-1}]$ [j] | $3.97\times10^{-15}$ | $1.21\times10^{-16}$ | $9.98\times10^{-16}$ | $9.92\times10^{-18}$ | $1.31\times10^{-15}$ | $1.74\times10^{-15}$ | $1.03\times10^{-14}$ |
| $\vert<g_B>\vert$ | $<6.3\times10^{-4}$ | $<5.6\times10^{-4}$ | $<3.8\times10^{-4}$ | $<3.7\times10^{-5}$ | $<2.1\times10^{-4}$ | $<1.0\times10^{-4}$ | $<1.2\times10^{-4}$ |

[a] Bar 87, [b] Cal 87, [c] Ell 87, [d] Kir 86, [e] Als 88, [f] Kli 86, [g] Mut 86, [h] Tom 87, [i] Gro 86, [j] Doi 87

Bounds on neutrino masses have been estimated from various kinds of experiments — single $\beta$ decay of tritium, electron capture of $^{163}$Ho, neutrino oscillations, the solar neutrinos, heavy neutrino searches — and from cosmology. Among them, the tritium decay experiments have set the smallest upper limits on the electron neutrino mass, $m_{\nu e} < 18$ eV [Fri 86], $m_{\nu e} < 29$ eV [Bow 86]. The ITEP group reported a finite neutrino mass from the tritium decay experiment, 17 eV $< m_{\nu e} <$ 40 eV [Bor 85,87], but critical comments were made on the analysis of this experiment [Ber 85]. The time-spectrum analyses of neutrinos from the recently observed supernova explosion [Hir 87, Bio 87] resulted in bounds on the electron neutrino mass around 20 eV (see also [Abb 88]). Thus at present, the $0\nu$ $\beta\beta$ decay experiments set the most stringent limit on the electron neutrino mass. As for the right-handed currents, the effective values of the mixing parameters deduced from the $0\nu$ $\beta\beta$ decay rates are more stringent by several orders of magnitude, compared to constraints obtained from other sources such as single nuclear $\beta$ decay, or decays of muons and kaons (for a detailed discussion see [Doi 85]).

## 7. Conclusion

The discovery of a non-zero neutrino mass would be an extemely profound example of 'new physics' beyond the simplest theories of grand unification. Neutrinoless $\beta\beta$ decay is at present the most sensitive test of the existence and mass of Majorana neutrinos. The most sensitive detector experiments currently yield an upper limit for the (electron) neutrino mass of about 1 eV. However, reliable deductions of neutrino masses from measured $0\nu$ $\beta\beta$ decay rates depend decisively on reliable calculations of nuclear matrix elements. In this paper the present status of such calculations is reviewed. It is shown that the long-standing discrepancy between experimental and calculated $2\nu$ $\beta\beta$ decay rates can be resolved by including particle-particle forces in a QRPA treatment. But, this progress has to be paid for by severe limitations in the accuracy with which $2\nu$ $\beta\beta$ decay rates now can be predicted. Only lower limits can still be reliably calculated. Also, for some nuclei there are limitations of the QRPA approach. We have, furthermore, given some general commutator relation argument that the $2\nu$ decay matrix elements should not vanish and that the vanishing of the $2\nu$ matrix element in the QRPA at a certain choice of $g_{pp}$ is an artifact of that model. Calculated rates for $0\nu$ $\beta\beta$ decay are found — in contrast to the $2\nu$ case — to be comparably insensitive to details of nuclear structure (various types of ground state correlations). This is of great importance: The search for $0\nu$ $\beta\beta$ decay remains the most powerful tool for investigating the mass of the (electron) neutrino.

In view of the important role neutrinos play in the framework of theories of grand unification and for astrophysics and cosmology, this should encourage further experimental efforts along this line. Next generation $\beta\beta$ decay experiments, particularly those using enriched $^{76}$Ge [Kla 86,87, Cal 86a] and $^{136}$Xe [Ale 88, Boe 88], could explore the neutrino mass down to $\sim 10^{-1}$ eV, corresponding to a see-saw model with a right-handed Majorana mass term of about 1 TeV which may be realized in some models based on $SU(2)_L \times SU(2)_R \times U(1)$.

References


Abb 88 L.F. Abbott, A. de Rújula and T.P. Walker Nucl. Phys. **B299** (1988) 734.

Ale 88 A. Alessandrello, A. Giuliani, E. Bellotti, D. Camin, O. Cremonesi, E. Fiorini, C. Liguori, S. Ragazzi, L. Rossi, P.P. Sverzellati and L. Zanotti, Nucl. Phys. **A478** (1988) 453c.

Als 88 M. Alston-Garnjost, B. Dougherty, R. Kenney, J. Krivicich, R. Tripp, H. Nicholson, S. Sutton, B. Dieterle, J. Kang and C. Leavitt, Phys. Rev. Lett. **60** (1988) 1928.

Ana 83 N. Anantaraman, H. Toki and G.F. Bertsch, Nucl. Phys. **A398** (1983) 269.

Avi 86 F.T. Avignone, III, R.L. Brodzinski, J.C. Evans, Jr., W.K. Hensley, H.S. Miley and J.H. Reeves, Phys. Rev. **C34** (1986) 666.

Bar 70 R.K. Bardin, P.J. Gollon, J.D. Ullman and C.S. Wu, Nucl. Phys. **A158** (1970) 337.

Bar 87 A.S. Barabash, V.M. Lobashev, V.V. Kuz'minov, V.M. Novikov, B.M. Ovchinnikov and A.A. Pomansky, Pisma w JETP **45** (1987) 171; A.S. Barabash, preprint ITEP, Moscow, 1987.

Ber 85 K.E. Bergkvist, Phys. Lett. **154B** (1985) 224; Phys. Lett. **159B** (1985) 408.

Bio 87 R.M. Bionta, G. Blewitt, C.B. Bratton, D. Casper, A. Ciocio, R. Claus, B. Cortez, M. Crouch, S.T. Dye, S. Errede, G.W. Foster, W. Gajewski, K.S. Ganezer, M. Goldhaber, T.J. Haines, T.W. Jones, D. Kielczewska, W.R. Kropp, J.G. Learned, J.M. Losecco, J. Matthews, R. Miller, M.S. Mudan, H.S. Park, L.R. Price, F. Reines, J. Schultz, S. Seidel, E. Schumard, D. Sinclair, H.W. Sobel, J.L. Stone, L.R. Sulak, R. Svoboda, G. Thornton, J.C. van der Velde and C. Wuest, Phys. Rev. Lett. **58** (1987) 1494.

Boe 88 F. Boehm, P.H. Fisher, H.E. Henrikson, M.Z. Iqbal, B.M.G. O'Callahgan, J.H. Thomas, H.T. Wong, E. Bovet, L.W. Mitchell, M. Treichel, J.-C. Vuilleumier, J.-L. Vuilleumier and K. Gabathuler, Proposal E88-01.1, 1988.

Boh 69 A. Bohr and B.R. Mottelson, Nuclear Structure (Benjamin, New York, 1969) Vol. I.

Boh 81 A. Bohr and B.R. Mottelson, Phys. Lett. **100B** (1981) 10.

Bor 85 S.D. Boris, A.I. Golutvin, L.P. Laptin, V.A. Lyubimov, V.V. Nagovitsyn, E.G. Novikov, V.Z. Nozik, V.A. Soloshchenko, I.N. Tikhomirov, E.F. Tret'yakov and N.F. Myasoedov, Pisma w JETP **42** (1985) 107; S. Boris, A. Golutvin, L. Laptin, V. Lubimov, V. Nagovizin, E. Novikov, V. Nozik, V. Soloshenko and E. Tretjakov, Phys. Lett. **159B** (1985) 217.

Bor 87 S.D. Boris, A.I. Golutvin, L.P. Laptin, V.A. Lyubimov, N.F. Myasoedov, V.V. Nagovitsyn, V.Z. Nozik, E.G. Novikov, V.A. Soloshchenko, I.N. Tikhomirov and E.F. Tret'yakov, Pisma w JETP **45** (1987) 267.

Bow 86 T.J. Bowler, J.F. Wilkinson, J.C. Browne, M.P. Maley, R.G.H. Robertson, P.A. Knapp and J.A. Helffrich, Proc. Int. Symp. on Weak and Electromagnetic Interactions in Nuclei (WEIN '86), Heidelberg, July 1-5, 1986, ed.

H.V. Klapdor (Springer, Heidelberg, 1986) p. 782.
Cal 86 D.O. Caldwell, R.M. Eisberg, D.M. Grumm, D.L. Hale, M.S. Witherell, F.S. Goulding, D.A. Landis, N.W. Madden, D.F. Malone, R.H. Pehl and A.R. Smith, Phys. Rev. **D33** (1986) 2737.
Cal 86a D.C. Caldwell, Proc. 12th Int. Conf. on Neutrino Physics and Astrophysics, Sendai, 3-8 June, 1986, eds. T. Kitagaki and H. Yuta, (World Scientific, Singapore, 1986) p. 77.
Cal 87 D.O. Caldwell, R.M. Eisberg, D.M. Grumm, M.S. Witherell, Phys. Rev. Lett. **59** (1987) 419.
Cha 85 D. Chang and R.N. Mohapatra, Phys. Rev. **D31** (1985) 1718.
Chi 80 Y. Chikashige, R.N. Mohapatra and R.D. Peccei, Phys. Rev. Lett. **45** (1980) 1926.
Chi 81 Y. Chikashige, R.N. Mohapatra and R.D. Peccei, Phy. Lett. **98B** (1981) 265.
Civ 87 O. Civitarese, A. Faessler and T. Tomoda, Phys. Lett. **194B** (1987) 11.
Dav 55 R. Davis, Phys. Rev. **97** (1955) 76.
Doi 81 M. Doi, T. Kotani, H. Nishiura, K. Okuda and E. Takasugi, Phys. Lett. **103B** (1981) 219; Prog. Theor. Phys. **66** (1981) 1739, 1765.
Doi 83a M. Doi, T. Kotani, H. Nishiura and E. Takasugi, Prog. Theor. Phys. **69** (1983) 602; **70** (1983) 1353.
Doi 83b M. Doi, M. Kenmoku, T. Kotani, H. Nishiura and E. Takasugi, Prog. Theor. Phys. **70** (1983) 1331.
Doi 85 M. Doi, T. Kotani and E. Takasugi, Prog. Theor. Phys. Suppl. **83** (1985) 1.
Doi 87 M. Doi, T. Kotani and E. Takasugi, preprint OS-GE-87-07, Osaka, 1987.
Edm 57 A.R. Edmonds, Angular Momentum in Quantum Mechanics (Princeton University Press, 1957).
Ell 87 S.R. Elliott, A.A. Hahn and M.K. Moe, Phys. Rev. Lett. **59** (1987) 2020.
Eng 88 J. Engel, P. Vogel and M.R. Zirnbauer, Phys. Rev. **C37** (1988) 731.
Fer 34 E. Fermi, Z. Phys. **88** (1934) 161.
Fir 49 E.L. Fireman, Phys. Rev. **75** (1949) 323.
Fri 75 H. Fritzsch and P. Minkowski, Ann. Phys. (N.Y.) **93** (1975) 193.
Fri 86 M. Fritschi, E. Holzschuh, W. Kündig, J.W. Petersen, R.E. Pixley and H. Stüssi, Phys. Lett. **173B** (1986) 485.
Fur 39 W.H. Furry, Phys. Rev. **56** (1939) 1184.
Gel 79 M. Gell-Mann, P. Ramond and R. Slansky, in Supergravity, ed. van Nieuvenhuizen and Freeman (North-Holland, Amsterdam, 1979).
Gel 81 G.B. Gelmini and M. Roncadelli, Phys. Lett. **99B** (1981) 411.
Gel 82 G.B. Gelmini, S. Nussinov and M. Roncadelli, Nucl. Phys. **B209** (1982) 157.
Gel 83 G.B. Gelmini, S. Nussinov and T. Yanagida, Nucl. Phys. **B219** (1983) 31.
Geo 74 H. Georgi and S.L. Glashow, Phys. Rev. Lett. **32** (1974) 438.
Geo 75 H. Georgi, in Particles and Fields, ed. C.E. Carlson (AIP, New York, 1975), p. 575.
Geo 81 H. Georgi, S.L. Glashow and S. Nussinov, Nucl. Phys. **B193** (1981) 297.
Gla 61 S.L. Glashow, Nucl. Phys. **22** (1961) 579.
Goe 35 M. Goeppert-Mayer, Phys. Rev. **48** (1935) 512.
Gro 83a K. Grotz, H.V. Klapdor and J. Metzinger, J. Phys. **G9** (1983) L 169.
Gro 83b K. Grotz, H.V. Klapdor and J. Metzinger, Phys. Lett. **132B** (1983) 22.
Gro 84 K. Grotz and H.V. Klapdor, Phys. Rev. **D30** (1984) 2095.
Gro 85a K. Grotz and H.V. Klapdor, Phys. Lett. **153B** (1985) 1.
Gro 85b K. Grotz and H.V. Klapdor, Phys. Lett. **157B** (1985) 242.
Gro 86 K. Grotz and H.V. Klapdor, Nucl. Phys. **A460** (1986) 395.
Gro 88 K. Grotz and H.V. Klapdor, Die Schwache Wechselwirkung in Kern-, Teilchen- und Astrophysik, Teubner Studienbuch, 1988.

Hal 67 J.A. Halbleib and R.A. Sorensen, Nucl. Phys. **A98** (1967) 542.

Hal 83 A. Halprin, S.T. Petcov and S.P. Rosen, Phys. Lett. **125B** (1983) 335.

Hax 81 W.C. Haxton, G.J. Stephenson, Jr. and D. Strottman, Phys. Rev. Lett. **47** (1981) 153.

Hax 82a W.C. Haxton, G.J. Stephenson, Jr. and D. Strottman, Phys. Rev. **D25** (1982) 2360.

Hax 82b W.C. Haxton, S.P. Rosen and G.J. Stephenson, Jr., Phys. Rev. **D26** (1982) 1805.

Hax 84 W.C. Haxton and G.J. Stephenson, Jr., Prog. Part. Nucl. Phys. **12** (1984) 409.

Hir 87 K. Hirata, T. Kajita, M. Koshiba, M. Nakahata, Y. Oyama, N. Sato, A. Suzuki, M. Takita, Y. Totsuka, T. Kifune, T. Suda, K. Takahashi, T. Tanimori, K. Miyano, M. Yamada, E.W. Beier, L.R. Feldscher, S.B. Kim, A.K. Mann, F.M. Newcomer, R. van Berg, W. Zhang and B.G. Cortez, Phys. Rev. Lett. **58** (1987) 1490.

Hol 81 K. Holinde, Phys. Rep. **68** (1981) 121.

Kir 67 T. Kirsten, W. Gentner and O.A. Schaeffer, Z. Phys. **202** (1967) 273.

Kir 68 T. Kirsten, O.A. Schaeffer, E. Norton and R.W. Stoenner, Phys. Rev. Lett. **20** (1968) 1300.

Kir 69 T. Kirsten and H.W. Müller, Earth Planet. Sci. Lett. **6** (1969) 271.

Kir 86 T. Kirsten, E. Heusser, D. Kaether, J. Oehm, E. Pernicka and H. Richter, Proc. Int. Symp. on Nuclear Beta Decays and Neutrino, eds. T. Kotani, H. Ejiri and E. Takasugi (World Scientific, Singapore, 1986) p. 81.

Kla 84 H.V. Klapdor and K. Grotz, Phys. Lett. **142B** (1984) 323.

Kla 86 H.V. Klapdor, Proc. Int. Symp. on Nuclear Beta Decays and Neutrino, eds. T. Kotani, H. Ejiri and E. Takasugi, (World Scientific, Singapore, 1986) p. 251.

Kla 87 H.V. Klapdor, Proposal, Internal Report, MPI H - 1987 - V17.

Kli 86 A.A. Klimenko, S.B. Osetrov, A.A. Pomansky, A.A. Smolnikov and S.I. Vasilyev, Proc. Int. Symp. on Weak and Electromagnetic Interactions in Nuclei (WEIN '86), Heidelberg, July 1-5, 1986, ed. H.V. Klapdor (Springer, Heidelberg, 1986) p. 701.

Lac 80 M. Lacombe, B. Loiseau, J.M. Richard, R. Vinh Mau, J. Cote, P. Pires and R. de Tourreil, Phys. Rev. **C21** (1980) 861.

Lan 81 P. Langacker, Phys. Rep. **72** (1981) 185.

Lan 86 P. Langacker, Proc. Int. Symp. on Weak and Electromagnetic Interactions in Nuclei (WEIN '86), Heidelberg, July 1-5, 1986, ed. H.V. Klapdor (Springer, Heidelberg, 1986) p. 879.

Lan 88 P. Langacker, Proc. Workshop on 'Neutrino Physics', Heidelberg, 20-22 Oct. 1987, eds. H.V. Klapdor and B. Povh (Springer, Heidelberg, 1988).

Led 78 Table of Isotopes, 7th ed., eds. C.M. Lederer and V.S. Shirley (Wiley, New York, 1978).

Lee 56 T.D. Lee and C.N. Yang, Phys. Rev. **104** (1956) 254.

Leu 83 C.N. Leung and S.T. Petcov, Phys. Lett. **125B** (1983) 461.

Mad 87 R. Madey, B.S. Flanders, B.D. Anderson, A.R. Baldwin, J.W. Watson, S.M. Austin, C.C. Foster, H.V. Klapdor and K. Grotz, preprint, 1987.

Maj 37 E. Majorana, Nuovo Cimento **14** (1937) 171.

Man 86 O.K. Manuel, Proc. Int. Symp. on Nuclear Beta Decays and Neutrino, eds. T. Kotani, H. Ejiri and E. Takasugi, (World Scientific, Singapore, 1986) p. 71.

Mey 86 H. Meyer, Proc. Int. Symp. on Weak and Electromagnetic Interactions in Nuclei (WEIN '86), Heidelberg, July 1-5, 1986, ed. H.V. Klapdor (Springer, Heidelberg, 1986) p. 846.

Mil 76 J.E. Miller and J.E. Spencer, Ann. Phys. (N.Y.) **100** (1976) 562.

Moh 80 R.N. Mohapatra and G. Senjanovic, Phys. Rev. Lett. **44** (1980) 912.

Moh 81a R.N. Mohapatra and G. Senjanovic, Phys. Rev. **D23** (1981) 165.

Moh 81b R.N. Mohapatra and J.D. Vergados, Phys. Rev. Lett. **47** (1981) 1713.

Moh 86 R.N. Mohapatra, Unification and Supersymmetry (Springer, New York, 1986).

Moh 88 R.N. Mohapatra, Proc. Workshop on 'Neutrino Physics', Heidelberg, 20-22 Oct. 1987, eds. H.V. Klapdor and B. Povh (Springer, Heidelberg, 1988).

Mut 86 K. Muto, Proc. Int. Symp. on Nuclear Beta Decays and Neutrino, eds. T. Kotani, H. Ejiri and E. Takasugi, (World Scientific, Singapore, 1986) p. 177; Proc. Int. Symp. on Weak and Electromagnetic Interactions in Nuclei (WEIN '86), Heidelberg, July 1-5, 1986, ed. H.V. Klapdor (Springer, Heidelberg, 1986) p. 668.

Mut 88a K. Muto and H.V. Klapdor, Phys. Lett. **201B** (1988) 420.

Mut 88b K. Muto, E. Bender and H.V. Klapdor, submitted to Phys. Rev. C.

Mut 88c K. Muto, E. Bender and H.V. Klapdor, to be published.

Nie 84 J.F. Nieves, Phys. Lett. **147B** (1984) 375.

PDG86 Particle Data Group, Phys. Lett. **170B** (1986) 1.

Pet 82 S.T. Petcov, Phys. Lett. **110B** (1982) 245.

Pri 69 H. Primakoff and S.P. Rosen, Phys. Rev. **184** (1969) 1925.

Pri 81 H. Primakoff and S.P. Rosen, Ann. Rev. Nucl. Part. Sci. **31** (1981) 145.

Rei 82 D.B. Reix, Phys. Lett. **115B** (1982) 217.

Rin 80 P. Ring and P. Schuck, The Nuclear Many-Body Problem (Springer, New York, 1980).

Sal 68 A. Salam, Proc. of the Eighth Nobel Symp., ed. N. Svartholm (Almqvist and Wiksell, Stockholm, 1968) p. 367.

Sch 80 J. Schechter and J.W.F. Valle, Phys. Rev. **D22** (1980) 2227.

Sch 82 J. Schechter and J.W.F. Valle, Phys. Rev. **D25** (1982) 2951.

Sch 84a K.W. Schmid, F. Grümmer, A. Faessler, Phys. Rev. **C29** (1984) 308.

Sch 84b K.W. Schmid, F. Grümmer, A. Faessler, Nucl. Phys. **A431** (1984) 205.

Sko 83 L.D. Skouras and J.D. Vergados, Phys. Rev. **C28** (1983) 2122.

Smi 73 D. Smith, C.E. Picciotto and D. Bryman, Phys. Lett. **46B** (1973) 157.

Smo 88 A.A. Smolnikov, private communication, May 1988.

Sol 76 V.G. Soloviev, Theory of Complex Nuclei (Pergamon, Oxford, 1976).

Tak 84 E. Takasugi, Phys. Lett. **149B** (1984) 372.

Tom 85 T. Tomoda, A. Faessler, K.W. Schmid and F. Grümmer, Phys. Lett. **157B** (1985) 4.

Tom 86 T. Tomoda, A. Faessler, K.W. Schmid and F. Grümmer, Nucl. Phys. **A452** (1986) 591.

Tom 87 T. Tomoda and A. Faessler, Phys. Lett. **199B** (1987) 475.

Tom 88 T. Tomoda, preprint, 1988, submitted to Nucl. Phys. **A**.

Tsu 84 T. Tsuboi, K. Muto and H. Horie, Phys. Lett. **143B** (1984) 293.

Val 83a J.W.F. Valle, Phys. Rev. **D27** (1983) 1672.

Val 83b J.W.F. Valle and M. Singer, Phys. Rev. **D28** (1983) 540.

Ver 81 J.D. Vergados, Phys. Rev. **C24** (1981) 640.

Ver 82 J.D. Vergados, Phys. Rev. **D25** (1982) 914.

Ver 83 J.D. Vergados, Nucl. Phys. **B218** (1983) 109.

Ver 86 J.D. Vergados, Phys. Rep. **133** (1986) 1.

Vin 79 R. Vinh Mau, Mesons in Nuclei, Vol. I., eds. M. Rho and D.H. Wilkinson (North-Holland, Amsterdam, 1979) p. 151.

Vog 85 P. Vogel and P. Fisher, Phys. Rev. **C32** (1985) 1362.

Vog 86 P. Vogel and M.R. Zirnbauer, Phys. Rev. Lett. **57** (1986) 3148.

Wap 85 A.H. Wapstra and G. Audi, Nucl. Phys. **A432** (1985) 1.

Wei 64 S. Weinberg, Phys. Rev. **133B** (1964) 1318; Phys. Rev. **134B** (1964) 882.

Wei 67 S. Weinberg, Phys. Rev. Lett. **19** (1967) 1264.

Wil 82 F. Wilczek, Phys. Rev. Lett. **49** (1982) 1549.

Wol 81 L. Wolfenstein, Nucl. Phys. **B185** (1981) 147.

Wu 57 C.S. Wu, E. Ambler, R.H. Hayward, D.D. Hopper and R.P. Hudson, Phys. Rev. **105** (1957) 1413.

Wu 85 H.F. Wu, H.Q. Song, T.T.S. Kuo, W.K. Chang and D. Strottman, Phys. Lett. **162B** (1985) 227.

Wyl 83 D. Wyler and L. Wolfenstein, Nucl. Phys. **B218** (1983) 205.

Yan 79 T. Yanagida, Proc. Workshop on Unified Theory and Baryon Number in the Universe, ed. Sawada and K. Sugimoto (KEK 1979).

Zam 82 L. Zamick and N. Auerbach, Phys. Rev. **C26** (1982) 2185.

# Neutrino Oscillations in Vacuum and Matter

*S.P. Mikheyev and A.Yu. Smirnov*

Institute for Nuclear Research, Academy of Sciences of the USSR,
60th October Anniversary prospect 7a, SU-117312 Moscow, USSR

## 1. Introduction

Neutrino oscillations in vacuum have been discussed first by Pontecorvo [1]. Strong transformations of mixed neutrinos in matter [2–4] have three aspects:
1) The detection of cosmic neutrinos is known to be a unique channel of information about the structure, energetics and evolution of stars as well as galaxies and the Universe. Experimental techniques for their detection are now developed very actively and first results have been obtained. The interpretation of the data should take into account possible effects of resonant neutrino oscillations: neutrinos produced in central regions of stars cross large layers of matter and the properties of neutrino fluxes may be changed drastically.
2) Changing of neutrino fluxes due to resonant oscillations depend on the difference of mass squared ($\Delta m^2 = m_1^2 - m_2^2$) and on neutrino mixing ($\sin^2 2\theta$) in a quite definite manner. Consequently, the search for effects of resonant oscillations involves measuring $\Delta m^2$ and $\sin^2 2\theta$. Moreover, the method is sensitive to regions of $\Delta m^2$ vs. $\sin^2 2\theta$ which are far beyond the possibilities of terrestrial experiments.
3) If effects of resonant oscillations are found (or if $\Delta m^2$ and $\sin^2 2\theta$ are fixed in other experiments), then it will be possible to make definite conclusions about the density profiles which are crossed by neutrinos.

In this paper the key points of neutrino resonant oscillations are considered in terms of (a) wave packets of neutrino eigenstates and (b) flavours (flavour decomposition) of neutrino eigenstates. Experiments that measure $\Delta m^2$ and $\sin^2 2\theta$ using matter filters such as the Sun, collapsing stars and the Earth are discussed.

## 2. Neutrino Mixing and Oscillations in Vacuum

### 2.1 Mixing

Mixing seems to be a fundamental property of Nature. For particles it means that the corresponding fields appear for different aspects (interactions) in different combinations. Consider flavour – mass mixing [5]:

$$\begin{aligned} \nu_e &= (\cos\theta)\nu_1 + (\sin\theta)\nu_2 \quad , \\ \nu_\mu &= (\cos\theta)\nu_2 - (\sin\theta)\nu_1 \quad . \end{aligned} \tag{1}$$

Here $\nu_e$ and $\nu_\mu$ are combinations which have definite weak interactions. They enter definite weak charge $(V-A)$ currents and are called "the states with definite ($e$ and $\mu$) flavours". $\nu_1$ and $\nu_2$ are the combinations with definite masses (roughly, with definite Higgs interactions), $\theta$ is the mixing angle.

One (but not unique) consequence of mixing is the generation of mixed states. Conditions are realized under which only one of combinations (1) is produced. The appearance of another one is suppressed either completely or partly.

The propagation of an unmixed state is known to be described by a wave packet with definite phase- and group velocities and with spatial length ($\tau$) determined by the generation conditions (dispersion of packets will be neglected). The propagation of a mixed state is described by two [in the case (1)] wave packets, which correspond to the $\nu_1$- and $\nu_2$-components of the mixed state. If the energy of the neutrino is large ($E \gg m_1, m_2$), then the amplitudes of the $\nu_1$- and $\nu_2$-waves are fixed by the coefficients in (1). For $\nu_e$ they are $\cos\theta$ and $\sin\theta$, respectively. Moreover, the $\nu_1$- and $\nu_2$-waves are in phase ($\Delta\varphi = 0$). In the $\nu_\mu$ state, the waves $\nu_1$ and $\nu_2$ have amplitudes $\sin\theta$ and $\cos\theta$ and opposite phases: $\Delta\varphi = \pi$ (minus sign in front of $\nu_1$) as shown in Fig. 1.

So mixed $\nu_e$ and $\nu_\mu$ are distinguished by a) amplitudes of $\nu_1$- and $\nu_2$-waves (admixtures of $\nu_1$ and $\nu_2$) and b) phase differences between $\nu_1$- and $\nu_2$-waves.

In the case of maximal mixing ($\theta = 45°$) admixtures of $\nu_1$ and $\nu_2$ are equal and $\nu_e$ and $\nu_\mu$ are distinguishable by phase difference only.

Properties of the mixed $\nu$ state, for example $\nu_e$, are the following:

(1) Mass composition: mass-spectroscopy experiments with a $\nu_e$ beam give the value $m_1$ in $N \cdot \cos^2\theta$ cases and value $m_2$ in $N \cdot \sin^2\theta$ cases ($N$ is the number of measurements);

(2) Coherence: the phase difference between $\nu_1$ and $\nu_2$ components is fixed and is the same for all production processes. The wave packets with lengths $\tau \ll 4\pi E/\Delta m^2$ will be considered, so that the phase difference along packets is practically the same. The opposite inequality ($\tau \gtrsim 4\pi/\Delta m^2$) means that the conditions of $\nu$ production permit the determination of the mass of produced neutrinos ($m_1$ or $m_2$) and, consequently, the coherence is destroyed. (This situation corresponds to averaging of oscillations).

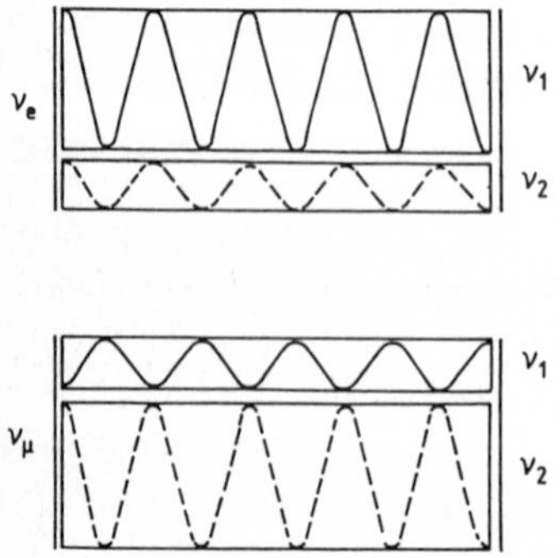


**Fig. 1.** The wave packets of mixed $\nu_e$ and $\nu_\mu$ states

## 2.2 Evolution of Mixed States. Oscillations

In the vacuum the states $\nu_1$ and $\nu_2$ with definite masses coincide with eigenstates of the Hamiltonian: $\nu_1$ and $\nu_2$ have definite energies and propagate independently. There is no $\nu_1 \leftrightarrow \nu_2$ transformation, and this determines the dynamics of evolution.

Due to the mass difference, $\nu_1$ and $\nu_2$ packets have different phase velocities ($v_{\rm ph} = E/k \simeq 1 + m^2/2k^2$), and

$$\Delta v_{\rm ph} = \frac{\Delta m^2}{2k^2} \quad , \tag{2}$$

where $k$ is the neutrino momentum. This results in a changing phase difference between $\nu_1$- and $\nu_2$-components during propagation of the mixed state:

$$\Delta\varphi(t) = \Delta v_{\rm ph} \cdot k \cdot t = \frac{\Delta m^2}{2k} \cdot t \quad , \tag{3}$$

where $t \cong x$ ($c = \hbar = 1$) is the time (distance).

The monotone increase in phase difference $\Delta\varphi$ leads to flavour oscillations in the propagating mixed neutrino. Oscillations can be traced very simply in the case of maximal mixing: $\nu_e$ and $\nu_\mu$ are distinguished by phase difference only. At the moment $t_\pi = 2\pi k/\Delta m^2$, an additional phase $\Delta\varphi = \pi$ appears in (3), and consequently $\nu_e$ transforms into $\nu_\mu$ completely. At $t_{2\pi} = 2t_\pi$, $\Delta\varphi = 2\pi$ and the mixed neutrino returns to the initial state. Thus the periodic transformations $\nu_e \to \nu_\mu \to \nu_e \to \ldots$ (oscillations) take place.

The distance over which the neutrino returns to the initial state is called the oscillation length. From the condition $\Delta\varphi(l_\nu) = 2\pi$ and (3), one has

$$l_\nu = \frac{4\pi k}{\Delta m^2} \quad . \tag{4}$$

## 2.3 Flavour Decomposition of $\nu$ Eigenstates. Parameters of Oscillations

In the general case (including both maximal and nonmaximal mixing) oscillations are the interference effect in $\nu_1$- and $\nu_2$-waves. The interference takes place between "parts" of the waves which have the same flavour. So to find the parameters of $\nu$ oscillations one needs to perform flavour decomposition of $\nu_1$ and $\nu_2$ waves: Relations (1) can be rewritten in the form

$$\begin{aligned} \nu_1 &= (\cos\theta)\nu_e - (\sin\theta)\nu_\mu \quad , \\ \nu_2 &= (\cos\theta)\nu_\mu + (\sin\theta)\nu_e \quad . \end{aligned} \tag{5}$$

This determines the flavours, that is $\nu_e$-, $\nu_\mu$-content of the $\nu$ eigenstates (mass states). The probability amplitude of finding $e$-flavour in $\nu_1$ is $\cos\theta$, and so on. According to (5), a $\nu_1$ wave can be decomposed into two waves with definite flavours and amplitudes, and

$$\nu_{1e} : \nu_{1\mu} = \cos\theta : \sin\theta \quad .$$

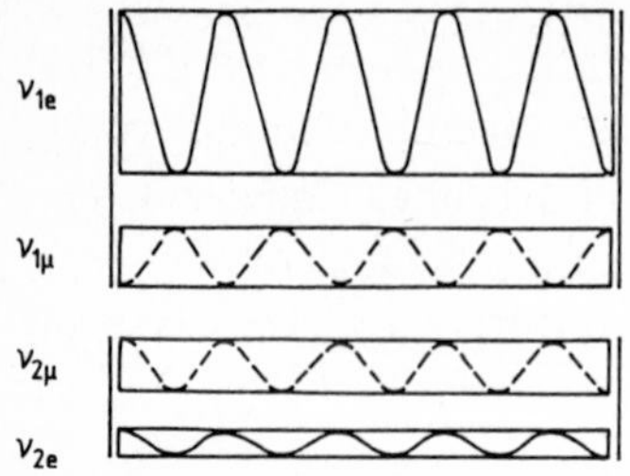


**Fig. 2.** The flavour decomposition of $\nu_1$ and $\nu_2$ for the $\nu_e$ state. Solid and dotted lines are waves with pure $e$- and $\mu$-flavours, respectively

Performing the flavour decomposition of $\nu_1$ and $\nu_2$ in the $\nu_e$ state leads to a description in terms of four waves, $\nu_{1e}$, $\nu_{1\mu}$, $\nu_{2e}$, $\nu_{2\mu}$, with definite flavours and phase velocities (see Fig. 2). From (1) and (5) one obtains the amplitudes of these waves as

$$\cos^2\theta, \quad \cos\theta\cdot\sin\theta, \quad \sin^2\theta, \quad \cos\theta\cdot\sin\theta \quad , \tag{6}$$

respectively. The phase velocities are

$$v_{\mathrm{ph}}(\nu_{1e}) = v_{\mathrm{ph}}(\nu_{1\mu}) = v_1 \quad ,$$
$$v_{\mathrm{ph}}(\nu_{2e}) = v_{\mathrm{ph}}(\nu_{2\mu}) = v_2 \quad .$$

Waves $\nu_{1e}$, $\nu_{2e}$ and $\nu_{2\mu}$ are in phase, while $\nu_{1\mu}$ has an opposite phase.

Interference of waves with the same flavour, that is $\nu_{1e}$ and $\nu_{2e}$ ($\nu_{1\mu}$ and $\nu_{2\mu}$), gives the oscillations of $e-$ ($\mu-$) flavour in the initially produced $\nu_e$-state.

Consider the oscillations of $e$ flavour. At the moments $t = l_\nu \cdot n$ ($n = 0, 1, 2\ldots$), $\nu_{1e}$ and $\nu_{2e}$ are in phase; constructive interference takes place and the total amplitude for $\nu_e$ is

$$A = \cos^2\theta + \sin^2\theta = 1$$

[see (6).] The corresponding probability is then

$$P^{\max} = |A|^2 = 1 \quad .$$

At times $t = l_\nu \cdot (1/2{+}n)$, $\nu_{1e}$ and $\nu_{2e}$ are in opposite phase (destructive interference) and the total amplitude is

$$A = \cos^2\theta - \sin^2\theta = \cos 2\theta \quad .$$

The probability of finding $\nu_e$ is now

$$P^{\min} = \cos^2 2\theta \quad .$$

So, the depth of oscillation is

$$A_p = P^{\max} - P^{\min} = \sin^2 2\theta \quad , \tag{7}$$

the averaged probability is

$$\overline{P} = \tfrac{1}{2}(P^{\max} + P^{\min}) = 1 - \tfrac{1}{2}\sin^2 2\theta \quad , \tag{8}$$

and the probability $P$ at an arbitrary time can be written as

$$P(t) = \overline{P} + \frac{1}{2} \cdot A_p \cdot \cos \frac{2\pi t}{l_\nu} \quad . \tag{9}$$

Here $l_\nu$ does not depend on mixing and has been determined in (4).

### 2.4 Remarks: From Vacuum to Matter

We summarize the results on vacuum oscillations in general form, which then permits the inclusion of matter effects:
1) Oscillations take place in states which do not coincide with eigenstates of the Hamiltonian ($\nu^H$). An oscillating state is the mixture of two (or more) $\nu^H$ eigenstates. Oscillations are the measure of incoincidence of a given $\nu$ state with the Hamiltonian eigenstate. The neutrino oscillates "around" the $\nu$-eigenstate.
2) In a given $\nu$ state the physical value $\hat{F}$ will oscillate, when the corresponding eigenstates $\nu_F$ do not coincide with eigenstates of the Hamiltonian.
3) To find the oscillation parameters one should represent the given $\nu$ state in terms of eigenstates of the Hamiltonian; perform flavour decomposition of $\nu_i^H$; consider the interference of waves with the same flavour from different $\nu_i^H$.
4) The vacuum is a special case: the eigenstates of the Hamiltonian are the states with definite masses, $\nu_i^H - \nu_i$. Flavours of $\nu_i^H$ are conserved and are fixed by $\theta$. Admixtures of $\nu_i^H$ in given $\nu(t)$ are conserved, too. If, initially, $\nu(0)$ coincides with flavour state $\nu_f$, then the admixtures are fixed by $\theta$. As a consequence the average probability and depth of oscillations are conserved: $\overline{P} = \overline{P}(\theta)$, $A_p = A_p(\theta) = \text{const.}$

## 3. Neutrino Oscillations in Matter

### 3.1 Refraction Indexes and Refraction Length. Effective Density

In matter, neutrino waves undergo scattering and absorption. If the width of the matter is sufficiently small, i.e. $d \ll l_a$ ($l_a$ is the absorption length), and the neutrino energy is small too, $s \ll G_F^{-1}$ ($s$ is the total centre of mass energy squared, $G_F$ is the Fermi constant), then the main effect is elastic forward-scattering, and the scattered waves sum coherently. As is well known from optics, the summation of such waves is equivalent to the appearance of a refraction index in the initial waves [1]:

$$(n_\alpha - 1) = \sum_i \frac{f_\alpha^i(0) \cdot N_i}{k^2} \quad , \tag{10}$$

$\alpha = e, \mu$. Here $f_\alpha^i(0)$ is the amplitude of $\nu_\alpha$ forward-scattering on the $i$-component of the matter, $N_i$ is the concentration of the $i$-component.

The refraction indexes change the phase velocities. If matter is nonsymmetric with respect to $\nu_e$ and $\nu_\mu$ (so that $n_e \neq n_\mu$), then an additional phase difference appears:

$$\Delta\varphi_m = k(n_e - n_\mu) \cdot t = \sum_i \frac{\Delta f^i(0) \cdot N_i}{k} \cdot t \quad . \tag{11}$$

There are no muons in most mediums, and this is the origin of $e - \mu$ nonsymmetry. Due to charged currents, $\nu_e$ and $\nu_\mu$ scatter on electrons differently:

$$f_e^e(0) - f_\mu^e(0) = \sqrt{2} G_F k \quad ,$$

and from (11) it follows that

$$\Delta\varphi_m = \sqrt{2} G_F N_e \cdot t \quad . \tag{12}$$

The distance over which an additional phase difference of $2\pi$ occurs is called "the refraction length" [2]. From (12) one has

$$l_0 = 2\pi (G_F N_e \sqrt{2})^{-1} \quad , \tag{13}$$

where $l_0$ determines the scale for which matter effects may become essential. The corresponding width of matter is

$$d_0 = l_0 \cdot \varrho \simeq 2\pi m_N / \sqrt{2} G_F \simeq 3.5 \cdot 10^9 \text{ g/cm}^2$$

($m_N$ is the nucleon mass). Its large value determines the field of applications, i.e. neutrino astrophysics and neutrino geophysics.
The refraction length may be rewritten as

$$l_0 = \frac{2\pi m_N}{G_F \varrho^{\text{eff}} \sqrt{2}} \quad ,$$

where $\varrho^{\text{eff}}$ is the effective density. For the $\nu_e - \nu_\mu$ channel, $\varrho^{\text{eff}} = \varrho Y_e$, where $Y_e$ is the number of electrons per nucleon. In general,

$$\varrho^{\text{eff}} = \sum_i \frac{\Delta f_i(0)}{G_F k \sqrt{2}} \cdot Y_i \quad ,$$

and the specification of a definite channel is contained in $\varrho^{\text{eff}}$.

### 3.2 Matter Effects and Oscillations

Inequality $f_e(0) \neq f_\mu(0)$ has two crucial consequences:
1) Appearance of an additional phase difference. This changes the oscillation length.
2) Transitions between neutrino mass states: $\nu_1 \leftrightarrow \nu_2$. Indeed, using (1) it can be shown that the amplitude of the $\nu_1 \leftrightarrow \nu_2$ transition is

$$f_{12}(0) = \sin\theta \cdot \cos\theta \cdot [f_e(0) - f_\mu(0)] \quad .$$

This means that mass states ($\nu_i$) are no longer eigenstates of the Hamiltonian in matter they have no definite phase or group velocities, and, moreover, the mass composition of the neutrino state may oscillate itself.

Eigenstates $\nu_{im}$ of the Hamiltonian in matter differ from $\nu_i$. The $\nu_{im}$ have definite phase velocities. Consequently, to find flavour oscillations, one should introduce neutrino mixing with respect to these new eigenstates $\nu_{im}$. The angle $\theta_m$ [which relates $\nu_e$ and $\nu_\mu$ with $\nu_{1m}$ and $\nu_{2m}$ by equations similar to (1)] is called the mixing angle in matter. $\theta_m \neq \theta$, i.e. matter changes the mixing, and $\nu_e$ and $\nu_\mu$ will oscillate "around" $\nu_{im}$. To find $\nu_{im}$ (in terms of $\nu_f$) the evolution equations and Hamiltonian for neutrinos in matter should be considered.

### 3.3 Evolution Equations, Hamiltonian and Eigenstates of the Hamiltonian in Matter

In matter, the phase difference is changed due to two reasons:
1) difference of masses of $\nu_1$ and $\nu_2$
2) difference of interactions of $\nu_e$ and $\nu_\mu$.

Both reasons can be taken into account in differential equations [2, 6]. In the vacuum, states with definite masses $\boldsymbol{\nu} = (\nu_1, \nu_2)$ propagate independently and the evolution equations for them $[i\dot{\nu}_i = E_i\nu_i \simeq (k + m_i^2/2k)]$ are

$$i\frac{d}{dt}\boldsymbol{\nu} = \begin{bmatrix} \frac{m_1^2}{2k} & 0 \\ 0 & \frac{m_2^2}{2k} \end{bmatrix} \boldsymbol{\nu} \quad . \tag{14}$$

The momentum term influences the phases of $\nu_i$ equally and therefore is omitted in (14). Substituting (5) in (14) we obtain the evolution equations for flavour wave functions, $\boldsymbol{\nu}_f = (\nu_e, \nu_\mu)$,

$$i\frac{d}{dt}\boldsymbol{\nu}_f = \hat{H}_f \cdot \boldsymbol{\nu}_f \quad , \tag{15}$$

with

$$\hat{H}_f = \frac{\Delta m^2}{2k}\begin{bmatrix} \cos 2\theta & -\sin 2\theta \\ -\sin 2\theta & -\cos 2\theta \end{bmatrix} \quad .$$

In matter, phases of $\boldsymbol{\nu}_f$ are changed due to the appearance of refraction indexes

$$d\boldsymbol{\nu}_f = -ik(n_f - 1)\boldsymbol{\nu}_f \cdot dt \equiv -iW_f \cdot \boldsymbol{\nu}_f \cdot dt \quad . \tag{16}$$

Adding these $d\boldsymbol{\nu}_f$ to (15), one finally has

$$i\frac{d}{dt}\boldsymbol{\nu}_f = \hat{H}\boldsymbol{\nu}_f \quad ,$$

$$\hat{H} = \hat{H}_f + \hat{W} = \begin{bmatrix} H_e & \overline{H}/2 \\ \overline{H}/2 & H_\mu \end{bmatrix} \quad . \tag{17}$$

Here $\hat{W} = \mathrm{diag}\,(W_e, W_\mu)$ and the elements of the Hamiltonian $\hat{H}$ are

$$H(\varrho) \equiv H_e - H_\mu = \frac{\Delta m^2}{2k}\cos 2\theta + \frac{\sqrt{2}G_F\varrho}{m_N} \quad , \quad \overline{H} = -\frac{\Delta m^2}{2k}\sin 2\theta \quad . \tag{18}$$

The eigenstates of the Hamiltonian $\nu_{im}$ are determined in the standard way from the diagonalization condition:

$$\boldsymbol{\nu}_f = \hat{S}(\theta_m)\cdot\boldsymbol{\nu}_m = \begin{bmatrix} \cos\theta_m & \sin\theta_m \\ -\sin\theta_m & \cos\theta_m \end{bmatrix}\boldsymbol{\nu}_m \quad ,$$

$$\hat{S}^+(\theta_m)\hat{H}\hat{S}(\theta_m) = \hat{H}_m^{\rm diag} = {\rm diag}(H_1^d,\, H_2^d) \quad . \tag{19}$$

$H_i^d$ are the eigenvalues of the Hamiltonian (energy levels in the system of two neutrinos). The mixing angle in matter ($\theta_m$) can be found from (19) and (17):

$$\tan 2\theta_m = \frac{\overline{H}}{H} \quad , \tag{20}$$

where the states $\nu_{im}$, the angle $\theta_m$ and $H_i^d$ are the analogies of $\nu_i$, $\theta$, and $m_i^2/2k$ for matter.

### 3.4 Oscillations in Matter. The Case of Constant Density

Propagation of mixed $\nu$-states is described by two wave packets, corresponding to $\nu_{1m}$ and $\nu_{2m}$. To find the characteristics of these oscillations, we perform flavour decomposition of the eigenstates.

In matter with constant density, the equations for $\nu_{im}$ are split:

$$i\dot{\nu}_{im} = H_i^d \nu_{im} \quad .$$

The $\nu_{im}$ evolve independently; there are no $\nu_{1m} \leftrightarrow \nu_{2m}$ transitions, and both flavours and admixtures of $\nu_{im}$ are determined by $\theta_m$. This means that the picture of $\nu$-oscillations is similar to that in the vacuum, but the parameters of the oscillations differ from vacuum ones. In $A_p$ and $\overline{P}$ (7, 8), $\theta$ should be replaced by $\theta_m$. The $\nu_{im}$ have definite phase velocities $v_{\rm ph} = H_i^d/k$, and the oscillation length is

$$l_m = \frac{2\pi}{\Delta H^d} = \frac{2\pi}{|H_1^d - H_2^d|} \quad . \tag{21}$$

### 3.5 Oscillations in Matter with Varying Density

Both $\theta_m$ and $H_i^d$ depend on density. Consequently, in matter with varying $\varrho$, both admixtures and flavours of neutrino eigenstates are changed. This results in changing (with time) $\overline{P}$, $A_p$ and $l_m$. There are two key points, which determine the oscillation picture in matter [3, 4]:

1) Resonance (or resonances, if the number of neutrinos is more than two) in the neutrino system.
2) Adiabaticity.

### 3.6 Resonance

Parameter $\sin^2 2\theta_m$ (which determines the depth of oscillations in the case $\varrho$ = const) can be written as

$$\sin^2 2\theta_m = \sin^2 2\theta \cdot R(\varrho) \quad ,$$

where $R(\varrho)$ is the resonant factor (see Fig. 3a). At

$$\varrho_R = -\frac{m_N \cos 2\theta \Delta m^2}{2\sqrt{2} G_F E} \quad , \tag{22}$$

$R(\varrho)$ reaches the maximum $R(\varrho_R) = (\sin 2\theta)^{-1}$, so that $\sin^2 2\theta_m(\varrho_R) = 1$. Equation (22) is called the resonant condition (see [3]) and the $\varrho_R$ (or $E_R$) which satisfy it are called the resonant densities (resonant energies). The resonance half width at half height is

$$\Delta\varrho_R = \varrho_R \tan 2\theta \quad . \tag{23}$$

The energy level difference at resonance is minimal (see Fig. 3b): $\Delta H^d = \Delta H_0 \sin 2\theta$, $(\Delta H_0 = \Delta m^2/2k)$ [8,9].

The physical sense of the resonance is described by the following:
1) At resonance, mixing is maximal and thus $\theta_m = 45°$. That is, in matter with resonant density the oscillations have maximal depth $A_p = 1$. If a neutrino beam with a continuous energy spectrum propagates through matter with a constant density,

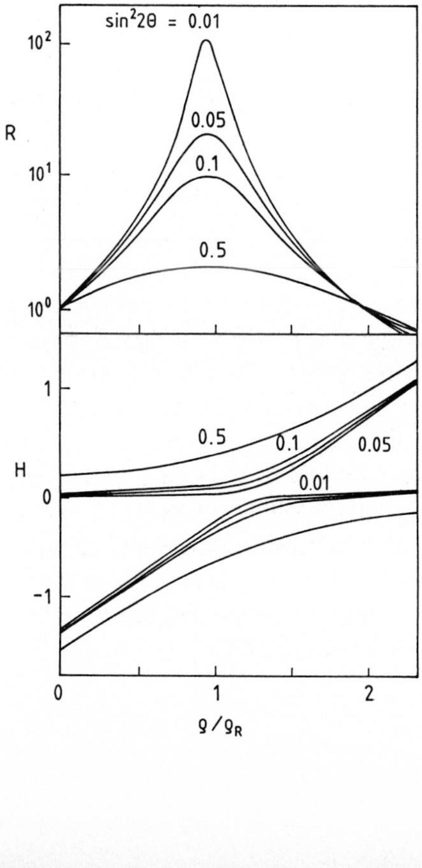


**Fig. 3. a)** Dependence of the resonant factor on effective density of matter for different values of $\sin^2 2\theta$. **b)** Dependence of the level energies in units $\Delta m^2 \cdot (2k \cos 2\theta)^{-1}$ on effective density for different $\sin^2 2\theta$

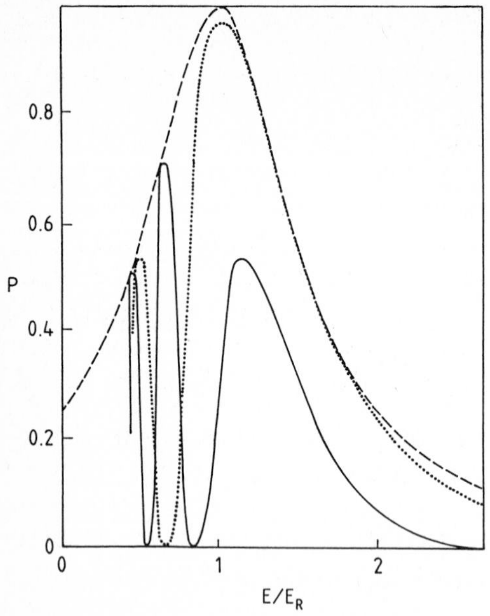


**Fig. 4.** Resonant enhancement of oscillations. The dependence of $\nu_e \to \nu_\mu$ probability in layers of matter with constant density and width $d$ on energy. The solid line corresponds to $d/l_0 = 1.8$, the dotted line to $d/l_0 = 0.9$, where $l_0$ is the refraction length and $\sin^2 2\theta = 0.25$. The dashed line is the dependence of $\sin^2 2\theta$ on $E/E_R$

then in the interval $(E_R - \Delta E_R) - (E_R + \Delta E_R)$ (where $\Delta E_R = E_R \tan 2\theta$) resonant enhancement of $\nu$-oscillations takes place. At the exit one has the oscillating curve shown in Fig. 4.

2) At resonance, for small $\theta$

$$l_\nu \cong l_0 \quad , \tag{24}$$

that is, the eigenfrequency of system (mixed neutrinos) $f_{\text{sys}} = 1/l_\nu$ coincides with the eigenfrequency of the external medium $f_{\text{ext}} = 1/l_0$.

3) At resonance

$$H_e = H_\mu \quad , \tag{25}$$

i.e. diagonal elements of the evolution matrix are equal. $H_e$ and $H_\mu$ are the energy levels of $\nu_e$ and $\nu_\mu$ in the case of zero mixing. So at resonance, energy levels coincide. Equality (25) has a very simple interpretation, which is based on the following analogy: The system of two mixed neutrinos $\nu_e$ and $\nu_\mu$ is similar to that of coupled oscillators. The effect of matter is equivalent to the changing of eigenfrequencies of these oscillators; moreover the change is different for $\nu_e$ and $\nu_\mu$. At $\varrho = \varrho_R$ the eigenfrequency of $\nu_e$ coincides with that of $\nu_\mu$, and consequently the vibration of one oscillator may be passed to another one completely.

In matter with varying density, $\varrho_R$ and $\Delta\varrho_R$ determine the density and the scale of densities, for which strong conversion of neutrinos takes place. The strongest transformations occur in the resonant layer $(\varrho_R - \Delta\varrho_R) - (\varrho_R + \Delta\varrho_R)$.

The conversion is basing essentially on the following points.

### 3.7 Flavour Changing of Neutrino Eigenstates

The mixing angle $\theta_m$ determines (as $\theta$ does in the vacuum) flavours of neutrino eigenstates. From (20), (18) and (22) one has

$$\tan 2\theta_m = \frac{\tan 2\theta}{1 - \varrho/\varrho_R} \quad . \tag{26}$$

When propagating in matter with varying $\varrho$, the $\nu_{im}$ change their flavours. If the density varies from $\varrho \gg \varrho_R$ to zero, then, according to (26), $\theta_m$ diminishes from $\pi/2$ to the vacuum value $\theta$. For small $\theta$ this means that the flavours of neutrino eigenstates change almost completely. This is illustrated in Fig. 5.
In resonance both flavours are represented in $\nu_{im}$ equally.

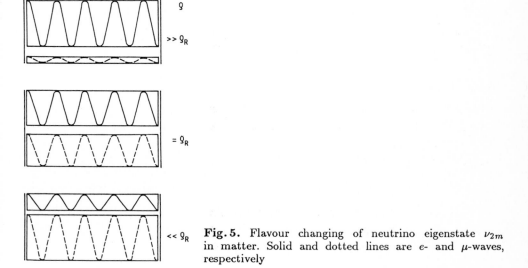

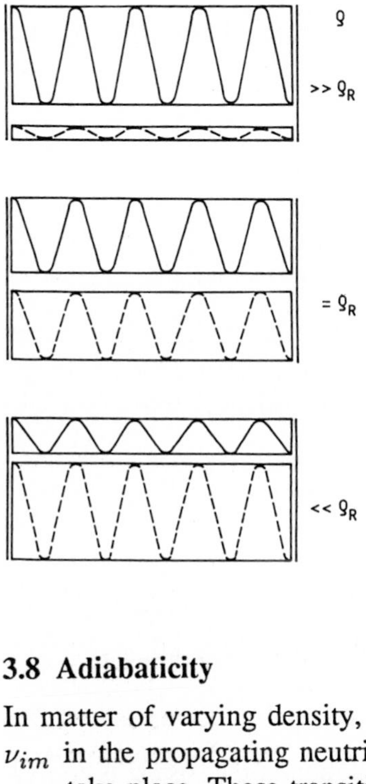


**Fig. 5.** Flavour changing of neutrino eigenstate $\nu_{2m}$ in matter. Solid and dotted lines are $e$- and $\mu$-waves, respectively

### 3.8 Adiabaticity

In matter of varying density, not only flavours of $\nu_{im}$, but also the admixtures of $\nu_{im}$ in the propagating neutrino are changed, that is, transitions between $\nu_{1m}$ and $\nu_{2m}$ take place. These transitions are governed by an adiabaticity condition, which is a condition for the rapidity of density changing.

Adiabaticity means that the system (mixed neutrino state) has time to be arranged to match changing external conditions (density). Concrete realization of adiabaticity for the mixed neutrino case is the following: If the density varies sufficiently slowly (see below), then the adiabaticity condition is fulfilled and the changing of admixtures of $\nu_{im}$ in $\nu(t)$ can be neglected. So in the adiabaticity regime, admixtures of $\nu_{im}$ in $\nu(t)$ are conserved and the flavours of $\nu_{im}$ are unambiguous functions of instantaneous values of $\varrho$. Consequently, properties of the $\nu$-state will follow changes in density, and the average probability $\overline{P}$ and the depth of oscillations $A_p$ turn out

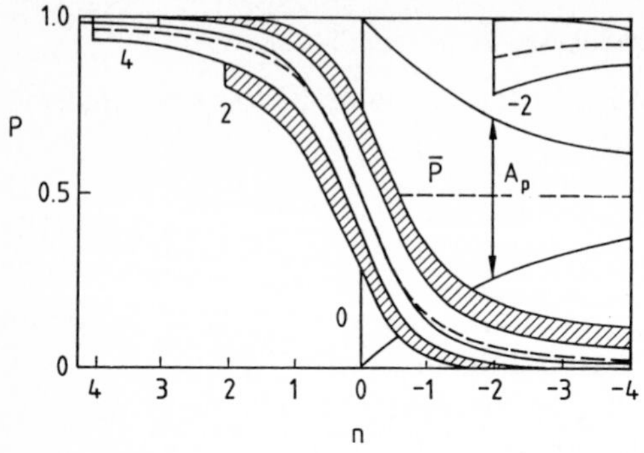


**Fig. 6.** The dependence of the average value of probability $\overline{P}$ and the depth of oscillations on $n = (\varrho - \varrho_R)/\Delta\varrho_R$ for different initial conditions $n_0$ in the adiabatic regime

to be functions of instantaneous values of $\varrho$ (see Fig. 6). [The admixtures of $\nu_{im}$ in $\nu(t)$ are fixed initially].

The adiabaticity condition has a very simple form at resonance (moreover it is at resonance that this condition is the most crucial):

$$2\Delta r_R \geq l_m^R \quad , \tag{27}$$

where

$$\Delta r_R = \left|\frac{d\varrho}{dr}\right|^{-1} \cdot \Delta\varrho_R = h \cdot \tan 2\theta$$

($h^{-1} = d \ln \varrho/dr$) is the spatial half width of the resonant layer, $l_m^R$ is the matter oscillation length at resonance. According to (27), at least one oscillation length should be obtained in the resonant layer. Condition (27) permits the introduction of the parameter of adiabaticity

$$\kappa_R = \frac{\Delta r_R}{l_m^R} = \frac{h \cdot \sin^2 2\theta}{l_\nu \cdot \cos 2\theta} \quad , \tag{28}$$

which determines the degree of adiabaticity violation. At $\kappa_R \geq 0.5$ oscillations occur in the adiabatic regime.

In the general case (for arbitrary values of $\varrho$) the adiabatic condition is

$$\dot{\theta}_m \ll \dot{\varphi}^\alpha \equiv \Delta H^d \quad , \tag{29}$$

which has a very simple geometrical interpretation. Evolution of the $\nu$-state is equivalent to a rotation of unit vector $\boldsymbol{\nu}$ in the complex space $\{\nu_e^R, \nu_e^{\mathrm{Im}}, \nu_\mu^R\}$ on the surface of the cone with axis $\nu_{1m}^R$ and cone angle $\theta_a$. The angle $\theta_a$ determines admixtures of $\nu_{im}$ in $\nu$, so that $\cos\theta_a$ gives the amplitude of a $\nu_{1m}$ wave, and $\sin\theta_a$ the amplitude of $\nu_{2m}$ (in vacuum $\theta_a = \theta$). This is shown in Fig. 7. The position of the cone is fixed by mixing angle $\theta_m$. Condition (29) means that the $\boldsymbol{\nu}$ vector should rotate around $\boldsymbol{\nu}_{1m}$ much faster than the axis of the cone does in flavour space.

If the density changes rapidly, then transitions $\nu_{1m} \leftrightarrow \nu_{2m}$ take place. Suppose the neutrino is on a definite level $H_2^d$ at $\varrho \gg \varrho_R$ (see Fig. 3). If the density diminishes slowly (adiabatically), then the system "inscribes" in the bend near resonance and

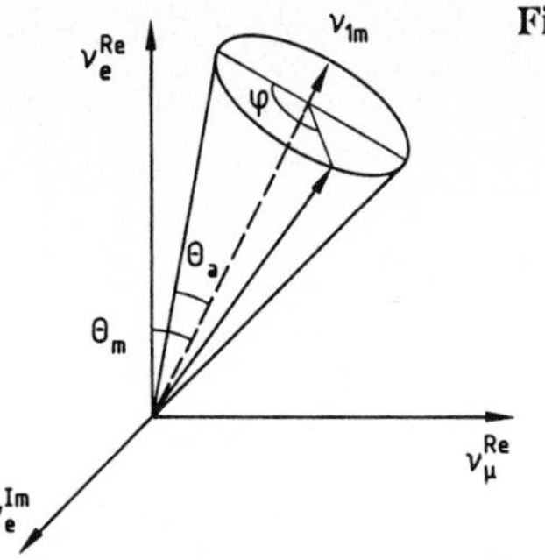


**Fig. 7.** Graphic representation of $\nu$ oscillations in matter

will stay on the same level $H_2^d$ [7, 8]. But if the density decreases rapidly (the system moves quickly), then with appreciable probability it jumps to the level $H_1^d$.

A naive estimate of the probability of the $\nu_{2m} \leftrightarrow \nu_{1m}$ transition ($P_{21}$) can be performed in the following way. The "kinetic" energy $E_K$ of the moving $\nu$-system along trajectory $H(\varrho)$ is proportional to $d\varrho/dt$, the scale of $\nu$ changing is determined by $\Delta\varrho_R$. Consequently from dimension consideration one has $E_K \sim (d\varrho_R/dt)/\Delta\varrho_R$. The smallest energy gap between levels is $U = \Delta H_R^d = 2\pi/l_m$. If $E_K > U$, then with $P_{12} \simeq 1$ the system goes from $H_2^d$ to $H_1^d$. If $E_K$ decreases, then $P_{21}$ decreases exponentially, i.e., $P_{21} = \exp(-CU/E_K)$, $C$ = const. A more careful consideration gives $C = \pi$ (integration should be done over the trajectories), and consequently [9, 10]

$$P_{21} = \exp\left(-\pi\frac{U}{E_K}\right) = \exp\left(-\pi^2\kappa_R\right) \quad . \tag{30}$$

The probability is determined by the parameter of adiabaticity. The expression (30) coincides with the well-known Landau-Zener result for the transition probability between two levels if the perturbation changes linearly with time [9, 10].

So, in the case of slow variations of density the neutrino state does not change the admixtures of $\nu_{im}$, but it changes its flavour according to changing $\nu$ eigenstates flavours. The system follows density variations. If the density varies very rapidly, then, conversely the neutrino state changes admixtures of $\nu_{im}$, but does not change its flavour. In the intermediate case, both admixtures of eigenstates $\nu_{im}$ and its flavours vary, so that $\nu(t)$ partly follows the density.

The strongest transformations of $\nu$ flavour take place in the adiabatic regime, when the initial density is large, i.e., $\varrho_0 \gg \varrho_R$. In this case $\theta_m^0 \cong \pi/2$ and the initial $\nu$ state (more definitely, $\nu_e$) coincides with $\nu_{2m}(t)$ throughout the propagation. If the final density is $\varrho_f \ll \varrho_R$, then at the exit one has $\nu \simeq \nu_{2m} \approx \nu_2$. Because the probability of finding $\nu_e$ in $\nu_2$ is $\sin^2\theta$, the suppression of the initial $\nu_e$ flux will be

$$P(\nu_e \to \nu_e) = \sin^2\theta \quad . \tag{31}$$

The smaller the vacuum mixing angle, the stronger the transformation that may be achieved compared with the vacuum case.

Note that when $\nu \approx \nu_{2m}$, the oscillation depth is practically zero (oscillations are the measure of incoincidence of a given $\nu$ with $\nu_{im}$). In the considered case the oscillation-free transformation of one neutrino type into another is realized. (For more details see [4] and references therein.)

# 4. Neutrino-Matter Oscillations Experiments

## 4.1 General Properties

The experimental scheme includes (a) a source of mixed neutrino, (b) a matter filter with known density distribution $\varrho(r)$ and (c) a neutrino detector (Fig. 8). The matter filter distorts the initial $\nu$ spectrum $F_0(E)$:

$$F(E) = P(E, \Delta m^2, \sin^2 2\theta, \varrho(r)) \cdot F_0(E) \quad ,$$

where $P$ is the suppression factor (distortion). By measuring the spectrum $F(E)$ at the exit one can extract the information on $\Delta m^2$ and $\sin^2 2\theta$.

For practical purposes two types of matter filters are interesting: (a) a filter with constant density (the density distribution in the Earth may be approximated by several layers with $\varrho = \text{const}$);
b) a filter with monotonous decreasing density (the Sun, collapsing stars). These density distributions are shown in Fig. 9.

Fig. 8

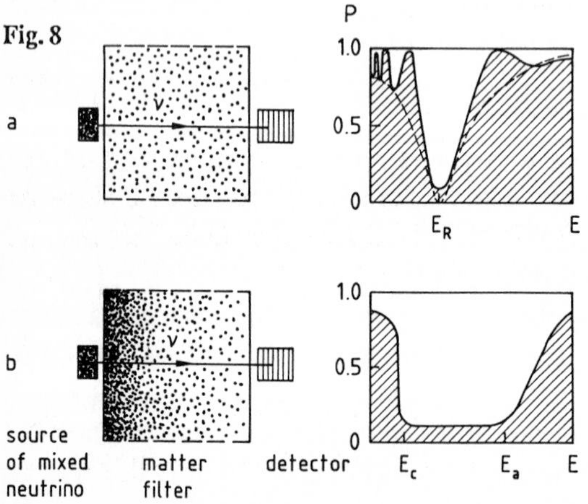


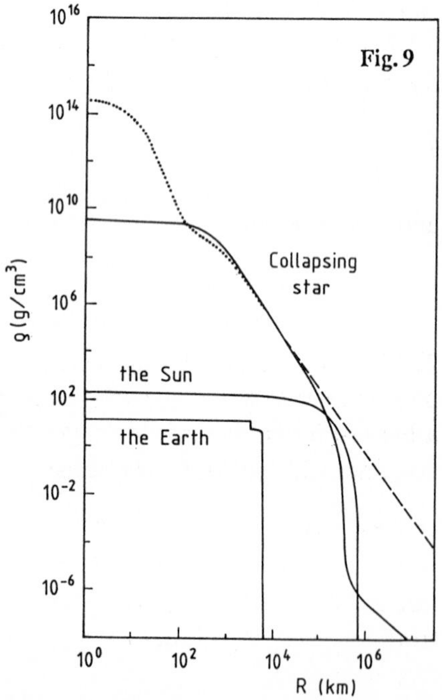


**Fig. 8.** Resonant $\nu$ oscillation experiments. The distortion of the initial $\nu$ spectrum **a)** with constant density filter; **b)** with monotonous decreasing density filter

**Fig. 9.** Density profiles of different matter filters. For collapsing stars the density distributions are shown: for progenitor with compact profile (*solid line*), for progenitor without sharp border between mantle and envelope (*dashed line*), for stars at $\nu$-opacity stage (*dotted line*)

Correspondingly there are two types of distortion factors:
1) Constant density filter, oscillation-like distortion. For fixed, $\theta$, $P$ (as a function of $E/\Delta m^2$) turns out to be the oscillating curve which inscribes into the resonant peak $\sin^2 2\theta_m(E)$. Its shape depends on

$$\frac{d}{l_m^R(\varrho)} \simeq \frac{d \cdot \tan 2\theta}{l_0(\varrho)} \quad ,$$

where $d$ is the width of filter and $l_0(\varrho)$ is the refraction length (see Fig. 4). At $d \cdot \tan 2\theta / l_0(\varrho) \ll 1$, there is no strong enhancement of oscillations, there are no oscillations of $P(E)$ in the resonant region, and the maximum of $P(E)$ is shifted to $E < E_R$. The frequency of $P(E)$ oscillations rises with $d/l_\nu^R$. In the limit $d/l_m^R \gg 1$ averaging over small intervals of $E$ gives

$$P(E/\Delta m^2) \simeq 1 - \sin^2 2\theta_m(E)/2 \quad .$$

Maximal filling of the resonant peak corresponds to

$$\frac{d}{l_m^R} = \frac{1}{2} \quad , \tag{32}$$

when the width of the layer is half of the resonant oscillation length. Under condition (32), the maximal effect of conversion integrated over energy interval $\Delta E \simeq 2\Delta E_R$ can be achieved.

Measuring $F(E)/F^0(E)$, one can restore the resonant curve (see Fig. 8). Then the position of resonance ($E_R$) determines $\Delta m^2$ according to (22):

$$|\Delta m^2| \cong \frac{2\sqrt{2} \cdot G_F \cdot \varrho}{m_N \cdot \cos 2\theta} E_R \quad , \tag{33}$$

and the width of resonance gives the mixing angle from

$$\tan 2\theta = \frac{\Delta E_R}{E_R} \quad . \tag{34}$$

2) Monotonous decreasing density, bath-like distortion. The dependence of the suppression factor on $E/\Delta m^2$ (averaged over one period of oscillation) has the shape of a bath ($\theta$ is fixed). The position of its left edge ($P = 1/2$) is determined by resonant condition (22) at maximal density:

$$(E/\Delta m^2)_c \cong \frac{m_N}{2\sqrt{2} \cdot G_F \cdot \varrho_c} \quad . \tag{35}$$

For $E < E_c$ there is no resonance in a given filter. The region where $P$ diminishes from its vacuum value $P = 1 - \sin^2 2\theta/2$ to $\sin^2 \theta$ is proportional to the width of resonance

$$\Delta E/E_c \sim 2\Delta E_R/E \simeq 2 \tan 2\theta \quad .$$

The energy of the right edge of the bath is fixed by adiabaticity condition $\kappa_R \simeq 1/2$ [see (28)]:

$$(E/\Delta m^2)_a = \frac{h(E) \cdot \sin^2 2\theta}{2\pi \cos 2\theta} \quad . \tag{36}$$

At $E > E_a$ adiabaticity is violated and $P$ rises with $E$. In the interval $(E/\Delta m^2)_c$–$(E/\Delta m^2)_a$ conditions for oscillation-free transformation are satisfied and on the bottom $P_b = \sin^2 \theta$ [see (31)]. Again, measuring the parameters of the bath $E_a$, $E_c$, $\Delta E/E_c$, $P_b$ (35, 36), one can find $\Delta m^2$ and $\sin^2 2\theta$.

For $\nu_e - \nu_\mu - \nu_\tau$ oscillations in the most interesting case (mass hierarchy), the distortion factor will be the superposition of two bath [11].
We now consider actual resonant oscillation experiments.

## 4.2 The Sun

The central part of the Sun is a source of electron neutrinos (see Fig. 10).

Resonant neutrino oscillations $\nu_e - \nu_\mu(\nu_\tau)$ as well as $\nu_e - \nu_s$ ($\nu_s$ is the sterile state) result in:
1) Suppression of the initial $\nu_e$ flux. Part of $\nu_e$ transforms into $\nu_\mu(\nu_\tau)$ or $\nu_s$.
2) Distortion of the energy spectrum. The suppression factor $P(E)$ has the bath-like shape. The relative position of the $\nu$ spectrum and bath on the energy scale depends on $\Delta m^2$. In contrast to astrophysical explanations of the observed low flux of solar neutrinos, resonant oscillations not only change relative $\nu$ fluxes from different reactions, but distort continuous spectra from elementary processes such as $pp$-reactions and $\beta$-decays of $B$, $^{13}N$, and $^{15}O$.

The region of strong conversion in an $\Delta m^2/E$ versus $\sin^2 2\theta$ plot is the triangle bounded by the resonant condition (35), the adiabaticity condition with $\kappa_R = 0.1$–$0.2$ [see (36), for the Sun $h(E) \simeq$ const.], and by $\sin^2 2\theta = 1$.
Let us consider existing data:
1) *Average* $^{37}$Ar *production rate in* Cl-Ar *experiments*. Resonant oscillations suppress the Ar production rate, because $\nu_\mu(\nu_\tau)$ with solar energies or $\nu_s$ do not transform Cl into Ar. This can explain Davis's data [12]. The suppression factor for the

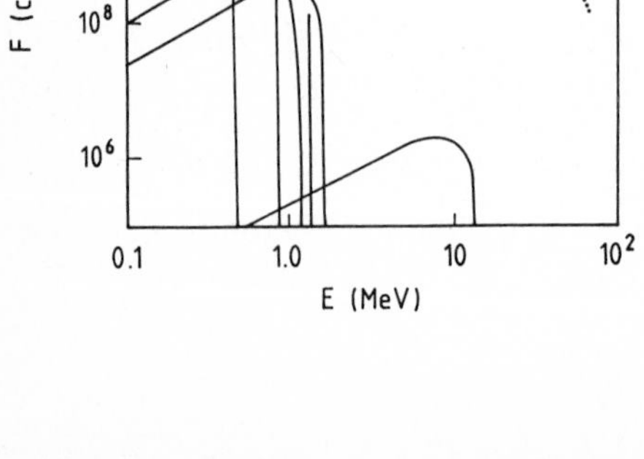

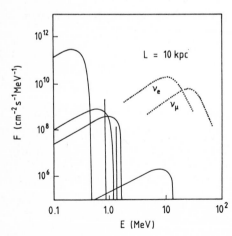

**Fig. 10.** $\nu$ spectra from different sources: *solid lines*, the Sun; *dotted lines*, collapsing stars

$^{37}$Ar production rate is

$$R_{\mathrm{Ar}} = \frac{Q_{\mathrm{Ar}}}{Q_{\mathrm{Ar}}^{\mathrm{SSM}}} = \frac{1}{Q_{\mathrm{Ar}}^{\mathrm{SSM}}} \int dE\, P(E) \cdot \sigma(E) \cdot F_0(E) \quad , \tag{37}$$

where $Q_{\mathrm{Ar}}^{\mathrm{SSM}}$ and $Q_{\mathrm{Ar}}$ are the production rates in the standard solar model without and with oscillations (for $Q_{\mathrm{Ar}}^{\mathrm{SSM}}$, $P = 1$), $\sigma$ is the cross-section, and $F_0(E)$ is the solar neutrino flux.

In the two-neutrino case the equation

$$R_{\mathrm{Ar}}(\Delta m^2, \sin^2 2\theta) = a = \text{const.}$$

determines the line of equal suppression [2] (isosnu line [13]) in the $\Delta m^2$ vs. $\sin^2 2\theta$ plot shown in Fig. 11. Values of $\Delta m^2$, $\sin^2 2\theta$ for which $R_{\mathrm{Ar}} = R_{\mathrm{Ar}}^{\mathrm{exp}} = 1/4 - 1/2$ are called "solutions" of the solar neutrino problem. The corresponding points are placed between isosnu lines $a = 1/4$ and $a = 1/2$. Depending on the regime of oscillations and on the distortion of energy spectra, one can single out the following types of solutions: adiabatic (A), nonadiabatic (NA), mixed (M), quasivacuum (Q), regeneration (R), (see Fig. 11).

The system of three neutrinos ($\nu_e - \nu_\mu - \nu_\tau$) has three resonances. For solar neutrinos only two of them are essential: the resonances in the $\nu_e - \nu_\mu$ and $\nu_e - \nu_\tau$ channels [11]. They are called $l$- (low) and $h$- (high) resonances, respectively. The $3\nu$ system is determined by two sets of parameters $(\Delta m^2, \sin^2 2\theta)_{l,h}$, which characterize $l$- and $h$-resonances. Consequently the $3\nu$ system is fixed (with respect to the $\nu_e \to \nu_e$ transition) by two points in the $\Delta m^2$ vs. $\sin^2 2\theta$ plot, that is, by positions of $l$- and $h$-resonances in this plot.

$3\nu$ solutions can be analyzed in terms of $2\nu$ ones: $l$- and $h$-resonances are placed on definite $2\nu$-isosnu lines, and there are two types of $3\nu$ solutions [11, 14]. 1) Mixed solution ($M_{3\nu}$): the $h$ resonance is on the adiabatic branch of the $2\nu$-isosnu line, the $l$ resonance is on the nonadiabatic one.

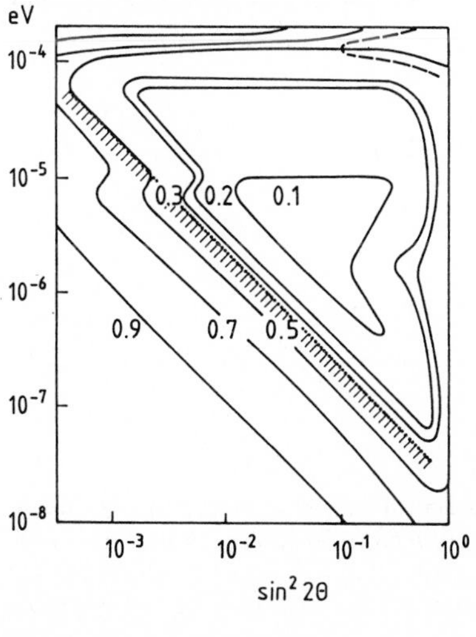


**Fig. 11.** Lines of equal suppression of $^{37}$Ar production rate due to $\nu_e - \nu_\mu(\nu_\tau)$ oscillations ($2\nu$-isosnu lines). Figures on the curves are values of suppression. The dotted line is the Kamiokande-II restriction, and the dashed line is the IMB restriction

2) Nonadiabatic $3\nu$ solution ($NA_{3\nu}$): both $l$- and $h$-resonances are on nonadiabatic branches (Fig. 11).

Rules can be formulated for the positions of $l$- and $h$-resonances [14]. In particular, to obtain a given suppression $R_{\mathrm{Ar}}$ one should place at least one resonance between $2\nu$-isosnu lines,

$$a_1 = \sqrt{R_{\mathrm{Ar}}} \quad \text{and} \quad a_2 = R_{\mathrm{Ar}} \quad , \tag{38}$$

and (38) determines isosnu strips on a $\Delta m^2$ vs. $\sin^2 2\theta$ plot.

If the suppression $R_{\mathrm{Ar}} = 1/4 - 1/2$ in Davis's experiment is due to resonant oscillations only, then according to (38) at least one resonance is between $2\nu$-isosnu lines $a = 0.25$ and $a = 0.7$ (Fig. 11).

2) *Limit on* $^8$B *Solar Neutrino Flux*. Resonant oscillations diminish the number of $\nu_{\odot}e \rightarrow \nu_{\odot}e$ events, because the $\nu_\mu(\nu_\tau)$ $e$-cross-section is 6-times smaller than that of the $\nu_e e$ ($\nu_s$ has no interactions with $e$). Suppression factors of $\nu_{\odot}e$-events $R_{\nu e}$ with threshold $E_e^{\mathrm{th}} = 9.5\,\mathrm{MeV}$ have been calculated for fixed values of $R_{\mathrm{Ar}}$ in Fig. 12. Note that $\nu e$ scattering with high $E_e^{\mathrm{th}}$ gives information which is essentially independent of the $^{37}$Ar data, the reasons being (a) there is an additional contribution to $R_{\mathrm{Ar}}$ from $^7$Be neutrinos, (b) cross-sections for Ar production and $\nu e$ scattering have different $E$ dependences, (c) there are contributions to $R_{\nu e}$ from $\nu_\mu(\nu_\tau)e$ scattering.

For the $3\nu$ case one can use $2\nu$ results [$2\nu$ lines of equal suppression $R_{\nu e}$ ($\Delta m^2$, $\sin^2 2\theta$) = const [15] and factorization of $R_{\nu e}$ : $R_{\nu e}^3 \cong R_{\nu e}^l \cdot R_{\nu e}^h$, where $R^{l,h}$ is the $2\nu$-suppression factor with parameters of $l$-, $h$-resonance].

The Kamiokande-II result $R_{\nu e}$ ($E > 9.5\,\mathrm{MeV}) \leq 0.53$ (90% Cl) [16] is not yet crucial for the choice of a definite solution of the solar neutrino problem (Fig. 12), it supports Davis's data, and moreover the obtained limit permits the exclusion of some

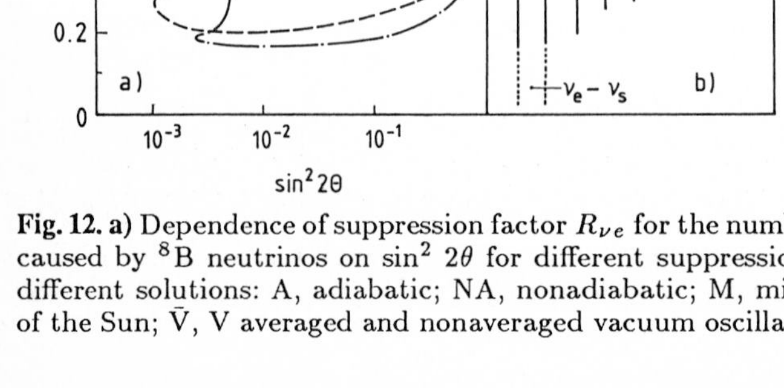

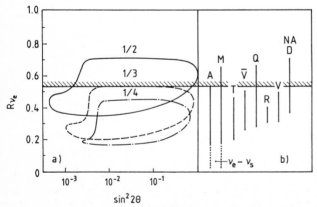

**Fig. 12. a)** Dependence of suppression factor $R_{\nu e}$ for the number of $\nu e - \nu e$ events ($E_e \geq 9.5\,\mathrm{MeV}$) caused by $^8$B neutrinos on $\sin^2 2\theta$ for different suppressions $R_{\mathrm{Ar}}$. **b)** Predictions of $R_{\nu e}$ in different solutions: A, adiabatic; NA, nonadiabatic; M, mixed; $\mathrm{T_c}$, low central temperature of the Sun; $\bar{\mathrm{V}}$, V averaged and nonaveraged vacuum oscillations; R, regeneration; D, $\nu$-decay

regions of $\Delta m^2$, $\sin^2 2\theta$ parameters in the nonadiabatic solution with $R_{\mathrm{Ar}} \geq 1/3$ (Fig. 11).

Further improvement of the Kamiokande results will give the following discrimination of solutions (see Fig. 12):

1) If $R_{\nu e} = 0.5-0.7$, then $2\nu$- and $3\nu$-nonadiabatic, $2\nu$- and $3\nu$-mixed, quasivacuum and $\nu$-decay solutions survive.
2) If $R_{\nu e} \simeq 0.25 - 0.5$, then nothing can be excluded.
3) If $R_{\nu e} \simeq 0.15 - 0.25$, then the solutions $2\nu$- and $3\nu$-mixed, adiabatic and low $T_c$ survive.
4) If $R_{\nu e} < 0.15$, then the flavour oscillations are excluded and only $2\nu$ mixed as well as adiabatic oscillations in sterile neutrinos remain.

Figure 13 shows the distortion of the solar $\nu_e$ spectrum for solutions of the solar neutrino problem as well as for solutions which are not related to resonant oscillations. Among the latter one has the "$T_c$" solution, which is based on a low central temperature of the Sun; averaged and nonaveraged vacuum solutions $(\overline{V}, V)$; and the $\nu$-decay solution. It follows from Fig. 13 that a complete solar neutrino spectroscopy, as well as some terrestrial experiments, in principle permit us to establish

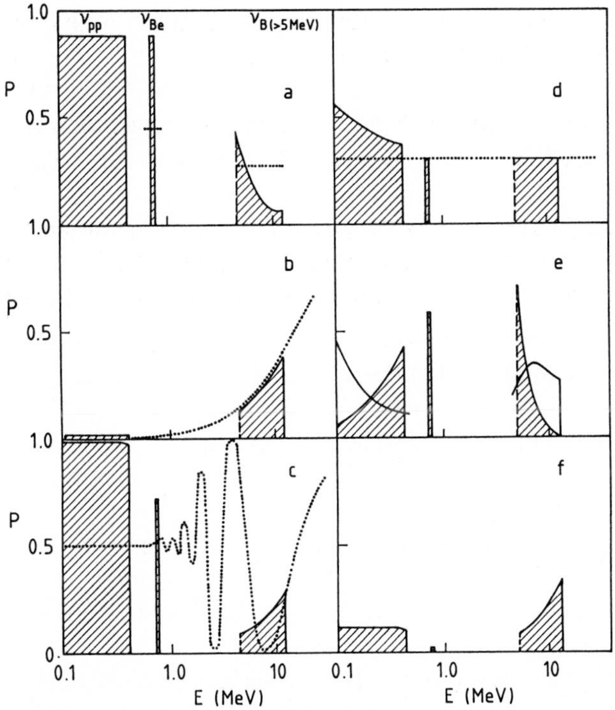


**Fig. 13a–f.** Distortion of the solar neutrino spectrum in different "solutions" of the solar neutrino problem **a** adiabatic, "$T_c$", **b** nonadiabatic $2\nu$, $\nu$-decay, **c** $2\nu$-mixed, nonaveraged vacuum oscillations, **d** quasivacuum, averaged vacuum oscillations, **e** $3\nu$-mixed, regeneration, **f** nonadiabatic $3\nu$. Solutions based on resonant oscillations are shown by *solid lines*, solutions which are not related to resonant oscillations, *dotted lines*. Regions of *pp*-, *Be*-, and *B*-neutrinos are *shaded*

whether resonant oscillations take place in the Sun. Moreover, in case of a positive answer this permits us to measure $\Delta m^2$ and $\sin^2 2\theta$.

A good approximation to complete spectroscopy will follow from (a) a series of radiochemical experiments with different thresholds (Cl-Ar, Ga-Ge, Li-Be, Br-Kr); (b) direct measurement of the $\nu$ spectrum by electronic methods ($\nu$ scattering on electrons, heavy water, $^{40}$Ar, $^{11}$B...); (c) geochemical experiments.

### 4.3 The Earth

Density profiles depend on the zenith angle ($\psi$) with which neutrinos cross the Earth. At $\cos\psi \lesssim 0.8$ neutrinos propagate in the mantle and the suppression factor has an oscillation-like form (see Sect. 4.1), with $\varrho = \overline{\varrho}_{\text{mantle}}$. At $\cos\psi \gtrsim 0.8$ resonances are essential both in the mantle and in the core, and the total suppression factor is the superposition of the two resonant curves.

The region of strong conversion is $E/\Delta m^2 = 10^6 - 10^7$ MeV/eV$^2$ and $\sin^2 2\theta > 10^{-2}$ [17, 15]. It is determined by resonant conditions for densities in the mantle and in the core, and by sizes of matter layers. The observable effects depend on properties of $\nu$-fluxes.

1) *Atmospheric neutrinos* [18]. The effects expected are
(a) distortion of the energy spectrum, (b) changing of the $\overline{\nu}/\nu$ ratio (c) changing of the angular dependences of $\nu$-fluxes and others.

The IMB collaboration [19] have compared the number of upgoing and downgoing events with $E \lesssim 350$ MeV induced by neutrinos. No oscillation effects have been observed and this permits the exclusion of the region $\Delta m^2 \simeq 1.1 \times 10^{-4}$ eV$^2$, $\sin^2 2\theta \gtrsim 10^{-1}$ (Fig. 11).

2) *Solar neutrinos*. Resonant oscillations of solar neutrinos in the Earth result in the partial regeneration of the $\nu_e$ flux for $\Delta m^2/E \simeq (2-7)\cdot 10^{-7}$ eV$^2$/MeV and $\sin^2 2\theta \gtrsim 10^{-2}$ ("regeneration solution"). Due to the rotation of the Earth, day-night and annual modulation are predicted [18, 15]. For $\sin^2 2\theta \lesssim 0.3$ the predicted strong season modulations are in contradiction with Davis's data [20], and this enables us to exclude the region $\Delta m^2 \simeq (1-3)\cdot 10^{-6}$ eV$^2$, $\sin^2 2\theta = (0.3-3)\cdot 10^{-7}$ [15].

Annual waves in the $^{37}$Ar production rate (latitude 43°), averaged over the day, for large values of $\sin^2 2\theta$ are shown on Fig. 14 [21]. The shifting of maxima in December is related to condition (32).

3) *Neutrino burst from collapsing stars*. Earth effects lead to the difference of properties of $\nu$-signals from the same collapse, measured at different installations. In particular differences in energy spectra, average $E$ and total $\nu$ luminosities are expected.

4) The search for resonant oscillation effects in neutrinos from accelerators is discussed [17].

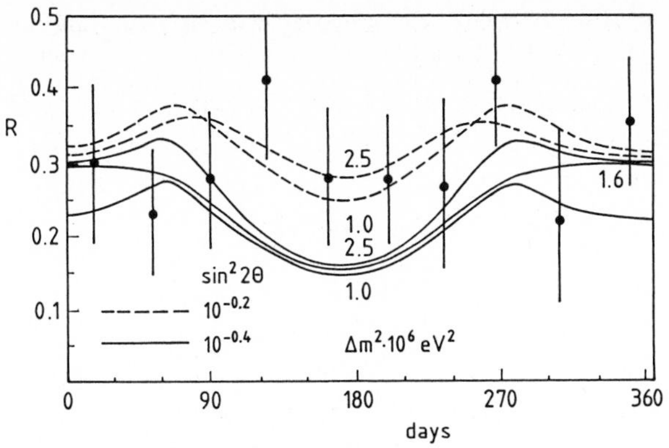


**Fig. 14.** Annual waves. Modulations of $^{37}$Ar-production rate by the Earth averaged over the day. Latitude 43°

## 4.4 Collapsing Stars

According to the standard picture the cores of collapsing stars are the source of all species of neutrinos $\nu_e$, $\nu_\mu$, $\nu_\tau$ and the corresponding antineutrinos. At the initial stage of collapse the $\nu_e$-flux from neutronization dominates. At the stage of $\nu$ opacity, comparable (but not equal) fluxes of $\nu_e$, $\nu_\mu$ and $\nu_\tau$ are emitted. It is claimed that the average energies of $\nu_\mu$ and $\nu_\tau$ are two-times larger than that of $\nu_e$. (Fig. 10).

For neutrinos of a particular type, the suppression factor has a bath-like shape. Due to the large central density ($\varrho_s \gg \varrho_\odot$), the size of the baths are much larger than for the Sun.

Regions of strong conversion on a $\Delta m^2$ vs. $\sin^2 2\theta$ plot are restricted by $\Delta m^2 \simeq 10^6 - 10^8\,\mathrm{eV}^2$ and $\sin^2 2\theta > 10^{-8} - 10^{-10}$ (for maximal $\Delta m^2$). The following observable effects are possible (see Fig. 15):

1) Suppression of $\overline{\nu}_e$ (or $\nu_e$) signal due to oscillations in the sterile state.
2) Interchange of $\nu_e$ and $\nu_\mu$ ($\nu_\tau$) spectra due to flavour oscillations in the $\nu$-opacity stage. If both $\nu_e$ and $\nu_\mu$ spectra are on the bottom of the bath, then symmetric exchange takes place (Fig. 15b). At the edges of the baths the interchange is non-symmetric and $\nu_e \to \nu_\mu$ dominates over $\nu_\mu \to \nu_e$, or vice versa.
3) Suppression of the $\nu_e$ peak from neutronization due to oscillations in flavour or sterile states (Fig. 15c).
4) Distortion of $\nu_e$ or $\overline{\nu}_e$ spectra.
5) Oscillation effects may be changed in time due to the $\varrho$ profile changing during a $\nu$ burst.

Possible observations of a $\nu$ burst related to SN1987A [22] can give restrictions on neutrino parameters:

1) If underground events are caused by $\overline{\nu}_e$ from gravitational collapse, then according to several criteria there is no strong suppression of the $\nu$ signal. In any case the region of $\Delta m^2$ vs. $\sin^2 2\theta$ restricted by the line $a = 0.1$ (Fig. 16a) can be excluded [23].

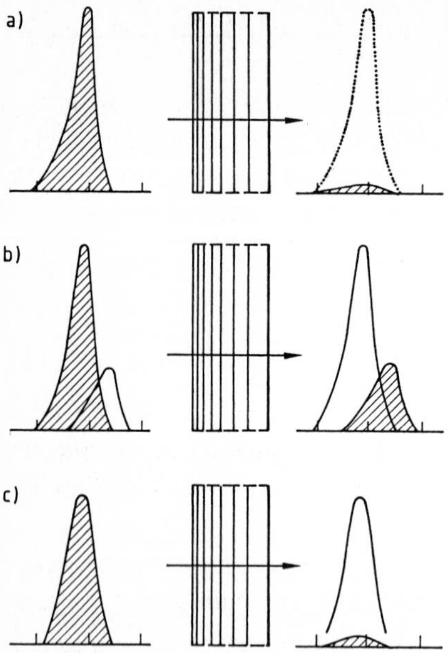


**Fig. 15.** The effects of resonant neutrino oscillations in a collapsing star. Shaded figures are $\nu_e$ spectra, *solid lines* are $\nu_\mu$ spectra, *dotted line* is the $\nu_s$ spectrum

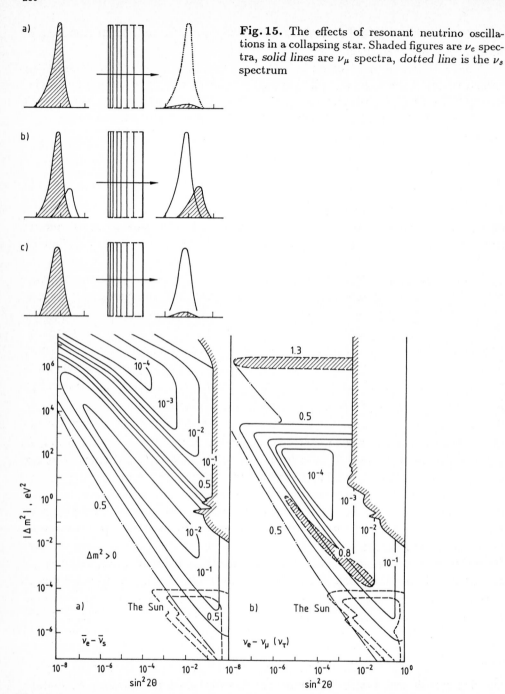

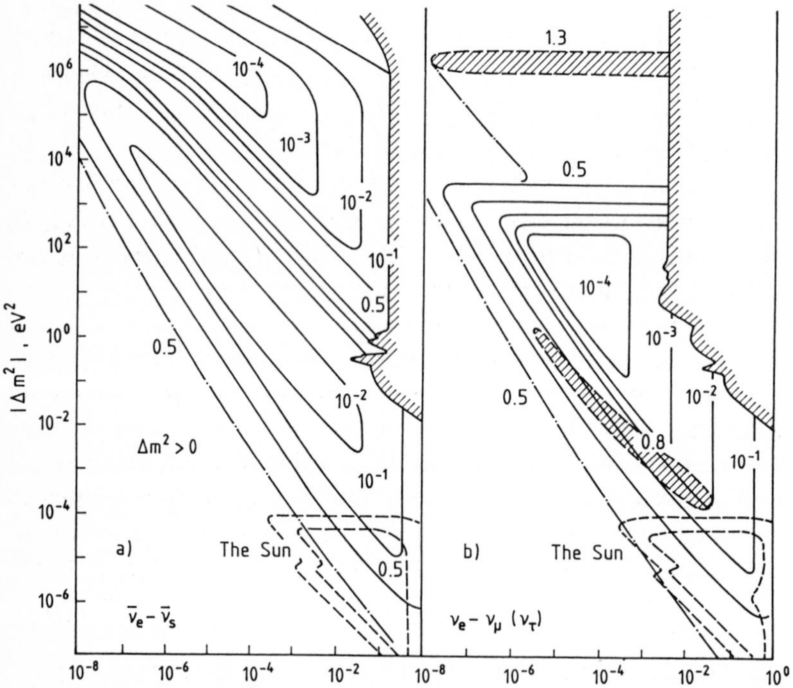


**Fig. 16. a)** Lines of equal suppression of $\overline{\nu}_e$-burst energy due to $\overline{\nu}_e - \overline{\nu}_s$ oscillations at the stage of $\nu$ opacity: *solid lines* represent a compact profile (see Fig. 9), *dashed-dotted line*, a profile with no sharp border between mantle and envelope. **b)** Lines of equal suppression of $\nu_e$ energy of neutronization peak due to $\nu_e - \nu_\mu$ $(\nu_\tau)$ oscillations. *Solid lines*, compact profile; *dashed-dotted*, noncompact one (Fig. 9). *Shaded regions* are the regions of nonsymmetric exchange of $\nu_e$ and $\nu_\mu$ spectra at the $\nu$ opacity stage ($R \gtrsim 1.3$ and $R \lesssim 0.8$)

Due to uncertainty in the distribution $\varrho(r)$ the edges of the excluded region may be shifted by an order of magnitude.
2) If the first two Kamiokande-II events [22] are due to neutronization neutrinos (which is extremely problematic), then there is no suppression of the neutronization peak. Consequently, the region of $\Delta m^2$ vs. $\sin^2 2\theta$ restricted by the line $a = 1/2$ (Fig. 16b) should be excluded [23, 24]. Solar neutrino spectroscopy may clarify this question.

## 5. Conclusion

1) Characteristics of the neutrino state such as flavour and mass composition may oscillate, that is, change in time periodically. Oscillations take place in mixed neutrino states, which are the combinations of two (several) eigenstates of the Hamiltonian ($\nu_i^H$).
2) Propagation of a mixed state is described by two (several) wave packets, corresponding to $\nu_i^H$. Packets have different phase velocities (determined by eigenvalues). Consequently, an additional phase difference appears between $\nu_i^H$ packets during neutrino propagation. A monotonous increasing phase difference results in oscillations. To find the characteristics of $\nu$ oscillations one should perform flavour decomposition of $\nu_i^H$ packets and consider the interference of waves with the same flavour, but from different $\nu_i^H$. The characteristics depend on (a) admixtures of $\nu_i^H$ in given $\nu$, (b) flavours of $\nu_i^H$, (c) phase velocities of $\nu_i^H$.
3) In the vacuum $\nu_i^H$ coincides with $\nu_i$, the states with definite masses. Both flavours of $\nu_i^H$ and admixtures of $\nu_i^H$ in $\nu(t)$ are conserved. If the initial $\nu$ state is $\nu_e$ or $\nu_\mu$ ($\nu_\tau$), that is the state with definite flavour, then both flavours and admixtures are determined by $\theta$. As a consequence the average probability and oscillations depth are constants and determined by $\theta$.
4) In matter, $\nu_i^H = \nu_{im}$ do not coincide with definite mass states. The flavours of $\nu_{im}$ depend on the density of matter, and change if neutrinos propagate through an inhomogeneous medium.
If the density changes slowly, so that the adiabaticity condition is fulfilled, then the admixtures of $\nu_{im}$ in a given $\nu(t)$ are fixed at the initial moment and are conserved during propagation. As a consequence, in the adiabatic regime $A_p$ and $\overline{P}$ are the functions of instantaneous value of density. The strongest conversion of neutrinos takes place in the resonant layer. If the adiabaticity is violated, then the admixtures of $\nu_{im}$ in given $\nu(t)$ are changed.
5) Resonant neutrino oscillations in matter offer unique possibilities in searching for neutrino masses and mixing. At the present time we have the following data which refer to matter oscillations:

- Measurements of the average $^{37}$Ar-production rates in the Cl-Ar experiment with solar neutrinos
- Search for annual modulations of the $^{37}$Ar-production rate

- Kamiokande-II limit on the $B$-solar neutrino flux
- Registration of up-going and down-going atmospheric neutrino events by the IMB collaboration
- Possible observation of a $\nu$ burst related to supernova 1987A in LMC

With these data we are at the beginning of a path which may lead to the establishing of the massiveness of neutrinos and, moreover, to the measuring of $\Delta m^2$ and mixing ($\sin^2 2\theta$). Also, "matter oscillation experiments" result in the exclusion of a very large and very important $\Delta m^2$ vs. $\sin^2 2\theta$ region.
6) Further progress in the field is related to the following experiments:

- Solar neutrino spectroscopy
- Searching for modulations of the solar neutrino flux
- Further observations of SN1987A, which will probably allow us to make a more accurate model of the star and to clarify the interpretation of underground events of February 23, 1987
- Searching for new $\nu$ bursts from gravitational collapses
- Underground and underwater experiments with atmospheric (and probably accelerator-produced) neutrinos
- Searching for vacuum oscillations and $\beta\beta$ decays, measuring of neutrino mass in $\beta$ decays.

# References

1. B. Pontecorvo: ZhETF **33**, 549 (1957); ibid. **34**, 247 (1958); ibid. **53**, 1717 (1967); B. Pontecorvo, S.M. Bilenky: Phys. Rep. **41C**, 225 (1978)
2. L. Wolfenstein: Phys. Rev. **D17**, 2369 (1978); ibid. **D20**, 2634 (1979)
3. S.P. Mikheyev, A.Yu. Smirnov: Sov. J. Nucl. Phys. **42**, 913 (1985)
4. S.P. Mikheyev, A.Yu. Smirnov: Sov. Phys. JETP **64**, 4 (1986); for review see S.P. Mikheyev, A.Yu. Smirnov: Usp. Fiz. Nauk. **153**, 3 (1987)
5. Z. Maki, M. Nagakawa, S. Sakata: Prog. Theor. Phys. **28**, 870 (1962)
6. V. Barger et al.: Phys. Rev. **D22** , 2718 (1980)
7. H. Bethe: Phys. Rev. Lett. **56**, 1305 (1986)
8. N. Cabibbo: Summary talk given at 10th Int. Conf. on Weak Interactions, Savonlinna, Finland, 1985
9. S.J. Parke: Phys. Rev. Lett. **57**, 1275 (1986)
10. W.C. Haxton: Phys. Rev. Lett. **57**, 1271 (1986)
11. T.K. Kuo, J. Pantaleone: Phys. Rev. Lett. **57**, 1805 (1986); Phys. Rev. **D35**, 3432 (1987);
S. Toshev: Phys. Lett. **185B**, 177 (1987);
A.Yu. Smirnov: Yad. Fiz. **46**, 1152 (1987) and others
12. R. Davis, Jr.: In Proc. of the 7th Workshop on Grand Unification/ICOBAN'86, ed. by J. Arafune (Toyama, Japan 1986) p. 237
13. S.J. Parke, T.P. Walker: Phys. Rev. Lett. **57**, 2322 (1986)
14. S.P. Mikheyev, A.Yu. Smirnov: Phys. Lett. **200B**, 560 (1988)
15. M. Cribier et al.: Phys. Lett. **182B**, 89 (1986)
J.N. Bahcall, J.M. Gelb, S.P. Rosen: Phys. Rev. **D35**, 2976 (1987)
16. M.B. Voloshin, M.I. Vysotsky, L.B. Okun: Zh. **91**, 754 (1986) Eksp. Teor. Fiz.
17. V.K. Ermilova, V.A. Tsarev, V.A. Chechin: Zh. Eksp. Teor. Fiz. Pis'ma Red. **43**, 353 (1986)

18. S.P. Mikheyev, A.Yu. Smirnov: In Proc. of the 6th Moriond Workshop Tignes, France, 1986, ed. by O. Fackler, J. Tran Thanh Van, p. 355
19. J.LoSecco et al.: Phys. Lett. **184B**, 305 (1987)
20. M.L. Cherry, R. Davis, Jr., K. Lande: Proc. of the ICRC, Moscow, 1987, ed. by A. Chudakov, v. 4, p. 332
21. S.P. Mikheyev, A.Yu. Smirnov: In Proc. of the Internat. Workshop on *Neutrino Physics*, Heidelberg 1987, Springer, Heidelberg 1988, ed. by H.V. Klapdor and B. Povh, p. 247
22. K. Hirata et al.: Phys. Rev. Lett. **58**, 1490 (1987); R.M. Bionta et al.: Phys. Rev. Lett. **58**, 1494 (1987); V.L. Dadykin et al.: Zh. Eksp. Teor. Fiz. Pis'ma Red. **45**, 464 (1987); E.N. Alexeyev et al.: Zh. Eksp. Teor. Fiz. Pis'ma Red. **45**, 461 (1987)
23. S.P. Mikheyev, A.Yu. Smirnov: Zh. Eksp. Teor. Fiz. Pis'ma Red. **46**, 11 (1987)
24. J. Arafune et al.: Phys. Rev. Lett. **59**, 1867 (1987); H. Minakata et al.: Tokyo Univ. preprint TMUP-HEL-87-03; T.K. Kuo, J. Pantaleone: Preprint PURD-TH-87-08 (1987); S.P. Rosen: Los Alamos preprint April 1987; D. Nötzold: Phys. Lett. **196B**, 315 (1987); P.O. Lagage et al.: Phys. Lett. **193B**, 127 (1987)

# Searches for Lepton-Flavour Violation

*P. Depommier*

Nuclear Physics Laboratory, University of Montreal, C.P. 6128,
Station "A", Montreal, Quebec, H3C3J7, Canada

## 1. Neutrino Physics, or No-Neutrino Physics?

It is certainly fair to say that understanding the nature of the neutrino is one of the most critical issues in particle physics, nowadays. In fact, a lot of experimental and theoretical work is being devoted to this problem. To study the properties of neutrinos there are essentially two different approaches. The first one makes use of neutrino (antineutrino) beams from high-energy accelerators and nuclear reactors, or takes advantage of the scarce neutrinos provided to us by nature (solar or cosmic neutrinos). The other approach consists in searching for processes where neutrinos do not appear: either they annihilate each other or they are replaced by some (more or less exotic) particle. For instance, neutrinoless double beta decay was proposed a long time ago as a probe of the Majorana nature of the neutrino. We now recognize the utmost importance of neutrinoless double beta decay in the context of the modern gauge theories and the quest for physics beyond the Standard Model: some fundamental issues are the problem of neutrino masses, the presence of right-handed charged weak currents, the existence of Goldstone bosons (majorons). Another well-known example is the neutrinoless process $\mu \rightarrow e\gamma$, which historically has led to the "two neutrino hypothesis", and is nowadays considered as one of the most powerful probes to search for physics beyond the Standard Model, as we shall see below.

It is therefore not surprising to have a chapter on (neutrinoless) lepton-flavour-violating processes in a book on Neutrino Physics. However, this contribution should not be considered as a complete review of the subject. The idea is rather to illustrate with selected examples the importance of the field in particle physics. I have to apologize to those whose work is not quoted.

## 2. The Muon Puzzle[1]

The problem arose soon after the discovery of the muon[2], as it was recognized[3] that this particle is not the meson predicted by Yukawa[4]. The existence of a new lepton, very similar to the electron but for its mass and instability, was very puzzling. One remembers the famous interrogation of Rabi: "Who ordered that?". Of course, at that time, it was not realized that the newly discovered particle was the precursor of a second generation of fermions. Then came the next question: how does the muon communicate with the electron? If the muon were some kind of heavy electron, it should easily decay via $\mu \rightarrow e\gamma$, or transform into an electron when captured by a nucleus (A,Z), giving rise to the process of *muon-to-electron conversion*, i.e. $\mu^-(A,Z) \rightarrow e^-(A,Z)$. Early experiments ruled out these two possibilities. Using cosmic rays, Hincks and Pontecorvo[5] established the first upper limit (of the order of one percent) on the $\mu \rightarrow e\gamma$ decay. Lagarrigue and Peyrou[6] showed that muon-to-electron conversion does not occur with a branching ratio greater than a few percent.

Later, it was established experimentally that muon decay is a three-body decay with neutral particles accompanying the electron[7]. The electron spectrum was analyzed according to the four-fermion theory developped by Michel[8]. It was assumed that two neutrinos were emitted along with the electron. If there is a conservation law for leptons, one must consider these two particles as being a neutrino and an antineutrino. However, the distinction between neutrino and antineutrino is still an open question, and it is only possible to say that the two neutral particles emitted in muon decay are neutrinos of different helicities.

The possibility that these two neutrinos mutually annihilate leads back to the occurrence of the $\mu \rightarrow e\gamma$ decay. Assuming that the weak interaction is transmitted by a vector boson W[9], one has to consider the following diagrams (Fig. 1):

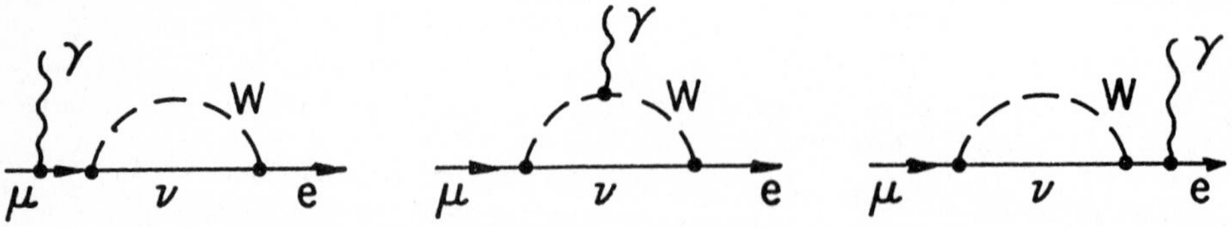


Fig. 1. Diagrams for the $\mu \rightarrow e\gamma$ decay. The neutrino emitted at the muon vertex is absorbed at the electron vertex. The photon is emitted from the charged current .

In spite of computational difficulties and ambiguities (in the absence of a good theory of weak interactions) it became clear that the theoretical expectation for the $\mu \rightarrow e\gamma$ branching ratio[10] (~ $10^{-4}$) exceeded the experimental upper limit by several orders of magnitude[11]. In fact, even before the advent of the meson factories the upper limit on the $\mu \rightarrow e\gamma$ decay had been brought down to the $10^{-8}$ scale[12]. This had led to the *two-neutrino hypothesis* [13]: the neutrino coupled to the electron and the one coupled to the muon are of different nature (they have different flavours). Like the electron and the muon, they carry different lepton numbers, electronic and muonic, and these partial lepton numbers are separately conserved (Table 1).

Table 1. Electronic and muonic lepton numbers ($L_e$ and $L_\mu$)

| | $e^-$ | $\nu_e$ | $\mu^-$ | $\nu_\mu$ | $e^+$ | $\bar{\nu}_e$ | $\mu^+$ | $\bar{\nu}_\mu$ |
|---|---|---|---|---|---|---|---|---|
| $L_e$ | +1 | +1 | 0 | 0 | -1 | -1 | 0 | 0 |
| $L_\mu$ | 0 | 0 | +1 | +1 | 0 | 0 | -1 | -1 |

This hypothesis was rapidly confirmed by experiment[14]: neutrinos (antineutrinos) from pion decay (of muonic flavour) do not produce electrons (positrons) when reacting with nucleons. As a result, there are two distinct lepton families, and no communication between them (Fig. 2).

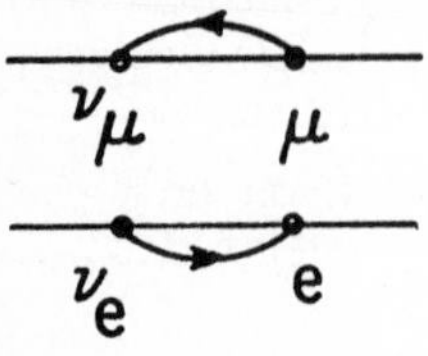


Fig. 2. In muon decay, partial lepton numbers are conserved. There is no communication between the two lepton families.

The conservation law considered above is the so-called "additive law". It forbids the processes $\mu \to e\gamma$, $\mu \to e\gamma\gamma$, $\mu \to 3e$, muon-to-electron conversion, etc.... As we shall see later there is also a multiplicative law.

## 3. Three Generations of Quarks and Leptons

In the hadronic sector, one was witnessing the discovery of strange particles, the advent of the SU(3)-based quark model, the prediction and discovery of charm. Quarks and leptons were recognized as being the building blocks of matter. Quarks were shown to differ from leptons in several respects (electric charges, confinement) and also because quarks mix via the Cabibbo matrix whereas leptons do not seem to mix. Later, this picture was extended with the discoveries of the $\tau$ lepton and b quark and the strong belief in the t quark. At present, we have strong evidence for three generations of quarks and leptons, but we do not know how many generations do exist. Again we observe that quarks mix via the Cabibbo-Kobayashi-Maskawa matrix whereas no lepton mixing has been observed yet. Very stringent constraints have been obtained on lepton-flavour mixing. The experimental situation forces us to conclude that there is a well-obeyed conservation law for partial lepton numbers, electronic, muonic and tauonic. But, do we have an explanation for such a conservation law?

## 4. The Standard Model and its Extensions

Nowadays, every particle physicist is familiar with the Standard Model of electroweak and strong interactions, based on the group product SU(3) X SU(2) X U(1). This model is still very successful[15] in explaining essentially all known experimental facts, with a minimal number of particles. In this model, neutrinos are massless (by construction), therefore there is no lepton mixing. With only one Higgs doublet, no heavy neutral leptons, no other particles, the model does not possess the ingredients which could lead to lepton-flavour violation. Therefore, there is a conservation law for partial lepton numbers, which happens to be accidental in the Standard Model. The $\mu \to e\gamma$ decay, muon-to-electron conversion and other similar processes should not exist, according to the Standard Model. Since they are not observed, why should one worry?

The answer lies in the future of the gauge theories of elementary particles and their interactions. It is obvious that the Standard Model can not be the final theory of nature, although it could be the correct low-energy limit of a more fundamental theory. The Standard Model is too arbitrary, it contains too many parameters and too many ad hoc assumptions. Many avenues are explored to build a more satisfactory theory: extension of the gauge group (additional fermions and gauge bosons), extension of the Higgs sector, left-right symmetric theories, grand unification, technicolor, supersymmetry, supergravity, superstrings, composite quarks and leptons, etc... It is a general feature of these theories that the conservation of partial lepton numbers must be violated at some level. In fact there are many mechanisms which can violate lepton flavour, and the difficulty is rather to explain why lepton-flavour conservation is so well respected in nature.There exists an abundant literature on the subject and I find it convenient to refer to the review papers by Scheck[16], Vergados[17], Costa and Zwirner[18].

A strong theoretical activity was triggered in the beginning of 1977 by a rumor that the $\mu \to e\gamma$ had been observed at SIN. It was suddenly realized that many models could predict a $\mu \to e\gamma$ branching ratio as large as $10^{-8}$ in a very natural way. The $\mu \to e\gamma$ and similar processes, $\mu \rightarrow e\gamma\gamma$, $\mu \rightarrow 3e$ and muon-to-electron conversion in a nucleus became the subject of intense experimental activities, mainly at the meson factories, LAMPF, SIN and TRIUMF.

The importance of these processes is easily understood. They do not exist in the Standard Model, therefore they open a window on the physics which lies beyond the Standard Model. Their discovery would be a major step forward and the beginning of a new era in theoretical and experimental particle physics. However, even if they are not found, upper limits on their branching ratios are extremely useful, since they put very stringent constraints on theoretical models. But why should one study all these processes and not just concentrate on the $\mu \rightarrow e\gamma$ decay? It is because these various muon-number violating processes are strongly model dependent. As an example, a model has been developped[19], which predicts that the $\mu \rightarrow e\gamma\gamma$ decay is faster than the $\mu \rightarrow e\gamma$ decay, something which is a priori not obvious. The model dependence of the $\mu \rightarrow 3e$ to $\mu \rightarrow e\gamma$ ratio has been studied by several authors[20][21]. This ratio can vary by several orders of magnitude. Therefore, it is important to search for all lepton-number-violating processes, because they can help in distinguishing between the numerous theoretical models.

One modest and straightforward extension of the Standard Model consists in giving small masses to the known neutrinos $\nu_e$, $\nu_\mu$ and $\nu_\tau$ (of course these masses must be compatible with our present experimental knowledge). Mixing between different lepton flavours can occur, giving rise to interesting phenomena like neutrino oscillations. But this is not sufficient to bring the rates of the lepton-number-violating decays to the level of experimental observation. In fact, assuming two generations for simplicity, the branching ratio of the $\mu \rightarrow e\gamma$ decay is given by[22]:

$$R(\mu \rightarrow e\gamma) = \frac{3\alpha}{32\pi} \sin^2\theta \cos^2\theta \frac{\left(m_1^2 - m_2^2\right)^2}{M_W^4}$$

where $m_1$ and $m_2$ are the neutrino masses (eigenvalues of the mass matrix), $\alpha$ the fine-structure constant, $M_W$ the W-boson mass and $\theta$ the mixing angle in the lepton sector. Using the experimental upper limit on the heaviest neutrino $\nu_\tau$ one calculates a branching ratio which is of the order of $10^{-18}$, not measurable. Of course, the situation would be different with heavy neutral leptons. Therefore, the $\mu \rightarrow e\gamma$ and similar processes are sensitive to new particles not contained in the Standard Model and to a range of high masses.

## 5. Searches for Lepton-Flavour Violation at Meson Factories[23]

These studies have been an important part of the mission of the meson factories, even if in the planning of these facilities the question of lepton-flavour violation was not so strongly emphasized. Experiments started in a rather modest way, using two sodium-iodide crystals to detect the electron and photon of the $\mu \rightarrow e\gamma$ decay. In fact, for a muon decaying from rest it is sufficient to measure the energies of these two particles. One unavoidable source of background is the radiative muon decay $\mu \rightarrow e\nu\nu\gamma$, due to the finite energy resolution of the detectors. Another important source of background comes from accidental coincidences between ordinary muon decay and radiative muon decay.

A measurement of the angle between electron and photon is redundant to define the kinematics but helps considerably in reducing the background. At SIN, the sodium-iodide crystals were equipped with a wire chamber for the electron and an hodoscope for the photon. However, such a device was limited to a sensitivity of $\sim 10^{-9}$ and therefore a second generation of experiments started with the construction of more sophisticated detectors. Let us consider three examples.

At LAMPF, the Crystal Box detector has achieved on the $\mu \rightarrow e\gamma$[24] and $\mu \rightarrow e\gamma\gamma$[25] the best upper limits (in the following, all upper limits correspond to a 90% confidence level, unless otherwise indicated):

$$BR(\mu \rightarrow e\gamma) \leq 4.9 \times 10^{-11}$$

$$BR(\mu \rightarrow e\gamma\gamma) \leq 7.2 \times 10^{-11}$$

The Crystal Box (Fig. 3) is a big modular sodium-iodide assembly (360 NaI crystals), providing a large solid angle (45% of $4\pi$) and high detection efficiency for the $\gamma$-rays. The positrons are tracked in a good-resolution cylindrical drift chamber arrangement (eight concentric shells). No magnetic field is applied. The apparatus also allows a good measurement of the angle between positron and $\gamma$-ray.

The first search for the $\mu \rightarrow 3e$ decay was performed by Korenchenko et al.[26], who established an upper limit of $10^{-9}$. Recently, at SIN, the SINDRUM1 detector has achieved the best upper limit on this decay[27]:

$$BR(\mu^+ \rightarrow e^+e^-e^+) \leq 1.0 \times 10^{-12}$$

SINDRUM1 (Fig. 4) consists of several cylindrical multiwire proportional chambers (C) placed in a magnetic field (M = magnet coils, S = solenoid) and surrounded by plastic scintillator hodoscopes (H) for triggering. Also represented are the preamplifiers (A), the light guides (L), photomultipliers (P), the muon beam (B) and the target (T). For a search of the $\mu \rightarrow 3e$ decay good timing resolution is essential as well as a precise reconstruction of the vertex.

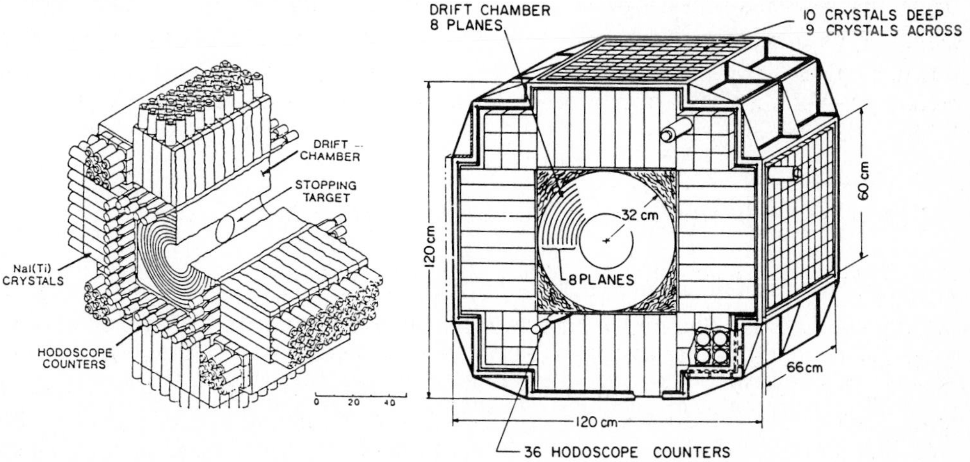


Fig. 3. Two views of the Crystal Box detector.

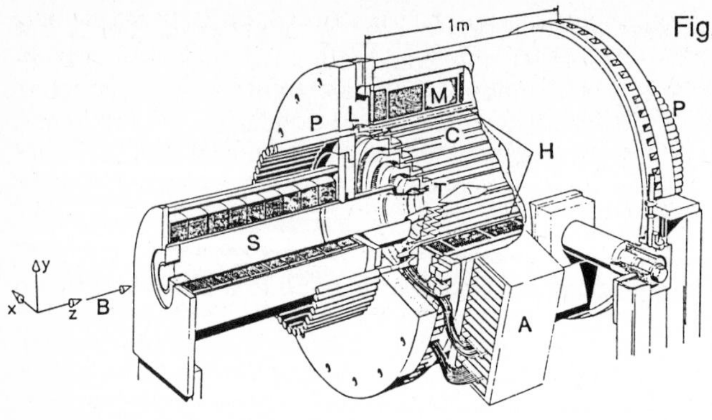


Fig. 4. The SINDRUM Detector.

At TRIUMF, the first operating *Time Projection Chamber* (Fig. 5) was built to study muon-to-electron conversion in Titanium[28]. It consists of a large gaseous volume where the ionization electrons drift towards planes equipped with sensitive wires. A position measurement on these wires defines two coordinates (x and y). The z coordinate is obtained by measuring the drift time of the ionization electrons, time zero being given by plastic-scintillator counters which signal that a charged particle has gone through the chamber. When studying muon-to-electron conversion one looks for the process in which the nucleus remains in its ground state. The signature is a

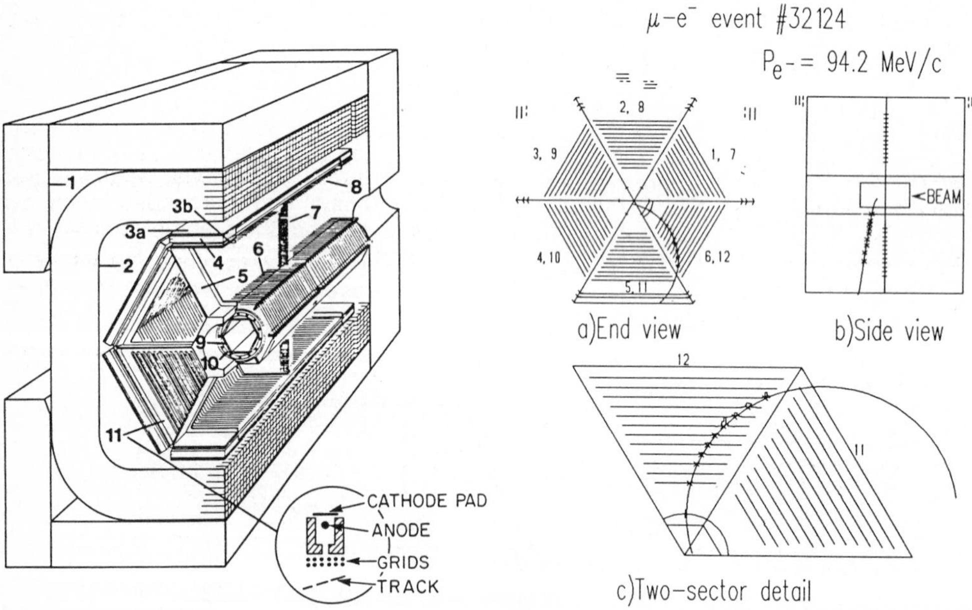


Fig. 5. Left: the TRIUMF Time-Projection-Chamber.

1: magnet iron; 2: coil; 3a and 3b: outer trigger scintillators; 4: outer trigger proportional counters; 5: support frame; 6: central electric field cage wires; 7: central high-voltage plane; 8: outer electric field cage wires; 9: inner trigger scintillators; 10: inner-trigger cylindrical proportional wire chamber; 11: end-cap proportional wire modules for track detection.

Fig. 5. Right: Electron trajectory in the TPC.

monoenergetic electron with a kinetic energy $E_e = m_\mu - B - R$, where B is the binding energy of the muonic atom and R the nuclear recoil energy (negligible). To a good approximation $E_e = m_\mu \sim 106$ [MeV]. This process has some advantages over the muon decays. The nuclear quarks act coherently, leading to an enhancement of the transition probability. The electron energy of 106 [MeV] is far above the end-point energy of the muon decay spectrum (~ 53 MeV). The background comes mainly from the muon decay in orbit: since the nucleus can participate in the momentum-energy sharing, the electron can (with a small probability however) take an energy up to 106 [MeV]. This contribution has been calculated theoretically [29]. Another source of background is due to muon radiative capture, i.e. $\mu^-(A,Z) \to \nu_\mu(A,Z-1)\gamma$. The photon can convert externally to give an asymmetric $e^+$-$e^-$ pair, and there is a certain probability that the electron takes up an energy close to 106 [MeV]. Other background sources are cosmic rays and residual pions in the muon beam. The best result on muon-to-electron conversion has been obtained with the TPC at TRIUMF[30]:

$$BR(\mu^- \to e^-) \leq 4.6 \times 10^{-12}$$

The same TRIUMF experiment has given an upper limit for the branching ratio of muon-to-positron conversion $\mu^-$ Ti $\to e^+$ Ca[31]:

$$BR(\mu^- \to e^+) \leq 9.0 \times 10^{-12} \qquad \text{for } E_e \geq 96 \text{ [MeV]}$$

and

$$BR(\mu^- \to e^+) \leq 1.7 \times 10^{-10} \qquad \text{(integrated over all positron energies).}$$

But this process is not coherent since there is some nuclear rearrangement, and this leads to many excited states in the final nucleus. Therefore the previous integrated limit is based on the use of a nuclear model (giant-resonance-excitation model) and it is subject to some ambiguity.

Fig. 5 shows an electron from $\mu$ decay in orbit. Fig. 5a) shows the trigger scintillators which have fired.

Let us note that the $\mu^- \to e^+$ conversion in a nucleus is very similar to the neutrinoless beta decay (Fig. 6).

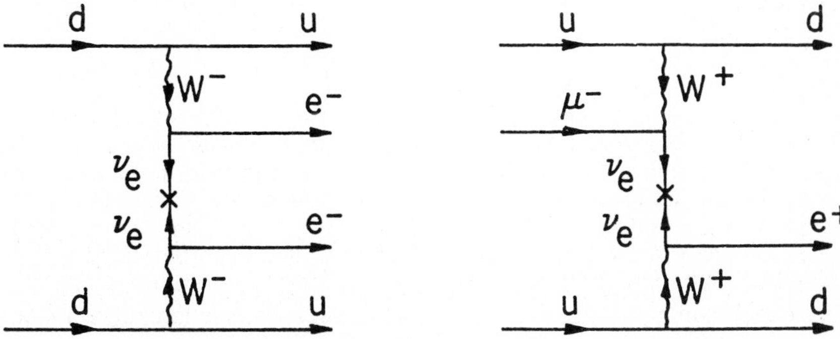


Fig. 6. Diagrams for the $\mu^- \to e^+$ conversion in a nucleus and the neutrinoless double-beta decay.

There is a possibility that the two neutrinos of normal muon decay are replaced by a neutral boson (massive or massless). Using a high-statistics positron spectrum from muon decay, obtained in their accurate measurement of the $\pi \rightarrow e\nu$ branching ratio, a TRIUMF group has searched for the exotic decay $\mu \rightarrow eX$, where X represents a neutral particle with a mass in the range 0 to 45 [MeV]. They have also analyzed other experiments and they present an upper limit for the decay $\mu \rightarrow eX$ in the mass range 0 to 104 [MeV][32].

So far none of these forbidden processes has been seen. They are still actively searched for, but how long can one still go? Obviously there is a hard limit which is probably not very far. Several on-going or proposed experiments are aiming at improvements in sensitivity by factors of 10 or even 100. The MEGA (MuEGAmma) experiment at LAMPF (a large U.S. collaboration) will search for the $\mu \rightarrow e\gamma$ decay at a level of 1-2 x $10^{-13}$ for the branching ratio[33]. The detector must combine extremely good energy resolution, high detection efficiency and the capability of handling high rates (the muon stopping rate will be 3 x $10^7$ [$s^{-1}$] at a duty cycle of 6%). The good energy resolution will be achieved by tracking the positrons and the electron-positron pairs from photon conversion in a large magnetic spectrometer. Four concentric conversion layers will allow a good detection efficiency for the photons. The very powerful multimicroprocessor system developped at Fermilab (Advanced Computer Program) will be used for data acquisition. Engineering runs should take place in the fall of 87 and the summer of 88. Data taking is expected in 89-90[34]. At SIN, there is a letter of intention[35] for the $\mu \rightarrow e\gamma$ decay, but more priority has been given to a new search for muon-to-electron conversion[36], using an upgraded version of the SINDRUM detector, SINDRUM2. The experimental difficulties are enormous. For instance, the pion contamination in the beam must be as low as $10^{-9}$. At high rates it is not possible to play the usual tricks to veto the beam pions, and one has to get rid of the pions in the muon beam itself.

Muonium ($M \equiv \mu^+e^-$) to antimuonium ($\overline{M} \equiv \mu^-e^+$) conversion is another interesting process. It is forbidden by the additive law of partial lepton numbers, but allowed by the multiplicative law. Here one defines parity-like numbers: +1 for electronic leptons and -1 for muonic ones. We now have fairly good data on muonium-to-antimuonium conversion, and several on-going or proposed experiments are still aiming at a better sensitivity. At LAMPF, a Heidelberg-Yale collaboration has searched for muonium-to-antimuonium conversion in vacuum[37]. Assuming that the process goes via a four-fermion interaction with a coupling constant $G_m$ they have obtained an upper limit for the ratio $G_m/G_F$:

$$G_m/G_F < 7.5 \text{ (90\% confidence level)}$$

($G_F$ is the usual Fermi constant as obtained from normal muon decay). At TRIUMF, an experiment uses a radiochemical technique and aims at a similar sensitivity[38]. Positive muons are stopped in $SIO_2$ powder where muonium is formed. It diffuses into vacuum, converts to antimuonium, producing a negative muon which is allowed to capture on Tungsten. This produces $^{184}Ta$, which has an 8.7 [h] half-life. The resulting activity is counted off-line in a low-background environment. It is important to note that there are several theoretical predictions giving $G_m/G_F$ values larger than one[39][40], claiming for new experimental results. The $\mu^+ \rightarrow e^+\overline{\nu}_e\nu_\mu$ decay is also forbidden by the additive law and allowed by the multiplicative law. It has been searched for at LAMPF[41]. The upper limit on the branching ratio is 5 X $10^{-2}$.

## 6. Searches in Other Sectors

Tau decays have been extensively studied at the DORIS II storage ring (DESY, Hamburg) with the ARGUS detector. These experiments have not yet reached the levels of sensitivity achieved at the meson factories but they are already providing very useful constraints on theoretical models. They are interesting because they are complementary to the searches at meson factories: they test lepton-flavour violation and lepton-number violation in a different sector (involving the third generation). Also, because of the higher mass of the $\tau$ lepton, they offer a larger variety of decays and they can test more violation mechanisms. In some models one introduces a *generation number* which characterizes each generation of quarks and leptons. In the $\tau^- \to \mu^- K^{*o}$ decay the generation number does not change since the decrease of -1 in the lepton sector is compensated by the increase of +1 in the quark sector (Fig. 7). Therefore, this decay is favored in certain models[42].

Using $e^+e^-$ annihilation at center of mass energies near 10 [GeV], the ARGUS collaboration has obtained the following results[43], which are compared to the MARK II results [44] (Table 2).

The MARK III Collaboration has also obtained upper limits on the decays $\tau \to eX^o$ and $\tau \to \mu X^o$, where $X^o$ is a light weakly interacting particle[45]. If $X^o$ is a massless Goldstone boson G, one gets the following limits (95% C.L.):

$$B(\tau \to \mu G)/B(\tau \to \mu \nu\nu) \leq 12.5\%$$

and:

$$B(\tau \to eG)/B(\tau \to e\nu\nu) \leq 4.0\%$$

from which one deduces a lower limit on the mass scale of the corresponding symmetry breaking:

$$M \geq 3 \times 10^6 \text{ [GeV]}.$$

Several groups have searched for the forbidden reaction $D^o \to \mu e$. The MARK II group has published an upper limit of $2.1 \times 10^{-3}$ for the branching ratio[47]. The ACCMOR group at CERN, using a 200 [MeV] negative pion beam incident on a fixed berylium target and high-quality microstrip silicon detectors, has obtained an upper limit of $9. \times 10^{-4}$[48]. An even better (and model-independent) upper limit has been obtained at SLAC by the MARK III group, i.e. $1.5 \times 10^{-4}$[49]. They used $e^+$-$e^-$ collisions at the $\psi$ (3770 [MeV]) resonance. More recently, preliminary results have been announced by

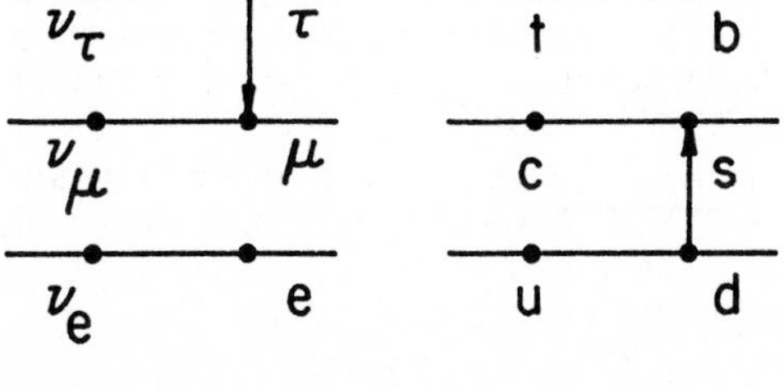


Fig. 7. The $\tau^- \to \mu^- K^{*o}$ decay

Table 2. Upper limits (90% C.L.) on forbidden $\tau$ decays[46]

| decay mode | ARGUS | | MARK II |
|---|---|---|---|
| $\tau^- \to e^-e^+e^-$ | $3.8 \times 10^{-5}$ | | $4.0 \times 10^{-4}$ |
| $\tau^- \to e^-\mu^+\mu^-$ | $3.3 \times 10^{-5}$ | | $3.3 \times 10^{-4}$ |
| $\tau^- \to \mu^-e^+e^-$ | $3.3 \times 10^{-5}$ | | $4.4 \times 10^{-4}$ |
| $\tau^- \to \mu^-\mu^+\mu^-$ | $2.9 \times 10^{-5}$ | | $4.9 \times 10^{-4}$ |
| $\tau^- \to \ell^{-+}\ell^{+-}\ell^-$ | $3.8 \times 10^{-5}$ | | |
| $\tau^- \to e^-\pi^+\pi^-$ | $4.2 \times 10^{-5}$ | | |
| $\tau^- \to \mu^-\pi^+\pi^-$ | $4.0 \times 10^{-5}$ | | |
| $\tau^- \to e^-\rho^0$ | $3.9 \times 10^{-5}$ | | $3.7 \times 10^{-4}$ |
| $\tau^- \to \mu^-\rho^0$ | $3.8 \times 10^{-5}$ | | $4.4 \times 10^{-4}$ |
| $\tau^- \to \ell^{-+}\pi^{+-}\pi^-$ | $6.3 \times 10^{-5}$ | | |
| $\tau^- \to e^-\pi^+K^-$ | $4.2 \times 10^{-5}$ | | |
| $\tau^- \to \mu^-\pi^+K^-$ | $1.2 \times 10^{-4}$ | | |
| $\tau^- \to e^-K^{*o}$ | $5.4 \times 10^{-5}$ | $(e^-K^o)$ | $1.3 \times 10^{-3}$ |
| $\tau^- \to \mu^-K^{*o}$ | $5.9 \times 10^{-5}$ | $(\mu^-K^o)$ | $1.0 \times 10^{-3}$ |
| $\tau^- \to \ell^{-+}\pi^{+-}K^-$ | $1.2 \times 10^{-4}$ | | |

ARGUS ($0.9 \times 10^{-4}$) and E691 ($0.8 \times 10^{-4}$)[50]. These $D^o$ decays imply a change in generation number both in the quark and lepton sectors.

They could be induced by massive leptoquarks (Fig. 8), which are predicted in a variety of unification models. In certain of these models, the leptoquarks couple "up-type" quarks (u, c, t) to charged leptons and "down-type" quarks (d, s, b) to neutral leptons. Therefore, the $D^o \to \mu e$ decay should be favored over the $K^o \to \mu e$ and $B^o \to \mu e$ decays. In any case, it is difficult to suppress all these decays simultaneously. Also, if Higgs bosons were responsible for lepton-flavour violation, they should be more strongly coupled to the charmed quarks than to the strange quarks. Therefore the

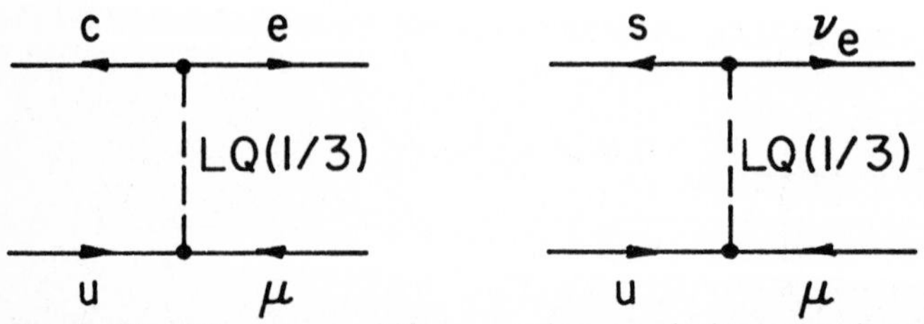


Fig. 8. The $D^o \to \mu e$ and $K^o \to \mu e$ decays via leptoquark exchange

forbidden $D^o$ decays could be more sensitive than the $K^o$ decays. There is also an upper limit on the $B^o \to \mu e$ decay, obtained at CLEO [51]:

$$BR(B^o \to \mu e) \leq 0.9 \times 10^{-4}$$

The kaon sector is presently being investigated very actively. Several experiments, at Brookhaven and KEK, aim at sensitivities of $10^{-10}$[52], $10^{-11}$[53] and $10^{-12}$[54] for the $K_L \to \mu e$ decay. Until recently the situation was somewhat confused. An experiment[55] gave an upper limit of $1.6 \times 10^{-9}$ on the $K_L \to \mu e$ branching ratio. But it failed to observe the $K_L \to \mu^+\mu^-$ decay and gave for its branching ratio an upper limit of $1.8 \times 10^{-9}$, a factor of 5 below the measured branching ratio which is $9.1 \times 10^{-9}$. The upper limit of $6 \times 10^{-6}$ quoted in the Table of Particle Properties[56] is a conservative estimate, but some experts think that the experiment should have seen the effect at the $10^{-8}$ level[57]. The recent preliminary result of the BNL-780 experiment has improved the situation. This experiment has obtained an upper limit[58]:

$$B(K^o{}_L \to \mu e) < 6.7 \times 10^{-9}$$

Assuming that the decay is mediated by a heavy boson of mass M, with the same coupling as the W, the branching ratio can be written as:

$$B(K^o{}_L \to \mu e) < 2.5 \times 10^{-3}\, M^{-4} \quad (M \text{ in TeV})$$

This provides a lower limit for M: $M > 25$ TeV.

On the $K \to \pi\mu e$ mode we still have the upper limits of Diamant-Berger et al.[59]:

$$B(K^+ \to \pi^+\mu^+e^-) \leq 5 \times 10^{-9} \qquad \text{and} \qquad B(K^+ \to \pi^{+-}\mu^{-+}e^+) \leq 7 \times 10^{-9}$$

A proposed BNL experiment aims at a sensitivity of $10^{-11}$[60]. It should be noted that searches for the $K \to \mu e$ and $K \to \pi\mu e$ are complementary. The first decay requires an axial-vector or a pseudoscalar coupling. The second requires a vector or a scalar coupling.

The $K^+ \to \mu^+\nu_e$ decay has been searched for[61] by looking for the $\nu_e$ interactions in a neon/hydrogen bubble chamber. The upper limit on the branching ratio is 0.4%.

These rare kaon decays will constitute an important part of the mission of the (proposed) future kaon factories[62].

## 7. A Very Arbitrary Selection of Theoretical Topics

There is a considerable theoretical litterature on the subject. Until 1986 the field has been well covered by the review papers mentioned above[63]. In fact, lepton-flavour-violating processes have been discussed abundantly and many mechanisms have been considered.

A straightforward generalization of the Standard Model introduces heavy neutral leptons. One keeps the general structure of the model, allowing for massive neutrinos[64] and mixing between different lepton flavours. Additional generations of quarks and leptons could provide these heavy neutrinos. Using the data on $\mu \to e\gamma$, etc... and muon-to-electron conversion, it is possible to put limits on heavy neutrino masses and mixing matrix elements[65]. Another possibility lies in theories with right-handed currents, where the right-handed charged lepton is put in a doublet with a heavy neutral partner. Heavy neutral leptons with masses in the multi-[GeV] range could bring the $\mu \to e\gamma$ decay to the level of observation[66]. Some models have also introduced heavy doubly-charged leptons[67].

It has been shown a long time ago that with two Higgs doublets lepton-flavour violation becomes possible[68]. One can also introduce other Higgs multiplets, as in the *Majoron Models* [69][70]. In fact, there are several good reasons to enlarge the Higgs sector. For instance, it is an economical way to give Majorana masses to the neutrinos[71]. It also happens naturally in left-right symmetric theories and there are interesting consequences for the various lepton-flavour violating decays[72] (Fig. 9). The left-right symmetric theories have obvious advantages: they can account for parity violation in the low-energy domain, they naturally give rise to neutrino masses. They are more aesthetic and less arbitrary than the purely left-handed theories. Bilenky and Petcov have discussed lepton-flavour violation in the Triplet Majoron Model[73]. One very interesting feature is the occurrence of the $\mu \to 3e$ decay with a branching ratio larger than that of the $\mu \to e\gamma$ and possibly close to the present experimental upper limit. Santamaria et al.[74] have studied the influence of the scalar particles of the model on the parameters of normal muon decay and have also considered the exotic forbidden decays $\mu \to e\theta\theta$, $\mu \to e\rho_L\rho_L$ and $\mu \to e\rho_L\theta$, where $\theta$ is the massless Goldstone boson associated with the breaking of lepton number, the Majoron, and $\rho_L$ a light scalar whose mass is proportional to the triplet vacuum expectation value. A model with three Higgs doublets has been studied by Cheng and Sher[75].

In several unification schemes one introduces leptoquarks which can also contribute to lepton-flavour violation (Fig. 10). Limits on leptoquark masses and couplings have been discussed by several authors[76][77].

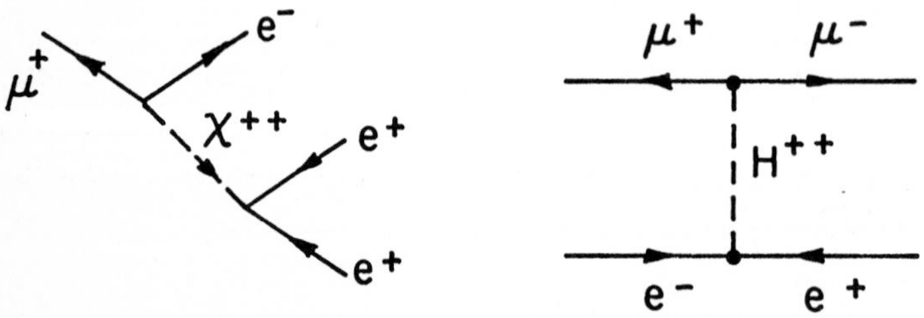


Fig. 9. The decay $\mu \to eee$ and muonium-antimuonium conversion via doubly-charged Higgs exchange

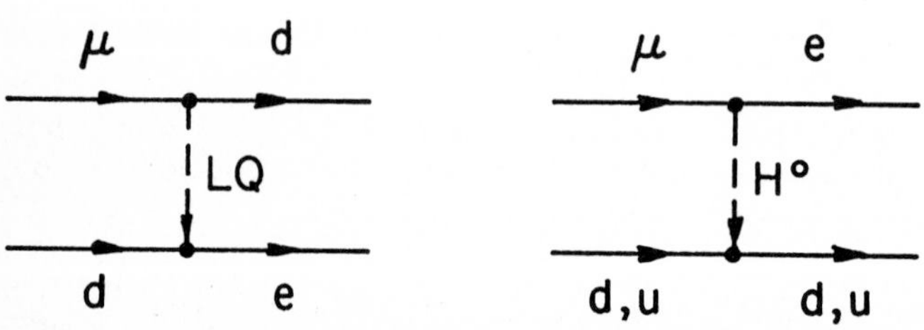


Fig. 10. Muon-to-electron conversion via leptoquark or Higgs exchange.

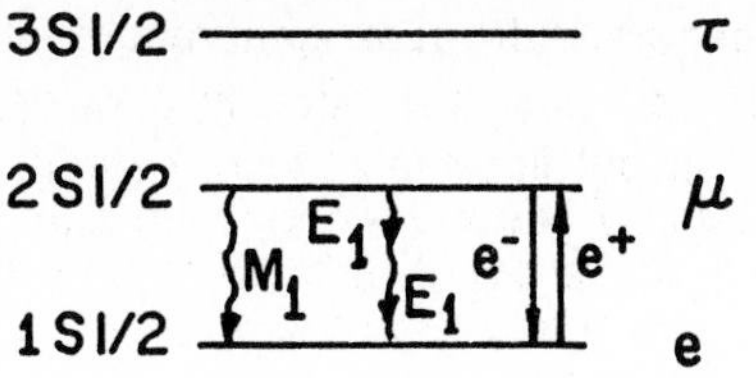


Fig. 11. The decays $\mu \to e\gamma$, $\mu \to e\gamma\gamma$ and $\mu \to eee$ in a composite model of leptons.

The recurrence of several quark and lepton generations strongly suggests that these particles are made of more elementary constituents[78]. In this context also the lepton-number-violating processes are important[79]. For example, it is conceivable that e, $\mu$ and $\tau$ are radial excitations ($1s_{1/2}$, $2s_{1/2}$ and $3s_{1/2}$) of a system of bound preons. In such a model the muon could decay by double $E_1$ transition to the electron rather than by a single $M_1$ transition (Fig. 11). The two-gamma transition involves an intermediate virtual P state. With the $\mu \to e\gamma\gamma$ it is possible to put stringent limits on the mass of the P states in a potential model of composite leptons[80].

Using the upper limits on the $\mu \to e\gamma$ and $\mu \to e\gamma\gamma$ branching ratios, Tomozawa[81] has obtained lower limits for the compositeness mass scale. He uses a potential of the form:

$$V(r) = V_C + V_\ell(E)$$

where $V_C$ is a Coulomb-like potential:

$$V_C = -\frac{\alpha_C}{r}$$

The potential $V_\ell(E)$ is energy- and angular momentum-dependent, and $\alpha_C$ is a dimensionless coupling constant (the analog of the fine-structure constant). The upper limit on the $\mu \to e\gamma\gamma$ decay provides the lower limit:

$$M\alpha_c = \left[\frac{\alpha^2 J}{12\pi}\frac{m_\mu}{\Gamma(\mu \to e\gamma\gamma)}\right]^{\frac{1}{4}}$$

where $\alpha$ is the fine-structure constant and $m_\mu$ the muon mass. The dimensionless parameter J satisfies the condition:

$$4.4 < J < 7.1$$

Using the upper limit $BR(\mu \to e\gamma\gamma) \le 7.2 \times 10^{-11}$ one obtains $M\alpha_C \ge 44$ TeV.

The $\mu \to e\gamma$ decay provides an independent limit:

$$\frac{M}{\alpha_c} = \frac{16}{81}\left[\frac{1}{2}\frac{\alpha m_\mu}{\Gamma(\mu \to e\gamma)}\right]^{\frac{1}{2}} m_\mu$$

Using the upper limit $BR(\mu \to e\gamma) \le 4.9 \times 10^{-11}$ one obtains $M/\alpha_C \ge 11 \times 10^7$ TeV.

Finally, the two results can be combined to get a limit which is independent of $\alpha_C$:

$$M = \frac{4}{9}\left[\frac{J\alpha^4 m_\mu^3}{48\pi\left[\Gamma(\mu \to e\gamma)\right]^2 \Gamma(\mu \to e\gamma\gamma)}\right]^{\frac{1}{8}} m_\mu$$

With the same experimental upper limits as above one obtains $M \geq 70 \times 10^3$ TeV. With the assumption $\alpha_C > \alpha$ the $\mu \to e\gamma$ decay gives the limit $M > 10^6$ TeV. Let us note that this limit is much higher than the one obtained from the anomalous magnetic moment of the electron (g - 2): $M \geq 10$ TeV. But it is, of course, model-dependent.

The existence of several quark and lepton families could be related to the existence of a new symmetry, called *family symmetry.* This symmetry is evidently broken, and if it is a global symmetry there should appear massless Goldstone bosons. They have received the name of *familons* [82]. These particles could be important for cosmology and could also serve as axions to solve the "Strong CP Problem"[83]. Familons are responsible for $s \to df$ and $\mu \to ef$ transitions and can be searched for in $K \to \pi f$ and $\mu \to ef$ decays. They would manifest themselves in the $\mu \to ef$ decay as a peak in the electron energy spectrum, at the end-point energy. The corresponding electrons would also have a longitudinal polarization less than one[84]. A nice experimental possibility results from the fact that with positive muons, at the end-point energy, no positrons are emitted at 180° with respect to the muon spin if one assumes V-A coupling (this suppression can be easily understood on the basis of helicity arguments). An experiment was done at TRIUMF to search for right-handed currents[85]. It has also been analyzed to extract an upper limit for familon emission. The result can be used to obtain a lower limit for the scale of family-symmetry breaking, which is $F \geq 9.9 \times 10^9$ [GeV]. Another limit has been obtained by a LAMPF group[86]. They have analysed their $\mu \to e\gamma$ data[87] to search for the $\mu \to e\gamma f$ decay. They obtain for F a lower limit of $2.9 \times 10^9$ [GeV]. A similar analysis has been performed by Bélanger and Ng[88]. In a more general way the role of various scalar particles in the $\mu \to e\gamma$ decay and other observables has been studied by Lopes and Martins Simoes[89].

*Horizontal gauge groups* have been introduced by several authors to connect the various generations of quarks and leptons[90]. Rare decays provide constraints on *horizontal gauge bosons* and their interactions[91]. Recently, horizontal gauge interactions have been discussed in the context of CP violation. Predictions for lepton-flavour violating decays of muons, kaons, D and B mesons have been made[92] and they show interesting rates.

Following the discovery of $e^+$-$e^-$ peaks in heavy ion collisions at GSI[93], the search for axions has been revived. Rare muon decays can be used for that purpose. The $\mu \to e\gamma\gamma$ data has been analyzed to search for $\mu \to ea$, followed by $a \to \gamma\gamma$[94]. More recently, at SIN, a measurement of the $\mu \to eee$ decay has been analyzed to put a limit on the decay $\mu \to ea$ followed by $a \to e^+e^-$. Combining this result with the study of the $\pi^+ \to e^+\nu_e e^+e^-$ decay, the SIN group can rule out the 1.7 [MeV] axion which had been postulated to explain the GSI results[95].

When discussing possible deviations from the Standard Model it is convenient to parametrize these deviations with a scale parameter $\Lambda$ which is simply related to the distance within which new interactions come into play[96][97]. Rates for the various forbidden muon decays can be expressed as a function of $\Lambda$ and useful lower limits on this parameter can be obtained. For instance, using the work of Buchmüller and Wyler[98] one calculates the following lower bounds on $\Lambda$ (in [TeV]): 14000, 175 and 340 from the $\mu \to e\gamma$, $\mu \to 3e$ decays and muon-to-electron conversion, respectively.

Majoron emission in $\mu$ and $\tau$ decays has been studied by Santamaria and Valle[99] in the context of spontaneous R-parity breaking in supergravity. Their main motivation was the solar neutrino problem and the generation of neutrino masses, but their model is also consistent with a branching ratio for the $\mu \to ef$ decay which is close to the present experimental upper limit.

*Supersymmetry* has been introduced to cure several problems in unification theories and also comes naturally from superstrings. Supersymmetric partners of ordinary particles can induce lepton-flavour violation (Fig. 12).

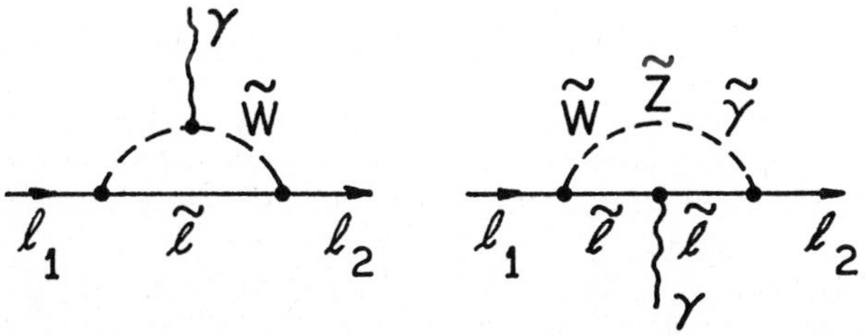

Fig. 12 Some diagrams for $\ell_1 \to \ell_2\gamma$ decay and and $Z^o$ decay via supersymmetric particles.

Rare processes can be used to put useful limits on flavour-violating interactions induced by supersymmetry[100]. Ellis and Nanopoulos have used the $\mu \to e\gamma$ decay to put constraints on slepton mass differences[101].

Supersymmetry and compositeness can be combined to yield supercomposite models, which are characterized by two different energy scales: $M_S$, where supersymmetry breaking occurs, and $M_C$, the scale of compositeness. Campbell and Peterson have studied the limits which can be placed on these energy scales by using experimental results on $\mu$-e universality (the $\pi \to e\nu/\pi \to \mu\nu$ ratio) and muon-to-electron conversion[102]. Using for the latter the TRIUMF result on Ti (upper limit of $4.5 \times 10^{-12}$) they obtain the constraint:

$$\frac{C}{4\pi^2}\frac{M_Z^2}{M_C M_S} \leq 1.7 \times 10^{-7}$$

where C is calculable in their model (0.143) and $M_Z$ is the $Z^o$ mass. Assuming that $M_S$ is not very different from the Fermi scale, they obtain for $M_C$ the bound:

$$M_C > 2.3 \times 10^6 \text{ [GeV]}$$

A slightly lower value is deduced from $\mu$-e universality. The authors conclude that under the assumption that supersymmetry breaking occurs at the Fermi scale the supercompositeness scale must exceed 100 TeV. In another paper, rare decays have been discussed in the context of an effective-Lagrangian study of three generations of supersymmetric composite quarks and leptons[103].

There is presently, among theorists, a great deal of enthusiasm for *superstring theories*, which look very promising and could, according to the optimists, lead to the "Theory of Everything". There is, however, a lot of skepticism about the possibility to test these theories in the low-energy domain, the one which is presently accessible to experiments, or will be in the near future. But one should recognize that some efforts have been made to derive low-energy phenomenology from the superstrings. There is still a considerable confusion on how to do that and there is no unique and universally accepted way to come down from the Planck mass to the laboratory energies. The superstring-inspired $E_6$ unification group has been largely exploited[104] (in fact it has always been an attractive possibility). It introduces a sector of particles associated with the 27 representation. In addition to the fermions of the Standard Model one has two Higgs doublets H and H', two colour field triplets D and $D^c$ and two neutral fermions: $\nu^c$ is a right-handed neutrino already present in SO(10) and N is an SO(10) singlet. New interactions appear, which can contribute to lepton-flavour violation. Campbell et al.[105] have made a complete analysis of the lepton-flavour-violating decays $K_L \to \mu e$, $K^+ \to \pi^+ \mu e$, $D^o \to \mu e$, $D^o \to \mu e X$, $B^o \to \mu e$, $\mu^-(A,Z) \to e^-(A,Z)$, $\mu \to 3e$, $\mu \to e\gamma$, together with other rare decays and other fundamental observables: mixing in the neutral K, D and B sectors, neutron dipole moment, etc... Whether evidence for superstrings can be obtained from lepton-flavour-violating processes represents a formidable challenge for theorists and experimentalists. The paper by Campbell et al. shows that the rare processes, especially the $\mu \to e\gamma$ decay and muon-to-electron conversion, provide powerful constraints on masses and coupling constants in the superpotential. The role of the $E_6$ exotic particles has also been discussed by Masiero et al.[106]. These particles can be active in such processes as the rare kaon decays, the $\mu \to e\gamma$ decay and muon-to-electron conversion, and the $\pi^o \to \mu e$ decay. Borzumati and Masiero[107] have studied renormalization effects in the scalar-lepton mass matrix in spontaneously broken N = 1 supergravity theories. They show that photino-lepton-slepton couplings can lead to appreciable rates for $\mu \to e\gamma$ and muon-to-electron conversion. In these two cases the present upper limits on rare muon processes can be used to put useful limits on the parameters of the models (masses and coupling constants). Lepton-number nonconservation in $E_6$ superstring models has been discussed by Ma[108]. Various scenarios are considered together with their phenomenological implications.

Interesting connections can be made between lepton-flavour violation and other fundamental observables. For instance, in a wide class of models[109], the $\mu \to e\gamma$ decay rate can be related to the magnitude of the electric dipole moment of the electron. The corresponding CP violation originates in the phases between several flavour-changing neutral currents[110]. The predicted upper limits on the electric dipole moments are on the $10^{-26}$ [e-cm] scale, still a factor of ~ 100 below the present experimental limit[111].

An interesting question is: will it be possible to observe lepton-flavour violation in the $Z^o$ decays at the forthcoming $e^+$-$e^-$ colliders (SLC, LEP)? Eilam and Rizzo[112] have considered the lepton-flavour-violating decay $Z^o \to \mu\tau$. Such a decay could result from the mixing of the new $E_6$ fermions with the ordinary quarks and leptons. Their conclusion is that the $Z^o \to e\mu$ has a too small branching ratio to be seen ($\leq 7.2 \times 10^{-13}$, as constrained by the $\mu \to eee$ decay) whereas there is still hope to see the $Z^o \to e\tau$ decay ($\leq 3.4 \times 10^{-4}$) and the $Z^o \to \mu\tau$ decay ($\leq 4.2 \times 10^{-4}$), where the constraints come from the upper limits on the $\tau \to 3\mu$, 3e, $2\mu e$ and $2e\mu$ decays.

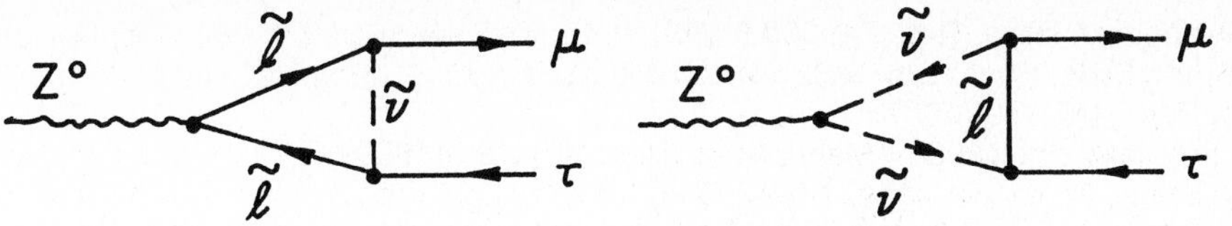


Fig. 13. Some diagrams for the $Z^o \rightarrow \mu\tau$ decay via sleptons.

The $Z^o \rightarrow \mu\tau$ decay via slepton mixing has been considered by Levine[113]. He uses softly-broken supersymmetry with flavour mixing in the supersymmetric sector (Fig. 13). The calculated branching ratio for the forbidden $Z^o \rightarrow \tau^+\mu^-$ decay is of the order of $10^{-6}$ to $10^{-7}$. With this mechanism alone there is no hope to observe this process. Bernabéu and Santamaria have studied lepton-flavour-violating decays of the $Z^o$ in the Scalar Triplet Model[114]. There is no GIM supression in their model and they find for the $Z^o \rightarrow \mu\tau$ and $Z^o \rightarrow e\tau$ decays the following branching ratios:

$$R(Z^o \rightarrow \mu\tau) \leq 5.9 \times 10^{-6}$$

$$R(Z^o \rightarrow e\tau) \leq 7.9 \times 10^{-6},$$

still difficult to detect.

Using a superstring-inspired model with new neutral fermions, Bernabéu et al.[115] have predicted, for the $Z^o \rightarrow \mu\tau$ decay, branching ratios in the $10^{-5}$ range, which could be accessible to experiments at SLC or LEP. But it should be noted that these predictions rest on the present experimental upper limits for the $\tau \rightarrow \mu\gamma$ and $\tau \rightarrow e\gamma$ decays, therefore one must exert some caution.

## 8. Conclusion

The experimental search for lepton-flavour-violating processes and the corresponding theoretical activity have played a useful role in the quest for "new physics", i.e. physics which lies beyond the Standard Model. These processes are sensitive to mass scales which can not be reached by the present and even future accelerators. The meson factories will soon complete they mission to search for rare muon and pion decays. The more distant future belongs to the kaon, charm and beauty factories, and to the large colliders. In view of the present experimental situation there is still some hope that lepton-flavour violation will be observed but it is also possible that it will escape all human efforts to detect it. Whatever the outcome is, lepton-flavour violation will remain as a beautiful chapter in the history of physics.

## References

1. See for instance S. Frankel, *"Rare and Ultrarare Muon Decays"*, in "Muon Physics", V.W. Hughes and C.S. Wu, Editors, Academic Press, New York (1975)
2. S.H. Neddermeyer and C.D. Anderson, Phys. Rev. 51, 884 (1937)
3. M. Conversi et al., Phys. Rev. 71, 209 (1947)
4. H. Yukawa, Proc. Phys. Math. Soc. Jpn 17, 48 (1935)
5. E.P. Hincks and B. Pontecorvo, Phys. Rev. 73, 257 (1948)
6. A. Lagarrigue and C. Peyrou, C.R. Heb. Acad. Sci. Paris 234, 873 (1952)

7. J. Steinberger, Phys. Rev. 74, 500 (1948)
8. L. Michel, Proc. Phys. Soc. A63, 514 (1950); see also C. Bouchiat and L. Michel, Phys. Rev. 106, 170 (1957)
9. R.P. Feynman and M. Gell-Mann, Phys. Rev. 109, 193 (1958)
10. G. Feinberg, Phys. Rev. 110, 1482 (1958)
11. S. Lokanathan and J. Steinberger, Phys. Rev. 98, 240A (1955)
12. S. Parker et al., Phys. Rev. 133, 768B (1964)
13. K. Nishijima, Phys. Rev. 108, 907 (1957)
J. Schwinger, Ann. Phys. 2, 407 (1957)
14. G. Danby et al., Phys. Rev. Lett. 9, 36 (1962)
15. P. Langacker, Proceedings of the 10th Workshop on Particles and Nuclei, Heidelberg, October 1987, Editor H.V. Klapdor, Springer Verlag
16. F. Scheck, Phys. Rep. 44, 187 (1978)
17. J.D. Vergados, Phys. Rep. 133, 1 (1986)
18. G. Costa and F. Zwirner, Nuovo Cimento Rivista 9, 1 (1986)
19. J.D. Bowman et al., Phys. Rev. Lett. 41, 442 (1978)
20. S.M. Barr and S. Wandzura, Phys. Rev. D16, 707 (1977)
21. W.J. Marciano and A.I. Sanda, Phys. Letters B67, 303 (1977)
22. S.M. Bilenky et al., Phys. Letters B67, 309 (1977)
23. R. Engfer and H.K. Walter, Ann. Rev. Nucl. Part. Sci. 36 , 327 (1986); see also the contributions of R. Engfer, R. Mischke and P. Depommier in "Proceedings of the International Symposium on Weak and Electromagnetic Interactions", Heidelberg, July 1986, Editor H.V. Klapdor, Springer Verlag; see also J. Schacher, Comments Nucl. Part. Phys., 8, 97 (1978)
24. R.D. Bolton et al., Phys. Rev. Lett. 56, 2461 (1986)
25. D. Grosnick et al., Phys. Rev. Lett. 57, 3241 (1986)
26. S.M. Korenchenko et al., Sov. Phys. JETP 43, 1 (1976)
27. U. Bellgardt et al., Nucl. Phys. B299, 1 (1988).
W. Bertl, Proccedings of the International Europhysics Conference on High-Energy Physics, Uppsala, Sweden, June 25 - July 1, 1987, Edited by O. Botner, Uppsala University, Vol. 1, p. 592
28. D. Bryman et al., in "The Time Projection Chamber", J.A. Macdonald, editor, AIP Conf. Proc. Number 108, American Institute of Physics (1983)
29. F. Herzog and K. Alder, Helv. Phys. Acta 53, 53 (1980)
30. S. Ahmad et al., Phys. Rev. Lett. 59, 970 (1987)
31. See reference 30
32. D.A. Bryman and E.T.H. Clifford, Phys. Rev. D57, 2787 (1986)
33. M.D. Cooper, Spokesman, LAMPF Experiment N^o^. 969 (1985)
34. M.D. Cooper, private communication
35. W. Bertl et al., *"Search for the decay $\mu \rightarrow e\gamma$,* Letter of intent for an experiment at SIN (February 1985)
36. A. Badertscher et al., *"Search for $\mu^- \rightarrow e^-$ conversion"*, Proposal for an experiment at SIN (November 1986)
37. B. Ni et al., Phys. Rev. 59, 2716 (1987)
38 A. Olin et al., TRIUMF- Research Proposal E304 (April 1987)
39. E. Derman, Phys. Rev. D19, 317 (1979)
40. A. Halprin, Phys. Rev. Lett. 48, 1313 (1982)
41. S.E. Willis, Phys. Rev. Lett. 44, 522 (1980)
42. Y. Ne'eman, Phys. Lett. B82, 69 (1976)
43. H. Albrecht et al., Phys. Letters B185, 228 (1987)
44. K.G. Hayes et al., Phys. Rev. D25, 2869 (1982)
45. R.M. Baltrusaitis et al., Phys. Rev. Lett. 55, 1842 (1985)

46. See reference 43
47. K. Riles et al., Phys. Rev. D35, 2914 (1987)
48. H. Palka et al., Phys. Letters B189, 238 (1987)
49. J.J. Becker et al., Phys. Letters B193, 147 (1987)
50. C. Grab, Proceedings of the International Europhysics Conference on High-Energy Physics, Uppsala, Sweden, June 25 - July 1, 1987, Edited by O. Botner, Uppsala University, Vol. 1, p. 349
51. P. Avery et al., Phys. Letters B183, 429 (1987)
52. M.P. Schmidt, W.M. Morse et al., *"A search for flavor-changing neutral currents $K_L \to \mu e$ and $K_L \to e^+ e^-$"*, BNL experiment AGS-780
53. T. Inagaki et al., KEK proposal 85-1 (1985)
54. S.G. Wojcicki et al., *"Study of very rare $K_L$ decays"*, BNL experiment AGS-791
55. A.R. Clark et al., Phys. Rev. Lett. 26, 1667 (1971)
56. Particle Data Group, Phys. Letters B170, 1 (1986)
57. D. Bryman, in *"New Frontiers in Particle Physics"*, Proceedings of the First Lake Louise Winter Institute, J.M. Cameron et al., Editors, World Scientific, Singapore (1986), p. 353
58. H.B. Greenlee et al., Phys. Rev. Lett 60, 893 (1988)
59. A.M. Diamant-Berger et al., Phys. Letters B62, 485 (1976)
60. M. Zeller et al., *"A Search for the Rare Decay $K^+ \to \pi^+ \mu^+ e^-$"*, AGS Experiment 777 (1982)
61. L. Lyons et al., Z. Phys. C10, 215 (1981)
62. P. Herczeg, Proceedings of the LAMPF II Workshop, 1980
H.K. Walter, in *"The Future of Medium- and High-Energy Physics in Switzerland"*, Les Rasses, Switzerland, May 17-18,1985; published by SIN, CH-5234, Villigen, Switzerland
63. See references 16, 17 and 18
64. R.E. Shrock, Phys. Rev. D24, 1232 (1981)
65. T. Numao, Phys. Rev. D34, 2900 (1986)
66. T.P. Cheng and L.F. Li, Phys. Rev. Lett. 38, 381 (1977); Phys. Rev. D16, 1425 (1977)
67. F. Wilczek and A. Zee, Phys. Rev. Lett. 38, 531 (1977)
68. J.D. Bjorken and S. Weinberg, Phys Rev. Lett. 38, 622 (1977)
69. G.B. Gelmini and M. Roncadelli, Phys. Letters B99, 411 (1981)
70. Y. Chikashige et al., Phys. Letters B98, 265 (1981)
71. See reference 69
72. R.N. Mohapatra and P.B. Pal, Phys. Lett. B179, 105 (1986)
73. S.M. Bilenky and S.T. Petcov, Rev. Mod. Phys. 59, 671 (1987)
74. A. Santamaria et al., Phys. Rev. D36, 1408 (1987)
75. T.P. Cheng and M. Sher, Phys. Rev. D35, 3484 (1987)
76. O. Shanker, Nucl. Phys. B206, 253 (1982)
77. W. Buchmüller and D. Wyler, Phys. Letters B177, 377 (1986)
78. L. Lyons, *"An Introduction to the Possible Substructure of Quarks and Leptons"*, in Prog. Nucl. Part. Phys., D. Wilkinson, Editor, Pergamon Press, Vol. 10, 227
79. V. Visnjic-Triantafillou, Phys. Letters B95, 47 (1980); Phys. Rev. D25, 248 (1982)
80. Y. Sirois, M.Sc. thesis, University of Montreal, unpublished
81. Y. Tomozawa, Phys. Rev. D25, 1448 (1982)
82. F. Wilczek, Phys. Rev. Lett. 49, 1549 (1982)
D. Reiss, Phys. Letters B115, 217, (1982)
G. Gelmini et al., Nucl. Phys. B219, 31 (1983)
83 D. Chang and G. Senjanovic, Phys. Letters B188, 231 (1987)
84. A.A. Anselm et al., Sov. J. Nucl. Phys. 41, 1060 (1985)

85. A. Jodidio et al., Phys. Rev. D34, 1967 (1986)
86. T. Goldman et al., Phys. Rev. D36, 1543 (1987)
87. See reference 24
88. G. Bélanger and J. Ng, TRIUMF Preprint TRI-PP-86-39 (June 1986)
89. J.H. Lopes and J.A. Martins Simoes, Phys. Rev. D35, 3428 (1987)
90. R.N. Cahn and H. Harari, Nucl. Phys. B176, 135 (1980)
91. O. Shanker, Nucl. Phys. B185, 382 (1981)
92. Wei-Shu Hou and A. Soni, Phys. Rev. D35, 2776 (1987)
93. T. Cowan et al., Phys. Rev. Lett. 56, 444 (1986)
94. See reference 19
95. R. Eichler et al., Phys. Lett. B175, 101 (1986)
96. W. Buchmüller and D. Wyler, Nucl. Phys. B268, 621 (1986)
97. I.I. Bigi et al., Phys. Letters B166, 238 (1986)
98. See reference 96
99. A. Santamaria and J.W.F. Valle, Lawrence Berkeley Laboratory Preprint LBL-23896 and University of Valencia Preprint FTUV-7/87; see also A. Santamaria and J.W.F. Valle, Phys. Letters B195, 423 (1987)
100. B.A. Campbell, Phys. Rev. D28, 209 (1983)
101. J. Ellis and D.V. Nanopoulos, Phys. Letters B110, 44 (1982)
102. B.A. Campbell and K.A. Peterson, Phys. Letters B192, 401 (1987)
103. R.R. Volkas and G.C. Joshi, Phys. Rev. D35, 1050 (1987)
104. See for instance J.W.F. Valle, Phys. Letters B196, 157 (1987) and references therein
105. B.A. Campbell et al., Int. Journal Mod. Phys. A2, 831 (1987)
106. A. Masiero et al., Phys. Rev. Lett. 57, 663 (1986)
107. F. Borzumati and A. Masiero, Phys. Rev. Lett. 57, 961 (1986)
108. E. Ma, Phys. Letters B191, 287 (1987)
109. S.M. Barr and A. Masiero, Phys. Rev. Lett. 58, 187 (1986)
110. See reference 97
111. Y. Kizukuri, Phys. Letters B185, 183 (1987)
112. G. Eilam and T.G. Rizzo, Phys. Letters B188, 91 (1987)
113. M.J.S. Levine, Phys. Rev. D36, 1329 (1987)
114. J. Bernabéu and A. Santamaria, Phys. Letters B197, 418 (1987)
115. J. Bernabéu et al., Phys. Letters B187, 303 (1987)

# Neutrino Physics and Supernovae: What have we learned from SN 1987A?

*W. Hillebrandt*

Max-Planck-Institut für Physik und Astrophysik,
Institut für Astrophysik, Karl-Schwarzschild-Str. 1,
D-8046 Garching, Fed. Rep. of Germany

Weak interaction processes leading to the production of neutrinos and their emission from collapsing stars and newly born neutron stars will be discussed. It will be demonstrated that theoretical predictions and expectations are in fairly good agreement with the luminosity and average energy of neutrinos observed from SN 1987A, provided they signalled the formation of a neutron star. Moreover, these neutrino observations offer a unique opportunity to study fundamental properties of neutrinos such as rest masses, decay modes, etc., and to compare them with results obtained from terrestrial experiments.

## 1. Introduction

On February 23, 1987, a supernova explosion in the Large Magellanic Cloud (LMC), a neighbor of our own galaxy, was observed. Spectra taken during the second night showed Balmer lines of hydrogen [1], indicating that the supernova was of type II. The position of the supernova coincided with that of a blue supergiant, Sanduleak - 69 202 [2]. When the supernova became weak in UV-radiation a few weeks after the explosion it became obvious that this star had indeed disappeared [3]. From its spectral class (B3) and its luminosity class (Ia) one could conclude the main sequence mass of the progenitor had been close to 20 solar masses ($M_\odot$) [4].

It had always been believed that the central cores of stars as massive as Sk - 69 202 should collapse to neutron star densities [5]. Most of the binding energy of a newly born neutron star was thought to be radiated away in form of neutrinos during the formation process [5]. Therefore, it did not come as a surprise that neutrinos were detected a few hours before the optical outburst [6] because theoretical models predicted that a few neutrinos should be detected by present neutrino detectors for a supernova at a distance of roughly 50 kiloparsec (kpc), appropriate for SN 1987A [7].

The arguments leading to this conclusion are fairly simple. Suppose that in the center of a collapsing star a neutron star of roughly 1.5 $M_\odot$ is formed. Initially this newly born neutron star will be very hot and, therefore, its gravitational binding energy will only be of the order of a few times $10^{52}$ erg. Some fraction of this energy may lead to the observed supernova outburst and mass ejection. The major part, however, has to be carried away by weakly interacting particles such as neutrinos. Moreover, during cooling and contraction the gravitational binding energy will encrease to a few times $10^{53}$ erg, the binding energy of a cold neutron star, and again the gain in binding will be radiated away by neutrinos. We thus can estimate that in total a few times $10^{53}$ erg in neutrinos will be emitted. During the very early cooling phase neutrinos will mainly be produced by thermal processes ($e^+ + e^- \rightarrow \bar{\nu} + \nu$, etc.), and thus all flavors of neutrinos and antineutrinos will carry roughly the same amount of energy. Assuming a typical value of $3 \times 10^{53}$ erg for their total energy and three different flavors, approximately $5 \times 10^{52}$ ergs can be expected in form of $\bar{\nu}_e$, with typical average energies of about 10 MeV. Water Cerenkov detectors such as the KAMIOKANDE experiment are mainly sensitive to $\bar{\nu}_e$ because they are detected via the reaction $\bar{\nu}_e + p \rightarrow n + e^+$, which has a cross section about

100 times the $(\nu, e)$ scattering cross section for 10 MeV neutrinos. From the numbers given above we expect $(6 \pm 4)$ counts in 2000 t of water for a collapse at a distance of 50 kpc, in fair agreement with the 12 (or 11) counts seen in the KAMIOKANDE detector.

Although these arguments seem to be very convincing, many questions remain open. We will, therefore, review in the following sections the basic ideas as well as the problems of the astrophysical models. Next the various neutrino detections will be discussed and finally implications for astrophysics and particle physics will be presented. A summary and conclusions follow.

## 2. Stellar Collapse and Supernova Explosions

Supernovae are rare events and recent estimates predict a rate of roughly one event every 40 ($\pm$ 20) years in a giant spiral galaxy like our own one [8], or even a rate that is lower by about a factor of two [9]. Among all supernovae roughly one half are of type II, and only those are thought to emit large amounts of neutrinos. Prior to SN 1987A there was no direct evidence that type II supernovae lead indeed to the formation of neutron stars, although several supernova remnants are known which do contain a neutron star (Crab, Vela, RCW 103, etc.). But in none of those cases do we know beyond doubt that the events were indeed of type II. Moreover, the remnant of an explosion of a rather massive star, Cas A, apparently does not contain a neutron star. The assumption, therefore, that massive stars ($M \gtrsim 8M_{\odot}$) are the progenitors of type II supernovae and neutron stars was mainly based on theoretical models of stellar evolution. On the other hand, a significant fraction of galactic supernovae may escape visual detection, as did Cas A some 350 years ago, and may only be discovered by their neutrino emission. It is interesting to note that if SN 1987A had exploded near the center of our own galaxy, it probably would have been seen as a neutrino source only. In the following subsections we will briefly discuss the theoretical models which have been developed over the last 20 years.

### 2.1 Pre-supernova Stellar Evolution

It is generally believed that stars less massive than approximately $8M_{\odot}$ on the main sequence end their lifes as white dwarfs by ejecting a large fraction of their mass by some "superwind" in the form of a planetary nebula [10]. These theoretical ideas are supported by estimates of white dwarf progenitor masses obtained from observations of star clusters [11]. Estimated birthrates of massive stars in the solar neighborhood are consistent with the assumption that all stars more massive than about 8 or 10 $M_{\odot}$ explode as type II supernova [12].

Stars can only collapse if their core mass exceeds the Chandrasekhar limit, given by

$$M_{CH} \simeq 5.72(y_e)^2 M_{\odot}, \tag{1}$$

where $y_e$ is the electron concentration, and if no significant amounts of burnable nuclear fuel (e.g. He, C, or O) are present in their cores. Again, this condition is fulfilled by massive stars, $M \gtrsim 8M_{\odot}$, for which stellar evolution models predict that they form degenerate or semi-degenerate cores of heavier elements. To be more precise, it has been shown that stars in the mass range from 8 $M_{\odot}$ to 10 or 12 $M_{\odot}$ form either O-Ne-Mg or Ne-Si cores with a mass close to $M_{CH}$ at the end of their hydrostatic evolution. These cores then contract due to electron captures, and nuclear burning incinerates a certain fraction of the core mass into iron-group elements [13]. More massive stars, on the other hand, burn all their nuclear fuel quietly and their iron-core masses range from 1.3 to 2.2 $M_{\odot}$ [14] (see also Fig. 1), the exact values depending on the main sequence mass, certain nuclear reaction rates, the way convection is modelled, etc.. The upper mass limit for this type of evolution is not well known and may range from about 50 $M_{\odot}$ to 100 $M_{\odot}$. Stars beyond this mass limit, which are extremely rare, will be disrupted by explosive oxygen-burning and can be ignored in our present discussion [15].

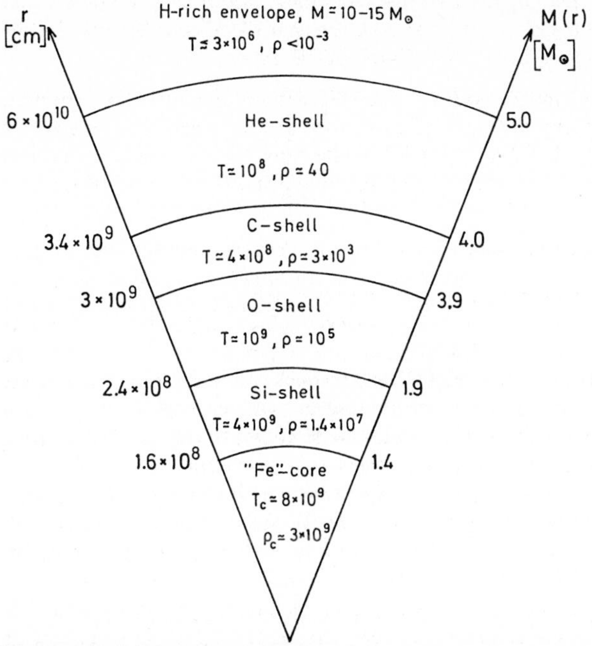


**Figure 1:** Schematic cross section of a massive star at the onset of collapse. $M(r)$ is the mass in units of solar masses inside the radius $r$. $T$ and $\rho$ are temperatures and densities in Kelvin and g cm $^{-3}$, respectively. The star has an onion-shell structure. The core, composed of iron-group elements, is surrounded by shells of silicon, oxygen, etc.. The iron core only collapses on a dynamical timescale

The properties of stellar cores at the onset of collapse can be summarized as follows: Central densities are typically around $10^{10}$g cm$^{-3}$, temperatures range from 8 to 10 times $10^9 K$, specific entropies are of the order of 1 $k_B$/nucleon, and core radii are typically a few $10^8$ cm. They become dynamically unstable because they have exhausted their nuclear fuel and electron captures on heavy nuclei and free protons reduce the pressure. There are, however, several problems and uncertainties which affect this general picture. The core mass and its entropy have turned out to be of crucial importance for the outcome of numerical simulations of stellar collapse. These quantities, in turn, depend on the way convective transport and mixing is modelled, on certain thermonuclear reaction rates, such as $^{12}C(\alpha,\gamma)^{16}O$, on weak interaction rates and neutrino losses, mass loss, etc.. So in conclusion, we are still far from knowing the necessary details with sufficient accuracy, even if we neglect effects that might be caused by rotation.

### 2.2. Some Remarks on the Equation of State

Computations of the equation of state (EOS) in the deep interior of a collapsing star are greatly simplified by the fact that strong and electromagnetic interactions are in equilibrium at temperatures above roughly $5 \times 10^9 K$. The task is, therefore, to minimize the free energy for given temperature, specific volume, and total number of neutrons and protons.

At sufficiently low densities, $\rho \lesssim 10^{12}$g cm$^{-3}$, nucleons and nuclei can be described as non-interacting Boltzmann-particles, and the free energy density of the nucleons and nuclei is given by

$$f_i = n_i\, k_B\, T \left\{ 1 + ln \left[ \frac{1}{n_i} \left( \frac{m_i\, k_B\, T}{2\pi\, \hbar^2} \right)^{3/2} \omega_i(T) \right] \right\} + m_i\, c^2\, n_i, \tag{2}$$

where $n_i$ denotes the number density of species $i$, $m_i$ is its mass, and $\omega_i$ the partition function, all other quantities having their usual meaning. Problems arise only because nuclear masses and partition functions have to be known for neutron-rich nuclei which are very unstable in the laboratory. Therefore extrapolations from known nuclei are commonly used in computations of the EOS of supernova matter (see, e.g., El Eid and Hillebrandt [16] or Mazurek et al. [17]).

At high density the Boltzmann-gas approximation breaks down, because nucleon-nucleon and nuleon-nucleus interactions become important and nuclei can no longer be treated as point-like particles. Consequently the Boltzmann-gas EOS has to be replaced by a self-consistent microscopic model.

A feasable way is to apply the temperature-dependent Hartree-Fock method [18] or simplifications of it [19]. The general approach is to put a fixed number of neutrons and protons into a spherical box of radius $R_c$, the so-called Wigner-Seitz cell, and to compute the free energy at a given temperature from a model Hamiltonian. An example of such a computation is shown in Figs. 2 and 3. It can be seen that at very high densities the most stable "nucleus" has a large mass ($A \simeq 560$), but that the pressure and therefore the adiabatic index is still dominated by contributions from relativistic electrons, which of course have to be added to the nucleonic part of the EOS.

At densities above nuclear matter density, "nuclei" dissolve into a homogeneous fluid of free neutrons and protons. Consequently, relativistic leptons are no longer the main contribution to the EOS, but nucleon-nucleon interactions in a Fermi fluid are dominant. Phenomenologically determined nucleon-nucleon forces will gradually lose reliability with increasing density and it is therefore not surprising that up to now the EOS at densities above, say, twice nuclear matter density is not well known and is subject to considerable dispute [20]. Fortunately, most EOS predict that the adiabatic index is of the order of 3 near nuclear matter density and thus the

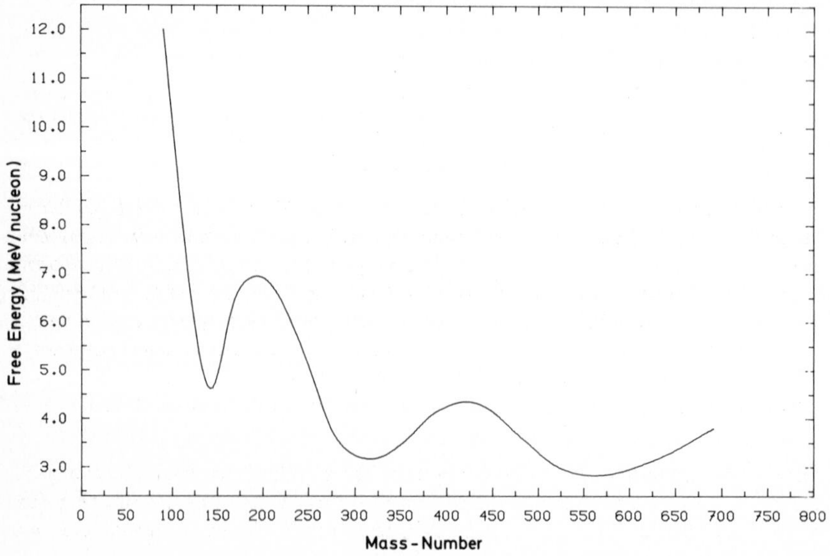


**Figure 2:** Free energy of a Wigner-Seitz cell versus mass number for fixed temperature (2.5 MeV), density ($6 \times 10^{13}$ g cm $^{-3}$) and electron concentration (0.3). The configuration with about 560 nucleons has the lowest free energy and is, therefore, the most stable one

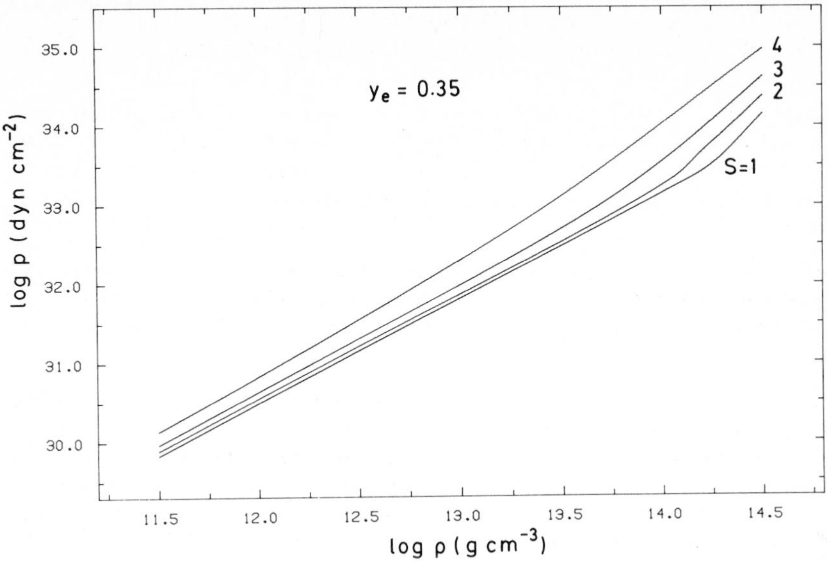


**Figure 3:** Pressure versus density for constant values of the entropy and an electron concentration of 0.35

collapsing core will, in general, not overshoot this density by much. As will be discussed later, however, the outcome of numerical simulations of stellar collapse may be very sensitive to the properties of the EOS just above nuclear matter density.

### 2.3. Weak Interaction Rates and Neutrino Transport

In general, in a collapsing star as well as during the explosion phase, weak interaction reactions such as

$$\begin{aligned} e^- + \left\{ \begin{matrix} p \\ A(Z,N) \end{matrix} \right\} &\rightarrow \left\{ \begin{matrix} n \\ A(Z-1,N+1) \end{matrix} \right\} + \nu_e \\ \nu_e + n &\rightarrow p + e^- \\ \nu_e + e^- &\rightarrow \nu_e + e^-, \quad \text{etc.} \end{aligned} \tag{3}$$

are not in equilibrium unless the density exceeds several times $10^{12}$g cm$^{-3}$. Of particular importance are reactions which change the electron concentration $y_e$, because, as we have seen, the electrons dominate the pressure during most of the collapse. The rates for electron captures on free protons can be computed numerically exactly (see, e.g., Bruenn [21]), but electron captures on nuclei are much more difficult. Present models are probably only accurate to within factors of about five [22]. Fortunately, most EOS predict a proton concentration sufficiently high, such that the change in $y_e$ is mainly due to electron captures on free protons. The same is not true for the entropy generation due to non-equilibrium weak interactions, as will be discussed below.

Other important processes include neutrino-absorption by free neutrons, neutrino-electron scattering, neutrino-nucleus coherent scattering and neutrino-neutron scattering. Again, the cross sections and reaction rates can be computed numerically exactly [21, 22], but in most numerical studies approximate values are used.

In a collapsing star neutrinos are mainly produced by electron captures because the electrons are degenerate and therefore the positron concentration is extremely low. We will find $\nu_e$'s only and their energy will be of the order of 2/3 of the electron Fermi energy ($\simeq$ 10 - 20 MeV). Neutrino opacities are dominated by coherent scattering off heavy nuclei, for which the cross section is given approximately by

$$\sigma \simeq 10^{-44} cm^2 \; N^2 \left[\frac{E_\nu}{\text{MeV}}\right]^2 , \tag{4}$$

where we have assumed a Weinberg angle $\sin^2 \theta_W = 0.25$ and $N$ is the neutron number of the nucleus.

Neutrinos are predominantly produced at densities between $10^{11}$ and $10^{12}$g cm$^{-3}$, the "deleptonization shell", where the typical nucleus has about 80 to 100 nucleons and 50 neutrons. The mean free path of the neutrinos is therefore

$$\ell_\nu \simeq \frac{1}{n_A \sigma} \simeq 10^7 \text{cm} \left[\frac{\rho}{10^{12} \text{g cm}^{-3}}\right]^{-1} \frac{A}{N^2} \left[\frac{E_\nu}{10 \text{MeV}}\right]^{-2} , \tag{5}$$

where $n_A$ is the density of an average heavy nucleus with mass number $A$. Here we have assumed that the concentration of free nucleons is small, which is well justified if the entropies per nucleon are less than about $2k_B$. From (5) we find that the matter becomes opaque for neutrinos once the density exceeds $\simeq 5 \times 10^{11}$g cm$^{-3}$. The diffusion time

$$\tau_{diff} \simeq \frac{d^2}{\frac{1}{3} c \ell_\nu} ; d \simeq 10^7 \text{cm} \tag{6}$$

is then of the order of $2s$ and thus much longer than the collapse time ($\simeq 10^{-3}s$). Consequently, neutrinos are trapped and will move with the matter. Neutrino-electron scattering, finally, will equilibrate the neutrinos energy distribution, and at densities beyond $10^{12}$g cm$^{-3}$ weak interactions also approach an equilibrium. Therefore, the entropy of stellar matter should be constant above $10^{12}$g cm$^{-3}$, and the collapse should be adiabatic unless shocks heat the matter.

The entropy generation during the equilibration phase can be estimated as follows [21]:

Before neutrino trapping, $(\nu_e, e)$-scattering as well as $\nu$-absorption can be neglected. The change in specific entropy is thus given by

$$k_B T \frac{ds}{dt} = (\mu_e + \mu_p - \mu_n - \langle E_\nu \rangle_{emit}) \left(-\frac{dy_e}{dt}\right) , \tag{7}$$

where the $\mu$'s denote chemical potentials of the respective particles and $\langle E_\nu \rangle_{emit}$ is the average neutrino energy in emission processes. For $e^-$-captures on free protons we have $\langle E_\nu \rangle_{emit} \simeq \frac{5}{6}\mu_e$ (since $\mu_e \gg k_B T$) and $e^-$-captures on heavy nuclei give approximately

$$\langle E_\nu \rangle_{emit} \simeq \frac{3}{5}(\mu_e + \mu_p - \mu_n - \Delta), \tag{8}$$

where $\Delta$ is the excitation energy, because part of the decay energy is stored in nuclear excited states. It is clear from (7) and (8) that the latter process always leads to an increase of the entropy, whereas $e^-$-captures on free protons will decrease the entropy. The net-effect is a very small increase.

After neutrino trapping the change of entropy is given by

$$k_B T \frac{ds}{dt} \simeq (\mu_e + \mu_p - \mu_\nu - \mu_n) \left(-\frac{dy_e}{dt}\right) , \tag{9}$$

which is always positive. From numerical models one obtains typically changes in $y_e$ of at most 0.1 and differences of the chemical potentials of less than 5 $k_B\ T$. The increase of the entropy is therefore at most 0.5 $k_B$ and the entropy always stays low [23].

After core-bounce the out-going shockwave increases the entropy to values of about (7 - 9)$k_B$. Heavy nuclei are dissociated into free neutrons and protons. This means that the neutrino scattering cross sections are significantly reduced and neutrinos can diffuse on timescales of the order of a few ms to the shock front, where they pile-up until the shock passes the neutrino sphere. But only those neutrinos which are created in the shocked gas can diffuse, because the diffusion time in the unshocked material is still of the order of several tenths of a second and thus much longer than the shock propagation time. Once the shock has passed the neutrino sphere most of these neutrinos are emitted on a hydrodynamical timescale ($\sim$ ms) [24](see also Fig. 4). In the shocked gas the temperature is sufficiently high such that the neutrino degeneracy is removed. This leads to an increase of $e^-$-captures on free protons and the matter becomes more neutron rich. As a result the electron chemical potential drops to values close to $k_BT$. But since the temperature is still high ($k_BT \gtrsim 1\,\mathrm{MeV}$) electron-positron pairs form in equilibrium and anti-neutrino producing reactions such as $e^+ + n \rightarrow \bar{\nu}_e + p$ and $e^+ + e^- \rightarrow \bar{\nu}_e + \nu_e$ become important. Moreover, electron-positron pairs will also decay into $\mu$-and $\tau$-neutrinos. We, therefore, expect that most of the energy of the proto-neutron star will be carried away by neutrinos of all flavours [24].

We conclude this section with some remarks on the way neutrino transport is usually treated in numerical simulations of stellar collapse and supernova explosions. During collapse and after core-bounce we will always find regions in the star where neutrinos are either streaming freely or diffusing outwards. So, in principle, we would have to solve the Boltzmann transport equation. This transport equation, however, is a set of complicated partial integro-differential equations [25] and, therefore, has never been used in core-collapse computations but only approximations to it.

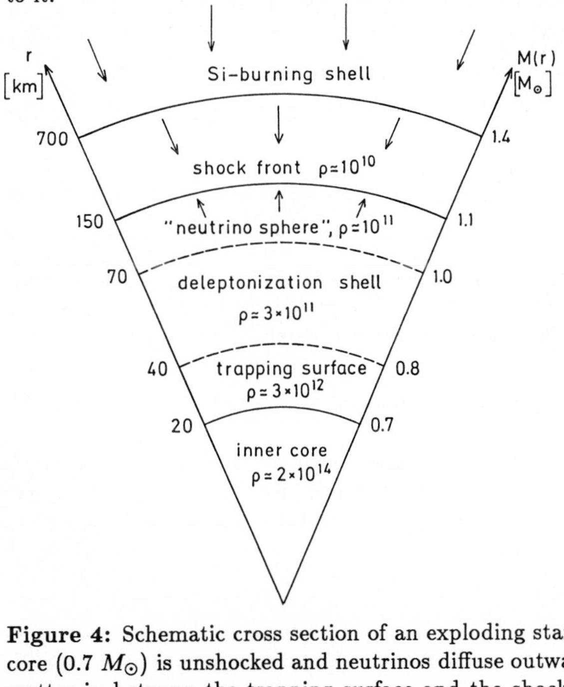


**Figure 4:** Schematic cross section of an exploding star a few ms after core-bounce. The inner core (0.7 $M_\odot$) is unshocked and neutrinos diffuse outwards on timescales of a few seconds. The matter in between the trapping surface and the shock front is losing neutrinos on dynamical timescales. Arrows indicate that the mass zones are moving outwards or inwards, respectively

In order to derive simple transport equations one may replace the Boltzmann equation by momentum equations

$$M_\nu^n := \frac{1}{2}\int_{-1}^{1} I_\nu\, \mu^n d\mu, n = 0,1,2,..., \tag{10}$$

where $\mu = \cos\theta$ and $I_\nu$ is the intensity of neutrinos of energy $\epsilon_\nu$. Then one obtains

$$M_\nu^0 = J_\nu =: \frac{1}{4\pi}\, G_\nu \quad \text{(neutrino energy)}, \tag{11}$$

$$M_\nu^1 = H_\nu =: \frac{1}{4\pi}\, F_\nu \quad \text{(neutrino energy flux)}, \tag{12}$$

$$M_\nu^2 = K_\nu \quad \text{(neutrino pressure)}. \tag{13}$$

The diffusion approximation follows if we use the closure conditions

$$M_\nu^3 \equiv 0, \quad K_\nu = J_\nu/3 = p_\nu. \tag{14}$$

In Lagrangian coordinates we then get

$$\dot{G}_\nu + \frac{1}{r^2}\frac{\partial(r^2 F_\nu)}{\partial r} + \frac{\dot{\rho}}{3\rho}\frac{\partial G_\nu}{\partial\, ln\, \epsilon_\nu} = c\Sigma_a^\nu G_\nu + q_\nu, \tag{15}$$

$$\frac{1}{c}\dot{F}_\nu + \frac{c}{3}\frac{\partial G_\nu}{\partial r} - F_\nu\left[\frac{2v}{cr} + \frac{2}{c}\frac{\dot{\rho}}{\rho}\right] - \frac{v}{cr}\frac{\partial}{\partial \epsilon_\nu}(\epsilon_\nu F_\nu) = -\Sigma_t^\nu F_\nu, \tag{16}$$

where $\Sigma_a^\nu$ is the macroscopic absorption cross section, $q_\nu$ is the spectral emissivity, and $\Sigma_t^\nu$ denotes the macroscopic transport cross section, which is equal to $(\lambda_\nu)^{-1}$ [26].

If, in addition, we assume a thermal distribution of neutrinos,

$$G_\nu = \text{const.}\epsilon_\nu^3(1 + \exp(\epsilon_\nu - \mu_\nu)/k_B T)^{-1}, \tag{17}$$

and drop terms of the order $\frac{v}{c}$ (non-relativistic limit), we obtain

$$F_\nu = -\frac{c}{3\Sigma_t^\nu}\frac{\partial G_\nu}{\partial r} \tag{18}$$

and

$$\dot{G}_\nu - \frac{1}{r^2}\frac{\partial}{\partial r}\left[r^2\frac{c}{3\Sigma_t^\nu}\frac{\partial G_\nu}{\partial r}\right] + \frac{\dot{\rho}}{3\rho}\frac{\partial G_\nu}{\partial\, ln\, \epsilon_\nu} = -c\Sigma_a^\nu G_\nu + q_\nu. \tag{19}$$

By multiplying (19) by $\epsilon_\nu^{-1}$ and integrating over $d\epsilon_\nu$ we get the usual diffusion approximation

$$\frac{\partial}{\partial t}(\rho y_\nu) - \frac{1}{r^2}\frac{\partial}{\partial r}\left[r^2\frac{c^2}{3\bar{\Sigma}_t}\frac{\partial}{\partial r}(\rho\, y_\nu)\right] = \text{sources and sinks.} \tag{20}$$

Here the mean free path is given by

$$(\bar{\Sigma}_t)^{-1} = \int_0^\infty (\Sigma_t^\nu)^{-1} d\epsilon_\nu / \int_0^\infty d\epsilon_\nu. \tag{21}$$

Since in thermal equilibrium $n_\nu \sim T^3 F_2(\mu_\nu/k_B T)$ and $g_\nu \sim T^4 F_3(\mu_\nu/k_B T)$, where the $F$'s denote relativistic Fermi-integrals, the neutrino energy flux can be computed once the particle flux is known.

The approximation described here has two major shortcomings. Firstly, at densities below $10^{12}$g cm$^{-3}$ neutrinos are not in thermal equilibrium and, secondly, the diffusion approximation breaks down at the neutrino-sphere, where the mean free path becomes comparable to the stellar radius. The second problem is usually circumvented by introducing a so-called flux-limiter which guarantees that for $\bar{\lambda} \gg \Delta r$ the free streaming limit is obtained. The first problem can only be solved by non-equilibrium transport models [26, 27]. It should be noted, however, that even the most elaborate transport schemes used in stellar collapse models are based on the diffusion equation and, therefore, may misrepresent the actual neutrino spectra. We will come back to this question later in Sect. 4.

## 2.4 Results of Some Numerical Models

The outcome of computer simulations of stellar collapse will depend on the pre-collapse stellar models, the adopted microphysics input data and the numerical methods. Given the fact that the various computations use different initial models, different equations of state, and/or different numerical schemes it is not surprising that the results also differ considerably. A simple estimate can clarify these difficulties. From the observations of SN 1987A we know that almost the entire binding energy of the newly born neutron star ($\gtrsim 10^{53}$ erg) was radiated away by neutrinos. The kinetic energy of the ejecta ($\simeq 10^{51}$ erg) as well as the energy in electromagnetic radiation ($\lesssim 10^{49}$ erg) were small corrections at the percent level or less. It is this small correction in the energetics of the outburst that numerical models try to determine.

There are, however, several features that are the same in all computations using the standard theory of weak interactions and non-rotating stellar models. During the collapse phase the entropy stays low ($S \leq 2k_B$) and, therefore, all models collapse to nuclear matter density. Moreover, about that fraction of the star corresponding to the Chandrasekhar mass ($M_{CH} \simeq 5.72 M_\odot\, y_e^2$) collapses homologously, $v = a(t)r$, where $a(t)$ is independent of $r$, in agreement with analytical considerations [28]. Consequently, a sonic point must exist and the matter outside the sonic point has supersonic velocities close to free-fall velocity. Since $M_{CH}$ is proportional to $y_e^2$ it will decrease during collapse. This decrease depends on the entropy of the initial model, on $e^-$-capture rates, the $\nu$-transport scheme, and the EOS, and turns out to be different in most computations [21, 29].

At core-bounce the central density of the star is in general only slightly higher than nuclear matter density ($\rho_c \gtrsim 3 \times 10^{14}$g cm$^{-3}$), unless rather soft EOS are used [30]. The inner core, defined by the condition that its velocity is subsonic, is stopped on a sound-crossing time ($\lesssim 1$ ms) once nuclei in the center of the star have dissolved into a homogeneous fluid of free nucleons and the EOS stiffens ($\gamma \simeq 2.5$ - $3$). Because the outer material is still falling with supersonic velocity, a shock must form near the sonic point at $M_{CH}$ (or a radius of about 20 km). The unshocked inner core cannot expand against the ram-pressure of the supersonic outer layers and achieves a hydrostatic equilibrium soon after bounce. From energy conservation one can estimate the energy that is put into the shock [31], wich is typically

$$E_{shock} \simeq E_B^{ic} \simeq (4 - 8) \times 10^{51}\text{erg}, \tag{22}$$

where $E_B^{ic}$ is the binding energy of the unshocked inner core, in agreement with the results of numerical simulations. Here, the main source of uncertainties is the stiffness of the EOS near nuclear matter density, but also $\nu$-transport can change these numbers considerably.

The outgoing shock wave is heavily damped by energy losses due to nuclear photodissociations, which cost about $8 \times 10^{18}$ erg/g, and neutrino losses once the shock has passed the neutrino sphere. Neglecting neutrino losses for a moment we find from (22) that at most 0.5 $M_\odot$ of heavy nuclei can be dissociated by the shock, and a necessary condition for successful propagation is

$$M_{"Fe"} - M^{ic} \lesssim 0.5 M_\odot, \tag{23}$$

where $M_{"Fe"}$ is the mass of the original iron core. Numerical models predict $y_e \lesssim 0.38$ and thus $M^{ic} \lesssim 0.7 M_\odot$. It follows immediately that only stellar models with iron-core masses less than 1.2 $M_\odot$ can lead to prompt explosions. Recent stellar evolution calculations [32] have shown that this condition limits the mass range of possible progenitor stars to values between 8 and 12 $M_\odot$ (or at most $15 M_\odot$) on the main sequence, and the only successful collapse computations have indeed been performed with stars in this mass range [29, 30].

In Figs. 5 and 6 some results of such a computation are shown. It should be noted, however, that in other simulations even the most favorable initial conditions do not lead to prompt explosions [33], probably because different equations of state and different $\nu$-transport schemes were used. Therefore we reach the conclusion that at present we cannot answer the question whether or not stars can explode by the core-bounce mechanism and have to keep this unpleasant situation in mind when we try to interpret the neutrino events detected from SN 1987A.

Among researchers in the field, however, there is general agreement that stars with iron cores more massive than about 1.4 $M_\odot$ cannot explode by the core-bounce mechanism. Conventional models of stellar evolution, therefore, exclude main sequence masses above about 15 $M_\odot$. Figures 7 and 8 show a typical example of a simulation where the shock stalled at about 1.2 $M_\odot$.

Because some supernova remnants (e.g., Cas A and Puppis A) show large oxygen overabundances indicating main sequence masses of at least 20 $M_\odot$, even prior to SN 1987A alternative explosion mechanisms had been searched for. A possibility which has been discussed extensively is the so-called delayed explosions model [34]. The idea is that a few hundred milliseconds after core-bounce energy transport by neutrinos may revive a stalled shock. At that time the edge of the unshocked inner core ($\simeq 0.7 M_\odot$) will be at a radius of about 20 km, and the density and temperature there are around $10^{12}$g cm$^{-3}$ and 2 to 3 MeV, respectively. The shock has changed to an almost standing accretion shock at 1.4 $M_\odot$ and a radius of about 500 km, the temperature and density just behind the shock being $10^7$g cm$^{-3}$ and 0.8 MeV, respectively. The shocked matter is irradiated by a neutrino flux originating from the neutrino-sphere at 30 to

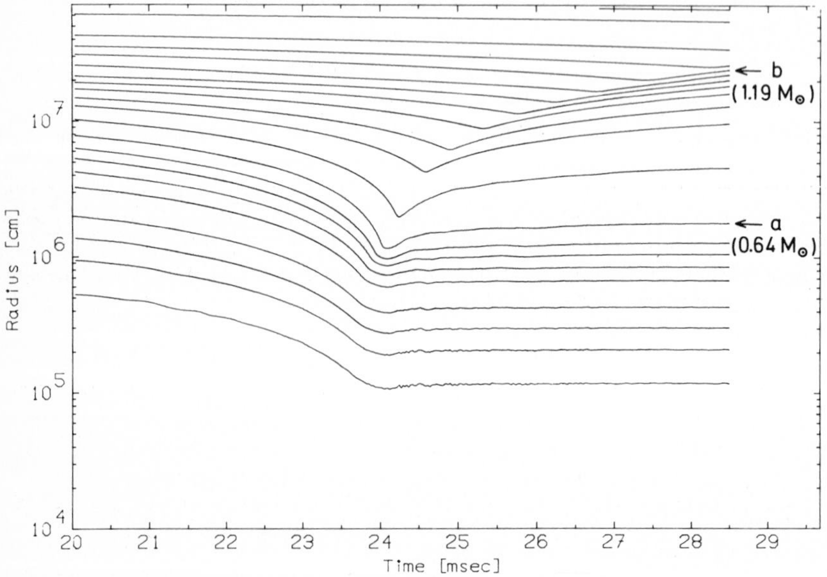


**Figure 5:** Radius versus time for various mass zones of a collapsing and exploding stellar model of 9 $M_\odot$. The unshocked inner core is labelled a, the first zone that reaches escape velocity, b

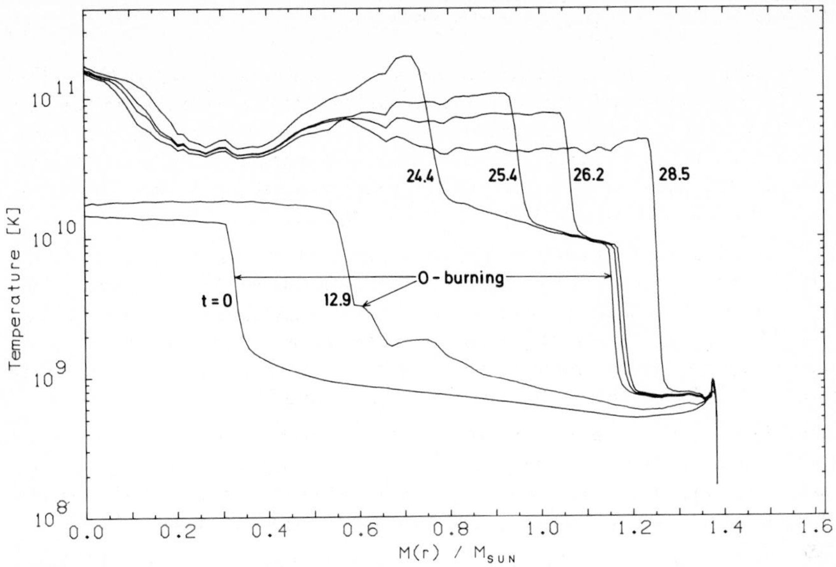


**Figure 6:** Snapshots of temperature versus mass in units of solar masses for the stellar model shown in Fig. 5. The curves are labelled with the time in ms measured from the beginning of the computations. The drop in temperature in the oxygen-burning shell is also indicated

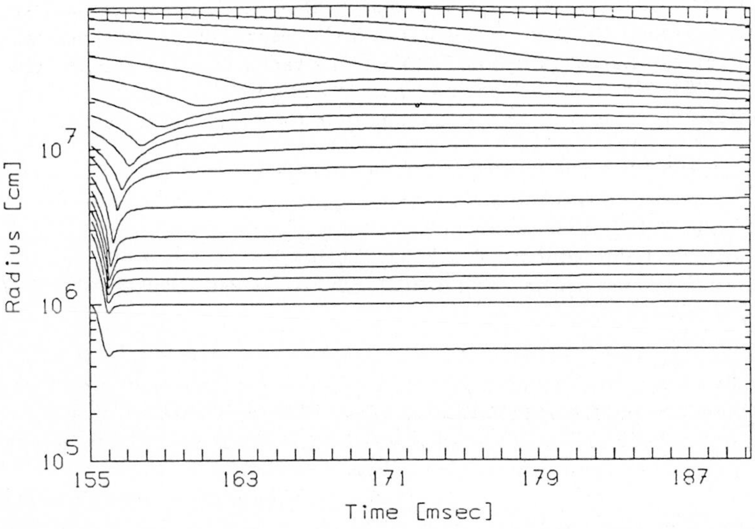


**Figure 7:** Same as Fig. 5 , but for a stellar model of 20 $M_{\odot}$. It is obvious that this time the shock stalls at about 300 km

70 km. Typical neutrino energies will be around 6 to 10 MeV. The energy gain per gramme of irradiated matter at radius $R$ much larger than $R_{\nu}$ of the neutrino sphere is given by

$$\dot{E}_{gain} = K_{abs} \frac{L_{\nu}}{4\pi R^2}, \tag{24}$$

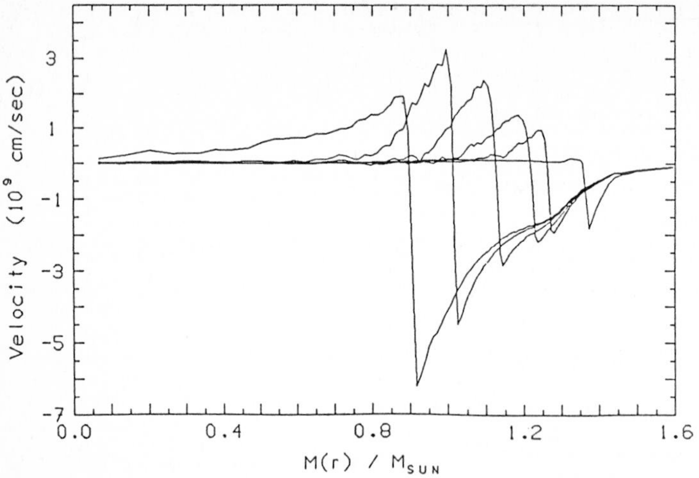


**Figure 8:** Snapshots of velocity versus mass for the model of Fig. 7. The curves, from the left, correspond to times 156.2, 156.6, 157.6, 159.9, 163.2, and 202.6 ms, respectively

where $L_\nu$ is the neutrino luminosity and $K_{abs}$ is the absorption opacity. Each gramme of matter will also lose energy by thermal emission of neutrinos, $\dot{E} \sim T^4 K_{emiss}$. Since the neutrino luminosity is proportional to $T^4 R_\nu^2$ and $K_{abs}$ and $K_{emiss}$ are proportional to $T_\nu^2$ and $T^2$, respectively, the net energy gain can be written as

$$\frac{dQ}{dt} = \frac{7}{16} ac\ K_{abs}\ T_\nu^4 \left[ \frac{1}{4} \left( \frac{R_\nu}{R} \right)^2 - \left( \frac{T}{T_\nu} \right)^6 \right], \tag{25}$$

where the temperatures are in MeV, $a = 1.37 \times 19^{26}$erg/c cm$^3$ MeV$^4$ and $K_{abs}$ in cgs units is $6 \times 10^{20}$ $\langle (\epsilon_\nu / m_e c^2)^2 \rangle$cm$^2$g$^{-1}$ [35]. The maximum matter temperature that can be obtained from this heating mechanism can be estimated from the equilibrium condition $dQ/dt = 0$, and we find roughly 1 MeV from typical stellar parameters. However, it is clear that the quantity of key importance is the neutrino temperature at the neutrino sphere, which, due to uncertainties in the flux-limited diffusion schemes, is not very well determined. Small changes in $T_\nu$ can change the net-heating considerably.

Wilson and Wilson et al. [34] have computed the hydrodynamic evolution of several stellar models for approximately 1s after core bounce and found explosions in all cases considered (see also Fig. 9). In most cases, however, the explosion energy ($\lesssim 4 \times 10^{50}$erg) was too low to account for typical type II supernova light curves and in particular for the rather high kinetic energy seen in SN 1987A. Only for rather massive stars, $M \gtrsim 25 M_\odot$, did explosive oxygen burning add enough energy to the explosion to explain a typical outburst. This problem can be understood from a simple argument. Neutrino heating proceeds on a time scale much longer than the hydrodynamical time scale. Once, due to neutrino heating, a sufficient overpressure has been built up behind the accretion shock the heated zones will start to expand and further heating will be turned off. The thermal energy needed to build up such an overpressure will be of the order of the binding energy of the overlaying material, i.e., a few times $10^{50}$erg. Consequently, one expects that the explosion energy is of the same order.

Recently Wilson (personal communication) repeated his computations with a slightly modified $\nu$-transport scheme and did not find explosions anymore. This, as well as results obtained by Hillebrandt [36], indicate that the delayed explosion mechanism may not work at all. So we

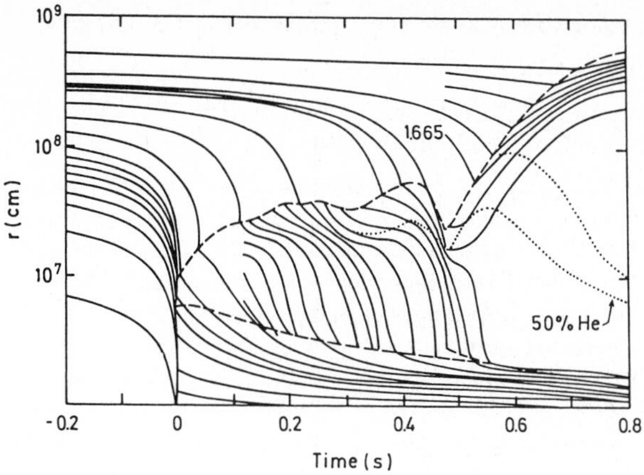


**Figure 9:** Radius versus time for selected mass zones of a model of 25 $M_\odot$ (from Wilson, ref. [34]). Time is measured from core bounce. The position of the shock front (upper dashed curve) as well as of the neutrino-sphere is also shown. The stalled shock is revived by neutrino heating 0.5s after bounce

are left with the problem that also (and maybe in particular) for massive stars, $M \simeq 20M_\odot$, the explosion mechanism is not well understood. It may well be that we have to invent more complicated scenarios in order to be able to solve these problems. For example, the neglect of rotation, magnetic fields and/or nuclear energy generation may be an oversimplification (see, e.g., [37, 38, 39]), or some of our basic assumptions concerning the EOS [30] or the theory of weak interactions are incorrect.

### 3. Neutrinos From Supernova 1987A

Two neutrino pulses were detected prior to the optical outburst of SN 1987A, at February 23.12 (UT) [40] and at February 23.32 (UT) [41, 42], respectively. There is also a report of neutrino events close to the second burst in the Baksan detector [43].

Two photographs taken at about February 23.44 (UT) [44] show the supernova at a visual magnitude of about $6^m.4$ only $2.8 \times 10^4 s$ and $1.1 \times 10^4 s$, respectively, after the neutrino events were discovered. This close correlation indicates that the neutrinos were indeed emitted from the exploding star.

The pulse seen in the Mont Blanc detector consisted of 5 events spread over $\Delta t \simeq 7s$ with measured positron energies between 5.8 and 7.8 MeV [45] (see also Table 1). From the second pulse, about 4.7 hours later, KAMIOKANDE detected 12 neutrinos spread over $\Delta t \simeq 13s$, whereas IMB saw 8 neutrinos with $\Delta t \simeq 6s$. The recorded electron (positron) energies in the second burst were significantly higher and ranged from 6 to 35 MeV (KAMIOKANDE) to 20 to 40 MeV (IMB) (see Table 1). It is interesting to note that the angular distribution of electrons (positrons) is not isotropic, as one would expect from the reaction $\bar{\nu}_e + p \rightarrow e^+ + n$. In particular, the high-energy events ($E_e \gtrsim 20$ MeV) seem to be strongly forward peaked away from the LMC. In the case of IMB this effect was not caused by the fact that about 25% of the photomultipliers were inoperative when the neutrinos were detected, as recent Monte Carlo simulations have shown [46]. Finally the Baksan group reported 5 events delayed by $30s$ relative to the IMB time, with energies from 12 to 24 MeV.

The detections of neutrinos from a type II supernova are the first proof that the cores of massive star do collapse to neutron star densities. However, certain aspects of the observed neutrino pulses are diffucult to understand. Therefore, we will discuss the experimental results in some detail in the following subsections.

### 3.1 Significance and Consistency of the Neutrino Detections

According to Hirata et al. [41], the rate at which events like the one observed in the KAMIOKANDE detector can arise from statistical fluctuations is of the order of one event every $7 \times 10^7$ yrs. There can be no doubt that the detected neutrinos were indeed emitted from SN 1987A. Although the statistical significance of the IMB detection is much less and gives a rate of one event of multiplicity 8 every 3 years from background noise, the close coincidence in time with the KAMIOKANDE detection is a strong argument in favor of the interpretation that these neutrinos also came from SN 1987A. Unfortunately, only a computer clock was used by the KAMIOKANDE group. The arrival time of the neutrinos in their detector, therefore, is uncertain by about $\pm 1$ minute. It is interesting to note that the Baksan detection lies within this window.

With respect to the first neutrino events, the conclusions are less certain. Aglietta et al. [40] state that their background rate of finding 5 positrons above threshold in a $10s$ interval is one every 1.5 yrs. This translates into a probability of $3 \times 10^{-4}$ of finding 5 events in a $10s$ interval in a 4 hrs window around the assumed time of the supernova explosion. Since at the time of the Mont Blanc detection no signal significantly above background was recorded by the (larger) KAMIOKANDE experiment, one cannot exclude the possibility that the first pulse was not real, but was due to random noise.

It is somewhat difficult to check the consistency of the various experiments from statistics alone, because they differ in their volume, their threshold energies, and in the way positrons are detected. Generally speaking, the count rate of the KAMIOKANDE detector should be higher than that of all other detectors if the average neutrino energy is between 8 and 15 MeV. As can be seen from Table 1 the average positron energy seen by the Mont Blanc detector was 6.7 MeV. Taking into account the sensitivity of the KAMIOKANDE detector at those energies ($\leq$ 30%) one would conclude that KAMIOKANDE should have seen about 20 events from the first burst. The fact that this was not the case is commonly used as an argument to omit the Mont Blanc events. On the other hand, we are dealing with the statistics of small numbers, and the count rate of Mont Blanc may have been accidentally high.

The same arguments hold for the Baksan experiment, which has a fiducial mass of 200 tons only (as compared to the 2140 tons of KAMIOKANDE) and a threshold energy of about 10 MeV. They found 1 event close to the Mont Blanc time, but 5 events near the KAMIOKANDE/IMB time. Again, the latter result is at best only marginal consistent with the KAMIOKANDE detections, because it requires a higher neutrino luminosity at the source.

IMB and KAMIOKANDE can be compared in a similar way. Taking into account the detection efficiency of IMB for events above 20 MeV and the fact that 25% of the photomultipliers did not work at the time of the neutrino burst, the total number of events above 20 MeV in the IMB detector would be $22 \pm 5$. Since the volume of KAMIOKANDE is about a factor of 2.3 smaller, one would expect $10 \pm 3$ events, which should be compared with the actual measurement of 4 $\pm$ 2 neutrinos.

So we reach the conclusion that although there are definitely inconsistencies among the various neutrino detections, the problem of the statistics of a small number of events does not allow us to rule out one or the other. The absence of a clear signal in the KAMIOKANDE detector during the first burst is certainly a strong argument against regarding this pulse to be real, but it is not a proof. It cannot be excluded from first principles that the neutrino energy in this burst was indeed low, in which case the higher detection efficiency of the Mont Blanc experiment at low energies could explain the observed number of events.

Table 1: **Properties of neutrino events detected from SN 1987A**

| Experiment | Event No. | Time (UT) (Feb. 23) | Electron energy (MeV) | Electron angle (degrees) |
|---|---|---|---|---|
| Mont Blanc | 1 | $2^h 53^m 36^s.79$ | 6.2 | - |
| | 2 | $40^s.65$ | 5.8 | - |
| | 3 | $41^s.01$ | 7.8 | - |
| | 4 | $42^s.70$ | 7.0 | - |
| | 5 | $43^s.80$ | 6.8 | - |
| KAMIOKANDE | 1 | $7^h 35^m 41^s.00$ (?) | $20.0 \pm 2.9$ | $18 \pm 18$ |
| | 2 | $41^s.11$ | $13.5 \pm 3.2$ | $40 \pm 27$ |
| | 3 | $41^s.30$ | $7.5 \pm 2.0$ | $108 \pm 32$ |
| | 4 | $41^s.32$ | $9.2 \pm 2.7$ | $70 \pm 30$ |
| | 5 | $41^s.51$ | $12.8 \pm 2.9$ | $135 \pm 23$ |
| | 6 | $41^s.69$ | $6.3 \pm 1.7$ | $68 \pm 77$ |
| | 7 | $42^s.54$ | $35.4 \pm 8.0$ | $32 \pm 16$ |
| | 8 | $42^s.73$ | $21.0 \pm 4.2$ | $30 \pm 18$ |
| | 9 | $42^s.92$ | $19.8 \pm 3.2$ | $38 \pm 22$ |
| | 10 | $50^s.22$ | $8.6 \pm 2.7$ | $122 \pm 30$ |
| | 11 | $51^s.43$ | $13.0 \pm 2.6$ | $49 \pm 26$ |
| | 12 | $53^s.44$ | $8.9 \pm 1.9$ | $91 \pm 39$ |
| IMB | 1 | $7^h 35^m 38^s.37$ | $41 \pm 7$ | $80 \pm 10$ |
| | 2 | $41^s.79$ | $37 \pm 7$ | $44 \pm 15$ |
| | 3 | $42^s.02$ | $28 \pm 6$ | $56 \pm 20$ |
| | 4 | $42^s.52$ | $39 \pm 7$ | $65 \pm 20$ |
| | 5 | $42^s.94$ | $36 \pm 9$ | $33 \pm 15$ |
| | 6 | $44^s.06$ | $36 \pm 6$ | $52 \pm 10$ |
| | 7 | $46^s.38$ | $19 \pm 5$ | $42 \pm 20$ |
| | 8 | $46^s.96$ | $22 \pm 5$ | $104 \pm 20$ |
| Baksan | 1 | $7^h 36^m 11^s.82$ | $12 \pm 2.4$ | - |
| | 2 | $12^s.25$ | $18 \pm 3.6$ | - |
| | 3 | $12^s.25$ | $23.3 \pm 4.7$ | - |
| | 4 | $19^s.51$ | $17 \pm 3.4$ | - |
| | 5 | $20^s.92$ | $20.1 \pm 4.0$ | - |

*Notes:* Mont Blanc data are taken from ref.[45]; the revised IMB data were presented by T. Haines at the Aspen Conf. on SN 1987A (Jan. 1988)

### 3.2 Neutrino Energetics

We will next assume as a working hypothesis that both neutrino pulses were real detections and will investigate the energetics of the events in order to check whether or not they are in conflict with astrophysical expectations. We will only present rough estimates here and will refer to more detailed discussions in passing.

We will first discuss the implications from the Mont Blanc detection. As can be seen from Table 1, two of the detected events had energies near threshold (around 5 MeV) and the first of them preceded the others by four seconds. We, therefore, will assume that the total number of events was 4 ± 2. Because of the much higher cross section of the ($\bar{\nu}_e$, $p$)-reaction compared to ($\nu_e$, $e$)-scattering we will assume that the incident reaction was $\bar{\nu}_e + p \rightarrow e^+ + n$, for which the cross section is $9.5 \times 10^{-44}$ $(E_\nu/\mathrm{MeV})^2$ cm$^2$, where $E_\nu$ is the incident neutrino energy in MeV. Note that for this reaction the incident neutrino energy will on average be only 10 to 20% higher than the observed positron energy. By taking the number of free protons in the Mont Blanc experiment ($\simeq 8 \times 10^{30}$) we can estimate the integrated flux at the detector and obtain $(5 \pm 3) \times 10^{12}$ $(E_\nu/\mathrm{MeV})^{-2}$ cm$^{-2}$. For the distance to the LMC ($\simeq$ 50 kpc) this corresponds to a total number of $\bar{\nu}_e$'s emitted from the source of about $(1.2 \pm 0.7) \times 10^{60}$ $(E_\nu/\mathrm{MeV})^{-2}$. The total energy is then $E_{tot} \simeq (1.9 \pm 1.1) \times 10^{54}$ erg $(E_\nu/\mathrm{MeV})^{-1}$. Assuming an average neutrino energy of 6 MeV in order to be consistent with the (non-)detection of KAMIOKANDE we end up with a total energy emitted in the form of $\bar{\nu}_e$'s of about $(3 \pm 2) \times 10^{53}$ erg. This number is in fact a lower limit because one probably has to go to even lower mean energies in order to satisfy the conditions set by the KAMIOKANDE experiment. If about the same energy was emitted in $\nu_e$'s, $\nu_{\mu,\tau}$'s, and $\bar{\nu}_{\mu,\tau}$'s, the total energy in the burst encreases by a factor of 6 and thus would be in conflict with the maximum binding energy of a neutron star ($\lesssim 6 \times 10^{53}$ erg) [47]. This problem can only be overcome if it is assumed that either the neutrino emission was beamed towards us or the luminosity of $\bar{\nu}$'s was higher than that of all other kinds. Both possibilities are not completely out of range, as will be discussed in Sect. 4.

An analysis similar to the one presented here has been performed by Sato and Suzuki [48] and many others [49] for the KAMIOKANDE and IMB data, assuming again that all neutrinos detected were $\bar{\nu}$'s. It was found that the KAMIOKANDE data can be fitted by a thermal neutrino spectrum corresponding to a temperature of (2.8 ± 0.3) MeV, whereas the IMB data require a significantly higher temperature of (4.6 ± 0.7) MeV, again indicating that the data are only marginally consistent. The total energy in three neutrino flavours is then $(2.9 \pm 0.6) \times 10^{53}$ erg for KAMIOKANDE and $(1.5 +1.2/-0.6) \times 10^{53}$ erg for IMB. If some of the events are considered to be noise, this energy estimate is reduced by a factor of about 2. If, on the other hand, some of the events are assumed to be caused by $(\nu, e)$-scattering, the energy in the burst would increase by about a factor of two. In any case, the data are consistent with the assumption that a neutron star of about 1.5 $M_\odot$ was born in SN 1987A and has radiated away a significant fraction of its binding energy in thermal neutrinos during the first few seconds of its life. It is also apparent, however, that this conclusion is still uncertain, because in the case of Mont Blanc we have to rely on the poor statistics of a few events and in the case of KAMIOKANDE and IMB we do not know for sure which (if any) of the detections were due to $(\nu, e)$-scattering.

## 4. A Comparison with Theoretical Predictions

### 4.1 The Standard Model

As we have discussed in Sect. 2 neutrinos are essentially trapped in a collapsing stellar core and, therefore, no measurable signal is expected from this early phase even of a typical galactic supernova. In fact, we can estimate from the discussion in Sect. 2 the expected neutrino flux in a more or less model independent way. During the core collapse phase the neutrino sphere is typically at a radius of 90 km and the neutrino temperature there is around 4 MeV. In the black body limit we obtain an electron neutrino luminosity of approximately $10^{53}$ erg s$^{-1}$ and a duration of $\Delta t \simeq 10^{-2} s$. Because in water Cerenkov detectors like the KAMIOKANDE-experiment these neutrinos can only be detected via neutrino-electron scattering, for which the cross section is of the order of $10^{-44}$ $(E_\nu/\mathrm{MeV})$ cm$^2$, a supernova at a distance of 3 kpc would give just 1 count in 2000t of water. Note that the distance to the galactic center is about 7.5 kpc.

After the shock front has passed the neutrino sphere the shocked matter becomes transparent to neutrinos. Electron captures on free protons in about 0.5 $M_\odot$ of the stellar material produce electron neutrinos with typical energies around 10 to 20 MeV. The total energy in neutrinos escaping on time scales of a few milliseconds from the star is of the order of $2 \times 10^{51}$ erg, the actual number depending on the strength of the shock. Therefore, the expected electron neutrino luminosity can be as high as $10^{54}$ erg s$^{-1}$. Again we can estimate the number of neutrinos which could be seen in existing neutrino detectors and find for the KAMIOKANDE experiment 60 counts $(\langle E_\nu \rangle / 10\ \mathrm{MeV})\ (D/\mathrm{kpc})^{-2}$ and a factor of 20 less for the Mont Blanc detector. Consequently, there is a certain chance that neutrinos from the deleptonization burst can be detected if the distance $D$ to a supernova is less than or of the order of a few kpc. On the other hand, it is very unlikely that any of those neutrinos have been seen in the KAMIOKANDE or IMB detector from SN 1987A ($D \simeq 50$ kpc). This result is in contradiction to interpretations which attribute the first two events detected by KAMIOKANDE to $\nu_e$'s emitted during this burst [48]. There remains the problem, however, that for both events the positron (electron?) was clearly directed away from the LMC, indicating $(\nu_e, e)$-scattering rather than $(\bar{\nu}_e, p)$-reactions.

After the deleptonization burst the electron concentration in certain mass zones behind the shock front has dropped to values of about 0.1 and consequently the electron chemical potential is close to $k_B T$ (see Sect. 2). Electron-positron pairs form in thermal equilibrium and reactions such as $e^+ + n \rightarrow \bar{\nu}_e + p$ or $e^+ + e^- \rightarrow \nu + \bar{\nu}$ produce neutrinos and anti-neutrinos of all flavors. Thus most of the thermal energy of the newly born neutron star is carried away by all kinds of neutrinos. Finally, once the temperature has dropped to values below 1 MeV URCA neutrinos and neutrinos from nucleon pair bremsstrahlung will take over, but their luminosity is too low to be detected by existing experiments. The time-scale over which we can hope to measure neutrinos is of the order of the neutrino duffusion time for the stellar core, i.e., a few seconds.

Unfortunately, only a few numerical studies have been performed which follow, moreover, the evolution over about $1s$ after core bounce only [50], and those computations dealing with the cooling of a newly born neutron star start from ad hoc initial models and are inconsistent with the results obtained from hydrodynamical simulations [51]. We can, therefore, give estimates of the expected neutrino flux only, and it is somewhat surprising that these estimates are in reasonable agreement with the detections from SN 1987A.

We expect from numerical simulations that for a time interval of about $0.5s$ the electron neutrino luminosity will be around $10^{52}$ to $5 \times 10^{52}$ erg s$^{-1}$ and that the electron antineutrino luminosity is about the same [50]. The luminosity in $\mu$-and $\tau$-neutrinos may be roughly a factor of two to four lower. Average energies will range from 10 to 12 MeV for $\nu_e$'s to 15 MeV for $\bar{\nu}_e$'s and 20 MeV for $\nu_\tau$'s. These numbers seem to be rather insensitive to the details of the numerical models and, therefore, may be typical. The total number of $\nu_e$'s and $\bar{\nu}_e$'s emitted from the newly born neutron star during this time ranges from a few times $10^{56}$ to $10^{57}$. The expected number of counts in the KAMIOKANDE experiment is thus $(6 \pm 4)\ (\langle E^{\nu_e} \rangle / 10\ \mathrm{MeV})\ (D/5\ \mathrm{kpc})^{-2}$ in $\nu_e$'s and $(60 \pm 40)\ (\langle E^{\bar{\nu}_e} \rangle / 10\ \mathrm{MeV})^2\ (D/5\ \mathrm{kpc})^{-2}$ in $\bar{\nu}_e$'s. For SN 1987A this translates into less than 0.1 counts of $\nu_e$'s and less than 2 counts of $\bar{\nu}_e$'s, whereas the actual count-rate for the first $0.4s$ was 4 (see Table 1). So, within the limited statistics of a few events, the agreement is quite good.

At even later times the predictions become extremely model dependent. They will, for example, be different for prompt and delayed explosions, the equation of state at high densities will play an important role, and the final mass of the neutron star will influence the neutrino fluxes and spectra. It is conceivable, however, that a neutrino luminosity of about $10^{52}$ erg s$^{-1}$ or somewhat higher can be sustained for a couple of seconds [51], and most of the neutrinos seen in the KAMIOKANDE experiment should have originated from this early cooling phase, provided the neutrinos have no restmass or their restmass is less than a few $eV/c^2$ [48]. However, it is difficult to understand why after about 10s neutrinos of rather high energies have still been seen from SN 1987A, since a typical neutrino diffusion time will be of the order of 2 to 3s only. Moreover, the large gap of about 7s between the arrival times of events number 9 and 10 (see table 1) in the KAMIOKANDE detector is difficult to understand. One is, therefore, tempted to assume that the last three events were noise and did not originate from SN 1987A. It is obvious, however, that while the standard model has no difficulties in explaining the KAMIOKANDE and IMB observations, it fails for LSD and Baksan, because the predicted average neutrino energies are too high and the luminosities are too low.

### 4.2 Non-standard Scenarios

If one insists that both neutrino pulses were real, a more complicated and speculative scenario has to be invented. Such a scenario has to explain the following facts:

1. The neutrino temperature of the first pulse was low ($T_\nu \lesssim 2$ MeV). Nevertheless, at least $2 \times 10^{53}$ erg have been emitted in the form of thermal electron neutrinos and anti-neutrinos, but significantly less in $\mu$- and $\tau$-neutrinos.

2. The second pulse was delayed by 4.7 hours relative to the first pulse.

3. The total energy in the second pulse was close to the binding energy of a cold neutron star of about 1.5 $M_\odot$.

It has been suggested that all three conditions can, in principle, be fulfilled if the first pulse signaled the formation of a meta-stable neutron star which later continued to collapse to either a black hole [52] or a strange matter configuration [53].

Let us first discuss condition 1. An upper limit of the electron neutrino luminosity can be obtained from the black-body limit which reads

$$L_\nu \lesssim 10^{49} \left[\frac{T_\nu}{\text{MeV}}\right]^4 \left[\frac{R_\nu}{10\text{km}}\right]^2 \ (\text{erg s}^{-1}), \tag{26}$$

where $T_\nu$ is the neutrino temperature in MeV and $R_\nu$ the radius of the neutrino sphere in km. From the Mont Blanc events we get $L_\nu \gtrsim 4 \times 10^{52}$ erg s$^{-1}$ and $T_\nu \lesssim 2$ MeV. Consequently the radius of the neutrino sphere has to be at least about 150 km. For a non-thermal neutrino spectrum the luminosity may exceed the limit given in (26) and thus the radius of the neutrino sphere may be somewhat smaller. In any case, it seems that the required radius of the neutrino sphere is significantly larger than that obtained from non-rotating core-collapse models ($\sim$ 30 to 80 km; see Sect. 2). One way out of this problem would be to assume that the stellar core was rapidly rotating in which case the neutrino "sphere" would become anisotropic, its "radius" might be larger and neutrinos could be transported more efficiently by large scale circular motions [54]. An anisotropy of the emitted neutrino pulse might also help to reduce the amount of energy needed in order to explain the Mont Blanc events, provided the pulse was beamed towards us.

In this scenario of a rapidly spinning proto-neutron star the time delay between the two neutrino bursts may also find a natural explanation. The collapse of the core of a 20 $M_\odot$ star, appropriate for the progenitor of SN 1987A, may lead to the formation of a rather massive neutron star, as was discussed in Sect. 2, and it is not excluded that its mass may exceed that of a cold non-rotating neutron star. Such a configuration would be stable until it has cooled down and/or has lost a significant fraction of its angular momentum. Loss of rotational energy by electro-magnetic dipole radiation can be estimated to be on typical time-scales of about

$$\tau \simeq 6 \times 10^4 \left[\frac{M}{M_\odot}\right] \left[\frac{P}{10^{-3}s}\right]^2 \left[\frac{B}{10^{13}\text{Gauss}}\right]^{-2} \left[\frac{R}{30\text{ km}}\right]^{-4} s, \tag{27}$$

where $P$ is the period of rotation, $B$ the magnetic field strength, and $M$ and $R$ are the mass and radius, respectively, of the newly born neutron star. From (27) we find that a magnetic field of the order of several $10^{13}$ Gauss will give the observed time scale, which seems to be somewhat high. If, on the other hand, the newly born neutron star was magnetically coupled to the expanding envelope lower values of $B$ will suffice.

A more difficult problem is to estimate the number of neutrinos and their energy which may be emitted during the collapse to a strange matter configuration or a black hole. Again we may use (26) to estimate the duration of the second neutrino pulse. Since we have an energy $E_\nu \gtrsim 4 \times 10^{52}$ erg in electron neutrinos and anti-neutrinos and the neutrino temperature was $\lesssim$ 4 MeV, we find $\tau \gtrsim$ 2s, which is much longer than the expected time-scale for the collapse to a black hole. But a neutron star collapsing to a black hole may not radiate neutrinos like a black body and we should replace (26) by the true Pauli-limit

$$L_\nu \lesssim 3.2 \times 10^{53} \left[\frac{\langle\nu\rangle}{20\mathrm{MeV}}\right]^4 \left[\frac{R_\nu}{10\mathrm{km}}\right]^2 (\mathrm{erg\ s^{-1}}), \tag{28}$$

where now $\langle\nu\rangle$ is the average neutrino energy. From (28) it is obvious that the observed amount of energy can, in principle, be emitted on time-scales of about a tenth of a second. Time scales of this order may result if a rapidly spinning object is collapsing to a black hole [55]. The observed dispersion of the neutrino signal should then be caused by other effects, e.g., a finite neutrino rest mass (see Sect. 5).

If, in contrast, the postulated second collapse proceeded to a strange matter configuration, the gain in binding energy might again be of the right order of magnitude, but now the time scale argument could be relaxed [56].

So in conclusion, it seems possible, though not easy, to invent an astrophysical scenario that can explain that two neutrino pulses may have been observed from SN 1987A. It is also obvious, however, that the most easy explanation of the neutrino events is to dismiss the Mont Blanc events as noise and to assume that the second pulse signaled the formation of a neutron star. Future observations will reveal which of the interpretations in the correct one. In fact, at present there is no direct evidence that a neutron star was born in SN 1987A.

## 5. Fundamental Properties of Neutrinos

The neutrino signal from SN 1987A has been used to place new constraints on fundamental properties of neutrinos, including rest masses, lifetimes, magnetic moments, mixing angles, etc..

In order to obtain an estimate of the electron neutrino rest mass we can proceed as follows. If neutrinos are emitted from a source at a distance $d$ with certain energy distribution on time scales short compared with the spread of the measured signal, a rest mass can be inferred from the relation [57]:

$$m_o = 4.28 \left[\frac{E_\nu^{min}}{7\mathrm{MeV}}\right] \left[\frac{d}{52\mathrm{kpc}}\right]^{-1/2} \sqrt{\Delta t \frac{\alpha^2}{\alpha^2 - 1}}, \tag{29}$$

where $\alpha = E_\nu^{max}/E_\nu^{min}$ is the observed energy band width of neutrinos. If we do not use a model for the intrinsic time spread of the signal, then (29) gives an upper limit for the rest mass. From the IMB and KAMIOKA date (Table 1) we obtain, allowing for possible errors in the energy determination, $m_o \leq 33eV/c^2$ and $15eV/c^2$, respectively, but the data are also compatible with a zero rest mass [58], in particular if we allow for the possibility that the last three KAMIOKANDE events as well as the last two IMB events, which were well separated from the main pulse, were caused by background noise. In any case, the supernova data are fully consistent with tritium decay experiments [59]. Therefore, only if it should turn out that SN 1987A left behind a black hole rather than a neutron star can we hope to get model-independent information on finite neutrino rest masses.

The fact that we have seen neutrinos from the supernova leads to a lifetime of the $\nu_e$ of

$$\tau \gtrsim 140h\left((m_o c^2)/(eV)\right) \tag{30}$$

in the co-moving frame of the neutrinos, but better limits exist for solar neutrinos [60]. The absence of a $\gamma$-ray burst associated with SN 1987A [61] can, however, be used to place limits on decay modes $\nu_H \to \nu_L + \gamma$ of heavy neutrinos into light ones which are essentially independent of the particular supernova model and apply to any massive weakly interacting particle emitted with a rate comparable to that of the $\bar{\nu}_e$'s. For temperatures below 5 MeV von Feilitzsch and Oberauer [62] obtain a lifetime larger than $10^{14}s\left((m_H c^2)/(eV)\right)$, which would also apply to heavy $\mu$-and $\tau$-neutrinos, provided $m_H - m_L \leq 2m_e$, for which $\nu_H \to \nu_L + \gamma$ is the only possible decay mode. For larger masses the limits are less stringent [63] because other decay modes become possible.

Most papers in which an attempt is made to place limits on other neutrino properties such as mixing angles [64] or magnetic moments [65] suffer from the fact that those estimates are based on the standard core collapse scenario and, therefore, have to be taken with some care. For example, if the first two of the KAMIOKANDE events are attributed to $\nu_e$'s from the deleptonization burst, absence of the Mikheyev-Smirnov-Wolfenstein effect [66] places very strong constraints on neutrino masses and mixing angles [64]. However, this assumption may be in conflict with the standard scenario, as was discussed in Sect. 4.

## 6. Summary and Conclusions

SN 1987A is the first clear case of a rather massive star that has undergone core collapse and has ejected its envelope in a type II supernova explosion. Such behavior was predicted by certain theoretical models, but direct observational evidence was not very strong. In fact, the rather high explosion energy as well as certain properties of the observed neutrino signals indicate that the explosion mechanism is not yet fully understood, whereas the blue nature of the progenitor star and the unusual light curve can be explained by rather standard stellar evolution models. Certainly, if a pulsar should show up in SN 1987, which is expected to happen in 1989, some of the uncertainties of the core collapse models will be removed. In particular we would know whether or not rotation played an important role, which, up to now, is still an open question. The presence of a neutron star in SN 1987A would also remove ambiguities in the interpretation of the neutrino data, but at present there is no direct evidence for a pulsar.

Theoretical models developed prior to the explosion of SN 1987A have predicted that the collapse of a massive star to neutron star densities will manifest itself in a fast burst of $\nu_e$'s of energies around 10 to 15 MeV, followed by significant emission of neutrinos and anti-neutrinos of all flavors over a timescale of several seconds. However, because the cross section for $(\bar{\nu}_e, p)$-reactions is about a factor of 100 higher than for $(\nu_e, e)$-scattering at the energies under consideration, only $\bar{\nu}_e$'s were expected to give a clear signal in existing neutrino detectors if the collapsing star was at a distance larger than about 5 kpc. The calculated neutrino luminosities and spectra are very sensitive to the astrophysical models and the physical input data, but the observed neutrino flux from SN 1987A indicates that the basic ideas are essentially correct. The observations have shown, in particular, that massive stars indeed do collapse, thereby emitting most of their gravitational binding energy in form of thermal neutrinos.

However, certain aspects of the neutrino events seen from SN 1987A are difficult to understand within the standard scenario. If indeed two neutrino bursts were emitted from SN 1987A, separated by almost five hours, this would call for significant modifications of the simple core collapse picture. Moreover, the apparent anisotropy of the positrons (electrons?) in both the KAMIOKANDE and IMB detectors is too large to be of purely statistical origin. It may indicate that the electron-neutrino luminosity of the supernova was much higher than predicted by the standard model. Other explanations are difficult to find unless one is willing to give up the standard Salam-Weinberg theory of weak interactions.

SN 1987A was just at the right distance to allow for model independent estimates of upper limits on the electron neutrino rest mass, which have turned out to be close to the best limits obtained so far from terrestrial experiments. Unfortunately, supernova models are still not good enough to improve those limits, and the same holds for most other estimates of fundamental properties of neutrinos. Future theoretical work as well as a detailed analysis of the information available in the electromagnetic spectrum may help to resolve some of the difficulties discussed here. In this respect, there is much more to be learnt from SN 1987A.

## References

[1] J.W. Menzies et al.: Mon. Not. R. Astr. Soc. 227, 39 (1987)

[2] R.M. West, A. Lauberts, H. Jørgensen, H.-E. Schuster: Astron. Astrophys. 177, L1 (1987)

[3] N. Panagia et al.: Astron. Astrophys. 117, L25 (1987)

[4] W.D. Arnett: Ap.J. 319, 136 (1987); W. Hillebrandt, P.Höflich, J.W. Truran, A.Weiss: Nature 327, 597 (1987); S.E. Woosley, P.A. Pinto, L. Ensman: Ap.J. 324, 466 (1987)

[5] H.A. Bethe, G.E. Brown, J. Applegate, J.M. Lattimer: Nucl. Phys. A324, 487 (1979); W. Hillebrandt: In Supernovae: A Survey of Current Research, ed. by M.J. Rees, R.J. Stoneham, NATO-ASI C90 (Reidel, Dordrecht 1982) p.123

[6] M. Aglietta et al.: Europhys. Lett. 3, 1315 (1987); R.M. Bionta et al.: Phys. Rev. Lett. 58, 1494 (1987); K. Hirata et al.: Phys. Rev. Lett. 58, 1490 (1987)

[7] J.N. Bahcall, S.L. Glashow: Nature 326, 476 (1987)

[8] G.A. Tammann: In Supernovae: A Survey of Current Research, ed. by. M.J. Rees, R.J. Stoneham, NATO-ASI C90 (Reidel, Dordrecht 1982) p. 371

[9] S. van den Bergh, R.D. Mc Clure, R. Evans: Ap.J. 323, 44 (1987)

[10] I. Iben, Jr.: Quark. J.R.A.S. 26, 1 (1985)

[11] V. Weidemann, D. Koester: Astron. Astrophys. 121, 77 (1983)

[12] G.E. Miller, J.H. Scalo: Ap.J. Suppl. 41, 513 (1979)

[13] K. Nomoto: In Stellar Nucleosynthesis, ed. by C. Chiosi, A. Renzini (Reidl, Dordrecht 1984) p. 205 and 238

[14] T.A. Weaver, S.E. Woosley, G.M. Fuller: In Numerical Astrophysics, ed. by J. Centrella, J. Le Blanc, R. Bowers (Jones & Bartlett, Portola Valley CA 1985) p. 374; J.R. Wilson, R.W. Mayle, S.E. Woosley, T.A. Weaver: In Proc. XII Texas Symp. on Relativistic Astrophys., Ann. NY Acad. Sci. 479, 267 (1986); K. Nomoto, M. Hashimoto: In Proc. Bethe Conference on Supernovae, ed. by G.E. Brown, Physics Reports, in press (1988)

[15] S.E. Woosley, T.A. Weaver: In Supernovae: A Survey of Current Research, ed. by M.J. Rees, R.J. Stoneham, NATO-ASI C90 (Reidel, Dordrecht 1982) p. 79 W.W. Ober, M.F. El Eid, K.J. Fricke: Astron. Astrophys. 119, 61 (1983)

[16] M.F. El Eid, W. Hillebrandt: Astron. Astrophys. Suppl. 42, 215 (1980)

[17] T.J. Mazurek, J.M. Lattimer, G.E. Brown: Ap.J. 229, 713 (1979)

[18] P. Bonche, D. Vautherin: Nucl. Phys. A372, 496 (1981); W. Hillebrandt, R.G. Wolff: In Nucleosynthesis, ed. by W.D. Arnett, J.W. Truran (University of Chicago Press, Chicago IL 1985) p. 131

[19] D.Q. Lamb, J.M. Lattimer, C.J. Pethick, D.G. Ravenhall: Phys. Rev. Lett. 41, 1623 (1978); M. Barranco, J.R. Buchler: Phys. Rev. C22, 1729 (1981) J.M. Lattimer: Ann. Rev. Nucl. Part. Sci. 31, 337 (1981)

[20] E. Baron, J. Cooperstein, S. Kahana: Nucl. Phys. A440, 744 (1985)

[21] S.W. Bruenn: Ap.J. Suppl. 58, 771 (1985)

[22] G.M. Fuller, W.A. Fowler, M.J. Newman: Ap.J. 252, 715 (1982); K. Takahashi, M.F. El Eid, W. Hillebrandt: Astron. Astrophys. 103, 358 (1978); J. Wambach: In Weak and Electromagnetic Interactions in Nuclei, ed. by H.V. Klapdor (Springer, Berlin Heidelberg 1986) p. 950

[23] H.A. Bethe: In Supernovae: A Survey of Current Research, ed. by M.J. Rees, R.J. Stoneham, NATO-ASI C90 (Reidel, Dordrecht 1982) p. 35 ; see also ref.[5]

[24] W. Hillebrandt, E. Müller: In Neutrino Physics and Astrophysics, ed. by K. Kleinknecht, E.A. Paschos (World Scientific, Singapore 1984) p. 229

[25] J.I. Castor: Ap.J. 178, 779 (1972)

[26] W.D. Arnett: Ap.J. 218, 815 (1977)

[27] R.B. Bowers, J.R. Wilson: Ap.J. 263, 366 (1982); see also ref.[21]

[28] P. Goldreich, S.V. Weber: Ap.J. 238, 991 (1980)

[29] W. Hillebrandt: In High Energy Phenomena Around Collapsed Stars, ed. by F. Pacini, NATO-ASI C195 (Reidel, Dordrecht 1985) p. 73

[30] E. Baron, H.A. Bethe, G.E. Brown, J. Cooperstein, S. Kahana: Phys. Rev. Lett. 59, 736 (1987)

[31] G.E. Brown, H.A. Bethe, G. Baym: Nucl.Phys. A375, 481 (1982)

[32] W. Hillebrandt, K. Nomoto, R.G. Wolff: Astron. Astrophys. 133, 175 (1984)

[33] J.R. Wilson: Talk presented at the Aspen Conference on SN 1987A (1988)

[34] J.R. Wilson: In Numerical Astrophysics, ed. by J. Centrella, J. Le Blanc, R. Bowers (Jones and Bartlett, Boston 1985) p. 422; H.A. Bethe, J.R. Wilson: Ap.J. 295, 14 (1985)

[35] J.M. Lattimer, A. Burrows: In Problems of Collapse and Numerical Relativity, ed. by D. Bancel, M. Signore, NATO-ASI C134 (Reidel, Dordrecht 1984) p. 147

[36] W. Hillebrandt: In Proc. 3rd Workshop on Nuclear Astrophysics, ed. by W. Hillebrandt, E. Müller, F.-K. Thielemann, MPA-Report 199, p. 59 (1985)

[37] E. Müller, W. Hillebrandt: Astron. Astrophys. 103, 358 (1981)

[38] J.M. Le Blanc, J.R. Wilson: Ap.J. 161, 541 (1970); E. Symbalisty: Ap.J. 285, 729 (1984)

[39] P. Bodenheimer, S.E. Woosley: Ap.J. 269, 281 (1983)

[40] M. Aglietta et al.: Europhys. Lett. 3, 1315 (1987)

[41] K. Hirata et al.: Phys. Rev. Lett. 58, 1490 (1987)

[42] R.M. Bionta et al.: Phys. Rev. Lett. 58, 1494 (1987)

[43] E.N. Alexeyev, L.N. Alexeyeva, I.V. Krivosheina, V.I. Volchenko: in "SN 1987A", ed. by I.J. Danziger, ESO, p. 237 (1987)

[44] R.H. McNaught: IAU Circular No. 4389 (1987)

[45] M. Aglietta et al.: In SN 1987A, ed. by I.J. Danziger, ESO, p. 207 (1987)

[46] R. Svoboda et al.: In SN 1987A, ed. by I.J. Danziger, ESO, p. 315 (1987)

[47] A. Burrows: In SN 1987A, ed. by I.J. Danziger, ESO, p. 315 (1987)

[48] K. Sato, H. Suzuki: Phys. Rev. Lett. 58, 2722 (1987)

[49] J.N. Bahcall, A. Dar, T. Piran: Nature 326, 135 (1987); A. Burrows, J. M. Lattimer: Ap.J. Lett. 318, L63 (1987); W.D. Arnett, J.L. Rosner: Phys. Rev. Lett. 58, 1906 (1987); A. De Rújula: Phys. Lett. 193, 525 (1987)

[50] R.W. Mayle, J.R. Wilson, D.N. Schramm: Ap.J. 318, 288 (1987); R.W. Mayle: Ph.D. Thesis, UC Berkeley, Lawrence Livermore Nat. Lab. preprint UCRL-53713 (1985)

[51] A. Burrows, J.M. Lattimer: Ap.J. 307, 178 (1986)

[52] W. Hillebrandt et al.: Astron. Astrophys. 180, L20 (1987); see also De Rújula [49]

[53] H.J. Haubold, B. Kaempfer, A.V. Senatorov, D.N. Voskresenski: Astron. Astrophys. 191, L22 (1988)

[54] E. Müller, W. Hillebrandt: Astron. Astrophys. 103, 358 (1981)

[55] D.M. Eardly: In Gravitational Radiation, ed. by N. Deruelle, T. Piran (North Holland, Amsterdam 1983) p. 257

[56] M. Takahara, K. Sato: in "Big Bang, Active Galactic Nuclei and Supernovae"; Proc. 20th Yamada Conference, ed. by S. Hayakawa and K. Sato, Univ. Tokyo, in press (1988)

[57] W. Hillebrandt et al.: Astron. Astrophys. 177, L41 (1987)

[58] W.D. Arnett, J.L. Rosner: Phys. Rev. Lett. 58, 1906 (1987); J.N. Bahcall, S.L. Glashow: Nature 326, 476 (1987); E.W. Kolb, A.J. Stebbins, M.S. Turner: Phys. Rev. D35, 3598 (1987)

[59] M. Fritschi et al.: Phys. Lett. B173, 485 (1086)

[60] G. Raffelt: Phys. Rev. D31, 3002 (1985)

[61] W.T. Vestraud, A. Gosh, E. Chupp: IAU-Circulars No. 4338, 4340 and 4365 (1987)

[62] F. von Feilitzsch, L. Oberauer: Phys. Lett. B200, 580 (1987)

[63] M. Takahara, K. Sato: Mod. Phys. Lett. A2, 293 (1987)

[64] D. Nötzold: Phys. Lett B 196, 315 (1987); T.K. Kuo, J. Pantaleone: Phys. Lett. B 198, 406 (1987)

[65] D. Nötzold: Preprint MPI-PAE/PTh 15/88 (1988)

[66] S.P. Mikheyev, A.Yu. Smirnov: Yad. Fiz 42, 1441 (1985); L. Wolfenstein: Phys. Rev. D17, 2369 (1978)

# Neutrinos in Cosmology

*G. Gelmini* *

SISSA/ISAS P.O. Box 586, I-34014 Trieste, Italy, and
ICTP, P.O. Box 586, I-34014 Trieste, Italy

## 1. Introduction

About neutrinos, we only know that the three left handed $\nu_{e_L}$, $\nu_{\mu_L}$, $\nu_\tau$ exist, that they have weak interactions and that they are much lighter than their corresponding charged leptons. They may be massive or massless, stable or not. If they are massive, they may be the left-handed part of the Majorana or Dirac fermions. If neutrinos are Majorana fermions, the right-handed neutrinos $\nu_R$ may or may not exist and may be heavy or light. If neutrinos are Dirac fermions, the $\nu_R$ must exist. Neutrinos may have new interactions, other than weak, which may allow them to annihilate or decay faster. Not just laboratory experiments can investigate these possibilities: because neutrinos were present in large numbers in the early universe, their properties affect the evolution and the present structure of the universe. Our knowledge of the history of the universe is thus used as a laboratory for placing limits on the mass, lifetime and number of flavours of neutrinos. Also astrophysics leads to limits on neutrinos, the recent observations of neutrinos from the supernova SN 1987A being an outstanding example. On the other hand, the knowledge of some properties of neutrinos may be crucial for cosmology and astrophysics. For example, the solution to a major cosmological problem would be given by a neutrino mass of around 30eV. This would mean that the "dark matter" consists of light neutrinos.

In the seventies the idea that neutrinos may have a mass started to be explored, in particular the cosmological consequences. Bounds on the masses, from the contribution of massive stable neutrinos to the density of the universe, were derived first. Then more complex bounds on unstable neutrinos were obtained from the effects of the decay products. Those tests derived for neutrinos are now routinely applied to any new proposed particle to limit its lifetime, mass, cosmological density, decay modes.

---

* On leave from Dept. of Physics, University of Rome II, Via O.Raimondo, 00173 Rome, Italy

Let us briefly review the features of standard cosmology that impose limits on neutrinos.The standard model of cosmology, the hot big bang, establishes that the universe is expanding from a state of extremely high density and temperature. The moment the expansion started, the moment of the big bang, is taken as the origin of the lifetime of the universe, t. This model provides a reliable description of the evolution of the universe from the epoch of primordial nucleosynthesis (at a time $t \sim 10^{-2}$ sec after the bang and temperature $T \simeq 10$MeV) until the present ($t_0 \simeq 1.5 \times 10^{10}$ y, $T_0 = 2.7$K). It rests on three major empirical pieces of evidence: the Hubble expansion, the cosmic blackbody microwave background radiation (MBR), and the relative abundance of light elements (that had to be synthesized in the early universe). Gamow proposed this model to explain the Hubble expansions, and indicated a MBR and the primordial synthesis of nuclei as necessary components [2]. The Hubble expansion of galaxies is the recession of galaxies (from us and from each other) with velocities proportional to their distances. The proportionality is given by the Hubble constant H (constant in space, but not in time). The value of H is now $H = h \times 100$ Km/(Mpc.sec), with h a number between 0.4 and 1 [3], that accounts for the uncertainty (astronomers measure distance in parsecs, pc, 1 pc = 3.26 light years). Astronomers observe that the redshift z of the spectra of distant galaxies

$$z = \frac{\lambda - \lambda_o}{\lambda} \tag{1}$$

($\lambda$ and $\lambda_0$ are the observed and emitted wavelengths) is proportional to their distance. The redshift is due to a Doppler effect: the universe is expanding around us. The "cosmological principle", which asserts that the universe is homogeneous and isotropic on large scales, then implies that the expansion is homogeneous: any distance r between two points increases with time as

$$r = R(t)\, r_0 \quad , \tag{2}$$

where $r_0$ is a constant and R(t) is the scale factor of the expansion. If R is chosen to be 1 now, $R_0 = 1$, then $r_0$ is the present distance between two points (subscript o usually denotes the value of the quantities today, in cosmology) and $R(t) = (1+z)^{-1}$. $r_0$ is also the distance measured in comoving coordinates, the coordinates that follow the expansion of the universe (as a grid painted on an expanding baloon). Deriving (2) with respect to time, Hubble's law is immediately obtained:

$$v = \dot{r} = \dot{R}R^{-1}\, r = Hr; \tag{3}$$

thus the Hubble "constant" is

$$H(t) \equiv \dot{R}R^{-1} \ . \tag{4}$$

Einstein's equations relate the value of H with the total density of the universe $\rho$. For a homogeneous and isotropic fluid,

$$H^2 = \frac{8\pi}{3} G\rho - \frac{k}{R^2} . \tag{5}$$

Here G is the Newton constant, the curvature constant $k = -1, 0, 1$ (any negative value can be brought to $-1$ and any positive to 1 by redefinition of coordinates) distinguishes a closed, flat or open universe. For $k = 0$, (5) gives the definition of the critical density $\rho_c$, whose value is now

$$\rho_c = 10.5\ h^2\ \mathrm{KeV/cm^3} . \tag{6}$$

$\rho_c$ is the density such that the universe is flat ( $\rho > \rho_c$ for a closed universe, which will re-collapse, and $\rho < \rho_c$ for an open universe, which will expand for ever). The energy density $\rho$ has three components: energy in relativistic particles (i.e. radiation) $\rho_r$, energy in non-relativistic particles (i.e. in matter) $\rho_m$, and energy in the vacuum (i.e. in a cosmological constant term) $\rho_v = \Lambda/8\pi G$, where $\Lambda$ is the usual cosmological constant. There is no evidence for a $\Lambda$-term, thus usually it is assumed $\Lambda=0$. Due to the different equations of state of matter and radiation, $\rho_r$ and $\rho_m$ decrease at a different rate as the volume increase with the Hubble expansion: $\rho_m \sim R^{-3}$ and $\rho_r \sim R^{-4}$. The number of free particles per unit volume decreases just as the volume increases, $n \sim R^{-3}$. While $\rho_m$ = mass $\times$ n, the energy of relativistic particles E decreases with R, $E = \lambda^{-1} \sim R^{-1}$ and $\rho_r = E \times n$. Thus, in the early universe ($t \leq 3 \times 10^2$ y, $T \gtrsim 10$ eV), the energy density of the universe was dominated by relativistic particle species, which were in thermal equilibrium with a temperature of $T \sim (t/\mathrm{sec})^{-1/2}$ MeV. For most of the history of the universe $T \sim R(t)^{-1}$. In a radiation dominated universe $R \sim t^{1/2}$, while $R \sim t^{2/3}$ for a matter dominated universe. These last relations are easy to remember by noticing that the lifetime of the universe is just $t = H^{-1}$ (except for a constant of order one) and $H \sim \sqrt{\rho}$ [by (5) with k=0]. We know observationally that k is approximately zero now, and in the early universe the curvature term ($\sim R^{-2}$) rapidly becomes irrelevant with respect to the density term ($\sim R^{-3}$ or $R^{-4}$). Thus $k = 0$ is always a good approximation.

Going back in time, the temperature, i.e. the mean energy of the particles in the universe, increases. When T is larger than the ionization energy of atoms, they cannot be stable; hence ions and electrons form a plasma. Going back further, nuclei are split into their components, further still and the universe contains only elementary particles. The moment when ions and electrons combine into stable atoms is called recombination or electromagnetic decoupling ($t \simeq 5 \times 10^5$y, $T \simeq 0.3$ eV). The last name refers to the fact that the universe becomes transparent to photons when the matter constituents become neutral (atoms). Photons that then interacted for the last time, became a background radiation, with a blackbody spectrum (because they were in equilibrium with matter before recombination). Due to the Hubble expansion those photons redshifted; thus the temperature of this radiation is now just 2.7 K = $2.3\ 10^{-4}$ eV (while it was approximately 0.3 eV when it was emitted) [4].

The existence of the 2.7 K blackbody background radiation is an essential piece of evidence, whose observation [5] confirmed the big bang model (against the steady-state model that also accounted for the Hubble expansion, but in which $\rho$ and H, and thus T, were constant in time). The blackbody character[4] and the high degree of isotropy [6] of this radiation give bounds on radiative neutrino decay modes. The isotropy of the MBR [6] $\Delta T/T < 10^{-4}$ (where $\Delta T$ is the difference of temperatures at two different points in the sky) is the best evidence of the isotropy and homogeneity of the universe, both in the past (at recombination and later) and at large scales now (larger than 100 Mpc). At scales smaller than 100 Mpc, the scale of superclusters of galaxies, the universe is now not homogeneous. There are stars, galaxies, clusters and superclusters of galaxies. Standard cosmology provides a general framework for understanding how structures were formed: small density inhomogeneities grew due to the effect of gravitation once the universe became matter dominated.

The third impressive piece of evidence, as already mentioned, is the nucleosynthesis of light elements. The cosmic abundance of $^4$He and of the trace elements D, $^3$He and $^7$Li are well accounted for, in terms of nuclear reactions that must have occurred at $t \simeq 10^{-2} - 10^2$ sec, $T \simeq$ 10-0.1 MeV [7,8].Nucleosynthesis gives bounds on any new particle that makes a non-negligible contribution to the energy density of the universe during the synthesis. It also provides bounds on radiative decays of neutrinos after the synthesis.

There is a most useful experimental bound on the present total energy density of the universe (for a universe either matter or radiation dominated) [9,10]:

$$(\Omega h^2)_0 \lesssim 1 \ . \qquad (7)$$

Here $\Omega = \rho/\rho_c$, is the energy density in units of $\rho_c$, and this limit can be obtained simply from the empirical bound on the lifetime of the universe, $t_0 \geq 1 \times 10^{10}$ y (as required by radioactive isotope dating [11]) in either a matter or radiation dominated universe. The values of H, $t_0$ and $\rho$ are not independent. The content of the universe determines the relation between them. Except in special models, the universe is matter dominated for most of its history. For a universe that is radiation dominated for most of the time (by relativistic neutrinos or Goldstone bosons, for example), the bound on $\Omega h^2$ is 0.3, slightly more restrictive than (7). For a higher minimum age $t_0 \geq 1.6\ 10^{10}$ y (as inferred from the age of the oldest stars [12]) the upper bound on $\Omega h^2$ becomes 0.2 and 0.05 for a matter or radiation dominated universe respectively. However, if the cosmological constant, i.e. the vacuum energy, is non-zero, these bounds are degraded. Direct observations indicate that $\Omega$ is of order one (between 0.1 and 1), which, when combined with the value of h, tells us that $\Omega h^2$ can not be larger than one. In the following we will consider a zero cosmological constant. $\Omega$ of order one is a very surprising result, unless a dynamical reason exists for this value, such as a period of inflation [13] in the early universe. In this case $\Omega$ should be exactly 1.

Another very surprising fact is that most of the mass of the universe does not consists of stars, gas or dust. The main mass component of the universe is "dark", it does not emit or absorb light and is revealed only through its gravitational effects [14]. A fascinating possibility is that this dark matter (D.M.) may be neutral elementary particles. From all the many D.M. candidates proposed in recent years [15], neutrinos are the only ones that definitely exist, and even so, we do not know if they are massive, as they should be.

In the following, we start by reviewing the implications of the condition (7) on the energy density in neutrinos, $\Omega_\nu \leq \Omega$. It provides an upper bound, $m_\nu < 100$ eV, on the mass of stable, light ( $m_\nu < 1$ MeV) neutrinos, a lower bound on the mass of heavy stable neutrinos ( $m_\nu >$ few GeV), and upper bounds on the lifetimes $\tau_\nu$. (By stable we mean $\tau > t_0$.) To apply this bound we need to compute the abundance of neutrinos in the universe. We call neutrinos, not only the three known left-handed $\nu_{e_L}$, $\nu_{\mu_L}$, $\nu_{\tau_L}$ but also the neutrinos of possible further generations, fourth etc., (light or heavy) and the possible (light or heavy) right-handed neutrinos. We will, then, briefly mention bounds derived from nucleosynthesis and bounds on radiative and invisible decay modes and annihilations. The last sections are devoted to neutrinos as the D.M.. Light or heavy neutrinos are good candidates for hot or cold D.M., respectively. The possibility of detecting cold D.M. from the halo of our galaxy, and the attempts under way to do so, will be our last topic.

This general review of the mentioned topics includes some recent developments, such as the following: With the experimental bounds known until recently, the neutrinos of the standard model (but massive) had the possibility of being unstable. A tau neutrino of mass around $m_\nu \simeq 50$ MeV could decay into $e^+e^-\nu$, with a lifetime of $\tau_\nu \simeq 10^2$ sec, within the standard electroweak model, respecting all known bounds. The observation that the neutrino burst from the supernova SN87A was not followed by photons seems to have closed this window. This window has, in any case, been closed by the new upper bound of 35 MeV on the $\nu_\tau$ mass [16]. The exclusion of this window is very important because it means that either all three neutrinos $\nu_e$, $\nu_\mu$, $\nu_\tau$ are lighter than 100 eV or they have non-standard interactions. With respect to invisible neutrino decay modes, the observation of neutrinos from the supernova does not preclude (against a naive belief) the decay of solar neutrinos while travelling to the earth as a solution of the solar neutrino problem. Light neutrinos, for sometime out of favour as the D.M., have been resurrected as possible D.M. candidates, not alone but with some other mechanism to help form galaxies, such as cosmic strings. Finally, interesting bounds on heavy neutrinos as components of the halo of our galaxy are being obtained, from both indirect and direct D.M. detection attempts.

## 2. Relic Abundances

In the big bang model, at high T left-handed neutrinos were in equilibrium with charged leptons and photons, due to their weak interactions. The number density of neutrinos $n_\nu$ is given, in equilibrium, by the Fermi-Dirac distribution, while the number density of photons $n_\gamma$ is given by the Bose-Einstein distribution. Thus, in equilibrium the ratio $n_\nu / n_\gamma$ is calculable:

- for relativistic neutrinos, $m_\nu < T$, $$\frac{n_\nu}{n_\gamma} = \frac{3}{4}\left(\frac{g_\nu}{2}\right) \quad , \tag{8}$$

- for non-relativistic neutrinos, $m_\nu > T$, $$\frac{n_\nu}{n_\gamma} = \left(\frac{\pi}{8}\right)^{1/2} \frac{g_\nu}{2\zeta(3)} \left(\frac{m_\nu}{T}\right)^{3/2} e^{-m_\nu/T} \,. \tag{9}$$

Here $g_\nu$ is the multiplicity of spin states, and $g_\nu=2$ for Weyl or Majorana neutrinos. Particles are in equilibrium when their interaction rate $\Gamma$ is larger than the expansion rate of the universe, which is the Hubble constant H or, equivalently, when their interaction time $\Gamma^{-1}$ is smaller than the age of the universe, that is approximately $H^{-1}$. When $\Gamma^{-1} < H^{-1}$ at high T and there is a temperature $T_{f.o.}$ such that $\Gamma^{-1} > H^{-1}$ for any smaller T, it is said that at that temperature $T_{f.o.}$ the particles "freeze-out" [17,18]. For any $T< T_{f.o.}$, the particles are out of equilibrium; they do not interact and they are neither created nor destroyed. The existence of equilibrium and then freeze out depends on the different behaviour of $\Gamma$ and H with T. For relativistic neutrinos (with full weak interactions, i.e. $\nu_L$), the interaction rate goes as $\Gamma \sim T^5$, while $H \sim T^2$, and $T_{f.o} \sim 1$ MeV. This is easy to see since the cross-section is $\sigma \sim T^2 G_F$, with $G_F$ the Fermi constant, while $n_\nu \sim T^3$ and the interaction rate is $\Gamma = n\sigma v$, where $v \simeq c = 1$. On the other hand $H \sim \sqrt{\rho_r G}$. By equating $\Gamma = H$ the $T_{f.o.}$ is obtained. The number per comoving volume of these light ($m_\nu < 1$ MeV) neutrinos is constant after the freeze-out. The number of photons, however, is not. When T drops below the mass threshold for the production of $e^+e^-$, these annihilate, increasing the number of photons, but not that of non-interacting particles such as neutrinos. While $T > 2m_e$, as many electrons are produced as annihilated, but for $T < 2m_e$ only the annihilation continues. So long as thermal equilibrium is maintained, the entropy per comoving volume $S \sim sR^3$ is constant. Considering that R is practically constant during the $e^+e^-$ annnihilation, then $s=(\rho+p)/T$ and the entropy per unit volume is conserved. Since for relativistic particles $p=\rho/3$, then $s = \rho/T \simeq g^* T^3$, where $g^*$ counts the contribution of relativistic particles to $\rho$; $g^*$ is the number of bosonic spin degrees of freedom plus (7/8) of the number of fermionic degrees of freedom. Thus $g^* = 2 + (7/8) \times 4 = 11/2$ at $T > 2m_e$, when photons and e+ and e- are in equilibrium, while $g^* = 2$ at $T < 2m_e$ (only photons). Thus, the temperature of the photons is increased by a factor of $(11/4)^{1/3}$, and the number of photons by a factor (11/4). Therefore the number of neutrinos (plus antineutrinos) of every light species with respect to photons is now smaller than (8):

$$n_{\nu_i} = (3/11)(g_{\nu i}/2) n_\gamma = (108/\mathrm{cm}^3)\,(g_{\nu i}/2) \quad . \tag{10}$$

The total contribution of the three light neutrinos to the density of the universe is now $\rho_\nu = \sum_i m_{\nu_i} n_{\nu_i} = \Omega_\nu \rho_c$; thus the sum of the masses is [19]

$$\sum_i \left[\frac{g_{\nu_i}}{2}\right] m_{\nu_i} = 97 \text{ eV } (\Omega_\nu h^2) \quad , \tag{11}$$

which runs over all stable neutrino species lighter than 1 MeV. This bound applies to the left-handed neutrinos $\nu_L$ with full weak interactions, i.e. $g_\nu = 2$. Even in the case of Dirac neutrinos, in the standard electroweak model the right-handed components $\nu_R$ are never in equilibrium for $m_\nu < 1$ MeV [18,20]. Since the $\nu_R$ do not have gauge interactions, they are mainly produced through Dirac mass insertions that flip $\nu_L$ into $\nu_R$ in gauge interactions of the $\nu_L$; thus the production rate is suppressed at least by $(m_\nu/T)^2$. In extensions of the standard model in which $\nu_R$ do have gauge interactions, the freeze-out occurs at $(T_{f.o.})_R > M_Z$ (the $Z^o$ boson mass), and thus $n_{\nu_R} < (n_{\nu_L}/27)$ [21]. The situation is different for heavy neutrinos ($m_\nu >>$ MeV), for which $g_\nu = 4$ if they are Dirac and $g_\nu = 2$ if they are Majorana.

For heavy neutrinos that become non-relativistic while annihilations are in equilibrium (i.e. $\Gamma_{annih.} > H$), the number of neutrinos per comoving volume decreases with the Boltzmann factor, $n_\nu R^3 \sim n_\nu / T^3 \sim e^{-m\nu/T}$ [see (9), where $n_\gamma \sim T^3$]. After annihilations cease (because $\Gamma_{annih.} < H$), $n_\nu/T^3$ departs from its exponential equilibrium function at $T_* = m_\nu/x_*$, and becomes constant soon after [17,22] (see Fig. 1). The final density $\rho_\nu$ is

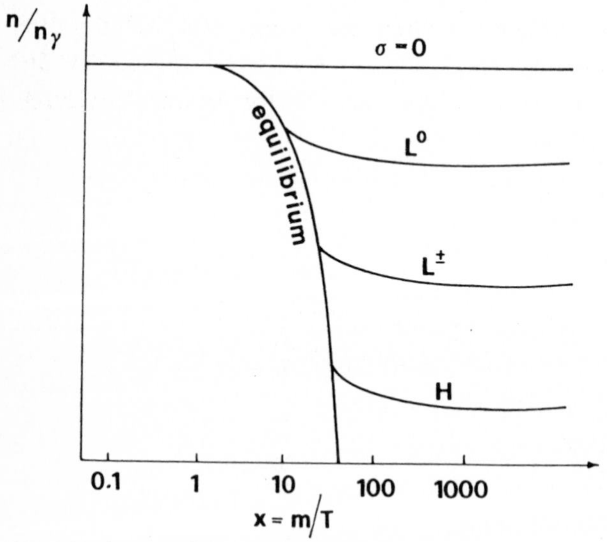


Fig. 1 Evolution of the ratio of the number of particles in the early universe n to the number of photons $n\gamma$. The ratio $n/n\gamma$ is constant for particles which become non-interacting while they are relativistic (the reheating of photons due to annihilations of other particles are not taken into account). This is the $\sigma=0$ line. $L^o$ indicates neutral, weakly interacting particles (this is the line for neutrinos). $L^\pm$ stands for charged leptons and H for hadrons. These particles go out of equilibrium at increasing values of $x=m/T$ : $x^*=20$ for $L^o$, $x^*=30$ for $L^\pm$, $x^* = 45$ for H. (Figure redrawn from [17])

proportional to the Boltzmann factor at $T_*$. Thus, an upper bound on $\rho_\nu$ implies a loweer bound on $m_\nu$. For weakly interacting relics $x_* \sim 20$, and for $\Omega_\nu = 0.1$ to 1, an $m_\nu$ in the few GeV range results [17, 22, 23]. The value of $x_*$ is larger for larger annihilation cross sections. Then, annihilations remain in equilibrium longer, and the final density for the same mass is smaller. This explains the difference between the relic abundances of heavy Dirac and Majorana neutrinos [24] (see Fig.2 re-drawn with data from [25]). The annihilation of non-relativistic Majorana neutrinos is suppressed, due to the presence of two identical particles in the initial state, since particle and antiparticle coincide, $\nu^c \equiv \nu$. Selection rules, then, imply that the annihilation occurs from an s-wave (with amplitude proportional to the mass of the annihilation products) or from a p-wave. This is because in the annihilations of Majorana neutrinos the vertices conserve chirality.If the masses in the final state are zero, chirality and helicity coincide, and the spin of the final state is $S_f = 1$. Therefore, the total angular momentum is $J_f = S_f + L_f \geq 1$, where $L_f$ is the final state orbital angular momentum. Thus $J_i = J_f \geq 1$, where i refers to the initial state. The initial wave function must be totally antisymmetric (for identical particles), and therefore $(-1)^{Li+Si+1} = (-1)$. Thus if $L_i = 0$, we should have $S_i = 0$, but $J_i = L_i + S_i$ must be 1 or larger, so that $L_i \geq 1$. If the mass of the final state fermions is $m_f \neq 0$, a mass insertion can change the helicity of a final fermion and $S_f = 0$ is possible. Hence $J_f \geq 0$ allows for $L_i = 0$, but the amplitude is proportional to $m_f$. Due to the suppression of annihilations,

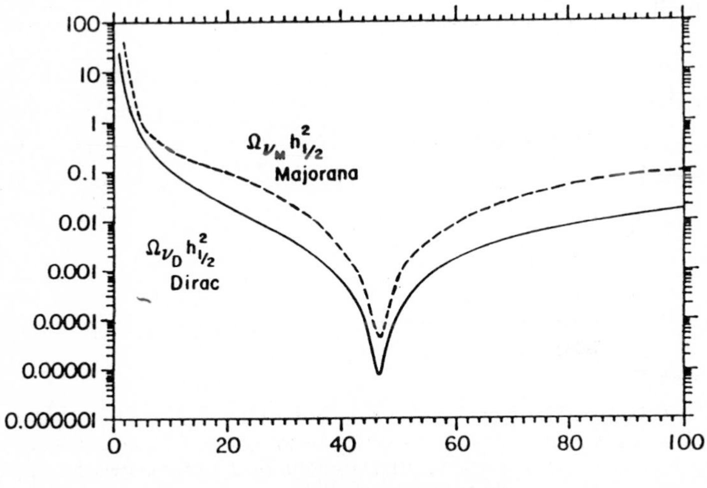


Fig. 2 The vertical axis shows $\Omega_\nu h^2_{1/2}$, where $h_{1/2}$ is the Hubble constant in units of 50 km/Mpc sec and $\Omega_\nu$ is the relic density for Dirac (solid curve) and Majorana (broken curve) neutrinos with $m_\nu$>1 MeV

the relic abundance of heavy Majorana neutrinos is larger than the abundance of Dirac neutrinos of the same mass [24]:

$$(\Omega_\nu h^2)^{\mathrm{DIRAC}}_{\mathrm{MAJORANA}} = \begin{Bmatrix} 5.4 \\ 18.0 \end{Bmatrix} \left(\frac{\mathrm{GeV}}{m_\nu}\right)^2 \left[1 + \frac{3}{25} \ln \left(\frac{m_\nu}{\mathrm{GeV}}\right)\right] \quad . \tag{12}$$

For $(\Omega_\nu h^2)_o < 0.25$ for example, $m_\nu^{\mathrm{DIRAC}} > 3.3$ GeV and $m_\nu^{\mathrm{MAJORANA}} > 6.8$ GeV [24]. The annihilation cross-section for heavy neutrinos has a resonance at $m_\nu = m_{z_0}/2$. Correspondingly $\rho_\nu$ has a dip. For even larger values of $m_\nu$ the cross section decreases with $m_\nu$, $\sigma_{annih} \sim m_\nu^{-2}$ (instead of $\sigma_{annih} \sim m_\nu^2/M_z^4$ for $m_\nu < M_z/2$). Accordingly, the relic density increases with $m_\nu$ (Fig. 2). For large values of $m_\nu$ other annihilation channels beyond $\nu\bar{\nu} \to f\bar{f}$ (such as $\nu\bar{\nu} \to W^+W^-$) become important (see Enqvist et al [26]). As a curiosity, note that the annihilation cross-section is also large for special values of $m_\nu$, where vector resonances such as $\psi$ and $\psi'$ are formed in the s-channel. Thus some special, discrete values of $m_\nu$ are also allowed (see Kane, Kani [26]).

Here we are assuming stable neutrinos. However, neutrinos of mass of order GeV or larger are very unlikely to be stable unless they carry an exactly conserved particle number. Unstable neutrinos may have masses in the range forbidden for stable neutrinos, between 100 eV and a few GeV. The condition (7) has to be applied to the present density of the decay products. If a heavy neutrino, whose contribution dominates the energy density of the universe decays, the decay products (D.P.) that are relativistic after the decay, become dominant. At the moment of the decay $(\Omega_\nu)_d = (\Omega_{D.P.})_d$ (subscript d meaning "at decay"). The universe passes from matter dominated to radiation dominated. Assuming that the universe is now still radiation dominated by the decay products, their density is now $(\Omega_{D.P.})_o = \Omega_\nu \times (R_d/R_o)$, where $\Omega_\nu$ is the density of the parent neutrino had it not decayed [since $(\Omega_{D.P.}) = (R_d/R_o)^4 \times (\Omega_{D.P.})_d$, while $\Omega_\nu = (R_d/R_o) \times (\Omega_\nu)_d$]. The conditions (7) and

$$(\Omega_{D.P.})_o \leq \Omega_o \tag{13}$$

lead to an upper bound on the lifetime into any mode [10, 27]. For $m_\nu < 1$ MeV this is

$$\tau_\nu < 5 \times 10^{14} \text{ sec } (\mathrm{KeV}/m_\nu)^2 (2/g_\nu)^2 (\Omega h^2/0.3)^2 \quad . \tag{14}$$

This bound is shown for all masses in Fig. 3 (large top left bottom right hatched area). The bound is more stringent if the universe again becomes matter dominated some time after the decay. This is why galaxy formation

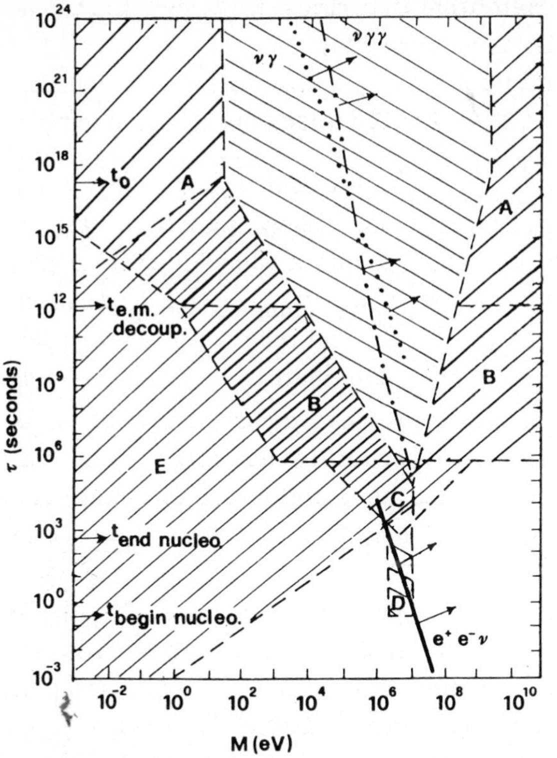


Fig. 3 In a lifetime τ versus neutrino mass M, excluded regions for any decay mode (hatched top left to bottom right) and for radiative decay modes (hatched bottom left to top right). The lines show respectively the lower lifetime limits for the modes νγ, νγγ and $\nu e^+e^-$ (M >1 MeV) in the standard electroweak model

provides a bound more stringent than (14) by a factor $10^5$ [28]. For $m_\nu < 1$ MeV this is

$$\tau_\nu < 2 \times 10^9 \ \text{sec} \ (\text{KeV}/m_\nu)^2 (2/g_\nu)^2 (\Omega h^2)^{3/2} \quad . \tag{15}$$

In fact in most models of galaxy formation, structures grow from originally small density fluctuations due to gravitation. This growth cannot happen in a radiation dominated universe. Thus, the contribution of relativistic decay products to the energy density should have ceased to be dominant at the moment density perturbations in baryonic matter started forming galaxies. This happens at recomination, that is [28],

$$(\Omega_{D.P.})_{\text{at e.m. decoupling}} < \Omega_{B \ \text{at e.m. decoupling}} \tag{16}$$

(and ever since, given that $\Omega_{D.P.} \sim R^{-4}$ and $\Omega_B \sim R^{-3}$). Some people do not consider this bound to be on the same footing as (13), since the problem of galaxy formation has not been solved, anyway.

## 3. Bounds from Primordial Nucleosynthesis

The light nuclei, mainly $^4$He, are formed using all the neutrinos available when the universe cools to $T \simeq 0.1$ MeV. At that moment the number per nucleon of photons that have enough energy to dissociate the nuclei falls below unity. The number of neutrons per proton is just $(n/p) = e^{-(m_n - m_p)/T}$, while protons and neutrons are kept in equilibrium by weak interaction. The n/p ratio freezes out when the rate of weak interactions becomes equal to, and then smaller than, H. H depends on the total energy density of the universe at that moment, $H \sim \sqrt{\rho}$. If $\rho$ is larger, the freeze out occurs at a higher temperature, the remaining n/p ratio is larger and so is the amount of $^4$He synthetized. This amount also increases with the number of baryons per photon, $\eta = n_B/n_\gamma$. Thus, in order to get bounds on any new contribution to $\rho$, such as new flavours of neutrinos $\rho_\nu$ , the abundances of at least two elements are needed (since $\eta$ also has to be fitted). Yang et al [8] concluded that one range of $\eta$ fixes three very different abundances ($^4$He, (D+$^3$He), $^7$Li), $\eta = 4\text{-}7 \times 10^{-10}$, which implies a present baryon density such that

$$0.014 < \Omega_B h^2 < 0.035 . \tag{17}$$

Recently, various alternative scenarios of nucleosynthesis have been proposed [29] in which the bounds on $\Omega_B$ are less stringent, but all of them have suffered partial failures up to now. Applegate et al. and Alcock et al. [29] consider a non-homogeneous nucleosynthesis, in which distinct regions are due to baryon segregation and neutron diffusion after the confinement phase transition of Quantum Chromo Dynamics. The problem with this approach is that several unknown QCD parameters must be suitably chosen to allow for the agreement with observations. Audouze et al. and Dominguez-Tenreiro [30] introduce a particle that decays into radiative modes after the nucleosynthesis. The decay products lead to photodissociation of $^4$He into D. The problem is that they cannot simultaneously account for the $^4$He , D and $^3$He abundances. Dimopoulos et al. [31] also introduce a particle that decays after the nucleosynthesis, but in this case there are hadronic decay products, which cause dissociation of $^4$He and production of light nuclei. The problem seems to be an overproduction of $^6$Li. Yang et al. find [8] that the number of flavours of 2-component relativistic neutrinos with full-strength weak interactions during the nucleosynthesis, i.e. with $m_\nu < 0.1$ MeV, cannot be larger than four ($N_\nu < 4$). If, during the nucleosynthesis, there are three 2-component

($g_\nu$=2) relativistic neutrino species and a fourth non-relativistic one, the fourth neutrino should have a mass $m_\nu$ > 20 MeV [32]. It is interesting to note that models with a nonzero vacuum energy density during the nucleosynthesis can accomodate $N_\nu$ = 5, or more [33].

Nucleosynthesis implies bounds on visible decays (which produce photons or charged particles) that occur after the synthesis. They alter the value of $\eta$ (since new $\gamma$'s are produced, $\eta$ during nucleosynthesis must be larger that $\eta$ today) and the high-energy decay products destroy the synthetized nuclei.

## 4. Bounds on Lifetimes

Unstable neutrinos that decay into visible channels, i.e. into charged particles (for $m_\nu$ > 1 MeV) or photons, are severely constrained by various observations (for a recent survey of bounds on radiative decays see [34]). Figure 3 (partially re-drawn from Kolb [35]) shows several of these bounds. The large top left to bottom right hatched region shows the lifetimes forbidden by the conditions (7) and (13), for any decay mode (it assumes the universe is radiation dominated by the decay products). The wider hatched top right to bottom left region is forbidden for visible modes. The blackbody microwave background radiation MBR consists of photons that last interacted at the moment of electromagnetic decoupling or recombination ($t \simeq 5 \times 10^5$ y, $T \simeq 0.3$ eV). At that moment ions and electrons, which had formed a plasma in which photons had a short mean free path, combined into atoms, and the universe became transparent to photons. Photons emitted after the e.m. recombination do not scatter and appear in the photon background. Because the number of relativistic neutrinos is close to the number of photons, when a neutrino species becomes non-relativistic ($m_\nu$ >T) its contribution to the energy density ($\rho_\nu = m_\nu n_\nu$) soon becomes dominant over the radiation density $\rho_r \simeq T n_\gamma$. Thus the radiative decay of a non-relativistic neutrino, after the recombination (in the region A) would produce a visible energy density larger than the MBR, in excess of observed limits [36,37]. The blackbody spectrum of the MBR must have been established before recombination, in the thermalization era (which started at $t \simeq 10^5$ sec, $T \simeq 5$ KeV). Any extra photons produced in this era would result in distorsions of the MBR (area B of Fig. 3) [36]. Photons produced before would have time to thermalize and become part of the blackbody spectrum. Areas C and D are excluded by nucleosynthesis: In the region C photons produced after the synthesis decrease $\eta$ too much [38], and in the region D the contribution of the heavy neutrino to the density of the universe during nucleosynthesis is too big, leading to overproduction of $^4$He [32]. The region E (of Fig. 3) is excluded by astrophysics: the neutrinos emitted by supernovas and white dwarfs decaying into visible modes would produce too large a background of high-energy photons [37,39]. These bounds apply to the decays $\nu \to \nu'\gamma$, $\nu \to \nu'\gamma\gamma$ and $\nu \to e^+e^-\nu'$ (the $e^+$ and $e^-$ end up producing energetic photons when annihilating or scattering with interstellar matter or radiation). These are the only modes allowed in the standard electroweak model. The lower bounds of the respective lifetimes of these modes in the standard electroweak model are shown in Fig. 3. The mode $\nu'\gamma\gamma$ [40] can be faster than $\nu'\gamma$ [41], because the former is not GIM suppressed, but both are far too slow. For $m_\nu > 1$ MeV, i.e. for the $\nu_\tau$, the dominant mode $e^+e^-\nu$ [42] seemed to be able to respect all bounds, for certain values of $m_\nu$ and $\tau$. There was a window [43] for tao neutrino masses between 40 and 70 MeV and lifetimes between $10^1$ and $10^3$ sec. The window was limited by astrophysical bounds (on the observation of $\gamma$-ray background radiation from supernovas [44]) and laboratory bounds (on the mass [45] and on mixing angles [46]).

The non-observation of photons following the burst of neutrinos from SN87A seems to have closed this window [47,48] (Takahara and Sato [47] arrive at this conclusion, provided the radius of the progenitor star was smaller than $10^{13}$ cm; now it is known to have been $3 \times 10^{12}$cm). This would be the case, anyway, due to the new upper bound of 35 MeV on the mass of the $\nu_\tau$ [16].

The decays of neutrinos into invisible particles, such as light neutrinos or Goldstone bosons, can be constrained only by dynamical arguments, such as conditions (13) and (16). The decay $\nu \to 3\nu'$, through gauge flavour changing neutral currents, F.C.N.C., or Higgs-mediated in extensions of the standard model, cannot satisfy the bound (13) without problems in avoiding conflict with laboratory experiments and having to suppress related radiative decays [49]. F.C.N.C. are possible in the standard model when Dirac and Majorana mass terms are both present. Harari and Nair [50] have recently tested many-neutrino models with the bound (13) and concluded that $\nu_\tau \to 3\nu'$ was allowed (through F.C.N.C.) for $m_\nu > 35$ MeV (a range of mass now excluded [16]).

The decays $\nu \to \nu'\phi$, with $\phi$ a Nambu-Goldstone boson, $m_\phi = 0$, may lead to very different lifetimes, according to the models. Lifetimes of a cosmological scale are obtained with singlet Majorons [51] or familons [52].

A Majoron is an N.G. boson that appears, together with Majorana neutrino masses (the reason for its name), when a global lepton number symmetry is spontaneously violated. There are two different models, the "singlet" [51] and the "triplet" Majoron [53,54]. Familons are N.G. bosons associated with the spontaneous breaking of a global interfamiliar (or "horizontal") symmetry. The breaking associated with singlet Majorons and with familons occurs at a large scale V.

The neutrino lifetime is of order

$$\tau = 10^{15} \sec \left[\frac{V}{10^{10}\ \mathrm{GeV}}\right]^2 \left(\frac{\mathrm{keV}}{m_\nu}\right)^3 . \tag{18}$$

Familons can also couple with equal strength to charged leptons and quarks. Thus, for familons there are lower bounds from astrophysics and particle physics on V. From energy loss in stars [55,56] into $\phi$'s, the most restrictive bound is $V>0.7 \times 10^{10}$ GeV [56]. Familons induce decays such as $\mu \to e\phi$, $\kappa \to \pi\phi$, $\tau \to e\phi$ that have not been observed. The bounds $V>2 \times 10^9$ GeV, $V>2 \times 10^{10}$ GeV and $V>1 \times 10^7$GeV are obtained, respectively [52]. That is, for familons the bound (13,14) can barely be respected, and the bound (15,16) could only be fulfilled in the odd case in which just the third generation could decay into the first, the second generation being stable (for familons with only flavour changing couplings the astrophysical bounds do not apply). The triplet Majoron [53,54] is the only model in which neutrinos have an interaction among themselves that is almost as strong as electromagnetism is among charged particles. In the original model neutrinos could not decay (just annihilate). In a slightly modified version [57], neutrinos may decay very fast. For example, solar neutrinos may decay while arriving at the earth. This mechanism has been proposed as a solution to the solar neutrino problem, with $\tau_\nu \simeq 5 \times 10^{-5}(m_\nu /\mathrm{eV})$ sec [58]. The observation of electron neutrinos from the supernova means that $\tau_\nu \simeq 5 \times 10^5(m_\nu /\mathrm{eV})$ sec . Still, due to mixings between flavour and mass neutrino eigenstates, both things are not incompatible if the mixing is large [59]. Taking for simplicity just two neutrinos, if $\nu_e = \nu_1 \cos\theta + \nu_2 \sin\theta$, and $\nu_2 \to \nu_1\phi$, ($\nu_1$ and $\nu_2$ are mass eigenstates), the solar neutrino solution requires $45^\circ \leq \theta \leq 90^\circ$ , while the signal from the supernova should have been the same for $\theta \leq 60^\circ$.Thus for $45^\circ \leq \theta \leq 60^\circ$ there is no incompatibility.

Not only unstable neutrinos, but also fast annihilating neutrinos can evade the mass bounds on standard stable neutrinos. In the triplet Majoron model, neutrinos annihilate completely into Majorons, leading to a neutrino-free universe [54,60]. A slower annihilation, mediated by a light massive Higgs field [61], may lead to any value of $\Omega_\nu$, with any value of $m_\nu$ [60].

## 5. Neutrinos as the Dark Matter

The "dark matter" (D.M.), the gravitationally dominant mass component of the universe, is only detected through its gravitational effects [14]. One of the most striking examples is given by the orbital velocity v of stars and gas clouds orbiting spiral galaxies, as a function of the distance r of these objects to the centre. The function v(r) is called the "rotation curve". If M(r) is the mass contained within the radius r, Kepler's third law says that $v(r)^2 = G\, M(r)/r$. The surprising fact is that v(r) is constant, and thus M(r)~r well outside the radius of the visible matter in the galaxy [62]. This indicates that most of the mass of spiral galaxies is in a dark, approximately spherical, halo, whose density falls off as $r^{-2}$ and extends to a distance more that 10 times the radius of the disk. The D.M. is seen at all cosmological scales. The light emitting regions account for $\Omega \simeq 0.01$, the haloes of galaxies for $\Omega \simeq 0.10$, at the largest scales (superclusters of galaxies) the detected D.M. density reaches $\Omega = 0.2 \pm 0.1$ (while inflation predicts that $\Omega = 1$!) [63].

Could the D.M. be dark, normal nucleonic matter? Big bang nucleosynthesis implies [see (17)] that $\Omega_B$ may be as large as 0.22 (for h = 0.4) [6]. Thus primordial nucleosynthesis does not preclude the possibility that the D.M. are baryons. There are, however, arguments against a universe composed only of baryonic matter [64].Cosmologists are examining the exciting possibility that the D.M. consists of neutral elementary particles. In this case, the structures in the universe (galaxies, clusters) start their formation as lumps of D.M., into which baryonic matter (atoms, molecules) fall, after the electromagnetic decoupling. Density perturbations in baryonic matter cannot grow before the e.m. decoupling (due to their interactions with photons).

Cosmologists classify the D.M. into three types, according to which structures form first (galaxies or clusters of galaxies): hot, warm, cold [65]. With hot D.M. large clusters form first. With cold D.M. ( and also warm, but barely) galaxies form first. Physical linear scales d, scale as $d(t) = R(t) d_c$ due to the Hubble expansion, where $d_c$ is the linear scale measured in comoving coordinates. Taking the scale factor now as $R(t_o) = 1$, the size $d_c$ of a given structure is always equal to the size it has now. Since in a radiation or matter dominated universe (see the introduction above), $R(t) \sim t^n$ with $n < 1$, the horizon ct grows with time even in comoving coordinates, encompassing more material as time passes. Gravitation can obviously act only within ct. Let us call $t_{GAL}$ the moment a perturbation of the size of a typical galaxy (which contains $10^{11}$ $M_{SUN}$, $10^{12}$ $M_{SUN}$ with the halo) is first encompassed by the horizon. The temperature at that moment turns out to be $T \simeq 1$ KeV. Hot D.M. particles are moving with velocity $\simeq$ c at $t_{GAL}$ and until much later. Galaxy size density fluctuations are erased, because every particle in the fluctuation moves from its original position to a distance much larger than the size of the fluctuation. The first fluctuation that survives has the size of the horizon when the hot D.M. particles become non-relativistic (and their velocities become very small). It is thus larger than a galaxy. Cold D.M. particles, on the other hand, move from their original positions in the fluctuations in a time $t_{GAL}$ a distance much smaller than $ct_{GAL}$. Thus galactic size fluctuations, and even smaller, survive.

Light neutrinos are hot D.M.. The structure that they form first has a mass

$$M = 3\times10^{15}\, M_{SUN}/\, (m_\nu/30\ \text{eV})^2 \quad , \qquad (19)$$

the mass contained in the horizon volume when neutrinos become non-relativistic.

With $m_\nu \simeq 30$ eV superclusters form first and galaxies form later by fragmentation. Numerical simulations showed that galaxies form too late in this scenario and light neutrinos fell in disfavour as D.M. candidates. The situation has recently changed. Large structures, at the scale of superclusters, have been observed: large voids, with galaxies in walls and filaments [66], large regions of the universe moving with large velocities [67]. Thus, neutrinos are reconsidered, because they naturally produce large structures. Some other mechanism, such as the presence of cosmic strings, is necessary to help galaxies form faster. Loops of cosmic strings would act as seeds for galaxies [68]. Cosmic strings are linear defects that may remain from the grand unified phase transition [69].

The mass of 30 eV is preferred for a light neutrino due to a kinematic constraint by Tremaine and Gunn [70]. They noticed that quantum mechanics implies that the number of D.M. particles which can fit within a gravitationally bound object is finite. Thus a minimum mass of the particles (corresponding to the maximal occupation number) is required to get a certain total mass. For rich clusters of galaxies this minimum mass is 5eV, for a large galaxy 30 eV (for dwarf galaxies it is 150eV, thus a 30 eV neutrino could not explain the dark matter there).

Particles of $m \simeq 1$ KeV are just becoming non-relativistic at $t_{GAL}$. Galaxies form first, but only just. They are warm D.M.. A neutrino of mass 1 keV must be unstable. The scenario of galaxy formation with unstable neutrinos

[71] has been analyzed at length and it does not work, even for lighter neutrinos, with mass closer to 30 eV [72]. The idea was that galaxies could be formed by heavy particles which decay at a later time. The decay products could produce an unclustered smooth background that would account for the non-observed contribution, 1- $\Omega_{observed}$ in such a way that $\Omega$ could be 1 now. The apparent discrepancy between $\Omega_{observed}$ and 1 is the basic motivation for these models. Besides several constraints, arising from the isotropy of the background radiation, stability of galaxies and clusters and peculiar velocities, there is a recent one based on the effect that the decay of most of the matter of a spiral galaxy would have on its rotation curve [72]. While the rotation curve would not be much affected near the centre, where the bulge and disk dominate, it should fall down outside, where the dominant matter component disappears through decay. Since only a few rotation curves show this behaviour (most stay flat or rise) a lower bound of 0.5 is obtained for the fraction of stable D.M. of the total D.M. [72]. This fraction is of order 0.05 in decaying neutrino models; thus they are ruled out (other mechanisms with unstable heavy D.M. are still viable [73], but this D.M. cannot be heavy neutrinos, due to the long lifetime required).

Cold D.M. particles are non-relativistic, since, much earlier than $t_{GAL}$, galaxies form first and aggregate into clusters and superclusters. Heavy neutrinos of $m_\nu \simeq$ few GeV have a relic abundance $\Omega \simeq 0.1$-1, and are therefore good cold D.M. candidates. Cold D.M. was definitely the preferred type of D.M. until the recent observation of the large scale structures mentioned above. Several ideas are in consideration to allow cold D.M. to account for the observations.

Thus either light 30 eV stable neutrinos (as hot D.M.) or heavy, few GeV, Dirac or Majorana stable neutrinos as cold D.M. are both good D.M. candidates. Light, relativistic neutrinos with a very short mean free path may behave as cold D.M., since the total distance they move by diffusion in a time t is much shorter than ct (i.e. they behave, in this respect, as non-relativistic particles). A variation of the triplet Majoron was recently proposed to obtain the strong interaction needed, with marginal success [74].

## 6. May We Detect D.M. Neutrinos from the Halo of our Galaxy?

We live in a spiral galaxy that, as all the others, has a dark halo, which contains most of its mass. The local density of our halo has been estimated to be 0.4 GeV/cm$^3$ (with a factor of two of uncertainty) and the characteristic velocity of the halo particles of the order $10^{-3}$ c [75]. Thus, the flux on earth is large, approximately $10^7$/cm$^2$ sec (m/1 GeV) for particles of mass m. The possibility of detecting 30 eV neutrinos with this velocity is out of reach at present. Heavy D.M. particles (of a few GeV mass) may, instead, be detected either indirectly, through their annihilation products, or directly, through the energy they may deposit in collisions within detectors.

Annihilations may occur in the halo [76]. If the D.M. particles can be trapped from the halo by the sun or the earth, annihilations would occur in the sun [77] or the earth [78]. The possibility of detecting the products of these annihilations has been extensively discussed recently. Heavy neutrinos annihilating in the halo may account for the entire spectrum of cosmic ray antiprotons (and contribute to the electrons and photons) [79]. Heavy Dirac or Majorana neutrinos with masses larger than 4 GeV would be trapped by the sun and then annihilate there, producing an excess of neutrinos coming from the sun. This is an active subject of research at the moment, because the data from large proton decay experiments (IMB, Frejus, Kamiokande) are at the point of providing interesting bounds [77,80]. If Dirac neutrinos are heavier than 12 GeV, they may be trapped by the earth and their annihilation would produce too many light neutrinos [78]. Particles lighter than those mentioned "evaporate": the rate at which they escape due to their thermal velocity is larger than the capture rate, and they never accumulate.

While the scattering cross sections of non-relativistic Dirac and Majorana neutrinos on protons are similar, those on heavier nuclei are very different. For a small momentum transfer, non-relativistic Dirac neutrinos have coherent interactions with heavy nuclei. The cross-section for a non-relativistic Dirac neutrino of mass m on a nucleus of mass M with Z protons and N nucleons is

$$\sigma^D_{SCATT} = \frac{G_F^M}{\pi} \frac{m^2 M^2}{(m+M)^2} T_{3L}^2 [(1-4\sin^2\theta_w)Z-N]^2 \quad , \tag{20}$$

where $T_{3L}$ is the third component of the weak isospin of the neutrino. Since $\sin^2\theta_w \sim 0.23$, the contribution of protons is negligible. The enhancement is by a large factor of order $10^3$ - $10^4$.

Majorana neutrinos have spin dependent interactions: the cross section is proportional to J(J+1), where J is the spin of the nucleus:

$$\sigma^{D}_{SCATT} = \frac{G^2}{\pi} \frac{m^2 M^2}{(m+M)^2} \lambda^2 J(J+1) \quad , \tag{21}$$

where $\lambda$ is a nucleus dependent factor. $\lambda J(J+1)$ is a number between 0.1 and 1 for most nuclei with nonzero spin [81]. Thus the spin-dependent cross-section is never much larger for a heavy nucleous than for a proton. This is why Majorana neutrinos in the halo cannot be captured by the earth. The difference in scattering cross sections are very relevant for direct detection. Interesting bounds are obtained on heavy Dirac neutrinos from germanium spectrometers, but not on Majorana neutrinos.

The idea of direct detection is that an incident halo particle could collide with a nucleus within a detector and transfer an energy that could be measured [81]. A recoiling nucleus in a crystal gives part of its kinetic energy to the lattice to produce phonon waves, which, after many scatterings within the crystal, thermalize, originating an increase in temperature. The detection of these effects requires cryogenic techniques still under development [82]. Another effect of a recoiling nucleus in a crystal is the production of ionization, the excitation of electrons. This mechanism allows the use of existing ultralow background germanium detectors to search for galactic D.M.

A search for D.M. was initiated with the Ge-detector of the PNL/USC collaboration [83]. (Similar results have been obtained by the UCSB/LBL collaboration with another Ge-detector [84].) The energy threshold of the detector was of 4 keV of electronic energy, corresponding to an initial nuclear recoil energy of 15 keV. The signal obtained when some energy is deposited on an electron is larger than when the same energy is deposited on a nucleus by a factor called R.E.F. in Ref. 83 (relative efficiency factor). The Ge-detectors were designed to detect the electrons emitted in double beta decay. The signal of a recoiling nucleus in these detectors had to be estimated. The R.E.F. is an energy-dependent quantity, which was determined, in [83] to be $\simeq$ 3.75 in the 10-100 keV region.

In order to obtain bounds in these experiments, the observed-interaction rate $R_o$ has to be compared with the predicted rate $R_p$ if the D.M. candidates under study are in the galactic halo. The predicted rate depends on the local density $\rho_{LOCAL}$ of the D.M. particle, its mass m, its cross-section $\sigma$ on Ge nuclei and the velocity, v, distribution in the halo: $R_p \simeq (\rho_{LOCAL}) \sigma v/m$. In [83], a Gaussian v-distribution was used with $v_{r.m.s.} = 250$ Km/sec and $v_{max} = 550$ km/sec. These are very conservative values. Also, to be conservative, it was assumed that the halo slowly rotates with a local velocity of 70 Km/sec in the direction of motion of the sun, like the galactic spheroid. The halo may not rotate at all, the rotation of the spheroid is taken as an upper bound. A non zero rotation velocity of the halo, reduces the velocities of the D.M. particles with respect to the earth, thus $R_p$ is reduced. After integrating over all velocities, an upper bound on $(\rho_{LOCAL}\sigma)$ is obtained by requiring $R_p \leq R_o$. By fixing $\sigma$ an area in $(\rho_{LOCAL}, m)$ space is excluded. For Dirac neutrinos this area is shown in Fig. 4. Dirac neutrinos with mass between 20 GeV and 2 TeV have been excluded as main components of the halo. These experimental bounds may reach at most to masses 8-10 GeV. There are no bounds on Majorana neutrinos because $\sigma$ is too small. Only 7.8% of the natural isotopic composition of Ge has a non-zero spin; besides, the cross-section is not enhanced by coherence. $R_o$ should improve by at least a factor $10^3$ before getting bounds on Majorana neutrinos. New cryogenic techniques will be necessary to reach smaller masses and to test particles with spin-independent interactions.

With a further assumption, bounds on the cosmological density of Dirac neutrinos may be obtained from the bounds on the local density [85]. With the assumption that the ratio of total D.M. mass over total baryonic mass in our galaxy $F_{GAL} = M_\nu/M_B$ has the same value in the whole universe, the bounds of Fig. 5 are obtained. This may be a good assumption for cold D.M. Fig. 5 shows the experimental upper limits on $F_{GAL}$ under two

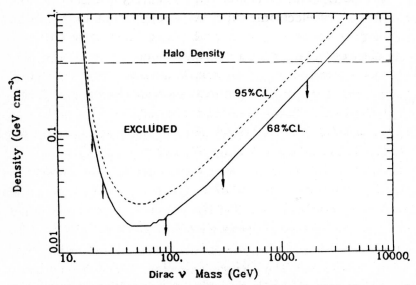

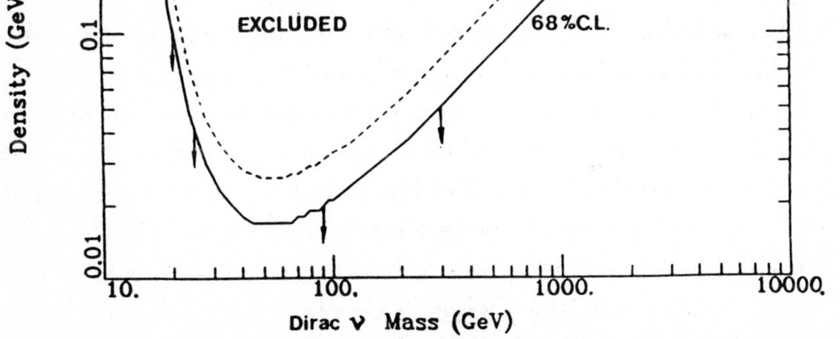


Fig. 4 Bounds obtained with the PNL/USB Ge-detector [83]. Maximum halo density of standard heavy Dirac neutrinos consistent with the observed count rate, at the 68% and 95% confidence levels

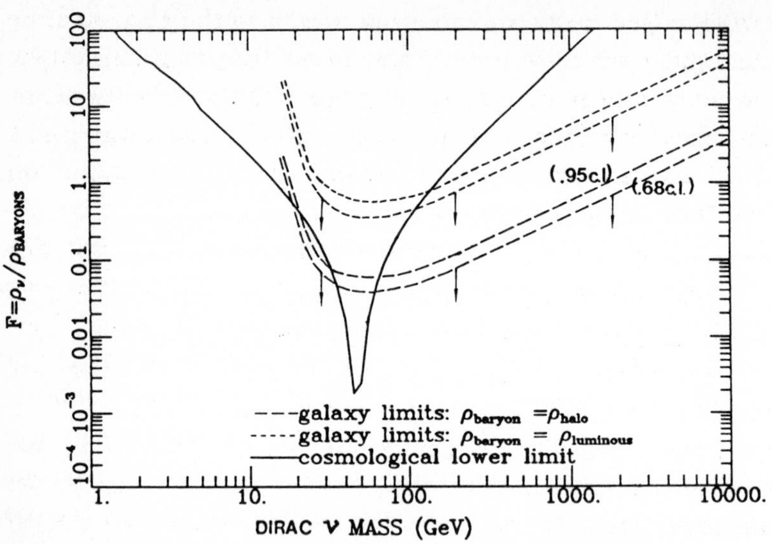


Fig. 5 Maximum ratio of total Dirac neutrino mass to total baryonic mass in our galaxy if (1) all the halo is baryonic and if (2) only the visible matter is baryonic. Assuming that the ratio is the same on a cosmological scale, it can be compared with the minimum cosmological ratio consistent with big bang nucleosynthesis (solid line) [85],[83]

assumptions: 1) the halo is baryonic. $M_B = M_{halo}$, and 2) $M_B$ includes only the observed luminous baryons (stars, gas, dust, etc.). These are to be compared with the fraction $F=\rho_\nu/\rho_B$ for standard Dirac neutrinos in the universe, where $\rho_\nu$ is the expected relic abundance. The minimum F expected has been computed by taking the maximum value for the cosmological baryonic density $\rho_B$ consistent with the bounds from nucleosynthesis (17). If the bounds on $F_{GAL}$ apply on F, then with the improvement in the data expected soon, the very existence in the universe of heavy stable Dirac neutrinos may be disproved (except for a range of masses around $m_Z/2$, due to the $Z^0$ resonance in the annihilation cross-section).

Finally, let us mention an amazing possibility: Could the solar neutrino problem be the first detected signature of D.M. in our galaxy? A decrease by 10% of the central temperature of the sun with respect to that expected in the standard model of the sun would account for the observed reduction of the predicted flux of solar neutrinos. Faulkner and Gilliland (in a paper unpublished in 1977, published in 1985 [86]) and independently Spergel and Press [87] proposed that one way the central solar temperature could be reduced is by improving the energy transport from the neutrino producing regions to the outer regions of the sun, with a population of weakly interacting particles. These particles, the "solar cosmions", if concentrated enough in the interior of the sun would not alter the outer photon producing region, which we see. Press and Spergel [88] proposed that these particles may have been trapped by the sun from the halo of our galaxy, if it consists of these particles. Thus, these cosmions would constitute the D.M. in our galaxy and simultaneously solve the solar neutrino problem. This scheme works only for particles with mass and scattering cross-sections in the sun over a small range: $m \simeq 5$ -10 GeV, $\sigma \simeq 10^{-36}$ cm$^2$. Besides, the annihilation of solar cosmions in the sun should be strongly suppressed [89]. Solar cosmions cannot be readily assigned to conventional D.M. candidates [89], whose properties do not satisfy all existing constraints [86 to 90]. However realistic models have been built [91 to 93]. In two of them the solar cosmion is a heavy neutrino of a fourth generation. The scattering cross-section due to the standard weak gauge interactions is too small. Thus, in one model the heavy neutrino has a large magnetic moment, so that its main interactions are electromagnetic. This candidate for solar cosmion is called "magnino" [92]. In order to get the large anomalous magnetic moment, the masses of the corresponding charged lepton of the same generation and new charged Higgs particles, must be very large, close to the magnino mass (less than 3 GeV larger). In the second model the solar neutrino is a normal neutrino, but the Higgs field is very light, with a mass of 400 MeV. Thus the main interactions of this neutrino are through Higgs exchange [93]. In both models annihilations are suppressed by having only neutrinos due to a primordial asymmetry (similar to the baryon asymmetry by which only baryons, and not antibaryons, exist now in the universe).

It is curious to realize that the heavy neutrinos of both these models might have escaped detection in accelerators experiments so far. They may be detected or rejected in direct detection experiments under study [94].

## References

1. For example S. Weinberg: Gravitation and Cosmology (Wiley, New York 1972)

2. G. Gamow: Phys. Rev. 70, 572 (1946); ibid. 74, 505 (1948); R. Alpher, H. Bethe, G. Gamow, Phys. Rev. 73, 803 (1948); R. Alpherand R.C. Herman: Phys. Rev. 75, 1089 (1949)

3. A.R. Sandage, G. Tammann: Nature 307, 326 (1984); G. de Vaucouleurs, G. Bollinger: Ap. J. 233, 433 (1979)

4. P.L. Richards: In Inner Space/Outer Space, ed. by E. Kolb et al. (University of Chicago Press, Chicago1986)

5. A. Penzias and R. Wilson: Ap. J. 142, 419 (1965); R. Dicke et al.: Ap. J. 142, 414 (1965)

6. D.T. Wilkinson: In Inner Space/Outer Space, ed. by E. Kolb et al. (University of Chicago Press, Chicago 1986)

7. R.V. Wagoner, W.A. Fowler, F. Hoyle: Ap. J. 148, 3 (1967)

8. J. Yang, M.S. Turner, G. Steigman, D.N. Schramm, K. A. Olive: Ap. J. 281, 493 (1984)

9. J. Bernstein, G. Feinberg: Phys. Lett 101B, 39 (1981)

10. P.B. Pal: Nucl. Phys. B227, 237 (1983)

11. F.K. Thielemann, J. Metzinger, H.V. Klapdor: Astr. Astrophys. 123, 162 (1983)

12. I. Iben, A. Renzini: Phys. Rep. 105C, 329 (1984)

13. A. Guth: Phys. Rev. D23 (1981); A. Linde: Phys. Lett. 108B, 389 (1982); A. Albrecht, P. Steinhardt, Phys. Rev. Lett 48, 1220 (1982)

14. See, for example, Dark Matter in the Universe, Proceedings of the International Astronomical Union Symposium No. 117, Princeton , June 1985 eds. J. Knapp and J. Kormendy (Reidel,1986)

15. see for example M. Turner in ref. 11

16. K. Schubert (ARGUS collaboration): Proceedings of the X Workshop on Particles and Nuclei, Neutrino Physics, Heidelberg, October 1987

17. For a review, G. Steigman: Ann. Rev. Nucl. Part. Sci. 29, 313 (1979)

18. Another review, A.D. Dolgov, Ya. B. Zeldovich: Rev. Mod. Phys. 53, 1 (1981)

19. R. Cowsik, J.McClelland: Phys. Lett. 29, 669 (1972) (however they assumed $g_V$=4)

20. F. Antonelli, R. Konoplich, D. Fargion: Lett. Nuovo Cim. 32, 289 (1981)

21. K. Olive, M. Turner: Phys. Rev. D25, 213 (1982)

22. S. Wolfram: Phys. Lett. 82B, 65 (1979)

23. B.W. Lee, S. Weinberg: Phys. Rev. Lett 39, 165 (1977); P. Hut: Phys. Lett. 69B, 85 (1977); K. Sato, M. Kobayashi: Prog. Theor. Phys. 58, 1775 (1977); M.I. Vysotskii, A.D. Dolgov, Ya. B Zel'dovich: JETP Lett. 26, 188 (1977)

24. E. Kolb, K. Olive: Phys. Rev. D33, 1202 (1986) and (erratum) Phys. Rev. D34, 2531 (1986) and references therein

25. T.K. Gaisser, G. Steigman, S. Tilav: Phys. Rev. D34, 2206 (1986)

26. G.L. Kane, I. Kani: Nucl. Phys. B277, 525 (1986)
K. Enqvist, K. Kainulainen and J. Maalampi: preprint HU-TFT-88-5

27. D. Dicus, E. Kolb, V. Teplitz: Phys. Rev. Lett 39, 169 (1977)

28. G. Steigman, M.S. Turner: Nucl. Phys. B253, 375 (1985)

29. J.H. Applegate, C.J. Hogan, R.J. Scherrer: Phys. Rev. D 35, 1151 (1985). C.A. Alcock, G.M. Fuller, G.J. Mathews: LLL preprint 1987

30. J. Audouze, D. Lindley, J. Silk: Ap. J. 293, L53 (1985); R. Dominguez-Tenreiro: Ap. J. 313, 523 (1987)

31. S. Dimopoulos, R. Esmailzadeh, L. Hall, G.D. Starkman: SLAC preprint PUB-4356, 1987

32. E. Kolb, R. Scherrer: Phys. Rev. D25, 1481 (1982)

33. K. Freese, F.C. Adams, J.A. Frieman, E. Mottola: Nucl. Phys. B287, 797 (1987)

34. M. Roos: Proceedings of the X Workshop on Particles and Nuclei, Neutrino Physics, Heidelberg October 1987

35 E. Kolb: Proceedings of the 1986 Theoretical Advanced Studies Institute, Santa Cruz, California, July 1986 (Fermilab preprint Conf. 86/146-A)

36. D. Dicus, E. Kolb, V. Teplitz: Ap. J. 221, 327 (1978)

37. R. Cowsick: Phys. Rev. Lett. 39, 784 (1977)

38. D. Dicus, E. Kolb, V. Teplitz, R. Wagoner: Phys. Rev. D17, 1529 (1978)

39. S.W. Falk, D.N. Schramm: Phys. Lett. 79B, 511 (1978)

40. J.F. Nieves: Phys. Rev. D28, 1664 (1983)

41. P.B. Pal, L. Wolfenstein, Phys. Rev. D25, 766 (1982)

42. K. Sato, M. Kobayashi, Prog. Theor. Phys. 58, 1775 (1977); E. Kolb, T. Goldman, Phys. Rev. Lett. 43, 897 (1979); R. Shrock, Phys. Rev. D24, 1232 (1981)

43. N. De Leener-Rosier, J. Deutsch, R. Prieels: Phys. Rev. Lett. 59, 1868 (1987) and references therein

44. M. Takahara, K. Sato: Phys. Lett. B174, 373 (1986)

45. H. Albrecht et al. (ARGUS collaboration): Phys. Lett 163B, 404 (1985)

46. N. De Leener-Rosier et al: Phys. Lett. B177, 228 (1986): G. Azuelos et al.:Phys. Rev. Lett. 56, 2241 (1986); F. Bergsma et al. (CHARM collaboration): Phys. Lett. 128B, 361 (1983)

47. M. Takahara, K. Sato: Modern Phys. Lett. A 2, 293 (1987)

48. A. Dar, S. Dado: preprint TECH-PH-40 (1987)

49. Y. Hosotani: Nucl. Phys. B191, 411(1981); M. Roncadelli, G. Senjanovich: Phys. Lett 107B, 59 (1981); A. Natale: Phys. Lett 141B, 323 (1984); P. Pal, Nucl. Phys. B227, 237 (1983); M. Gronau, R. Yahalom, Phys. Rev. D30, 2422 (1984)

50. H. Harari, Y. Nair: SLAC preprint PUB-4224 (1987)

51. Y. Chicashige, R.N. Mohapatra, R.D. Peccei: Phys. Lett. 98D, 265 (1981) and Phys. Rev. Lett. 45, 1926 (1980)

52. D.B. Reiss: Phys. Lett. 114B, 217 (1982); F. Wilczek; Phys. Rev. Lett. 49, 1549 (1982); G. Gelmini, S. Nussinov, T. Yanagida: Nucl. Phys. B219, 31 (1983); D. Dicus, V. Teplitz: Phys.Rev. D28, 1778 (1983); H.Y. Cheng: Indiana University preprint IUHET-129 (1987)

53. G. Gelmini, M. Roncadelli: Phys. Lett. 99B, 411 (1981)

54. H. Georgi, S.L. Glashow, S. Nussinov: Nucl. Phys. B193, 297 (1981)

55. D. Dicus, E. Kolb, V. Teplitz, R. Wagoner: Phys. Rev. D23 839 (1987); M. Fukugita, S. Watamura, M. Yoshimura, Phys. Rev. D26, 1840 (1982); N. Iwamoto, Phys. Rev. Lett. 53, 1198 (1984); G. Raffelt, Phys. Rev. D34, 3927 (1986)

56. D.S.P. Dearborn, D.N. Schramm, G. Steigman: Phys. Rev. Lett. 56, 26 (1986). For a recent review: H.Y. Cheng, Indiana preprint IUHET-129 (1987)

57. J.W.F. Valle: Phys. Lett. 131B, 87 (1983); G. Gelmini, J.W.F. Valle: Phys. Lett. 142B, 181 (1984)

58. J.N. Bahcall, S.T. Petcov, S. Toshev, J.W.F. Valle: Phys. Lett. 181B, 369 (1986); J.N. Bahcall, N. Cabibbo, A. Yahill: Phys. Rev. Lett. 28, 316 (1972).

59. J. Frieman, H. Haber, K. Freese: SLAC preprint, PUB-4261 (1987)

60. E. Kolb, M. Turner: Phys. Lett. 159B, 102 (1985)

61. P.B. Pal, Phys. Rev. 30D, 2100 (1984)

62. S.M. Faber, J.S. Gallagher: Ann. Rev. Astron. and Astrophys. 17, 135 (1979); A. Bosma; Astron. J. 220, 1825 (1981); V.C. Rubin, W.K. Ford, N. Thonnard: Ap. J. 238, 471 (1980); D. Burstein, V.C. Rubin: Astroph. J. 297, 423 (1985)

63. P.J. Peebles, Ap. J. 284, 439 (1984)

64. Lectures of J. Silk at First ESO-CERN School on Particle Physics and Cosmology, Erice, Italy, January 1987

65. S.J.Bond, A. Szalay: Ap. J. 274, 443 (1983); J.R. Primak, G.R. Blumenthal: in Formation and Evolution of Galaxies and Large Structures in the universe Moriond conference 1983, ed. by J. Andouze andJ. Tran Thanh Van (Reidel, Dortrecht) 1983

66. V. De Lapparent, M. Geller, J. Huchra: Ap. J. 302, L1

67. A. Dressler et al.: Ap. J. 313, L37 (1987)

68. R. Brandenberger, N. Keiser, N. Turok: DAMTP preprint, June 1987; R.Brandenberger, N. Kaiser, D. Schramm, N. Turok: FERMILAB preprint, Pub. 87/126-A; E. Berschinger, P. Watts, MIT preprint 1987

69. A. Vilenkin: Phys. Rep. 121, 263 (1985) and references therein

70. S.Tremaine, J. Gunn: Phys. Rev. Lett. 42, 407 (1979)

71. M. Turner, G. Steigman, L. Krauss: Phys. Rev. Lett. 52, 2090 (1984); G. Gelmini, D. Schramm, J.W.F. Valle, Phys. Lett. 146B, 311 (1984)

72. See R. Flores, G. Blumenthal, A. Dekel and J. Primack: Nature 323, 781 (1986), R. Flores: XXIII Int. Conf. on H.E.P., Berkeley, July 1986 and references therein

73. K. Olive, D. Seckel, E. Vishniac: Ap. J. 292, 1 (1985)

74. G. Raffelt, J. Silk: Phys. Lett. 192B, 65 (1987)

75. J. Caldwell, J. Ostriker: Ap. J. 251, 61 (1981); J. Bahcall, R. Soneira: Ap. J. Suppl. 55, 67 (1984), ibid. 44, 73 (1980); J. Bahcall, M. Schmidt, R. Soneira: Ap. J. 265, 730 (1983). See useful discussions in M. Turner: Phys. Rev. D33, 889 (1986): A.K. Drukier, K. Freese, D. Spergel: Phys. Rev. D33, 3495 (1986)

76. J. Silk, M. Srednicki: Phys. Rev. Lett. 53, 624 (1984); J. Hagelin, G.L. Kane: Nucl. Phys. B263, 399 (1986); M. Srednicki, S. Theisen, J. Silk: Phys. Rev. D33, 2079 (1986); M. Turner: Phys. Rev. D34, 1921 (1986)

77. J. Silk, K. Olive, M. Srednicki: Phys. Rev. Lett. 55, 257 (1985); M. Srednicki, K. Olive, J. Silk: Nucl. Phys. B279, 804 (1987); J. Hagelin, K.W. Ng, K. Olive: Phys. Lett. 180B, 375 (1986); K.W. Ng, K.Olive, M. Srednicki: Phys. Lett. 188B, 138 (1986)

78. K. Freese: Phys. Lett. B167, 295 (1986); L. Krauss, M. Srednicki, F. Wilczek: Phys. Rev. D33, 2079 (1986)

79. S. Rudaz, F. Stecker: Lab. for High Energy Astroph. NASA/Goddard Space Flight Center preprint 87-015 (1987)

80. K. Olive and M. Srednicki, in preparation (private communication)

81. M.W. Goodman, E. Witten: Phys. Rev. D31, 3059 (1985); I. Wasserman,; Phys. Rev. D33, 2071 (1985); A.K. Drukier et al. in ref. [64]. See also A. Drukier, L. Stodolski: Phys. Rev. D30, 2295 (1984)

82. P.F. Smith, review at 2nd ESO/CERN Symposium on Cosmology, Astronomy and Fundamental Physics (ESO, Garching, March 1986), Rutherford Laboratory preprint RAL-886-029

83. S.P. Ahlen, F.T. Avignone III, R.L. Brodzinski, A.K. Drukier, G. Gelmini, D.N. Spergel: Phys. Lett. 195B, 603 (1987)

84. D.O. Caldwell, R.M. Eisberg, D.M. Grumm, M.S. Witherell, F.S. Goulding, A.R. Smith, Santa Barbara preprint, USCB-HEP-87-3

85. G. Gelmini et al., Quarks and Galaxies workshop, L.B.L. Berkeley, California, 1986; Enrico Fermi Institute preprint EFI 86-77. (Computations done with the collaboration of ref. 83)

86. J. Faulkner, R. Gilliland: Ap. J. B299, 663 (1985); see also G. Steigman, C. Sarazin, H. Quintana, J. Faulkner: Ap. J. 83, 1050 (1978)

87. D.N. Spergel, W.H. Press: Ap. J. 294, 663 (1985)

88. W.H. Press, D.N. Spergel: Ap. J. 296, 679 (1985)

89. L.M. Krauss, K. Freese, D.N. Spergel, W.H. Press: Ap. J. 299, 1001 (1985).

90. R.L. Gilliland, J. Faulkner, W.H. Press, D.N. Spergel: Ap. J. 306, 703 (1986).

91. G. Gelmini, L. Hall, M.J. Lin: Nucl. Phys. B281, 726 (1987)

92. S. Raby, G.B. West: Nucl. Phys. B292, 793 (1987) and Los Alamos preprints LA-UR-87-1255 and LA-UR-87-1734 (1987)

93. S. Raby, G.B. West: Los Alamos preprint LA-UR-87-3664 (1987)

94. B. Sadoulay, J. Rich, M. Spiro, D.O. Caldwell: Lawrence Berkeley Lab. preprint, LBL-23779; B. Sadoulay: LBL preprints 23098, 23468, 23469 and references therein

# Index of Contributors